Authors

Stanley A. Smith teaches at Loyola College in Baltimore, Marylan[...] served as Coordinator, Office of Mathematics (K–12) for Baltimore County Public Schools, Maryland. He has taught junior high school mathematics and science and senior high school mathematics. He earned his M.A. degree at the University of Maryland. Mr. Smith was named Outstanding Mathematics Educator by the Maryland Council of Teachers of Mathematics in 1987. He is co-author of *Addison-Wesley Essentials of Mathematics* (1992), *Addison-Wesley Consumer Mathematics* (1989), *Addison-Wesley Algebra* (1992), and *Addison-Wesley Algebra and Trigonometry* (1992).

Charles W. Nelson is Professor and Associate Chair at the department of mathematics at the University of Florida. He earned his Ph.D. from the University of Maryland and taught at Purdue University before going to the University of Florida. A former high school mathematics teacher, Dr. Nelson has been active in organizing workshops for secondary and middle school mathematics teachers. He has co-authored mathematics texts from the elementary to the college levels, including several high school and college geometry texts.

Roberta K. Koss is Mathematics Department Chair at Redwood High School in Larkspur, California. She received her B.A. from the University of California, Berkeley and her M.S. from Holy Names College. In 1986 she was a fellow at the Woodrow Wilson Institute on geometry. She was a California state finalist for the Presidential Award for Excellence in the Teaching of Mathematics in 1987 and 1988, and in 1990 was the recipient of the California Educator Award. She is a member of the NCTM Commission on Teaching Standards for School Mathematics, and is a member of the Mathematics Advisory Committee for the California Assessment Program. In addition, she is serving on the Adolescence and Young Adulthood/Mathematics Standards Committee for the National Board of Professional Teaching Standards.

Mervin L. Keedy is Professor of Mathematics Emeritus at Purdue University. He received his Ph.D. at the University of Nebraska, and formerly taught at the University of Maryland. He has also taught mathematics and science in junior and senior high schools. Professor Keedy is the author of many books on mathematics. Most recently he is co-author of *Addison-Wesley Algebra* (1992) and *Addison-Wesley Algebra and Trigonometry* (1992).

Marvin L. Bittinger is Professor of Mathematics Education at Indiana University-Purdue University at Indianapolis. He earned his Ph.D. at Purdue University. Dr. Bittinger has authored numerous mathematics books at the high school and college levels. He is co-author of *Addison-Wesley Algebra* (1992) and *Addison-Wesley Algebra and Trigonometry* (1992).

Consultants and Reviewers

Bridget A. Arvold
Blacksburg High School
Blacksburg, Virginia

William B. Duffie, Ph.D.
Steinmetz High School
Chicago, Illinois

Cynthia Felton
Steinmetz High School
Chicago, Illinois

Adrienne Roth Hanus
Taft High School
Chicago, Illinois

Adrienne A. Kapisak
Gateway Senior High School
Monroeville, Pennsylvania

Jo Anne J. Martin
Cypress Creek High School
Houston, Texas

Cheryl A. Milliman
Hempstead High School
Dubuque, Iowa

Jane Swaggerty
Southwest High School
San Antonio, Texas

Michael Duckworth
Eastside High School
Gainesville, Florida

Essie H. Espinosa
East Chambers Independent School
    District
Winnie, Texas

David Hammett
South Cobb High School
Smyrna, Georgia

Kay Hodges
John Tyler High School
Tyler, Texas

Dr. D. J. Lefstad
Lake Whitney, Texas

Clementene D. Mathis
Cooper High School
Abilene, Texas

Barbara A. Schlachter
Osborne High School
Detroit, Michigan

Contents

Symbols   xiii

Arithmetic Review   xiv

1 Exploring Geometry

1-1   Geometry in the World Around Us   1
- *Enrichment*   4

1-2   Terms and Symbols in Geometry   5

1-3   Geometric Figures and Measurement   9
- *Enrichment*   14
- *Algebra Connection*   15
  Distances on the Number Line

1-4   Greek Constructions   16
- *Discover*   20
- *Geometry in Design*   21
  Tiles

1-5   Introduction to Transformations   22

1-6   Using Inductive Reasoning   25
- *Enrichment*   28
- *Problem-Solving Strategies*   29
  Look for a Pattern

Chapter 1 Review   30

Chapter 1 Test   31

2 Organizing Geometry

2-1   Deductive Reasoning   33
- *Visualization*   37

2-2   Definitions   38
- *Algebra Connection*   43
  Adding Positive and Negative Numbers

2-3   Postulates   44
- *Problem Solving*   48

2-4   Theorems   49
- *Algebra Connection*   53
  Subtracting Positive and Negative Numbers

2-5   Conditional Sentences   54
- *Geometry in Communication*   58
  Advertisements

Chapter 2 Review   59

Chapter 2 Test   60

3 Distance and Angle Measure

3-1 Coordinates and Distance 63
  • *Visualization* 66
3-2 Segments and Rays 67
  • *Enrichment* 71
3-3 Coordinate Geometry 72
  • *Geometry in Communication* 77
    Line Graphs
3-4 Angles 78
  • *Algebra Connection* 82
    Solving Equations
3-5 Angle Measurement 84
  • *Algebra Connection* 88
    Order of Operations and Solving Equations
3-6 Adding Angle Measures 89
  • *Problem Solving* 94
3-7 Greek Constructions—Bisecting and Copying Angles 95
  Chapter 3 Review 99
  Chapter 3 Test 100

4 Angle Relationships

4-1 Congruent Angles and Segments 103
  • *Algebra Connection* 108
    Combining Like Terms and Solving Equations
4-2 Complementary and Supplementary Angles 109
  • *Discover* 114
  • *Problem-Solving Strategies* 115
    Write an Equation
4-3 Adjacent Angles and Linear Pairs 116
  • *Problem Solving* 120
  • *Algebra Connection* 121
    Solving Equations with Variables on Both Sides
4-4 Vertical Angles 122
  • *Enrichment* 127
  • *Geometry in Nature* 128
    Spirals
4-5 Right Angles and Perpendicular Lines 129
  • *Visualization* 135
4-6 Reflections 136
  • *Discover* 141
  Chapter 4 Review 142
  Chapter 4 Test 144
  Cumulative Review Chapters 1–4 146

5   Triangles and Congruence

5-1	Classifying Triangles	149
	• *Enrichment*	153
	• *Geometry in Design*	154
	Using Equilateral Triangles	
5-2	Symmetry	155
5-3	Congruent Triangles	158
	• *Visualization*	162
5-4	Reading and Making Drawings	163
	• *Problem-Solving Strategies*	166
	Make a Drawing	
5-5	SAS, SSS, and ASA Congruence	167
5-6	Showing Triangles Congruent	173
	• *Enrichment*	177
5-7	Showing Corresponding Parts Congruent	178
5-8	The AAS and HL Postulates	182
	• *Algebra Connection*	188
	Number Properties	
	Chapter 5 Review	189
	Chapter 5 Test	190

6   Triangle Relationships

6-1	Isosceles Triangles	193
	• *Enrichment*	197
	• *Problem-Solving Strategies*	198
	Use a Flowchart	
6-2	Exterior Angles of a Triangle	200
	• *Visualization*	203
6-3	The Opposite Parts Theorem	204
6-4	The Triangle Inequality	208
	• *Algebra Connection*	212
	Inequalities on a Number Line	
6-5	Inequalities in Two Triangles	213
6-6	Concurrent Lines in Triangles	217
	• *Discover*	221
	• *Geometry in Science*	222
	The Centroid	
	Chapter 6 Review	223
	Chapter 6 Test	224

7 Parallel Lines

7-1 Parallel and Skew Lines   227
- *Geometry in Art*   232
  Perspective Drawing
7-2 Transversals and Angles   234
- *Algebra Connection*   238
  Solving Equations Involving Parentheses
7-3 When Are Lines Parallel?   239
- *Enrichment*   243
7-4 The Parallel Postulate   244
- *Enrichment*   248
7-5 The Angles of a Triangle   249
- *Problem Solving*   254
7-6 Translations   255
- *Visualization*   260
Chapter 7 Review   261
Chapter 7 Test   262

8 Quadrilaterals

8-1 Classifying Quadrilaterals   265
- *Enrichment*   270
- *Problem-Solving Strategies*   271
  Draw a Venn Diagram
8-2 Properties of Parallelograms   272
- *Visualization*   278
- *Geometry in Art*   279
  Computer Graphics
8-3 Determining Parallelograms   280
- *Enrichment*   284
8-4 Special Parallelograms   285
- *Enrichment*   290
8-5 Trapezoids   291
- *Discover*   297
- *Statistics Connection*   298
  Collecting and Displaying Data
Chapter 8 Review   299
Chapter 8 Test   300
Cumulative Review Chapters 5–8   302

9 Similarity and Scale Change

9-1 Ratio 305
    • *Geometry in Nature* 309
    The Golden Ratio

9-2 Proportion and Scale 310
    • *Visualization* 315

9-3 Similarity and Size Transformations ... 316
    • *Enrichment* 320

9-4 Similar Triangles 321
    • *Problem Solving* 325

9-5 AA Similarity 326
    • *Enrichment* 330

9-6 SAS and SSS Similarity 331
    • *Discover* 335
    • *Algebra Connection* 336
    Radicals

Chapter 9 Review 337
Chapter 9 Test 338

10 Using Similar Triangles

10-1 Similar Right Triangles 341

10-2 The Pythagorean Theorem 346
    • *Visualization* 350
    • *Probability Connection* 351
    Probability of a Simple Event

10-3 The Converse of the Pythagorean Theorem ... 352

10-4 The Isosceles Right Triangle 356
    • *Problem Solving* 359

10-5 The 30°–60° Right Triangle 360
    • *Enrichment* 364
    • *Geometry in Science* 365
    Making Measuring Instruments

Chapter 10 Review 366
Chapter 10 Test 367

11 Polygons

11-1 Identifying Polygons 369
    • *Enrichment* 373
11-2 Diagonals of a Polygon 374
    • *Enrichment* 377
    • *Problem-Solving Strategies* 378
      Simplify the Problem
11-3 Perimeter 379
11-4 Angles of a Polygon 383
11-5 Similar Polygons 387
    • *Visualization* 389
11-6 Regular Polygons 390
    • *Enrichment* 394
    • *Statistics Connection* 395
      Mean and Median
11-7 Tessellations 396
    • *Geometry in Design* 400
      Tessellations with Translations
    Chapter 11 Review 401
    Chapter 11 Test 402

12 Area of Polygons

12-1 The Meaning of Area 405
    • *Visualization* 409
    • *Algebra Connection* 410
      Exponents
12-2 Area of Rectangles 411
    • *Enrichment* 415
    • *Problem-Solving Strategies* 416
      Make an Organized List
12-3 Area of Parallelograms 417
12-4 Area of Triangles 421
    • *Enrichment* 425
12-5 Area of Trapezoids 426
    • *Geometry in Technology* 430
      Computer-Assisted Design
    Chapter 12 Review 431
    Chapter 12 Test 432
    Cumulative Review Chapters 9–12 434

13 Circles

13-1 Circles and Chords — 437
    • *Problem Solving* — 440

13-2 More About Chords — 441
    • *Discover* — 445

13-3 Tangents and Secants — 446
    • *Enrichment* — 450

13-4 Angles and Arcs — 451
    • *Visualization* — 456

13-5 Rotations — 457
    • *Geometry in Design* — 461
    Tessellations with Rotations

13-6 Inscribed Angles — 462
    • *Discover* — 466

13-7 Inscribed Polygons — 467

13-8 Circumference of Circles — 470
    • *Probability Connection* — 475
    Probability and Pi

13-9 Area of Circles — 476
    Chapter 13 Review — 481
    Chapter 13 Test — 482

14 Space Figures

14-1 Polyhedrons — 485
    • *Enrichment* — 488

14-2 Prisms and Their Surface Area — 489
    • *Visualization* — 493

14-3 Pyramids and Their Surface Area — 494
    • *Geometry in Art* — 499
    More on Perspective Drawing

14-4 Volume of Prisms — 500
    • *Enrichment* — 504

14-5 Volume of Pyramids — 505
    • *Algebra Connection* — 509
    Scientific Notation

14-6 Space Figures with Curved Surfaces — 510
    • *Visualization* — 513

14-7 Surface Area of Cylinders and Cones — 514
    • *Enrichment* — 517

14-8 Volume of Cylinders and Cones — 518

14-9 Surface Area and Volume of Spheres — 523
    • *Problem Solving* — 527

14-10 Surface Areas and Volumes of Similar Solids — 528
    Chapter 14 Review — 533
    Chapter 14 Test — 534

15 Coordinate Geometry

15-1	The Distance Formula	537
	• *Enrichment*	540
	• *Algebra Connection*	541
	Multiplying and Dividing with Negative Numbers	
15-2	Slope	542
	• *Geometry in Science*	546
	Numerical Control	
15-3	The Midpoint Theorem	547
	• *Enrichment*	549
15-4	Graphing Equations	550
	• *Visualization*	552
15-5	Lines and Equations	553
	Chapter 15 Review	556
	Chapter 15 Test	557

16 Trigonometric Ratios

16-1	Trigonometric Ratios	559
	• *Visualization*	562
16-2	The Sine, Cosine, and Tangent Ratios	563
	• *Geometry in Science*	567
	The speed of Raindrops	
16-3	Using the Sine, Cosine, and Tangent Ratios	568
	• *Problem Solving*	571
16-4	Solving Triangle Problems	572
	• *Probability Connection*	576
	Compound Probability	
	Chapter 16 Review	577
	Chapter 16 Test	578
	Cumulative Review Chapters 13–16	579

Table of Squares and Square Roots	582
Table of Trigonometric Ratios	583
Tables of Times and Measures	584
Postulates	586
Theorems	589
Glossary	597
Selected Answers	607
Dot Paper	635
Milestones in Mathematics	641
Index	645

Symbols

$\overleftrightarrow{AB}$, ℓ	line containing A and B, line ℓ	π	pi
$\overline{AB}$	line segment with endpoints A and B	(a,b)	ordered pair consisting of x-coordinate a and y-coordinate b
$\overrightarrow{AB}$	ray with endpoint A and containing B	$P(a,b)$	point P located by coordinates a and b
AB	distance between A and B, length of $\overline{AB}$	$\sin A$	sine of $\angle A$
		$\cos A$	cosine of $\angle A$
$\mathcal{K}$	plane K	$\tan A$	tangent of $\angle A$
$\triangle ABC$	triangle with vertices A, B, and C	n-gon	polygon with n sides
$\angle ABC$	angle with sides $\overrightarrow{BA}$ and $\overrightarrow{BC}$	p	perimeter
$\angle B$	angle with vertex B	s	length of side of square
$m\angle ABC$	measure of $\angle ABC$	A	area
$^\circ$	degree(s)	b	base length
$\cong$	is congruent to	h	height
$\not\cong$	is not congruent to	a	apothem
$\perp$	is perpendicular to	ℓ, w	length, width
$\not\perp$	is not perpendicular to	B	base area
$\parallel$	is parallel to	V	volume
$\nparallel$	is not parallel to	k	slant height
$\leftrightarrow$	corresponds to	r	radius
⌐	shows a right angle	d	diameter, distance
⨯ ⨯	shows congruence	C	circumference
$P \rightarrow Q$	P implies Q	m	slope
A'	A prime	mm	millimeter(s)
$\square$	parallelogram	cm	centimeter(s)
$\triangle$s	triangles	m	meter(s)
$<$	is less than	km	kilometer(s)
$>$	is greater than	in.	inch(es)
$\leq$	is less than or equal to	ft	foot (feet)
$\geq$	is greater than or equal to	yd	yard(s)
$|x|$	absolute value of x	mi	mile(s)
$\sqrt{}$	square root of	cm^2	square centimeter(s)
$\sim$	is similar to	m^2	square meter(s)
$\approx$	is approximately equal to	in.^{2}	square inch(es)
$\odot$	circle	ft^2	square foot (feet)
$\overset{\frown}{AB}$	arc with endpoints A and B	cm^3	cubic centimeter(s)
$\overset{\frown}{ACB}$	arc with endpoints A and B and containing C	m^3	cubic meter(s)
		in.^{3}	cubic inch(es)
$m\overset{\frown}{AB}$	measure of $\overset{\frown}{AB}$	ft^3	cubic foot (feet)

Arithmetic Review

Write each as an improper fraction.

1. $1\frac{1}{3}$ **2.** $2\frac{1}{2}$ **3.** $4\frac{3}{4}$ **4.** $3\frac{2}{5}$

5. $2\frac{5}{6}$ **6.** $6\frac{2}{3}$ **7.** $10\frac{1}{2}$ **8.** $8\frac{1}{4}$

9. $4\frac{3}{10}$ **10.** $3\frac{3}{8}$ **11.** $4\frac{3}{7}$ **12.** $6\frac{7}{10}$

Multiply. Reduce to lowest terms.

13. $8 \times \frac{1}{4}$ **14.** $\frac{2}{3} \times 6$ **15.** $\frac{1}{3} \times \frac{2}{5}$ **16.** $\frac{3}{4} \times \frac{1}{2}$

17. $\frac{2}{5} \times \frac{1}{4}$ **18.** $\frac{5}{6} \times \frac{3}{5}$ **19.** $\frac{2}{3} \times \frac{3}{2}$ **20.** $\frac{3}{8} \times \frac{4}{9}$

21. $\frac{1}{5} \times 2\frac{1}{2}$ **22.** $4\frac{2}{3} \times \frac{2}{7}$ **23.** $1\frac{1}{3} \times 3\frac{3}{4}$ **24.** $4\frac{1}{8} \times 1\frac{1}{9}$

Divide. Reduce to lowest terms.

25. $\frac{3}{8} \div 3$ **26.** $\frac{5}{6} \div 5$ **27.** $\frac{3}{4} \div \frac{1}{4}$ **28.** $\frac{7}{8} \div \frac{1}{8}$

29. $\frac{3}{8} \div \frac{3}{4}$ **30.** $\frac{2}{3} \div \frac{1}{6}$ **31.** $\frac{5}{8} \div \frac{5}{6}$ **32.** $\frac{9}{10} \div \frac{3}{8}$

33. $3\frac{1}{3} \div \frac{5}{8}$ **34.** $4\frac{1}{2} \div \frac{3}{4}$ **35.** $5 \div 1\frac{1}{2}$ **36.** $6 \div 2\frac{2}{3}$

Add. Reduce to lowest terms.

37. $\frac{3}{5} + \frac{1}{5}$ **38.** $\frac{2}{7} + \frac{4}{7}$ **39.** $\frac{1}{4} + \frac{1}{2}$ **40.** $\frac{2}{3} + \frac{1}{6}$

41. $\frac{2}{3} + \frac{3}{5}$ **42.** $\frac{3}{4} + \frac{2}{3}$ **43.** $3\frac{3}{8} + 1\frac{1}{2}$ **44.** $5\frac{2}{5} + 1\frac{3}{10}$

45. $5\frac{2}{3} + 6\frac{1}{2}$ **46.** $2\frac{7}{8} + 4\frac{1}{12}$ **47.** $\frac{1}{2} + \frac{1}{8} + \frac{3}{4}$ **48.** $\frac{5}{12} + \frac{1}{6} + \frac{2}{3}$

Subtract. Reduce to lowest terms.

49. $\frac{5}{6} - \frac{1}{6}$ **50.** $\frac{4}{5} - \frac{2}{5}$ **51.** $\frac{3}{4} - \frac{1}{2}$ **52.** $\frac{5}{6} - \frac{2}{3}$

53. $\frac{4}{5} - \frac{3}{4}$ **54.** $\frac{5}{7} - \frac{2}{3}$ **55.** $3\frac{5}{8} - 2\frac{1}{2}$ **56.** $7\frac{8}{9} - 1\frac{1}{3}$

57. $8 - 4\frac{2}{5}$ **58.** $6 - 5\frac{3}{4}$ **59.** $7\frac{1}{8} - 3\frac{3}{4}$ **60.** $4\frac{2}{3} - 2\frac{3}{4}$

Express as a decimal. Use a bar to show repeating decimals.

61. $\frac{1}{2}$ **62.** $\frac{1}{4}$ **63.** $\frac{1}{3}$ **64.** $\frac{3}{4}$

65. $\frac{3}{5}$ **66.** $\frac{2}{3}$ **67.** $\frac{3}{8}$ **68.** $\frac{3}{10}$

69. $\frac{7}{20}$ **70.** $\frac{5}{8}$ **71.** $\frac{12}{25}$ **72.** $\frac{5}{6}$

Express as a fraction in lowest terms.

73. 0.75 **74.** 0.3 **75.** 0.8 **76.** 0.45

77. 0.5 **78.** 0.125 **79.** 0.12 **80.** 0.625

Multiply.

81. 0.4×32 **82.** 0.18×4 **83.** 0.7×0.3
84. 0.88×10 **85.** 0.6×7.1 **86.** 26.4×0.1
87. 0.63×100 **88.** 2.4×0.03 **89.** 0.01×9.8

Divide.

90. $0.6 \div 3$ **91.** $0.8 \div 2$ **92.** $5 \div 0.5$
93. $8 \div 0.2$ **94.** $0.52 \div 0.2$ **95.** $4.6 \div 2$
96. $7.3 \div 10$ **97.** $4.8 \div 0.4$ **98.** $0.7 \div 100$

Round to the given units.

99. 0.81 (tenths) **100.** 4.7 (whole number)
101. 0.038 (hundredths) **102.** 9.86 (tenths)
103. 25.19 (whole number) **104.** 9.782 (hundredths)

Express each decimal as a percent.

105. 0.01 **106.** 0.05 **107.** 0.25 **108.** 0.35
109. 0.4 **110.** 0.9 **111.** 0.015 **112.** 0.302

Express each fraction as a percent.

113. $\frac{1}{100}$ **114.** $\frac{3}{4}$ **115.** $\frac{1}{2}$ **116.** $\frac{21}{100}$

117. $\frac{4}{5}$ **118.** $\frac{7}{10}$ **119.** $\frac{16}{25}$ **120.** $\frac{7}{50}$

Express each percent as a decimal.

121. 4% **122.** 12% **123.** 29% **124.** 2%
125. 80% **126.** 30% **127.** 0.5% **128.** 2.4%

1 Exploring Geometry

The Epcot Center in Florida contains many geometric structures.
How many common geometric figures can you find in the above photograph?

1-1 Geometry in the World Around Us

Objective: Intuitively recognize common geometric shapes.

The world is a wonderland of geometric shapes and figures. Geometric shapes are found both in nature and in human activity. We use geometry whenever we work with the size, shape, or position of objects or their parts.

Astronomers use geometry to measure distances in space, artists use geometry to create balanced drawings, architects use geometric drawings in the plans of buildings, and highway builders use geometry to plan the layout of roads.

The automobile is designed using geometry. What size wheels will lift the car's body the appropriate distance off the ground? What size should the pistons be in order to fit the engine cylinders? What shapes work best for tire treads? What kinds of devices will keep the front wheels parallel?

Explore

List the names of as many shapes as you can recognize in the pictures on this page and on the facing page.

Here are some common geometric shapes that you may have found on this page or elsewhere.

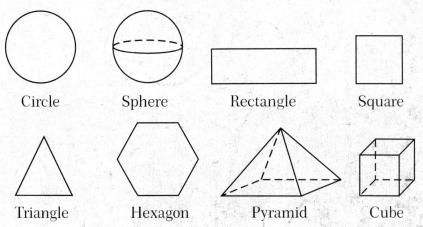

Circle	Sphere	Rectangle	Square
Triangle	Hexagon	Pyramid	Cube

Example **1.** What do a circle and a sphere have in common? What is different about them?

In common: Both are round.

Different: A circle is flat, and a sphere is a space figure.

Try This... a. What do a triangle and a square have in common? What is different about them?

b. What do a triangle and a pyramid have in common? What is different about them?

This is a picture of a solid figure that is made from cubes. There are several other ways that we can look at this picture. Complex models of this type are often used for technical drawings.

Examples 2. How many cubes are in the figure above?

Although we cannot see all the cubes, we can figure out the total number by counting the cubes in each group shown. There are 11 cubes in all.

3. Draw the figure as it would look if you were directly above it. This is called the *top view.*

4. Draw the figure as if you were directly in front of it. This is called the *front view.*

Try This... c. Draw the *right side view* of the figure.

Discussion Look around the classroom. What geometric shapes can you identify? Where do you encounter geometry in your everyday life? Where might your parents encounter geometry in their work?

Exercises

Practice

Identify the following shapes by name, if possible. Indicate those you have never heard of before today.

1.

2.

3.

4.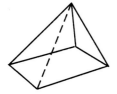

5.

6.

7. What do a square and a cube have in common? What is different about them?

8. What do a square and a rectangle have in common? What may be different about them?

9. Study the figure made from cubes.
   a. How many cubes are in this figure?
   b. Draw the top view.
   c. Draw the front view.
   d. Draw the right side view.

10. Study the figure made from cubes.
   a. How many cubes are in this figure?
   b. Draw the front view.
   c. Draw the right side view.

Top view

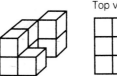

Extend and Apply

For each shape, give an example of where it occurs in nature or in the world.

11. Circle

12. Square

13. Triangle

14. Angle

15. Pentagon (5-sided figure)

16. Cone

17. Sphere

18. Rectangle

19. Star

20. Collect from magazines some pictures that illustrate geometric shapes in nature.

21. When sand falls through an hourglass, what shape appears in the lower section?

22. When a book is partially opened, what shape is suggested by the front and back covers?

23. When a carrot is cut straight through, what shape is suggested by the cut surface?

24. Which of these shapes is suggested by the cut surface if a carrot is cut through with a slant cut?

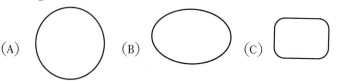

(A)　　　(B)　　　(C)　　　(D) None of these

Use Mathematical Reasoning

25. Draw a cube figure that shows 12 cubes, where the front view and the right side view are the same.

26. Explain why these cannot be views of the same figure.

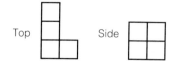

Top　　　Side

Mixed Review

Add.

27. $\frac{1}{8} + \frac{5}{8}$　　**28.** $\frac{3}{16} + \frac{7}{16}$　　**29.** $\frac{9}{10} + \frac{1}{10}$　　**30.** $2\frac{1}{4} + 1\frac{1}{2}$

Round to the nearest hundredth.

31. 3.267　　**32.** 150.991　　**33.** 11.455　　**34.** 0.005

Express as a decimal.

35. $\frac{1}{20}$　　**36.** $\frac{5}{8}$　　**37.** 80%　　**38.** 125%

Express as a percent.

39. 0.7　　**40.** 1.75　　**41.** $\frac{1}{10}$　　**42.** $\frac{3}{4}$

Enrichment

DNA, or deoxyribonucleic (de·ox′·e·ri·bo·nu·cle′·ic) acid, is considered the building block of all living things. Ask your science teacher or librarian for a picture of a DNA molecule. Its shape is formally known as a *double helix*.

Describe the geometric shape that you observe.

Does this shape remind you of any familiar shapes or objects?

1-2 Terms and Symbols in Geometry

Objective: Become familiar with basic vocabulary and symbols of geometry.

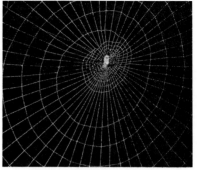

In Lesson 1 we saw many familiar geometric shapes that occur in the world. To explore relationships among geometric figures, we will need to consider some terms and symbols. Three of the most basic terms of geometry are **point, line,** and **plane.**

Name	Figure	Symbol	Understanding
Point	*A* •	*A*	The smallest thing in geometry. It has no size. We represent it by a dot.
Line *AB*, line *BA*, or line ℓ		$\overleftrightarrow{AB}$, $\overleftrightarrow{BA}$, or ℓ	A set of points that is straight and continues without gaps indefinitely in each direction. It has no width.
Plane *ABC*, or plane ℛ		plane *ABC*, or ℛ	A flat surface that extends in all directions indefinitely. It has no thickness.

Examples Tell whether a point, a line, or a plane is suggested by each object.

1. A parking lot   A flat surface, it suggests a plane.

2. The edge of a   It suggests a line.
   chalkboard

3. The tip of a pencil   It suggests a point.

Try This... Tell whether a point, a line, or a plane is suggested by each object.

a. A basketball court   **b.** The tip of an ice pick

c. A tightly stretched cord   **d.** A beam of light

A **segment** consists of two points of a line and all points between them. The two points are the **endpoints** of the segment.

We write $\overline{BC}$ or $\overline{CB}$ to name this segment.

A **ray,** or half-line, consists of one point of a line and all points of the line in one direction. The point is the endpoint of the ray.

There are two rays shown in the figure: the rays QP and QR. We can write them as $\overrightarrow{QP}$ and $\overrightarrow{QR}$. Rays are always named endpoint first, and the arrow symbol points right.

Examples Write each of the following using symbols.

4. This is the line FD or the line DF, so we can write $\overleftrightarrow{FD}$ or $\overleftrightarrow{DF}$.

5. This is the segment PQ or the segment QP, so we can write $\overline{PQ}$ or $\overline{QP}$.

6. This is the ray VX, so we write $\overrightarrow{VX}$.

7. A segment with endpoints G and H We write $\overline{GH}$ or $\overline{HG}$.

Try This... Write each of the following using symbols.

e.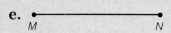

f.

g. The line that contains points E and F

h. The ray KJ

Discussion   What are some real-world illustrations of the idea of a point? a line? a plane? Our world is referred to as three-dimensional. Discuss the number of dimensions suggested by a point, a line, and a plane.

Exercises

Practice

Tell whether a point, a line, or a plane is suggested by each object.

1. A football field
2. A tight telephone wire
3. The edge of a window
4. The tip of a ball point pen
5. The tip of a pair of scissors
6. A guy wire for a TV tower
7. A tennis court
8. A sharpened pencil end
9. A tightly stretched tape measure
10. The bottom of an ice cream cone

Write each of the following using symbols.

11. The segment with endpoints T and S

12.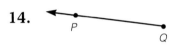

13.

14.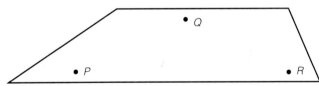

15. The line that contains the points D and E

16.

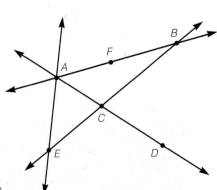

Extend and Apply

17. Draw $\overline{PQ}$.

18. Draw $\overrightarrow{ST}$.

19. Draw a segment with G and H as endpoints.

Use the figure to the right for Exercises 20–27.

20. Name four points.
21. Name three lines.
22. Name three rays.
23. Name two segments that contain point B.
24. Name two lines that contain point A.
25. Name a plane.
26. Name the point that is between A and B.
27. Name the point that is common to three lines.

28. Which is *not* a segment shown in the figure?

(A) $\overline{EB}$ (B) $\overline{FD}$

(C) $\overline{BD}$ (D) $\overline{CA}$

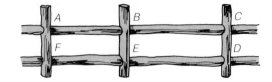

Use Mathematical Reasoning

29. Use the idea of a segment to describe a triangle.

30. Use the idea of a segment to describe a six-sided figure.

31. Draw a figure with points B, C, D, E, F, and G that shows lines $\overleftrightarrow{CD}$, $\overrightarrow{BG}$, and $\overleftrightarrow{EF}$, with the point C on all three lines.

How many *different* lines can you create that contain at least two points?

32.   **33.**   **34.**

Mixed Review

Subtract.

35. $\dfrac{3}{4} - \dfrac{1}{4}$ **36.** $\dfrac{15}{16} - \dfrac{11}{16}$ **37.** $3\dfrac{3}{4} - 2\dfrac{1}{2}$

38. $1\dfrac{1}{2} - \dfrac{3}{4}$ **39.** $5\dfrac{1}{8} - \dfrac{1}{4}$

40. What do a triangle and a rectangle have in common? What is different about them?

Express as a fraction in lowest terms.

41. 0.25   **42.** 0.6   **43.** 90%   **44.** 35%

45. Express $\dfrac{2}{3}$ as a decimal to the nearest hundredth.

46. Express $\dfrac{2}{3}$ as a percent to the nearest tenth of a percent.

What is the next number in the pattern?

47. 1   2   4   8   ?

48. 2.3   2.5   2.7   2.9   ?

49. 10   11.5   13   14.5   ?

50. 8   4   12   6   ?

51. 1   3   6   10   ?

52. 81   27   9   3   ?

53. 12   34   56   78   ?

1-3 Geometric Figures and Measurement

Objective: Measure segments and classify angles.

Explore

Use a ruler to measure the length of your right shoe.

Discuss What units did you use to measure the shoe? If you chose to use a specific mark on the ruler, how exact was the measurement? Did you estimate or round? Did you use a fraction or decimal in your numerical answer? Compare with your classmates.

When investigating geometric figures, it is often necessary to use a ruler to measure lengths. Many rulers are marked off in customary units (inches) on one side and in metric units (centimeters) on the other side. We can measure or draw using either side.

To measure a segment, line up the 0 end of the ruler against one endpoint. Then read the length at the other endpoint. On the customary side, every $\frac{1}{16}$ inch is marked off. Using this ruler we can measure to the nearest $\frac{1}{16}$ inch. We say that the **precision** of this ruler is $\frac{1}{16}$ inch. The smaller the unit of measure, the more precise the ruler. For some applications we may require very precise readings; for others we may use less precision.

Example   1. Measure $\overline{AB}$ using customary units. Use the precision of the ruler.

The endpoint B is closest to the $3\frac{11}{16}$ in. mark. To the nearest $\frac{1}{16}$ inch, $\overline{AB}$ measures $3\frac{11}{16}$ inches.

Examples **2.** Find the length of $\overline{AB}$ to the nearest $\frac{1}{4}$ in., to the nearest $\frac{1}{2}$ in., and to the nearest inch.

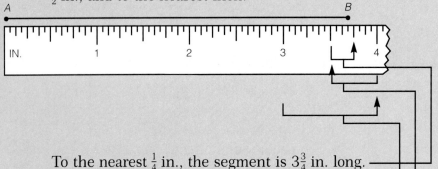

To the nearest $\frac{1}{4}$ in., the segment is $3\frac{3}{4}$ in. long.
To the nearest $\frac{1}{2}$ in., the segment is $3\frac{1}{2}$ in. long.
To the nearest inch, the segment is 4 in. long.

3. Measure $\overline{AB}$ using metric units. Use the precision of the ruler.

The other side of the ruler is marked off in tenths of a centimeter, or millimeters. The endpoint B is closest to the 9.4 cm mark. To the nearest tenth of a centimeter, $\overline{AB}$ measures 9.4 cm.

Try This... Measure each segment using both customary and metric units. Use the precision of the ruler.

a.

b.

Example **4.** Draw a segment that is $3\frac{5}{16}$ in. long.

Locate the $3\frac{5}{16}$ in. mark by counting 5 marks past the 3 in. mark. Then draw a segment from the 0 end to the $3\frac{5}{16}$ in. mark.

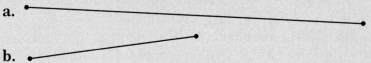

     c. $2\frac{2}{16}$ in.   **d.** $5\frac{3}{16}$ in.   **e.** 6.8 cm

An **angle** is another familiar geometric term. An angle is formed by two rays with a common endpoint, called the **vertex.**

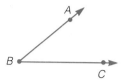

The angle shown is denoted $\angle B$, or $\angle ABC$. Note that the middle letter is the vertex.

The usual unit of angle measure is the **degree.** One degree is $\frac{1}{360}$ of a complete circle. Here are some angles and their measures.

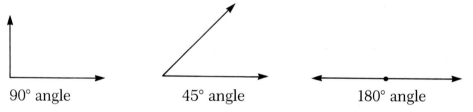

90° angle   45° angle   180° angle

An angle measuring 90° is called a **right angle,** and an angle measuring 180° is called a **straight angle.** Corners of cards or papers form right angles, so they can be used to determine right angles. The edge of a paper can be used to determine whether an angle is a straight angle.

Examples Use a card or a paper to determine whether the angle shown is a right angle, a straight angle, or neither.

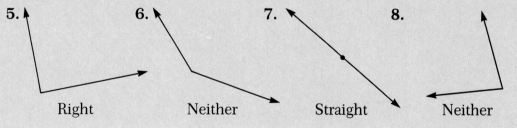

5. Right   **6.** Neither   **7.** Straight   **8.** Neither

Try This... Use a card or a paper to determine whether the angle shown is a right angle, a straight angle, or neither.

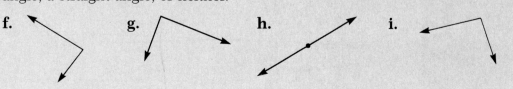

f.   **g.**   **h.**   **i.**

Angles that measure less than 90° are called **acute angles.**
Angles between 90° and 180° are called **obtuse angles.**

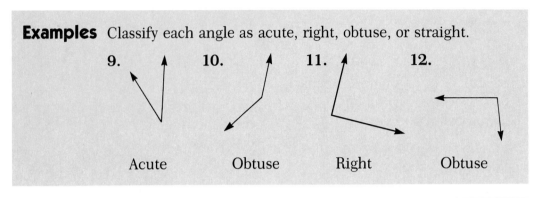

Examples Classify each angle as acute, right, obtuse, or straight.

9.   10.   11.   12.

Acute   Obtuse   Right   Obtuse

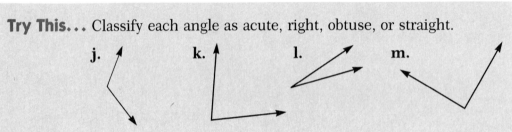

Try This... Classify each angle as acute, right, obtuse, or straight.

j.   k.   l.   m.

Discussion   Which side of the ruler gives the more precise
measurement, the customary side or the metric side? Why?

Exercises

Practice

Measure each segment below using both customary and
metric units. Use the precision of the ruler.

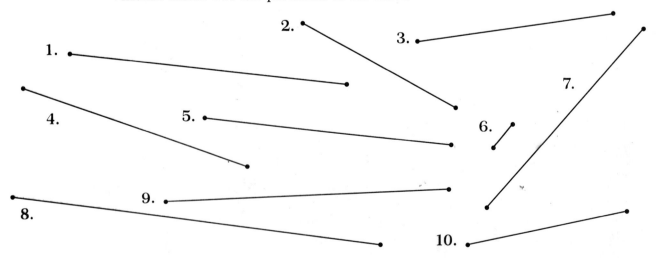

Draw segments with these measures.

11. $2\frac{9}{16}$ in.　　**12.** $8\frac{1}{16}$ in.　　**13.** 9.6 cm　　**14.** 12.3 cm　　**15.** 2.9 cm

16. $10\frac{11}{16}$ in.　　**17.** $4\frac{1}{2}$ in.　　**18.** $5\frac{1}{4}$ in.　　**19.** 22.4 cm　　**20.** $3\frac{1}{8}$ in.

Classify each angle as acute, right, obtuse, or straight.

21. 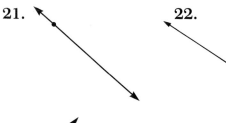　　　　**22.** 　　　　**23.**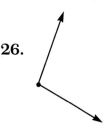

24.　　　　**25.**　　　　**26.**

What is the precision of the ruler?

27.

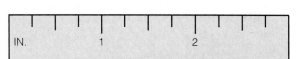

28.

Extend and Apply

29. How much has the child grown?

(A) $2\frac{3}{4}$ in.　　　(B) $2\frac{1}{4}$ in.

(C) $3\frac{1}{4}$ in.　　　(D) $3\frac{3}{4}$ in.

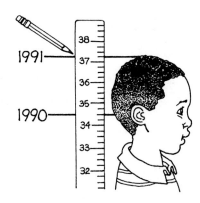

30. Estimate these lengths.
  a. The height of the classroom doorway
  b. The height of your desk
  c. The width of your desk top
  d. The distance from the front of the classroom to the back

31. Classify each angle formed as acute, right, obtuse, or straight.
  a. The corner of a piece of paper
  b. Your thumb and index finger open as wide as possible
  c. Your elbow bend when your arm is extended
  d. Your elbow bend when your arm is bent as far as possible

Use Mathematical Reasoning

Measure, then estimate the distance in miles.

32. Clinton to Bass

33. Avon to Delton

34. Fline to Empire

35. a. How far is Bass from Fline?

b. What is the *shortest route* available from Bass to Fline? What is the distance along this route?

c. Suppose it takes 2 minutes per mile to travel between cities, plus 10 minutes to go *through* a city. How long does it take to travel from Bass to Fline? How long would it take if a *direct* route were available?

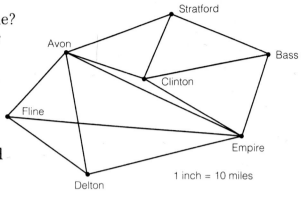

1 inch = 10 miles

Mixed Review

Tell whether a point, a line, or a plane is suggested by each object.

36. A cable   **37.** A grain of sand   **38.** A mirror

39. The corner of a room where two walls and the ceiling meet

Convert between units.

40. 4 ft = _____ in.   **41.** 276 in. = _____ ft   **42.** 42 ft = _____ yd

43. 6 cm = _____ mm   **44.** 500 cm = _____ m   **45.** 6 km = _____ m

46. 4 mi = _____ ft   **47.** 13,200 ft = _____ mi   **48.** 3.5 m = _____ cm

Enrichment

Make the objects as shown from cardboard strips and paper fasteners.

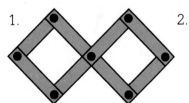

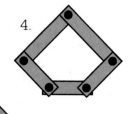

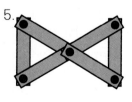

Which of these objects are *rigid*—that is, have shapes that cannot be changed without force? How can the others be made rigid?

Algebra Connection

Distances on the Number Line

Positive and negative numbers describe situations that are opposites of each other.

$-3°$, three degrees *below* zero $+3°$, three degrees *above* zero

This number line shows examples of opposites.

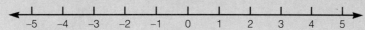

Samples   **1.** What number is suggested by "a loss of $15."
-15 A loss suggests a negative number.

2. What is the opposite of -7?
7 is the opposite of -7.

3. What is the opposite of 8?
-8 is the opposite of 8.

The **absolute value of a number** is its distance from 0 on the number line.
We use the symbol $|n|$ to represent "the absolute value of n."

Samples   **4.** The absolute value of 4 is 4, as 4 is 4 units from 0.
We write $|4| = 4$.

5. The absolute value of -4 is 4, as -4 is 4 units from 0.
We write $|-4| = 4$.

6. The absolute value of 0 is 0, as 0 is 0 units from 0.
We write $|0| = 0$.

The absolute value of a negative number is its opposite. The absolute value of any nonnegative number is just the number itself.

Problems   Name the number that is suggested by each situation.

1. 450 ft below sea level
2. A gain of 10 yards
3. 4 seconds before liftoff

4. A debt of $15.25
5. A drop of $2\frac{5}{8}$
6. A 5000-point bonus

Give the opposite of each number.

7. -20
8. 32
9. -2.4
10. -4001
11. $1\frac{1}{2}$

Find the absolute value.

12. $|3|$
13. $|-7|$
14. $|0|$
15. $|-12|$
16. $|-23|$

17. $|-56|$
18. $|8.6|$
19. $|-100|$
20. $|-0.1|$
21. $\left|-3\frac{3}{4}\right|$

1-4 Greek Constructions

Objective: Use a compass and straightedge to copy a segment, bisect a segment, and copy a triangle.

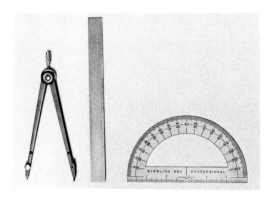

In geometry we will often use drawings to help us picture shapes and make discoveries. Straightedges, rulers, protractors, and compasses are the drawing tools we will use.

The ancient Greeks were particularly interested in drawings that could be made with only a compass and a straightedge. Such drawings are called **constructions.**

The following example shows how to construct a copy of a segment.

Example   1. Copy $\overline{AB}$ using a compass and a straightedge.

Step 1
Use a straightedge to draw a segment that is longer than $\overline{AB}$. Mark a point C near one end of the segment.

Step 2
Open your compass so that it fits $\overline{AB}$.

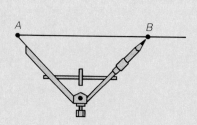

Step 3
Keep the same compass setting, place the point on C, and make a short arc at D.

$\overline{CD}$ is a copy of $\overline{AB}$.

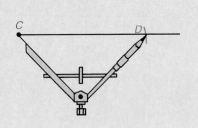

Try This... a. Copy $\overline{MN}$ using a compass and a straightedge.

The **midpoint** of a segment is a point halfway between its endpoints. To **bisect** a segment, we find its midpoint.

We can bisect a segment with a compass and a straightedge.

Example **2.** Bisect $\overline{AB}$ using a compass and a straightedge.

Step 1
Open the compass to more than half the length of $\overline{AB}$. Place the point at A and draw an arc as shown in the picture.

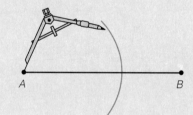

Step 2
Keep the same setting. Place the compass point at B and make an arc that intersects the other.

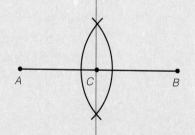

Step 3
Join the points of intersection using a straightedge. Point C is the midpoint of $\overline{AB}$.

Try This... b. Draw a segment about 7 cm long. Bisect it using a compass and a straightedge.

A **triangle** consists of three segments joined at their endpoints. Each corner of the triangle is a *vertex*. Triangle ABC is denoted △ABC.

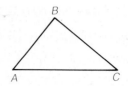

Example   **3.** Use a compass and a straightedge to construct a copy of △PQR.

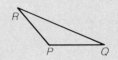

Step 1
Copy $\overline{PQ}$ using a straightedge and a compass. Label the endpoints A and B.

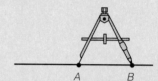

Step 2
Open the compass to fit $\overline{QR}$. Then place the point on B and draw an arc as shown.

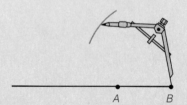

Step 3
Open the compass to fit $\overline{PR}$. Then place the point on A and make a second arc that crosses the first. The two arcs cross at C.

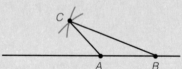

Step 4
Use the straightedge to draw $\overline{AC}$ and $\overline{BC}$. △ABC is a copy of △PQR.

Try This... c. Use a compass and a straightedge to construct a copy of △ABC.

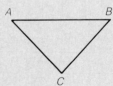

d. Use a compass and a straightedge to construct a triangle with sides the same length as these segments.

M ——————————— N

N ————————————— T

T ————————————————— M

Which do you think is more accurate: finding the midpoint of a segment with a compass and a straightedge or with a ruler? Why?

Exercises

Practice

Copy each segment using a compass and a straightedge.

1. •————————• 2. •—————•

Use a ruler to draw each segment. Copy the segments using a straightedge and a compass.

3. $\overline{RS}$, 6 cm long

4. $\overline{PQ}$, 8 cm long

5. $\overline{TV}$, 3 cm long

6. $\overline{QT}$, 9 cm long

Use a compass and a straightedge.

7. Draw a segment about 8 cm long. Bisect it.

8. Draw a segment about 5 cm long. Bisect it.

9. Draw a segment about 12 cm long. Bisect it.

10. Copy △PQR.

11. Copy △ABC.

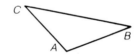

12. Construct a triangle with sides the same length as these segments.

————————————

————————————

————————————————

Construct a triangle with sides having these lengths.

13. 3 cm, 5 cm, 4.5 cm

14. 2.5 cm, 3.5 cm, 5 cm

Extend and Apply

15. A triangular structure was to be constructed with sides of length $1\frac{1}{3}$ yd, $\frac{3}{4}$ yd, and $2\frac{2}{3}$ ft. When the structure was completed, the sides measured 48 in., 27 in., and 32 in. Which statement is true?
    (A) All of the side lengths were incorrect.
    (B) Two of the side lengths were incorrect.
    (C) One of the side lengths was incorrect.
    (D) The structure was constructed correctly.

The segments below have lengths of a and b as shown. Use a compass to construct segments of the following lengths.

—————— a ——————   ——————————— b ———————————

16. $2a$ **17.** $a + b$ **18.** $b - a$
19. $2a + b$ **20.** $3a + b$ **21.** $3a + 2b$

Use Mathematical Reasoning

22. Suppose you had only a compass. Explain how you could bisect a segment by trial and error.

23. Can you use a compass and a straightedge to construct a triangle with sides of lengths 4 cm, 4 cm, and 8 cm? Why or why not?

Mixed Review

24. A _____ consists of two points of a line and all points between them. The two points are the _____ of the segment.

25. A half-line with one endpoint is called a _____ .

26. A flat surface that extends indefinitely in all directions is a _____ .

27. To find the midpoint of a segment, _____ the segment.

◇ **28.** Find $|-20|$ ◇ **29.** Give the opposite of -20.

Discover

Fold an $8\frac{1}{2}$-in. by 11-in. piece of paper in half crosswise. Unfold the paper. What does the fold represent?

Label one endpoint of the fold A and the other B. Now fold the paper in half the other way. What does this fold represent?

Label the endpoints C and D.

Look at both folds. What is the measure of $\overline{AB}$? What is the measure of $\overline{CD}$?

Label the point where $\overline{AB}$ intersects $\overline{CD}$, O.

Measure $\overline{AO}$ and $\overline{OB}$. What are their lengths?
Measure $\overline{CO}$ and $\overline{OD}$. What are their lengths?

What can you conclude about $\overline{AB}$ and $\overline{CD}$?
What is another name for O?

Tiles

Tiles contained some of the earliest geometric designs. Tiles have been found that date back 5200 years, to as early as the cultures of the Tigris, Euphrates, and Nile rivers. One of the simplest tile designs was based on the square.

1. Draw a square with sides 12 cm long. Use your ruler to find the midpoint of each side.

2. Connect consecutive midpoints.

3. Repeat this procedure for the smaller square. Continue this pattern.

4. Color or shade your drawing to form an interesting tile design.

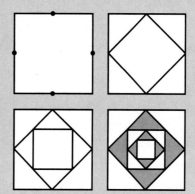

Problems

1. Make a square design exactly like the one made above. Change the design by drawing the diagonals of each of the squares. Color or shade the drawing to make an attractive design.

2. Make a new square-based design by joining the midpoints of the opposite sides of a square and then making a simple square design inside each of the four new squares.

3. Make a new square-based design by rotating squares of the same size. Follow the pattern shown below.

4. Make your own original square-based design.

5. This tile design, which uses pentagons (five-sided figures), is based on the simple square design. Try to visualize how this was done and repeat the design.

1-5 Introduction to Transformations

Objective: Identify how reflections, translations, and rotations can change a figure.

Many artists have used mathematics in creating their art. The early Greeks decorated their temples with frieze designs. They took a simple shape such as ⌐

and created a design by *sliding,* or **translating,** and repeating the same figure.

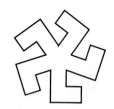

They found they could *flip,* or **reflect,** and repeat the shape to create another interesting design.

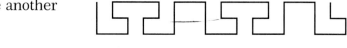

By *turning,* or **rotating,** the shape, yet another design is produced.

A **transformation** describes some kind of a change in a figure. Flips, slides, and turns are transformations that move figures to new positions. The resulting shape and location of the transformed figure is its **image.**

Examples Describe how the first figure has been transformed into the second figure.

1.

Rotated

2. Reflected

3. Translate, or slide, the figure to the new point and sketch its image.

4. Reflect, or flip, the figure over the given line.

Try This... **a.** Describe how the first figure has been transformed into the second figure.

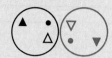

Copy each figure and sketch the image.

b. Reflect the figure over the line and draw its image.

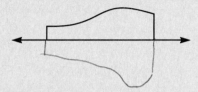

c. Translate the figure to the new point and sketch its image.

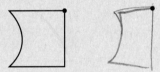

Discussion How can you reverse a transformation? That is, what must be done to an image to make *its* image the original figure?

Exercises

Practice

Describe how the first figure has been transformed into the second figure.

1.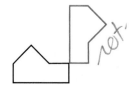

2.

3.

4.

Copy each figure and sketch its image.

5. Reflect over the given line.

6. Translate to the new point.

7. Translate to the new point.

8. Reflect over the given line.

Extend and Apply

9. Reflect the figure over the given line to find the hidden message.

10. This is a _____ .
   (A) translation   (B) reflection
   (C) rotation   (D) None of these

11. Use the edge of a paper to write the top, bottom, or side of a hidden message. Then try to read the message by placing the edge of a mirror against it.

12. Find examples of rotations, translations, or reflections used in wallpaper or other designs.

Use Mathematical Reasoning

13. Suppose the figure is rotated 180° clockwise about the point, translated 1 in. to the right, rotated 180° again, and finally reflected across $\overline{AB}$. What is the result?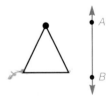

Mixed Review

14. Draw a segment that is 6.5 cm long.

15. Use a compass and straightedge to copy the segment from Exercise 14.

16. Use a compass and straightedge to bisect the copy from Exercise 15.

Round to the nearest tenth.

17. 8.2497   **18.** 1.04   **19.** 1.96   **20.** 0.005

21. What number is suggested by a reduction in capacity of 14 gallons.

1-6 Using Inductive Reasoning

Objective: Use inductive reasoning to reach conclusions and to provide counterexamples to disprove statements.

Explore

Try the following with several types of triangles.

1. Draw a large triangle using a straightedge.

2. Cut out the triangle. Then tear off the corners.

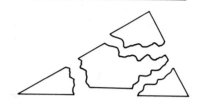

3. Place the corners together.

Discuss What does the bottom edge of the angles of the triangle appear to form? What does this suggest about the sum of the measures of the angles of a triangle?

When we reach conclusions by observing figures and studying patterns as we did in the Explore above, we are using *inductive reasoning*. **Inductive reasoning** is reaching a conclusion on the basis of a series of examples.

The Explore suggests that the sum of the measures of the angles of a triangle is 180°. We can repeat this exploration many times and find that this is always true.

We will see how inductive reasoning can be used to reach a conclusion about the *medians* of a triangle.

A **median** of a triangle is a segment from a vertex to the midpoint of the opposite side.

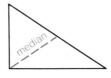

Example   **1.** Draw a large triangle. Draw its medians and see what happens.

Step 1
Use a whole sheet of paper to draw a large triangle.

Step 2
Bisect one of the sides. Then join the midpoint with the opposite vertex. This segment is a median.

Step 3
Draw the other two medians in the same way. What do you notice about the medians?

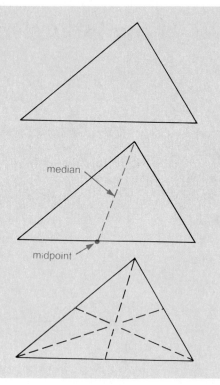

Try This... **a.** Draw two more triangles and their medians. What do you observe? Compare your results with those of several classmates.

When you use inductive reasoning, you can rarely look at all possible examples. Thus, you cannot really be sure that a statement based on inductive reasoning is always true. In fact, if you find only one example that disproves the statement, the statement is false. An example showing that a statement is not always true is called a **counterexample.**

Examples Suggest a possible counterexample for each conclusion.

2. All compact discs play for 45 minutes.

Possible counterexample: a CD that plays for 50 minutes

3. All mammals have two or four legs.

Possible counterexample: A dolphin is a mammal without legs.

4. Any three segments can form a triangle.

Possible counterexample: ●—● ●—● ●————●

Discussion Suppose you cannot find a counterexample for a conclusion reached by inductive reasoning. Does this guarantee that the conclusion is true?

Exercises

Practice

Use a ruler or a compass and a straightedge. Draw the figures described. Study the angles formed and look for patterns. What do you observe?

  1. Three triangles, each with all sides the same length

  2. Three triangles, each with two sides the same length

  3. Three triangles, each with no sides the same length

  4. Draw a square. Find the midpoints of the sides of the square and label them P, Q, R, S. Draw $\overline{PQ}$, $\overline{QR}$, $\overline{RS}$, and $\overline{SP}$. What do you observe?

Suggest a possible counterexample for each conclusion.

  5. All horror movies involve vampires.

  6. All rectangles are squares.

  7. All roses are red.

  8. All birds can fly.

  9. All pizzas contain pepperoni.

10. All pencils have 6 sides.

11. Dividing two numbers always results in a different number.

12. All Supreme Court justices are men.

13. All candy contains sugar.

14. All prime numbers are odd.

Extend and Apply

15. Use the corner-tearing method from the Explore to investigate the sum of the measures of any quadrilateral (four-sided figure).

16. Explain how the median experiment from Example 1 could be done by folding paper.

17. Which is a possible counterexample to the statement "All flizzles are carnaples"?

    (A) A carnaple that is not a flizzle

    (B) A flizzle that is not a carnaple

    (C) A carnaple that is a flizzle

    (D) A flizzle that is a carnaple

Use Mathematical Reasoning

18. Draw several different-shaped triangles and their medians. Look at the distance from a corner to the point of intersection of the medians. How does this distance compare to the distance from the point of intersection to the opposite side?

Mixed Review

19. What is the precision of the ruler?

20. Use symbols to name the ray containing *A* that has an endpoint at *B*.

Tell whether the figures are related by reflection, rotation, or translation.

21.   **22.**   **23.**   **24.**

What is the next number in the pattern?

25. $\dfrac{1}{4}$ $\dfrac{1}{2}$ $\dfrac{3}{4}$ 1 ?

26. What is the absolute value of the opposite of 7?

Enrichment

Make a list of statements made in product claims for which you think you can find counterexamples. Look in newspapers, your mailbox, on cereal boxes, or wherever you find product claims.

Why is it difficult to suggest a counterexample for claims such as the following:

The most popular car in its class
Preferred by more people than any top-selling brand
The number one pain reliever
Audiences have been captivated by *Leave Them Cold.*

Problem-Solving Strategies

Look for a Pattern

Problems in geometry can often be solved by using problem-solving strategies, some of which may already be familiar to you. Some common strategies are: *Make a Drawing, Make an Organized List, Write an Equation, Guess-Check-Revise, Look for a Pattern, Work Backward, Simplify the Problem, Make a Table, Make a Model,* and *Use Logical Reasoning.*

Many problems can be solved in more than one way. To solve some problems, we can use *inductive reasoning* by finding a pattern. When we find a pattern, it allows us to make a prediction.

Sample Predict the number of stars there will be in A_{40}.

A_1 ☆☆

A_2 ☆☆☆
         ☆☆☆

A_3 ☆☆☆☆
         ☆☆☆☆
         ☆☆☆☆

A_4 ☆☆☆☆☆
         ☆☆☆☆☆
         ☆☆☆☆☆
         ☆☆☆☆☆

In A_1, there is 1 row of 2 stars. In A_2, there are 2 rows of 3 stars. In A_3, there are 3 rows of 4 stars. In A_4, there are 4 rows of 5 stars.

Arrangement	1	2	3	4
Number of stars	$1 \times 2 = 2$	$2 \times 3 = 6$	$3 \times 4 = 12$	$4 \times 5 = 20$

It looks like for arrangement number n, there are n rows of $n + 1$ stars. Continuing the pattern, for arrangement 40 there will be 40 rows of 41 stars, or $40 \times 41 = 1640$ stars.

Problems

1. How many cubes would be needed for 10 steps?
2. How many different squares are in this 4 × 4 square?
3. How many segments make a pattern 15 triangles long?
4. How many cuts must be made to get 64 small cubes from the larger cube?

Chapter 1 Review

1-1

1. What do and ⬡ have in common? What is different about them?

2. Study the figure made from cubes.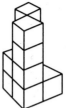
   a. How many cubes are used in this figure?
   b. Draw the top view.
   c. Draw the front view.
   d. Draw the right side view.

   Front

1-2   Tell whether a point, a line, or a plane is suggested by each object.

3. A ramp   4. A tape measure

Write each of the following using symbols.

5. The segment with endpoints K and S

6.

1-3   Draw segments with these measures.

7. $3\frac{5}{16}$ in.   8. $5\frac{5}{8}$ in.   9. 12.1 cm   10. 9.7 cm   11. 0.7 cm

Classify each angle as acute, right, obtuse, or straight.

12. ⌐   13. ∠   14. ⟋

1-4

15. Use a compass and a straightedge to copy $\overline{AB}$ and bisect the copy.

16. Use a compass and a straightedge to construct a copy of $\triangle PQR$.

1-5

17. Describe how the first figure has been transformed to the second figure.

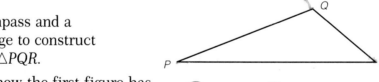

1-6   Sketch the image of the given figure.

18. Rotate about the point.   19. Reflect over the given line.

1-7   Suggest a counterexample.

20. All triangles have three different side lengths.

Chapter 1 Test

1. What do a cone and a pyramid have in common? What is different about them?

2. Study the figure made from cubes.
   a. How many cubes are used in this figure?
   b. Draw the top view.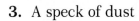
   c. Draw the front view.
   d. Draw the right side view.

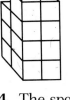

Tell whether a point, a line, or a plane is suggested by each object.

3. A speck of dust

4. The spot where a spinning top touches the floor

Write each of the following using symbols.

5. The line containing points C and B

6.

Draw segments with these measures.

7. $1\frac{13}{16}$ in.   8. $4\frac{3}{8}$ in.   9. 13.8 cm   10. 5.5 cm   11. 5.5 in.

Classify each angle as acute, right, obtuse, or straight.

12.   13.   14.

15. Use a compass and a straightedge to copy $\overline{LM}$ and bisect the copy.

16. Use a compass and a straightedge to construct a copy of $\triangle MNO$.

17. Describe how the first figure has been transformed to the second figure.

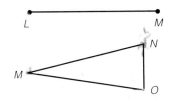

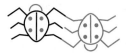

Sketch the image of the given figure.

18. Translate to the new point.

19. Reflect over the given line.

Suggest a counterexample.

20. All lines can be named exactly two ways.

2 *Organizing Geometry*

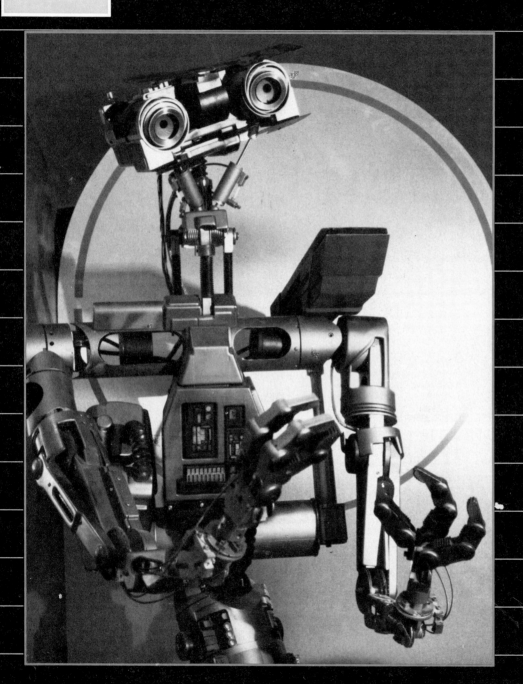

The commands to program a robot must be ordered correctly. In the same way, the concepts of geometry can be organized so that they build logically.

2-1 Deductive Reasoning

Objective: Use deductive reasoning to reach conclusions.

Joan constructed a figure with five sides, a **pentagon.** Maryann wanted to know the sum of the measures of its angles and was going to use a protractor to measure each of the angles. Joan said she knew the sum of the measures of all five angles without measuring them because the sum is related to the angles of a triangle. Can you guess how Joan found the sum of the pentagon's angle measures?

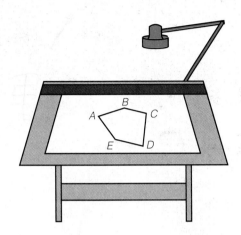

In Chapter 1 you used *inductive reasoning* to conclude that the angle measures of any triangle add up to 180°. Your conclusion was based on several observations when you tore off the corners of triangles and fitted them together.

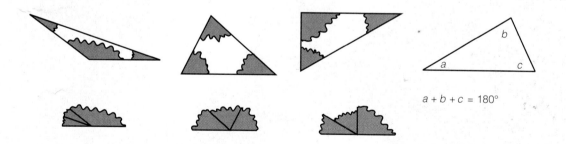

$$a + b + c = 180°$$

In this lesson we will look at another way of reaching a conclusion. Instead of basing the conclusion on a series of observations, we will use **deductive reasoning.** Deductive reasoning allows us to reach a conclusion based on given assumptions and rules of logic.

In the following example we begin by assuming that the angle measures of any triangle add up to 180°. Then we use deductive reasoning to draw a conclusion about the angle measures of any four-sided figure.

Example   **1.** Start with the assumption that the measures of the angles of any triangle add up to 180°. Use this assumption to reach a conclusion about the total angle measure of a four-sided figure, which is called a **quadrilateral.**

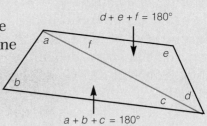

Draw a quadrilateral and one of its diagonals as shown.

Notice that two triangles are formed. Their angles combine to form the angles of the quadrilateral. So the total for the quadrilateral is 180° + 180°, or 360°.

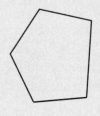

$d + e + f = 180°$

$a + b + c = 180°$

Try This...   **a.** Start with the assumption that the measures of the angles of any triangle add up to 180°. Use this assumption to reach a conclusion about the total angle measure for the five-sided figure (pentagon) in the lesson opener.

Here are some other examples of deductive reasoning.

Examples   Use the following assumptions to reach a conclusion.

   2. Assumptions: (i) All citizens of Dallas are Texans.
                              (ii) All Texans are Americans.

        Conclusion: All citizens of Dallas are Americans.

   3. Assumptions: (i) All Californians are skiers.
                              (ii) Mark is a Californian.

        Conclusion: Mark is a skier.

   4. Assumptions: (i) If you shop at Joe's Best Buy, you will save money.
                              (ii) You shop at Joe's Best Buy.

        Conclusion: You will save money.

Try This... Use the following assumptions to reach a conclusion.

b. Assumptions: (i) All rectangles are quadrilaterals.
(ii) All squares are rectangles.

c. Assumptions: (i) If a number has the factor 2, then it is an even number.
(ii) The number 346 has the factor 2.

d. Assumptions: (i) John arrived at school one hour before Harry.
(ii) Harry arrived at school at 9 a.m.

Discussion How does deductive reasoning differ from inductive reasoning?

Exercises

Practice

Use the following assumptions to reach a conclusion.

1. (i) All elephants are mammals.
(ii) All mammals have hair.

2. (i) All jumbos are mumbos.
(ii) All mumbos are gumbos.

3. (i) If it rains, then the grass will grow.
(ii) It rains.

4. (i) If you study hard, then you will succeed.
(ii) You study hard.

5. (i) All Greeks are good mathematicians.
(ii) Pythagoras was Greek.

6. (i) All squares are rectangles.
(ii) The diagonals of a rectangle bisect each other.

7. (i) Jody lives a mile and a half farther from school than Jing does.
(ii) Jing lives 5 miles from school.

8. (i) Paul is definitely taller than Karl is.
(ii) Karl is 5 ft 9 in. tall.

9. (i) No juniors ordered a yearbook.
(ii) Carrie is a junior.

Start with the assumption that the measures of the angles of any triangle add up to 180°. Use this assumption to draw conclusions about the total angle measure of the following figures.

10. Six-sided figure (hexagon)

11. Seven-sided figure (heptagon)

12. Eight-sided figure (octagon)

Extend and Apply

Detectives often use deductive reasoning in their work. As the famous detective Sherlock Holmes would say, "It's elementary, my dear Watson." Deduce as much information as you can from each set of assumptions in Exercises 13 and 14.

13. (i) Marty bowled from 7:00 to 10:00 on Wednesday evening.
    (ii) At 8:30 on Wednesday evening, Marty's car was seen running a red light.

14. (i) Mrs. Jones came home at 4:30 p.m. to find the cookie jar empty although she had filled it that morning.
    (ii) Her husband came home for lunch.
    (iii) Mrs. Jones's daughter said the cookie jar was empty when she came home after school at 3 p.m.
    (iv) Mrs. Jones's son came home at 4 p.m.

In Exercises 15 and 16, use the assumptions to reach a conclusion.

15. (i) The angle measures of a triangle add up to 180°.
    (ii) One angle of the triangle measures 90°.

16. (i) The angle measures of a triangle add up to 180°.
    (ii) One angle of the triangle measures 90°.
    (iii) The other two angles have the same measure.

17. Given that all widgets are doodads and all snickets are widgets, which of the following must be true?
    (A) All widgets are snickets.   (B) All doodads are snickets.
    (C) All doodads are widgets.   (D) All snickets are doodads.

Use Mathematical Reasoning

18. Suppose a figure has n sides. Try to discover a relationship between the number of sides and the total angle measure of the figure. (Hint: Consider figures with three, four, five, and six sides and look for a pattern.)

Mixed Review

Suggest a possible counterexample for each statement.

19. All tables have four legs.

20. No apples are yellow.

21. All cars have exactly two doors.

22. Every resident of California lives in Los Angeles.

Name the number suggested by each situation.

23. 27,325 ft above sea level

24. Owing $125.37

25. A 5-yd penalty

26. Winning $1000

Interpreting figures is an important part of understanding mathematics. In geometry, we often use flat, two-dimensional drawings to represent three-dimensional solids. In such figures, dotted or dashed lines usually represent "hidden" segments. The following problems will give you practice in using these drawings.

Use the figure at the right for Problems 1–4.

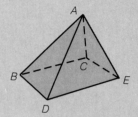

1. Which point is meant to be farthest from the viewer?

2. Which point is meant to be closest to the viewer?

3. Which segments are "hidden" from the viewer?

4. Which of the drawings below shows a different view of the same figure?

(A) 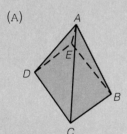　　(B) 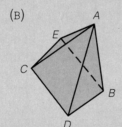　　(C) 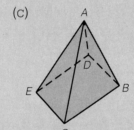　　(D)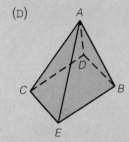

When a solid object is cut by a plane, the resulting figure is called its **cross section.** For example, slicing an apple in half results in a cross section like the figure at the right. For Problems 5–6, match each solid sliced by a plane with its cross section.

5.

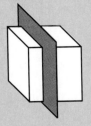

(A) 　　(B) 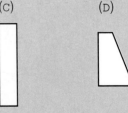　　(C)　　(D)

6.

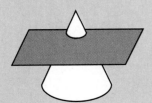

(A)　　(B) 　　(C) 　　(D)

2-2 Definitions

Objective: Use the definitions of collinear, noncollinear, coplanar, noncoplanar, *and* intersection, *and recognize good definitions.*

Over two thousand years ago, a mathematician named Euclid organized all of the geometric ideas known at that time. Little is known about Euclid's life, except that he taught in Alexandria, Egypt. The sequence of ideas in this textbook is quite similar to that developed by Euclid in his book, known as Euclid's *Elements*.

Recall from Chapter 1 that we already have an intuitive understanding of geometric terms such as *point, line,* and *plane*. We can use these basic undefined terms to give definitions for other terms.

Definitions **Space** is the set of all points.
Collinear points are points that are on the same line.
Noncollinear points are points that are not all on the same line.

Point *A*, point *C*, and point *D* are collinear because they all lie on line *m*. Point *A*, point *B*, and point *C* are noncollinear because they are not all contained in any one line.

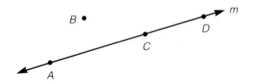

Example 1. Name a set of collinear points and a set of noncollinear points.

R, *S*, and *T* are collinear. A straightedge can be placed along these points to check this.

R, *S*, and *P* are noncollinear.

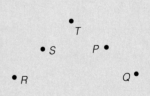

Try This... **a.** Name another set of collinear points and another set of noncollinear points.

H ere are some additional basic definitions that we will use throughout the book.

Definition   **Coplanar points** are points that are in the same plane.

Think of the bees as points and the table as a plane. The bees on the table are coplanar.

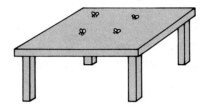

Definition   **Noncoplanar points** are points that are not all in the same plane.

In this picture, the bees are noncoplanar points.

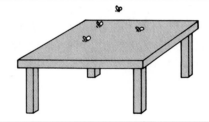

Definition   The **intersection** of two figures is the set of points that they have in common.

In this figure, line *m* and line *n* intersect at point *A*. Planes $\mathcal{R}$ and S intersect in line ℓ.

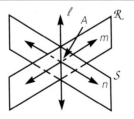

Example   **2.** Consider this TV antenna tower. Name a set of four coplanar points. Name a set of four noncoplanar points.

C, D, E, and *J* are coplanar.
C, D, E, and *F* are noncoplanar.

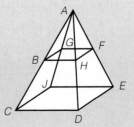

Try This... **b.** Name another set of four coplanar points and another set of four noncoplanar points.

Explore

1. Choose a point on a sheet of paper and name it *P*.
2. Use your ruler to find at least a dozen points that are each 3 cm from *P*.
3. What kind of figure do you get if you connect these points?

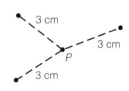

You have discovered that the set of all points 3 cm from a given point is a circle. We call a set of points satisfying a specified condition a **locus** of points.

When we write a definition of a geometric figure, we give a name to the figure and then describe the figure so that:

(1) every member of the group of figures we are defining fits the description, and
(2) no other geometric figure fits that description.

So, we can define a circle as the locus of points in the plane that are a given distance from a given point. This is a good definition because (1) every circle, regardless of its size, fits this description, and (2) no other figure fits the description.

Example 3. What is wrong with the following definition?

"Definition": A <u>triangle</u> is a figure formed by joining noncollinear points with segments.

The definition does not satisfy (2). Other figures, such as squares, also fit the description.

Correct Definition: A **triangle** is a figure formed by joining three noncollinear points with three segments.

Try This... What is wrong with the following definitions?

c. "Definition": <u>Coplanar</u> points are points that are all in a horizontal plane.

d. "Definition": A <u>circle</u> is the locus of points 3 cm from a given point.

Discussion Select an object in the classroom and define the object. Compare your definition with those of your classmates. Can there be more than one correct definition?

Exercises

Practice

Name two examples from everyday life for each of the terms in Exercises 1–3.

1. Point   **2.** Line   **3.** Plane

4. Draw three collinear points.

5. Name an everyday example of three collinear points.

6. Draw three noncollinear points.

7. Name an everyday example of three noncollinear points.

8. Name an everyday example of coplanar points.

9. Name an everyday example of noncoplanar points.

Use the figure at the right to determine whether each statement is true or false.

10. Point A, point B, and point C are collinear.

11. Point C, point B, and point D are collinear.

12. Point A, point B, and point D are noncollinear.

13. Line ℓ and line m intersect at point B.

14. $\overleftrightarrow{AC}$ and $\overleftrightarrow{BC}$ intersect at point C.

15. All points on line n are collinear.

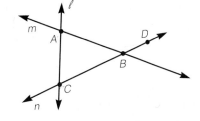

Consider the tent at the right for Exercises 16–19.

16. Name a set of three collinear points.

17. Name a set of three noncollinear points.

18. Name a set of four coplanar points.

19. Name a set of four noncoplanar points.

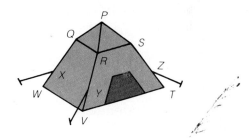

Identify the following sets of points as coplanar or noncoplanar.

20. A, B, E, and D **21.** A, C, F, and D

22. A, H, C, and B **23.** D, G, C, and A

24. D, G, F, and A **25.** H, B, F, and D

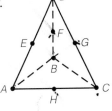

What is wrong with each of the following definitions?

26. A triangle consists of three different segments.

27. It consists of three different segments joined at their endpoints.

28. A triangle consists of three or more different segments joined at their endpoints.

Extend and Apply

Use the figure at the right for Exercises 29–33.

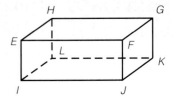

29. Which of the following is the intersection of $\overline{HL}$ and $\overline{LK}$?
(A) Point L (B) Point K
(C) Point H (D) Point I

30. What is the intersection of $\overline{EF}$ and $\overline{FG}$?

31. What is the intersection of $\overline{EI}$ and $\overline{IJ}$?

32. Which line is the intersection of plane $EFJI$ and plane $FGKJ$?

33. What is the intersection of plane $EHGF$, plane $EHLI$, and plane $EFJI$?

Use Mathematical Reasoning

In Exercises 34–36, describe the characteristics that separate the members from the non-members. Write a definition for the members.

34.   Members   Non-members

35.   Members   Non-members

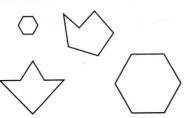
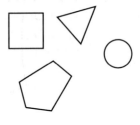

36.   Members   Non-members

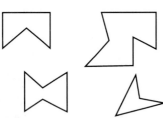
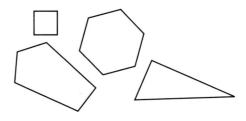

Mixed Review

Use the following assumptions to reach a conclusion.

37. (i) The angle measures of any quadrilateral add up to 360°.
    (ii) A square is a quadrilateral.

◈ Give the opposite of each number.

38. 14   **39.** −201   **40.** −57.3   **41.** 99

Algebra Connection

Adding Positive and Negative Numbers

The Lincoln High School football team gained 5 yd,
lost 7 yd, and gained 3 yd. What was their net gain?   5 + (−7) + 3 = ?

Addition of positive and negative numbers can be shown by moving
on a number line. Move to the right if the number added is positive;
move to the left if the number added is negative.

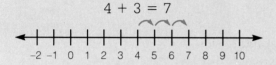
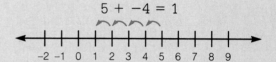

You can also use the following rules for finding the sum of any two numbers.

To add numbers with like signs, add their absolute values, and
if the numbers are positive, the sum is positive.
if the numbers are negative, the sum is negative.

For example, −2 + −3 = −5
            13 + 8 = 21

To add numbers with unlike signs, subtract their absolute values, and
if the number farther from 0 is negative, the sum is negative.
if the number farther from 0 is positive, the sum is positive.

For example, 8 + −3 =   Subtracting absolute values: 8 − 3
            5   The sum is positive, since 8 is farther from
                           0 than −3 is.

            13 + −15 =   Subtracting absolute values: 15 − 13
            −2   The sum is negative, since −15 is farther
                           from 0 than 13 is.

Problems   Add.

1. −4 + −8   2. −5 + −14   3. −23 + −3   4. −17 + −11
5. −3 + −21   6. −29 + −6   7. −22 + −36   8. −11 + −33
9. 8 + −5   10. 12 + −4   11. 23 + −18   12. 37 + −25
13. −7 + 18   14. −12 + 38   15. −6 + 25   16. −27 + 54
17. 9 + −15   18. 12 + −26   19. 14 + −41   20. 29 + −34
21. −34 + 17   22. −26 + 3   23. −65 + 32   24. −54 + 13
25. −2 + 14   26. −14 + −38   27. 23 + −6   28. −61 + 16

2-3 Postulates

Objective: Recognize postulates in geometric and everyday situations.

Steve and Lee play a tic-tac-toe tournament using the following rules.

1. One player uses Xs, the other player uses Os.
2. Players take turns placing their mark in any empty box.
3. The winner is the first player to get three of his or her marks in a row.
4. The first player to win three games is the champion.

In the same way that people playing a game must agree beforehand on the rules, mathematicians have agreed to begin the study of geometry with certain statements that they assume to be true. The statements that mathematicians assume true are called **postulates.**

We will do the same. We will select some postulates at the beginning and assume that they are true. When you read a postulate, you may start by thinking, "We will assume that…"

Postulate 1   (We will assume that…) Given any two points, there is exactly one line containing the two points.

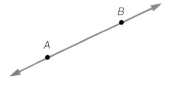

Postulate 2   (We will assume that…) Given any three noncollinear points, there is exactly one plane containing them.

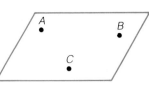

Postulates 1 and 2 tell us that two points determine a line and that three noncollinear points determine a plane. For example, in the figure, we say that points A and B determine line ℓ. The noncollinear points A, B, and C determine plane $\mathcal{R}$.

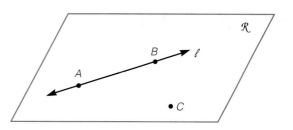

The next postulate assures us that there are differences among lines, planes, and space.

Postulate 3 (We will assume that...) Any line contains at least two points. Any plane contains at least three noncollinear points. Space contains at least four noncoplanar points.

Examples Which postulate is suggested by each statement?

1. The camera tripod does not wobble.   Postulate 2
2. A taut clothesline forms a straight line.   Postulate 1

Try This... Which postulate is suggested by each statement?

a. Space is not flat like a plane.
b. A bracing wire forms a straight line.
c. A three-runner iceboat rests on the ice without wobbling.

Example 3. How many lines are determined by three noncollinear points, A, B, and C? Draw a picture to explain your answer.

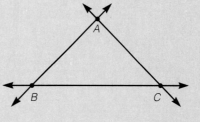

Each pair of points determines a line. Thus, three lines are determined.

Try This... d. How many lines are determined by four points, R, S, T, and U, no three of which are collinear? Draw a picture to explain your answer.

Here are two more postulates describing the relationships
among points, lines, and planes.

Postulate 4 (We will assume that . . .) If two points lie in a plane,
then the line containing them is in
the plane.

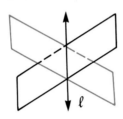

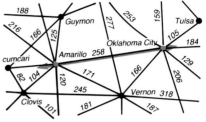

Postulate 5 (We will assume that . . .) If two planes intersect, then
their intersection is a line.

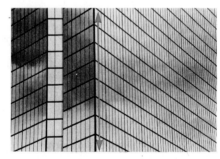

Example 4. Tell which postulate is
suggested by the photo.

Postulate 5
(If two planes intersect,
then their intersection
is a line.)

Try This. . . e. Tell which postulate is suggested
by the following fact. A crack
between two points on a sheet of
glass lies entirely within the
glass.

Exercises

Practice

In Exercises 1–12, tell which postulate is suggested by each statement.

1. A three-legged stool sits level on a floor.
2. A rope stretched tight at a building site forms a straight line.
3. A line is not like a point.
4. A surveyor's transit tripod does not wobble.
5. A gardener uses a stretched cord to make straight rows.
6. Space consists of more than one point.
7. A groundskeeper uses two stakes and a string to mark a straight line for a baseline.
8. Points R and S are contained in $\overleftrightarrow{AB}$. Points R and S are also contained in $\overleftrightarrow{CD}$. Thus, $\overleftrightarrow{AB}$ and $\overleftrightarrow{CD}$ are the same line.
9. Three noncollinear points G, H, and K are in plane $\mathcal{M}$. Points G, H, and K are also in plane $\mathcal{P}$. Thus, planes $\mathcal{M}$ and $\mathcal{P}$ are the same plane.
10. Point A is in plane $\mathcal{R}$ and point B is in plane $\mathcal{R}$. Therefore, $\overleftrightarrow{AB}$ is in plane $\mathcal{R}$.
11. The points common to two intersecting planes form a line.
12. If two points are in a plane, then the line containing the points lies in the same plane.

Tell which postulate is suggested by each picture.

13.

14.

15.

16.

17. How many different lines are determined by points E, F, G, and H? Draw a picture to explain your answer.

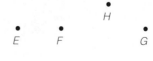

18. How many different lines are determined by points I, J, K, and L? Draw a picture to explain your answer.

Extend and Apply

Complete each statement.

19. If two planes intersect, then their intersection is a _____ .

20. A statement that mathematicians assume to be true is called a _____ .

21. If point D and point E lie in plane M, then line _____ lies in plane M.

22. The edge of a room formed by the ceiling and a wall suggests Postulate _____ .

23. Three noncollinear points always determine _____ .
  (A) a line   (B) a plane   (C) two planes   (D) three lines

Use Mathematical Reasoning

Complete each statement with *always*, *sometimes*, or *never*.

24. Two points _____ determine a line.

25. A plane _____ contains at least three noncollinear points.

26. Two points are _____ noncollinear.

27. Three points are _____ collinear.

Mixed Review

28. Name a set of three collinear points.

29. Name a set of three noncollinear points.

30. What is the intersection of $\overleftrightarrow{CE}$ and $\overleftrightarrow{FG}$?

31. What is the intersection of $\overline{EF}$ and $\overline{FG}$?

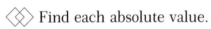

Find each absolute value.

32. $|43|$ **33.** $|0|$ **34.** $|-51|$ **35.** $|17.1|$

Problem Solving

In this lesson you found that three noncollinear points determine three lines, and that four points, no three of which are collinear, determine six lines. Use the strategies *draw a picture* and *look for a pattern* to investigate the following problem.

How many lines are determined by fifteen points, no three of which are collinear?

3 points-3 lines   4 points-6 lines   5 points-? lines   6 points-? lines

2-4 Theorems

Objective: Apply basic theorems to geometric and everyday situations.

Explore

1. Fold a sheet of paper to make a straight crease.

2. Unfold the paper and make another crease that crosses the first.

Discuss Is it possible to make the two creases so that they intersect in more than one point? State a conclusion about intersecting lines.

In the Explore you discovered that lines cannot intersect in more than one point. Using deductive reasoning, we could prove this fact. When definitions and postulates can be used to prove a statement, that statement is called a **theorem.** When you read a theorem, you may start by thinking, "We can prove that..."

Theorem 2.1   (We can prove that...) If two lines intersect, then they intersect in exactly one point.

Although we will not formally prove theorems, we will often give a **convincing argument** or demonstration for them. The Explore above is a demonstration for Theorem 2.1.

Here are some additional theorems that follow from our postulates and definitions.

Theorem 2.2   (We can prove that...) If a line and a plane intersect and the line is not in the plane, then they intersect in a point.

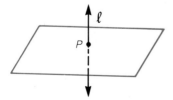 

Theorem 2.3 (We can prove that . . .) If ℓ is a line and P is a point not on the line, then ℓ and P are contained in exactly one plane.

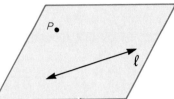

The following Examples show how these theorems can be used to explain geometric facts.

Examples Tell which theorem explains each statement.

1. A line not in a plane cannot contain more than one point of the plane.

   Theorem 2.2 (If a line and a plane intersect and the line is not in the plane, then they intersect in a point.)

2. A line and a point not on the line cannot be contained in more than one plane.

   Theorem 2.3 (If ℓ is a line and P is a point not on the line, then ℓ and P are contained in exactly one plane.)

3. Two intersecting lines, m and n, cannot look like this:

   Theorem 2.1 (If two lines intersect, then they intersect in exactly one point.)

Try This. . . Tell which theorem explains each statement.

a. A line not contained in a plane does not intersect the plane in two places.

b. Two lines cannot intersect in two different points.

c. A line cannot intersect a plane like this:

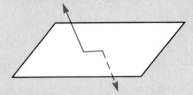

Discussion Give a convincing argument for Theorem 2.2. You may use inductive reasoning, deductive reasoning, drawings, or models.

Exercises

Practice

Tell which theorem is suggested by each picture.

1.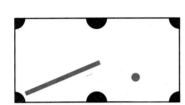

2.

3.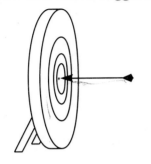

4.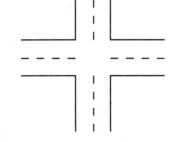

Tell which theorem explains each statement in Exercises 5–8.

5. A line and a point not on the line are in only one plane.

6. It is impossible for two different lines to have two points in common.

7. Two lines cannot look like this:

8. A line and a plane cannot look like this:

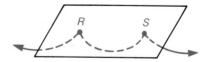

Extend and Apply

Determine whether each statement in Exercises 9–12 is true or false.

9. Two lines can intersect in more than one point.

10. If m is a line and Q is a point not on the line, then there is a plane that contains m and Q.

11. When a line is not in a plane but the line intersects the plane, the line and plane intersect in a point.

12. If point A is not on line ℓ, then A and ℓ can be contained in two different planes.

13. A helicopter pilot knows that Lakeview Hospital is located at the intersection of two straight streets. When she flies over the intersection of the streets, she concludes that the hospital must be directly below because the streets do not intersect anywhere else. Which theorem is she using?

14. Angela uses a hammer to drive a nail into the wall. Which of the following postulates or theorems explains the resulting hole in the wall?
  (A) If two lines intersect, then they intersect in exactly one point.
  (B) Given any two points, there is exactly one line containing the two points.
  (C) If a line and a plane intersect and the line is not in the plane, then they intersect in a point.
  (D) Any plane contains at least three noncollinear points.

Use Mathematical Reasoning

Complete each statement with *always*, *sometimes*, or *never*.

15. Two lines _____ intersect.

16. If two lines intersect, they _____ intersect in a point.

17. A line and a point not on the line are _____ contained in two planes.

18. If line ℓ is not in plane $\mathcal{P}$ but it intersects plane $\mathcal{P}$, then the intersection is _____ a point.

19. A line and a plane _____ intersect.

Mixed Review

Use the figure at the right for Exercises 20–29. Determine whether each statement is true or false.

20. Points *A*, *B*, and *C* are collinear.

21. Points *A*, *B*, and *P* are collinear.

22. Points *A*, *P*, and *C* are noncollinear.

23. Line *r* and line ℓ intersect at point *P*.

At which point do the following lines intersect?

24. $\overleftrightarrow{AB}$ and $\overleftrightarrow{CD}$

25. Line ℓ and line *n*

26. Line *m* and $\overleftrightarrow{BD}$

What type of angle (acute, right, obtuse, or straight) do the following angles appear to be?

27. ∠1   **28.** ∠2   **29.** ∠3

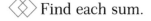 Find each sum.

  30. $12 + -7$ **31.** $-8 + -6$ **32.** $-14 + 17$ **33.** $45 + -58$

Algebra Connection

Subtracting Positive and Negative Numbers

The temperature during the night fell from 10°F to −5°F.
How many degrees did the temperature change?   10 − (−5) = ?

Subtraction is used when we want to know the difference of two numbers.
We can use a number line to see how far apart the numbers are.

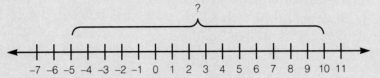

We can see that the temperature dropped 15°.   10 − (−5) = 15

Compare these results.   7 + −3 = 4   7 − 3 = 4

8 + −5 = 3   8 − 5 = 3

10 + −2 = 8   10 − 2 = 8

These examples indicate that subtracting a number and
adding the opposite of the same number yield the same
answer. Try more examples. We can conclude by inductive
reasoning that this is correct. In fact, we can subtract both
positive and negative numbers by *adding the opposite*.

$$a - b = a + (-b)$$

Samples   Subtract.

1. 7 − 3 = 7 + (−3) = 4   **2.** 2 − 5 = 2 + (−5) = −3

3. 4 − (−5) = 4 + 5 = 9   **4.** −5 − (−9) = −5 + 9 = 4

Problems   Subtract.

1. 2 − 7	**2.** 4 − 9	**3.** 7 − 16	**4.** 0 − 6
5. 12 − 23	**6.** 8 − 25	**7.** 11 − 32	**8.** 1 − 19
9. 3 − (−6)	**10.** 7 − (−9)	**11.** 10 − (−2)	**12.** 1 − (−14)
13. 23 − (−12)	**14.** 17 − (−1)	**15.** 21 − (−9)	**16.** 0 − (−12)
17. −7 − (−9)	**18.** −4 − (−2)	**19.** −8 − (−4)	**20.** −9 − (−3)
21. −11 − (−3)	**22.** −23 − (−16)	**23.** −16 − (−3)	**24.** −41 − (−41)
25. −7 − 23	**26.** 14 − 23	**27.** −12 − 26	**28.** 71 − 100

2-5 Conditional Sentences

Objective: Identify the hypothesis and the conclusion of conditional sentences, and write the converses of conditional sentences.

The advertisement says, "If you want a brighter smile, then brush with Goldmate." Sentences like these are called **conditional sentences.** It is important to understand how they are used in mathematics as well as in everyday life.

In mathematics, many postulates and theorems are stated as conditional sentences. We say they are in **if-then** form. Look for examples of if-then sentences in the postulates and theorems of Lessons 2-3 and 2-4.

Here is a new theorem. It is stated in if-then form.

Theorem 2.4   If two lines, ℓ and m, intersect, then the two lines are contained in exactly one plane.

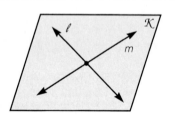

The "if" part of the sentence is called the **hypothesis.** The "then" part of the sentence is called the **conclusion.**

$$\text{If } \underset{\uparrow}{P}, \text{ then } \underset{\uparrow}{Q}$$
$$\text{hypothesis} \qquad \text{conclusion}$$

In Theorem 2.4, the hypothesis is "two lines, ℓ and m, intersect." The conclusion is "the two lines are contained in exactly one plane." Notice that the words "if" and "then" are not part of the hypothesis and conclusion.

For "If P, then Q" we may use symbols to write "$P \rightarrow Q$," which is read "P implies Q."

Examples Identify each hypothesis and conclusion.

1. If <u>it snows today</u>, then <u>school will close early</u>.
         hypothesis   *conclusion*

2. If <u>the figure has four sides</u>, then <u>it is a quadrilateral</u>.
          hypothesis   *conclusion*

Try This... Identify each hypothesis and conclusion.

a. If you go out in the rain, then you will get wet.

b. If a number is odd, then it is not even.

Conditional sentences may be stated in different ways.
Sometimes the "then" is omitted.

Examples Identify each hypothesis and conclusion.

3. If <u>you go shopping</u>, <u>I will go with you</u>.
       hypothesis   *conclusion*

4. If <u>two lines intersect</u>, <u>they intersect in exactly one point</u>.
     hypothesis   *conclusion*

Try This... Identify each hypothesis and conclusion.

c. If Jamal works hard, he will succeed.

d. If she earns enough money, she will go to the concert.

Sometimes the "if" part (the hypothesis) is written at the end
of the conditional sentence. In this case the word "then" is omitted.

Example 5. Identify the hypothesis and conclusion.
   <u>Sue will buy the car</u> if <u>she earns enough money</u>.
       conclusion   *hypothesis*

Try This... e. Identify the hypothesis and conclusion.

You will get wet if you go out in the rain.

If we interchange the two parts of a conditional sentence, we obtain a new conditional sentence called its **converse.**

Conditional: If P, then Q. $(P \rightarrow Q)$

Converse: If Q, then P. $(Q \rightarrow P)$

Examples Write the converse of each sentence.

6. If a figure is a triangle, then it has three sides.
Converse: If a figure has three sides, then it is a triangle.

7. If Tim lives in Dallas, then he lives in Texas.
Converse: If Tim lives in Texas, then he lives in Dallas.

Try This... Write the converse of each sentence.

f. If a figure is a square, then it has four sides.

g. If Hal is 18 or older, then he can vote.

Discussion Look again at the conditional sentences above. If a conditional is true, is its converse always true? How are the conditional and its converse for Goldmate toothpaste related? Make up other examples of true conditional sentences with true converses and with false converses.

Exercises

Practice

Identify the hypothesis and the conclusion in each sentence. Then write the converse of the conditional.

1. If a number is odd, then it is not divisible by 2.
2. If the grass is green, then there are no weeds.
3. If water freezes, then the temperature is low.
4. The bread tears if the butter is hard.
5. If Rita saves enough money, then she will buy the dress.
6. If the sun is out, then Roberto will go to the beach.
7. If you change the oil, your car runs better.
8. I will wash the car if you help me.
9. If the birds are flying south, then winter is coming.

Identify the hypothesis and the conclusion in each sentence.
Then write the converse of the conditional.

10. If you water the flowers, they grow.　**11.** It is easy if you practice.

12. If $2x = 10$, then $x = 5$.　**13.** If $3y + 2 = 17$, then $y = 5$.

Extend and Apply

14. Which of the following is the same as the conditional, "If you want the best, then buy Cougar running shoes"?
 (A) If you buy Cougar running shoes, then you want the best.
 (B) Buy Cougar running shoes if you want the best.
 (C) Cougar running shoes are the best.
 (D) If you buy Cougar running shoes, then you will get the best.

Rewrite each statement in "if-then" form.

15. Spare the rod and spoil the child.

16. Too much pepper ruins the sauce.

17. All squares are rectangles.

18. All right angles have a measure of 90°.

19. An apple a day keeps the doctor away.

20. Absence makes the heart grow fonder.

Use Mathematical Reasoning

Determine whether each conditional is true or false. Then
decide whether the converse is true or false.

21. If a figure is a square, then it has four sides.

22. If a figure has three sides, then it is a triangle.

23. If points A, B, and C lie on a line, then they are collinear.

24. If two lines intersect, then the lines intersect in one point.

25. If four points are coplanar, then the points are collinear.

Mixed Review

Tell which theorem is suggested by each situation.

26. A pencil and a penny lying on a table

27. Crossed swords　　　　**28.** A dart in a dartboard

Use the following assumptions to reach a conclusion.

29. (i) Fran eats Crunchies cereal.
 (ii) If you eat Crunchies cereal, you will have a good morning.

30. (i) No sophomore went to the picnic.
 (ii) Jane went to the picnic.

◇ Subtract.

31. $8 - 15$　　　　**32.** $12 - (-9)$　　　　**33.** $-3 - 1$

Geometry In Communication

Advertisements

Influenced by the wording of advertisements, consumers may shop in a new store or try a new product. Many advertisements in magazines and newspapers contain examples of conditional sentences.

Do your teeth look dull? Have they lost their sparkle? Then read on!!!

If you've always wanted a beautiful smile, then brush with **BRIGHTWHITE**. **BRIGHTWHITE** will bring back the sparkle in your smile.

Sample   Identify the conditional sentence in this advertisement. Find the hypothesis and the conclusion. Then write its converse.

If you've always wanted a beautiful smile, then brush with BRIGHTWHITE.

Write down the conditional sentence.

If <u>you've always wanted a beautiful smile</u>,
hypothesis
then <u>brush with BRIGHTWHITE</u>.
conclusion

Identify the hypothesis and the conclusion.

If you brush with BRIGHTWHITE, then you've always wanted a beautiful smile.

Write the converse.

Identify the conditional sentence in each advertisement. Find the hypothesis and conclusion for each sentence. Then write its converse.

1.

If you're out running a couple of miles a day, then you're on your way to building a better body. And what better way to show off all your hard work than in **RUNTEX** shorts and tops?

For the look and comfort that all runners want, buy the **RUNTEX** label.

2.

Rub-A-Tub-Tub.

That's our motto.

If you're tired of scrubbing tubs clean, then use **SCRUBEASE**. Gently rub **SCRUBEASE** on your tub and then rinse off with water. There's no scrubbing with **SCRUBEASE**.

Soon Rub-A-Tub-Tub will be your motto, too.

3.

I bet you haven't tasted a green bean this fresh since you were a kid on your grandfather's farm. If freshness is what counts, then shop at the FARM MART. All our produce arrives daily, fresh from local farmers.

Chapter 2 Review

Use the following assumptions to reach a conclusion. *2-1*

1. (i) All fish have gills.
   (ii) All trout are fish.

2. (i) None of the freshmen went to the game.
   (ii) Jim is a freshman.

3. (i) If you save your money, you can afford that stereo.
   (ii) You save your money.

4. Start with the assumption that the measures of the angles of any triangle add up to 180°. Use this assumption to draw a conclusion about the total angle measure of a nine-sided figure (nonagon).

Use the figure at the right for Problems 5–8. *2-2*

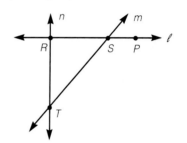

5. Name a set of three collinear points.

6. Name a set of three noncollinear points.

7. What is the intersection of line m and line n?

8. What is the intersection of $\overleftrightarrow{SP}$ and $\overleftrightarrow{RT}$?

Use the figure at the right for Problems 9–11.

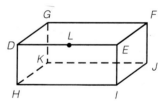

9. Name a set of three collinear points.

10. Name a set of four coplanar points.

11. Name a set of four noncoplanar points.

12. What is wrong with this definition? A circle is the locus of points 5 in. from a given point.

Which postulate is suggested by each statement or picture? *2-3*

13. A three-legged table does not wobble.

14. A bricklayer uses a cord stretched between two nails to make a straight row.

15.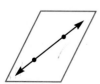

16.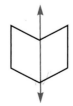

17. How many different lines are determined by three collinear points?

2-4 Tell which theorem explains each statement.

18. A line not in a plane cannot intersect the plane in two points.

19. When you drive a nail into a board, it makes only one hole.

Tell which theorem is suggested by each picture.

20.  **21.**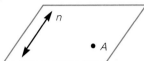

2-5 Identify the hypothesis and the conclusion. Then write the converse.

22. If the rain falls, then weeds grow.

23. If two lines are parallel, then they do not intersect.

24. If there is lightning, I am staying inside.

25. Frank mows the lawn if his sister helps him.

Chapter 2 Test

Use the following assumptions to reach a conclusion.

1. (i) All birds have feathers.
(ii) All ostriches are birds.

2. (i) No one who lives in Texas lives in Denver.
(ii) Juana lives in Texas.

3. (i) If you buy XYZ vitamins, you will be healthy.
(ii) Alberto buys XYZ vitamins.

4. Start with the assumption that the measures of the angles of any triangle add up to 180°. Use this assumption to draw a conclusion about the total angle measure of a seven-sided figure (heptagon).

Use the figure at the right for Problems 5–8.

5. Name a set of three noncollinear points.

6. Name a set of three collinear points.

7. What is the intersection of line m and line n?

8. What is the intersection of $\overleftrightarrow{AC}$ and $\overleftrightarrow{DE}$?

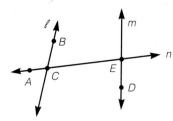

9. What is wrong with this definition? It is the locus of points a given distance from a given point.

Use the figure at the right for Problems 10–12.

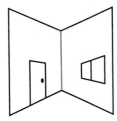

10. Name a set of four coplanar points.

11. Name a set of three collinear points.

12. Name a set of four noncoplanar points.

Which postulate is suggested by each statement or picture?

13. The edge of a cardboard box is a straight line.

14. A three-legged chair sits level on a tile floor.

15.

16.

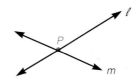

17. How many different lines are determined by three noncollinear points?

Tell which theorem explains each statement.

18. Two different lines cannot contain the same two points.

19. If point A is not on line m, then m and A are contained in exactly one plane.

Tell which theorem is suggested by each picture.

20.

21.

Identify the hypothesis and the conclusion. Then write the converse.

22. If two lines intersect, then they have a point in common.

23. If the sky is blue, then the sun shines.

24. Corliss will drive if we go to the beach.

25. If this is Monday, I have a Spanish class.

3 Distance and Angle Measure

These cranes are being used in the construction of a bridge. The measure of their angles with the horizontal determines their height above the water.

3-1 Coordinates and Distance

Objective: Name coordinates of points on a line and use coordinates to find distances.

Vietta wanted to measure her science textbook to make a cover for it, but she was only able to find a broken yardstick. She placed the yardstick so that the 17-in. mark was at the top of the book and the 27-in. mark was at the bottom of the book. When she determined that the book was 10 in. tall, Vietta was using some of the ideas of this lesson.

To begin, we start with a line and apply the idea of a ruler extending forever in both directions. This creates a **number line.**

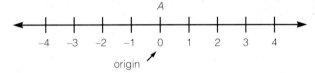

The number that corresponds to a point on the line is called the **coordinate** of that point. The point with coordinate zero is called the **origin.** Point A is the origin on the above number line.

Examples Use the number line below.

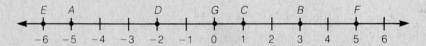

1. Name the coordinates of points A and C.
   The coordinates of A and C are -5 and 1, respectively.
2. Which point has the coordinate 3?
   Point B has the coordinate 3.

Try This... Use the number line above.

a. Name the coordinates of points D and G.

b. Which point is the origin?

Notice how the paper clip is 5 cm long regardless of how we place the ruler. This is related to the following postulate:

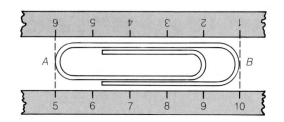

Postulate 6 **The Distance Postulate**

To every pair of distinct points A and B, there corresponds a positive number, AB, called the distance between the points.

We can find distances on a number line by subtracting coordinates. When subtracting coordinates, order makes a difference. On the number line in Examples 1 and 2,

$$BC = 3 - 1 = 2,$$

but changing the order gives

$$BC = 1 - 3 = -2$$

Because we want distance to be a positive number, we take the absolute value of the difference. Then the order of subtraction does not matter.

Examples Find these distances on the number line.

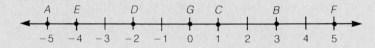

3. FB $FB = |5 - 3| = |2| = 2$
       or $FB = |3 - 5| = |-2| = 2$

4. AB $AB = |3 - (-5)| = |8| = 8$
       or $AB = |-5 - 3| = |-8| = 8$

Try This... Find these distances on the number line above.

    c. EF **d.** GF

    e. DB **f.** EG

Discussion Give examples of how number lines are used in sports. When might the Distance Postulate be used in each case?

Exercises

Practice

Use the number line below for Exercises 1–22.

Name the coordinate of each of these points.

1. F   **2.** G   **3.** H   **4.** J   **5.** K

6. E   **7.** A   **8.** T   **9.** R   **10.** P

Name the points that have these coordinates.

11. −4   **12.** 8   **13.** −9   **14.** −3   **15.** 3

Give the letter of the point named by each coordinate to solve the coded messages in Exercises 16 and 17.

16. $\underline{}\ \underline{}\ \underline{}\ \underline{}\ \underline{}\ \ \underline{}\ \underline{}\ \underline{}\ \underline{}\ \underline{}\ \underline{}\ \underline{}\ \underline{}$
$\ \ \ 8\quad -4\quad 8\quad -8\quad -7\quad\ \ 4\quad -7\quad 2\quad 10\quad -7\quad 0\quad -10\quad -3$

17. $\underline{}\ \underline{}\ \underline{}\ \underline{}\ \underline{}\ \ \underline{}\ \underline{}\ \underline{}\ \underline{}\ \underline{}\ \underline{}\ \underline{}\ \underline{}$
$\ \ \ 7\quad 2\quad -6\quad -7\quad -6\quad\ \ 10\quad -7\quad -1\quad -1\quad -2\quad 4\quad -7\quad -1$

$\underline{}\ \underline{}\ \underline{}\ \ \underline{}\ \underline{}\ \underline{}$
$-2\ -10\ -7\quad 5\quad -9\quad 9$

Find these distances on the number line.

18. AB   **19.** BC   **20.** CD   **21.** DE   **22.** EF

Extend and Apply

23. Point F has coordinate −3. Point G has coordinate 5. Which of the following is FG?

(A) 3   (B) 8   (C) −8   (D) 2

Name the coordinate of each point on the thermometer.

24. A   **25.** B   **26.** C   **27.** D

Determine whether each statement is true or false.

28. For any points A and B on a line, AB is the same distance as BA.

29. When we find the distance between two points on a number line, the distance may be a negative number.

30. If point P has coordinate 7 and R has coordinate −2, then PR = 5.

31. The origin on a number line has the coordinate 0.

Use Mathematical Reasoning

Tanya places a surveyor's steel tape measure along the front of her yard and then plants some bushes along the tape so that they are equally spaced. Given the coordinates of two of the bushes, find the coordinates of the others.

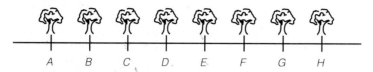

A   B   C   D   E   F   G   H

32. $A = 2$ and $B = 4$ **33.** $B = 4$ and $D = 10$

34. $E = 17$ and $A = 1$ **35.** $C = 13$ and $H = 43$

36. $A = 3$ and $H = 38$ **37.** $A = 6$ and $G = 33$

38. The coordinate of X is -3 and $XY = 15$. What is the coordinate of Y?

39. The coordinate of X is 7 and $XY = 7$. What is the coordinate of Y?

40. The coordinate of Y is 1 and $XY = 2$. What is the coordinate of Y?

Mixed Review

Write the converse of each sentence.

41. If it is 11:00, then you are late.

42. If $\overrightarrow{AB}$ and $\overrightarrow{AC}$ are rays, they have a common endpoint.

43. The ice melts if it is too hot.

◈ Subtract.

44. $12 - 15$ **45.** $-3 - 0$ **46.** $-16 - 5$ **47.** $-4 - (-6)$

Visualization

Which of the patterns below can be folded into a cube?

(A)

(B)

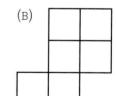

(C)

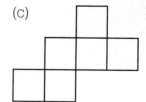

3-2 Segments and Rays

Objective: Recognize subsets of lines, such as segments, rays, and opposite rays, and find midpoints of segments.

Norman wants to hang a painting so that its left edge is 5 feet from the corner of the room. The painting is 18 inches wide. Where should he drive the nail in the wall? The answer to this question is related to the midpoint of a segment.

In Chapter 1 we described segments and rays. We will now state these descriptions as definitions, and add the definition of *opposite rays*.

Term	Symbol	Definition	Picture
Segment	$\overline{AB}$ or $\overline{BA}$	The set of points of a line containing A, B, and all points between A and B. A and B are called the **endpoints** of the segment.	
Ray	$\overrightarrow{AB}$	The subset of a line $\overleftrightarrow{AB}$ that contains A and all points on the same side of A as B. Point A is the **endpoint** of $\overrightarrow{AB}$.	
Opposite rays		Given a line $\overleftrightarrow{CB}$ with A between C and B, the rays $\overrightarrow{AB}$ and $\overrightarrow{AC}$ are called opposite rays.	

Examples Lines $\overleftrightarrow{AC}$ and $\overleftrightarrow{DE}$ intersect at B.

1. What is the endpoint of $\overrightarrow{BA}$?
   Point B is the endpoint of $\overrightarrow{BA}$.

2. Name $\overrightarrow{DE}$ a different way.
   $\overrightarrow{DB}$ is the same ray.

3. Name all the pairs of opposite rays shown.
   $\overrightarrow{BA}$ and $\overrightarrow{BC}$; $\overrightarrow{BD}$ and $\overrightarrow{BE}$

Try This... Lines $\overleftrightarrow{AE}$ and $\overleftrightarrow{DB}$ intersect at C.

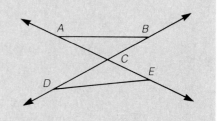

a. What is the endpoint of $\overrightarrow{DB}$?

b. Name $\overrightarrow{EA}$ a different way.

c. Name all the pairs of opposite rays shown.

Example   4. On $\overleftrightarrow{AD}$, what points do $\overline{AC}$ and $\overline{BD}$ have in common?

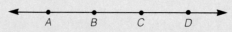

All of the points of $\overline{BC}$ are common to both $\overline{AC}$ and $\overline{BD}$.

Try This... d. Refer to $\overleftrightarrow{AD}$ above. What points do $\overrightarrow{BD}$ and $\overrightarrow{DA}$ have in common?

Using the Distance Postulate, we can talk about the length of a segment. We will now use the concept of distance to define the *midpoint* of a segment.

Definition M is the **midpoint** of segment $\overline{AB}$ whenever M is between A and B, and $AM = MB$. The midpoint M is said to **bisect** $\overline{AB}$.

Every segment has exactly one midpoint. The definition of a midpoint does not say this, but it can be proven.

Theorem 3.1   Every segment has exactly one midpoint.

Here is a way to find the coordinate of a midpoint. For any segment $\overline{AB}$ whose endpoints have coordinates a and b, the coordinate of midpoint M is

$$\frac{a + b}{2}$$

That is, we simply find the *average* of the coordinates of the endpoints.

The following Examples illustrate how to find the midpoint of a segment.

Examples For each segment, find the coordinate of the midpoint.

5.

A   M   B
2   8

The coordinate of the midpoint is $\dfrac{2 + 8}{2} = \dfrac{10}{2} = 5$.

6.

A   M   B
-8.5   7.9

The coordinate of the midpoint is $\dfrac{-8.5 + 7.9}{2} = -0.3$.

We can use a calculator as follows:

8.5 $\boxed{+/-}$ $\boxed{+}$ 7.9 $\boxed{=}$ $\boxed{\div}$ 2 $\boxed{=}$ $\boxed{\hspace{3cm} -0.3}$

Try This... For each segment, find the coordinate of the midpoint.

e. $\overline{AB}$

A   M   B
-5   1

f. $\overline{GH}$

G   M   H
-4.52   7.14

Discussion For which of the symbols AB, $\overline{AB}$, $\overrightarrow{AB}$, and $\overleftrightarrow{AB}$ can we change the order of the letters without changing what the symbol represents?

Exercises

Practice

Use the figure at the right for Exercises 1–7.

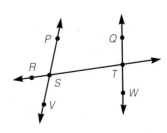

1. What is the endpoint of $\overrightarrow{TQ}$?
2. What is the endpoint of $\overrightarrow{ST}$?
3. Name $\overrightarrow{TR}$ a different way.
4. Name $\overrightarrow{PV}$ a different way.
5. Name all pairs of opposite rays shown.
6. What points do $\overrightarrow{WQ}$ and $\overrightarrow{TW}$ have in common?
7. What points do $\overline{RS}$ and $\overrightarrow{ST}$ have in common?

For each segment, find its length and the coordinate of its midpoint.

8.
```
A       M       B
●———————●———————●
0               26
```

9.
```
A       M       B
●———————●———————●
-4              0
```

10.
```
A     M     B
●—————●—————●
5           18
```

11.
```
A       M       B
●———————●———————●
-9.5            -3.5
```

12.
```
A         M         B
●—————————●—————————●
-6.75               5.32
```

13.
```
A         M         B
●—————————●—————————●
-10.08              7.72
```

Extend and Apply

14. The coordinates of R and T are as shown and S is the midpoint of $\overline{RT}$. What is the coordinate of S?
    (A) -2 (B) 2 (C) 0 (D) 1

```
R       S       T
●———————●———————●
-2              4
```

Find the coordinate of the midpoint for each of the following.

15.

16.
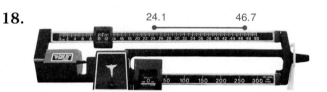

17.

18.

19. Kim wants to hang a 4-ft-wide painting so that its left edge is 7 ft from the edge of the wall. How far from the edge of the wall should she place the nail?

20. Draw points R and S. Then draw $\overrightarrow{RS}$ and $\overrightarrow{SR}$. What set of points do $\overrightarrow{RS}$ and $\overrightarrow{SR}$ have in common?

21. Draw three collinear points L, M, and N, where M is between L and N. Then draw $\overrightarrow{NL}$ and $\overrightarrow{MN}$. What set of points do these rays have in common?

22. Draw three collinear points A, B, and C, where B is between A and C. Then draw $\overrightarrow{BA}$ and $\overrightarrow{BC}$. What set of points do these rays have in common?

Use Mathematical Reasoning

23. Suppose the coordinate of point D is 5 and the coordinate of point F is 28. E is the midpoint of $\overline{DF}$ and G is the midpoint of $\overline{DE}$. Find the coordinate of G.

```
D   G   E           F
●———●———●———————————●
```

Complete each sentence with *always*, *sometimes*, or *never*.

24. A ray _always_ has an endpoint.

25. A ray _never_ has two endpoints.

26. A segment _A_ has two endpoints.

27. Lines _N_ have endpoints.

28. Opposite rays _A_ form a line.

29. The coordinate of the midpoint of a segment is _____ negative.

30. The coordinate of the midpoint of a segment is _____ zero.

31. The coordinate of the midpoint of a segment is _____ positive.

32. The length of a segment is _N_ negative.

33. The length of a segment is _A_ positive.

34. A segment _N_ has more than one midpoint.

Mixed Review

Tell which theorem or postulate is suggested by each situation or picture.

35. The tip of a pencil makes one dot on a piece of paper.

36. A three-legged chair does not wobble.

37.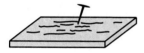

38.

Use the figure at the right for Exercises 39–41.

39. Name two sets of three collinear points.

40. What is the intersection of $\overleftrightarrow{JK}$ and $\overleftrightarrow{LM}$?

41. What is the intersection of $\overline{JK}$ and $\overline{KM}$?

 Find each absolute value.

42. $|-13|$ **43.** $|-5.3|$ **44.** $|7.1|$ **45.** $|123|$

3-3 Coordinate Geometry

Objective: Find the coordinates of a point on a graph and find its coordinates after a reflection or translation.

The city map at the right shows a building located at the intersection of 34th Street and Seventh Avenue. We can locate a point on a plane in much the same way as we located this building on the map.

Just as a point on a line can be named by its coordinate, a point on a plane can be named by a pair of numbers called its **coordinates.**

We begin with two lines, the **x-axis** and the **y-axis,** which intersect at right angles at a point O, called the **origin.** The x-axis and the y-axis determine the **x-y plane.**

Point A has coordinates $(4, 3)$. The first number, 4, is the **x-coordinate** and the second number, 3, is the **y-coordinate.** The x-coordinate tells the distance right (positive) or left (negative) from the vertical axis. The y-coordinate tells the distance up (positive) or down (negative) from the horizontal axis. We use the notation $P(a, b)$ to denote the point P with coordinates (a, b).

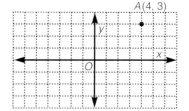

Example 1. Plot the points $A(-2, 3)$, $B(3, -5)$, and $C(-3, 0)$.

For $A(-2, 3)$, the x-coordinate, -2, is negative. We move 2 units left of the vertical axis. The y-coordinate, 3, is positive. We move 3 units up from the horizontal axis.

The graphs of $B(3, -5)$ and $C(-3, 0)$ are shown.

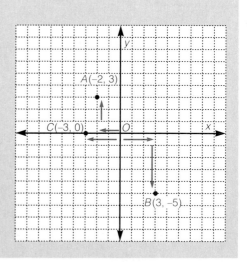

Try This... Use graph paper. Draw and label an x-axis and a y-axis. Plot these points.

    a. $M(4, 2)$ **b.** $P(5, 0)$

    c. $D(-3, 5)$ **d.** $R(-2, -6)$

    e. $K(3, -1)$ **f.** $W(0, -5)$

Example   **2.** Find the coordinates of the points E and S.

Point E is 2 units to the left of the vertical axis and 4 units down from the horizontal axis. Its coordinates are $(-2, -4)$.

The coordinates of point S are $(5, -3)$.

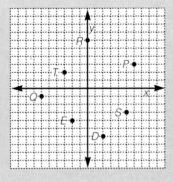

Try This... g. Find the coordinates of points T, Q, R, P, and D in the graph of Example 2.

In Chapter 1 we looked at some transformations, including reflections (or flips) and translations (or slides). We will see how these transformations move points in the x-y plane.

Explore

1. Mark the points $A(5, 3)$, $B(3, 0)$, and $C(-4, -2)$ on a piece of graph paper.
2. Fold the graph paper back along the y-axis. Then use the point of your compass or pencil to carefully punch a hole through the paper at these three points.
3. Open the paper and label the three marks on the other side of the y-axis as A', B', and C', as shown.
4. How far is A from the y-axis? How far is A'? How is the y-axis (the fold line) related to each of the pairs of points?

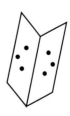

You might have said that point B was flipped over the y-axis to B'. So, a reflection over a line moves the figure the same distance on the opposite side of the line.

Examples **3.** Find the reflection of the point $(2, 3)$ when reflected over the x-axis.

The point $(2, 3)$ is 3 units from the x-axis. Its reflection is 3 units from the x-axis on the other side of the x-axis. This is the point $(2, -3)$.

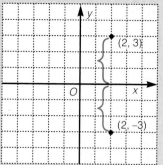

4. Find the reflection of the point $(1, 5)$ when it is reflected over the y-axis.

The point $(1, 5)$ is 1 unit from the y-axis. Its reflection is 1 unit from the y-axis on the other side of the y-axis. This is the point $(-1, 5)$.

Try This... **h.** Find the reflection of the point $(2, 3)$ when reflected over the y-axis.

i. Find the reflection of the point $(4, 5)$ when reflected over the x-axis.

When a point is translated vertically or horizontally, its x- or y-coordinate changes as illustrated in the next example.

Example **5.** Find the translation of the point $(4, -1)$ when translated upward 6 units.

We move 6 units upward to $(4, 5)$.

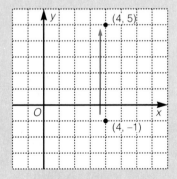

Try This... **j.** Find the translation of the point $(-2, 2)$ when translated 4 units to the left.

Discussion Describe a transformation that will move the point $(2, 5)$ to the point $(2, -5)$. Is there more than one answer?

Exercises

Practice

Use graph paper. Draw and label an x-axis and a y-axis.
Plot these points.

1. $N(5, 2)$ **2.** $T(6, 4)$ **3.** $P(-3, 1)$

4. $L(-4, 2)$ **5.** $C(2, -3)$ **6.** $A(3, -5)$

7. $Q(-2, -4)$ **8.** $W(-7, -5)$ **9.** $H(0, 6)$

10. $S(0, -4)$ **11.** $J(6, 0)$ **12.** $M(-5, 0)$

Find the coordinates of points A, B, C, D, and E on the graphs
below.

13.

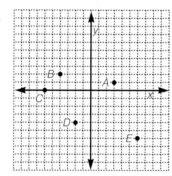

14.

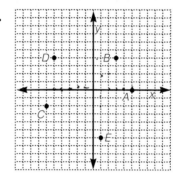

Find the coordinates of the reflection of
each point when reflected over the x-axis.

15. A **16.** B

17. C **18.** D

19. E **20.** F

Find the coordinates of the reflection of
each point when reflected over the y-axis.

21. A **22.** B

23. C **24.** D

25. E **26.** F

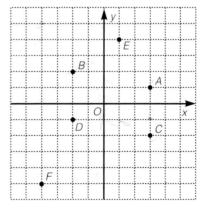

27. Find the coordinates of the translation of point A when
translated downward 5 units.

28. Find the coordinates of the translation of point C when
translated 3 units to the right.

29. Find the coordinates of the translation of point E when
translated upward 1 unit.

30. Find the coordinates of the translation of point F when
translated 5 units to the left.

Extend and Apply

31. The coordinates of the reflection of $(2, -3)$ when reflected over the y-axis are ____.

(A) $(-2, -3)$ (B) $(2, 3)$ (C) $(-2, 3)$ (D) $(2, -3)$

32. Suppose a point is located on the x-axis. What happens when the point is reflected over the x-axis? Find the coordinates of the reflection of the point $(3, 0)$ reflected over the x-axis.

33. Suppose a point is located on the y-axis. What happens when the point is reflected over the y-axis? Find the coordinates of the reflection of the point $(0, -5)$ reflected over the y-axis.

Use graph paper to plot a point that satisfies each of the following conditions.

34. The x-coordinate is 3 greater than the y-coordinate.

35. The x-coordinate and the y-coordinate are the same.

36. The x-coordinate is twice the y-coordinate.

37. The product of the coordinates is -24.

38. The sum of the coordinates is 12.

Use Mathematical Reasoning

39. Graph 12 points such that the sum of the coordinates of each point is 8. What do you observe?

40. Graph 12 points such that the difference between the coordinates of each point is 2. What do you observe?

Mixed Review

Use the figure at the right for Exercises 41–45.

41. Name $\overrightarrow{PC}$ a different way.

42. What is the endpoint of $\overrightarrow{PB}$?

43. Name all pairs of opposite rays shown.

44. What points do $\overline{AE}$ and $\overline{PC}$ have in common?

45. What points do $\overline{AP}$ and $\overline{PE}$ have in common?

For each segment, find its length and the coordinate of its midpoint.

46.

47.

 Add.

48. $-6 + 13$

49. $-7 + -7$

50. $5 + -4$

51. $18 + -19$

52. $-15 + -2$

53. $9 + -12$

Line Graphs

A common use of coordinate geometry is in the creation of line graphs. Line graphs are useful when a large amount of information is to be communicated visually.

The graph at the right shows how a population of bacteria changes after exposure to light for a week. The *x*-axis tells how many days have elapsed since the start of the experiment. The *y*-axis tells the number of bacteria present.

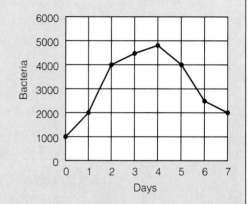

The coordinates of points on the graph give information about the population of bacteria. For example, the point with coordinates (0, 1000) tells us that on Day 0, the start of the experiment, 1000 bacteria were present.

Problems

Use the above graph.

1. About how many bacteria were present after 6 days?

2. On which days were about 4000 bacteria present?

3. After how many days was the bacteria population at its peak?

4. At the end of the experiment, how many more bacteria were present than at the start?

Students at Springfield High School were asked how many hours they work at weekend jobs. The line graph at the right shows the results.

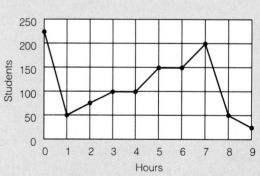

5. What does the point with coordinates (5, 150) represent?

6. How many students work 4 hours on weekends?

7. About how many students do not work on weekends?

8. Among students who work on weekends, what is the most common number of hours worked?

9. How many students work 8 or more hours on weekends?

3-4 Angles

Objective: Identify angles and parts of angles, and name angles in different ways.

The structure shown contains many angles. Recall that we considered angles briefly in Chapter 1. We will now look at angles in greater detail.

> **Definition** An **angle** is a figure consisting of two rays with a common endpoint. If the rays are opposite rays, the angle is called a *straight angle*.

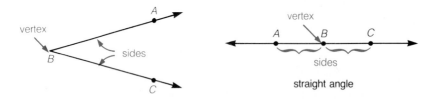

The angles above may be named $\angle ABC$ or $\angle CBA$. The middle letter *always* names the vertex. If there are no other angles with vertex B, we may write $\angle B$.

We sometimes use numerals to name angles.

$\angle 1$ is $\angle ABE$ (or $\angle ABD$)
$\angle 2$ is $\angle EDF$
$\angle 3$ is $\angle EBC$ (or $\angle DBC$)

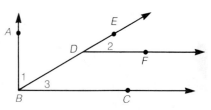

The following Examples show how angles may be named in different ways:

> **Example** Identify all the angles and name them in as many ways as possible.
>
> **1.** There is one angle that may be named $\angle PQR$, $\angle RQP$, or $\angle Q$.

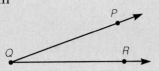

Examples

2. There are three angles:

   $\angle ADB$ or $\angle BDA$ or $\angle 1$,
   $\angle BDC$ or $\angle CDB$ or $\angle 2$,
   and $\angle ADC$ or $\angle CDA$.

   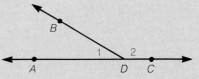

   (We do not use $\angle D$ because it could name any of the three angles.)

3. Although the sides of an angle are rays, the sides of $\triangle ABC$ each determine a ray. Thus we refer to $\angle ABC$, $\angle BCA$, and $\angle CAB$. We can also name these angles $\angle CBA$, $\angle ACB$, and $\angle BAC$, or $\angle B$, $\angle C$, and $\angle A$.

   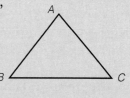

Try This... Identify all the angles and name them in as many ways as possible.

a.

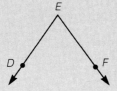

b.

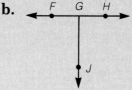

Examples Identify the vertex and the sides of each angle.

4. Vertex: Y
   Sides: $\overrightarrow{YX}$ and $\overrightarrow{YZ}$

5. All three angles in the bicycle figure have vertex E.
   Sides of $\angle DEG$: $\overrightarrow{ED}$ and $\overrightarrow{EG}$
   Sides of $\angle FEG$: $\overrightarrow{EF}$ and $\overrightarrow{EG}$
   Sides of $\angle DEF$: $\overrightarrow{ED}$ and $\overrightarrow{EF}$

   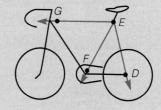

Try This... Identify the vertex and the sides of each angle.

c.

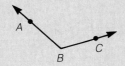

d.

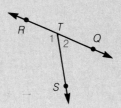

Sometimes we refer to the **interior** of an angle. Points P, Q, and R are in the interior of $\angle ABC$.

Points X, Y, and Z are *not* in the interior of $\angle ABC$. They are in the **exterior** of $\angle ABC$.

The points *on* the angle itself are not in the interior or the exterior. For example, point C is on $\angle ABC$. Straight angles have no interior or exterior.

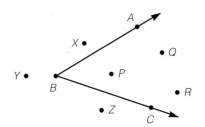

Examples 6. Which points are in the interior of $\angle BAC$?

Points Q, S, and T are in the interior of $\angle BAC$.

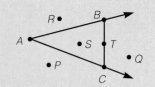

7. Which points are in the exterior of $\angle BAC$?

Points P and R are in the exterior of $\angle BAC$.

Try This... e. Which points are in the interior of $\angle DFE$?

f. Which points are in the exterior of $\angle DFG$?

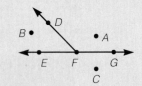

Discussion Why do you suppose we do not identify an interior or an exterior for straight angles?

Exercises

Practice

In Exercises 1–6, name in two ways each angle shown. Identify the vertex and the sides.

1.
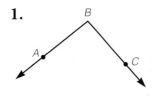

2.
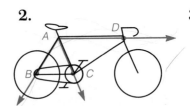

3.
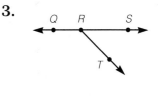

4.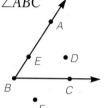

5.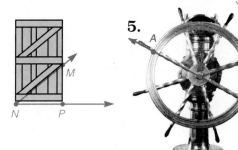

6.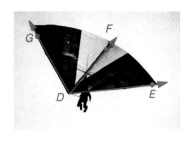

Draw and label each of these angles.

7. $\angle ABD$ **8.** $\angle FEG$ **9.** $\angle PRQ$ **10.** $\angle HSR$

Identify the points in the interior, in the exterior, and on each named angle.

11. $\angle ABC$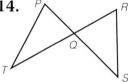

12. $\angle MNP$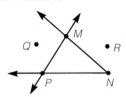

13. $\angle XYZ$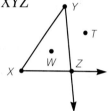

Extend and Apply

Identify the angles that can be named using only the vertex letter.

14.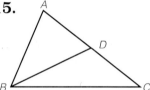

15.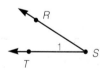

16.

17. Which of these is *not* a name for the angle at the right?

(A) $\angle RST$ (B) $\angle T$

(c) $\angle TSR$ (D) $\angle 1$

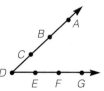

18. Draw and label an angle that could correctly be named $\angle ABC$ or $\angle DBF$.

Use Mathematical Reasoning

19. In how many different ways can the angle shown be named using three letters?

Mixed Review

Suggest a possible counterexample for each statement.

20. All geometric figures can be drawn using only straight lines.

21. All angles are acute or obtuse.

◇ Add.

22. $19 + -31$ **23.** $-45 + 52$ **24.** $-4 + -12$ **25.** $14 + -14$

Algebra Connection

Solving Equations

Addition and subtraction are inverse operations. That is, each undoes the other.

$$7 + 3 - 3 = 7$$

Likewise, multiplication and division are inverse operations.

$$8 \div 2 \times 2 = 8$$

An equation is like a balance. If you do the same operation on both sides of the equation, you keep the balance.

Samples   **1.** Solve $x + 42 = 63$.

$$\begin{aligned} x + 42 &= 63 \qquad \text{We want the variable } x \text{ by itself.}\\ -42 &= -42 \qquad \text{Subtracting 42 undoes adding 42.}\\ \hline x &= 21 \qquad \text{Subtracting 42} \end{aligned}$$

The answer is 21.
We can check to see if the answer is correct by replacing x with 21 in the original equation.

$$\begin{aligned} x + 42 &= 63\\ 21 + 42 &= 63\\ 63 &= 63 \qquad \checkmark \quad \text{The answer checks.} \end{aligned}$$

2. Solve $p - 10 = 14$.

$$\begin{aligned} p - 10 &= 14 \qquad \text{We want the variable } p \text{ by itself.}\\ +10 &= +10 \qquad \text{Adding 10 undoes subtracting 10.}\\ \hline p &= 24 \qquad \text{Adding 10} \end{aligned}$$

The answer is 24.
Replacing p with 24 will show that the answer checks.

3. Solve $3t = 12$.

$$\begin{aligned} 3t &= 12 \qquad \text{We want the variable } t \text{ by itself.}\\ \frac{3t}{3} &= \frac{12}{3} \qquad \text{Dividing by 3 undoes multiplying by 3.}\\ t &= 4 \qquad \text{Dividing by 3} \end{aligned}$$

The answer is 4.
Replacing t with 4 will show that the answer checks.

Sample **4.** Solve $\frac{y}{22} = 10$.

$$\frac{y}{22} = 10 \qquad \text{We want the variable } y \text{ by itself.}$$

$$\frac{y}{22} \cdot 22 = 10 \cdot 22 \qquad \text{Multiplying by 22 undoes dividing by 22.}$$

$$y = 220$$

The answer is 220.
Replacing y with 220 will show that the answer checks.

Problems Solve and check.

1. $x + 28 = 82$

2. $x + 13 = 67$

3. $x + 129 = 162$

4. $y + 103 = 174$

5. $y + 91 = 303$

6. $y + 37 = 222$

7. $306 = x + 72$

8. $195 = y + 81$

9. $100 = x + 73$

10. $x - 83 = 29$

11. $x - 36 = 390$

12. $x - 6 = 39$

13. $y - 113 = 41$

14. $y - 81 = 38$

15. $y - 136 = 32$

16. $294 = y - 45$

17. $39 = x - 29$

18. $364 = y - 21$

19. $n + 34 = 69$

20. $31 = n - 17$

21. $n + 78 = 126$

22. $396 = n - 500$

23. $n + 76 = 189$

24. $n - 58 = 23$

25. $72 = n + 34$

26. $n - 91 = 101$

27. $39 = n - 28$

28. $p + 65 = 103$

29. $29 = p - 321$

30. $p + 48 = 129$

31. $3x = 18$

32. $7x = 91$

33. $112 = 4x$

34. $6x = 54$

35. $72 = 8x$

36. $9x = 126$

37. $3y = 78$

38. $44 = 11y$

39. $2y = 116$

40. $\frac{y}{4} = 23$

41. $\frac{y}{8} = 34$

42. $29 = \frac{y}{3}$

43. $\frac{x}{7} = 19$

44. $\frac{x}{6} = 45$

45. $143 = \frac{y}{3}$

46. $\frac{p}{2} = 38$

47. $34 = \frac{p}{7}$

48. $\frac{p}{5} = 27$

49. $12r = 120$

50. $36 = 3r$

51. $155 = \frac{r}{2}$

52. $16z = 144$

53. $9z = 63$

54. $205 = \frac{z}{5}$

55. $10x = 140$

56. $\frac{y}{13} = 13$

57. $312 = 3r$

58. $\frac{y}{25} = 1$

59. $10 = \frac{x}{7}$

60. $139 = \frac{p}{5}$

3-5 Angle Measurement

Objective: Measure angles with a protractor, draw angles with a given measure, and classify angles according to their measures.

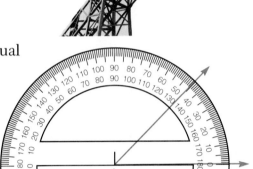

Angles are such an important part of the world around us that we need to be able to measure them as accurately as we can measure segments.

Just as we can measure the length of a segment with a ruler, we can use a **protractor** to find the measure of an angle. Recall from Chapter 1 that the usual unit of angle measure is called a *degree*.

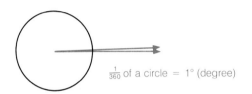

$\frac{1}{360}$ of a circle = 1° (degree)

Vertex

Postulate 7 The Angle Measure Postulate

For each angle there is exactly one number n, called its measure, where $0° < n \le 180°$.

For the measure of $\angle ABC$ we write "$m\angle ABC$." We can find the measure of an angle by using a protractor as shown below.

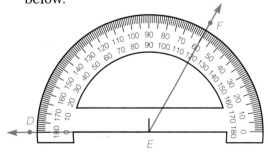

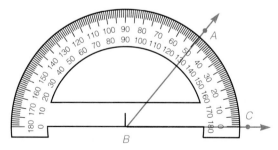

Read the inner scale. (Since the angle is obtuse, we expect the measure to be between 90° and 180°. Therefore, it makes sense to use the numbers on the inner scale.) The measure of $\angle DEF$ is 120°. We write $m\angle DEF = 120°$.

Read the outer scale. (Since the angle is acute, we expect the measure to be between 0° and 90°. Therefore, it makes sense to use the numbers on the outer scale.) The measure of $\angle ABC$ is 50°. We write $m\angle ABC = 50°$.

Examples Read the measure of each angle.

1. ∠ABC, ∠ABD, ∠ABE
m∠ABC = 30°, m∠ABD = 70°, m∠ABE = 120°

2. ∠PQR, ∠PQS
m∠PQR = 40°, m∠PQS = 130°

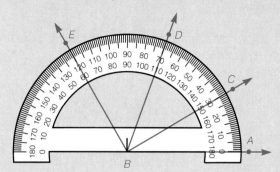

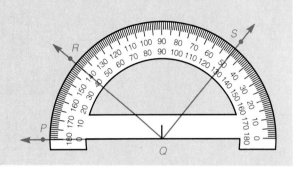

Try This... Read the measure of each angle.

a. ∠PQR, ∠PQS, ∠PQT

b. ∠ABC, ∠ABD

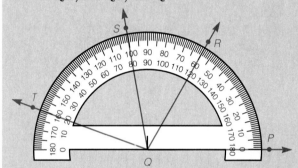

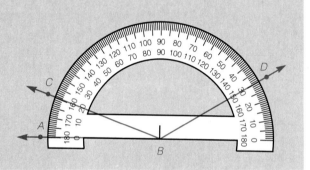

Example **3.** First estimate and then use a protractor to measure ∠JKL.

Estimate: The angle is acute and appears to be about 60°.

Step 1
Place the protractor so that the vertex mark is at K and the bottom edge is along $\overrightarrow{KL}$.

Step 2
Find the point where $\overrightarrow{KJ}$ intersects the edge of the protractor.

Step 3
Read the measurement. Since ∠JKL is acute, the numbers on the outer scale make sense.
m∠JKL = 52°.

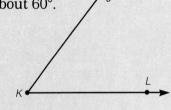

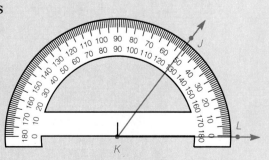

First estimate and then use a protractor to measure these angles.

c.

d.

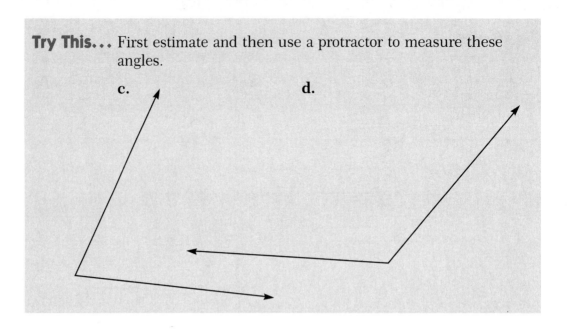

We can use a protractor and a straightedge to draw an angle with a specified measure.

Example 4. Draw an angle with measure 60°.

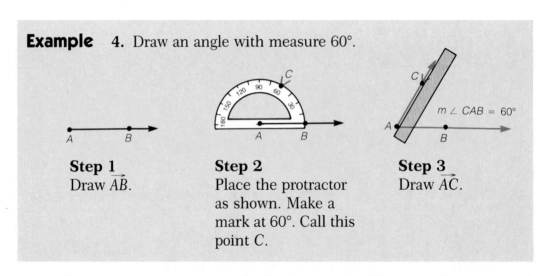

Step 1
Draw $\overrightarrow{AB}$.

Step 2
Place the protractor as shown. Make a mark at 60°. Call this point C.

Step 3
Draw $\overrightarrow{AC}$.

Try This... Draw angles with these measures.

    e. 70°   **f.** 130°   **g.** 85°

Discussion Try to find an example in your classroom of an angle that you think measures approximately 60°. Do the same for a 90° angle and a 150° angle. Which angle was easiest to find? Why do you think this is?

Exercises

Practice

Read the measure of each angle.

1. $\angle ABD$
2. $\angle CBD$
3. $\angle ABE$

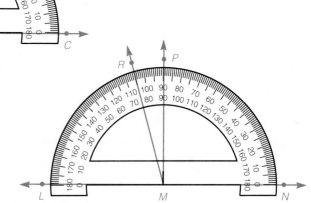

4. $\angle PMN$
5. $\angle RML$
6. $\angle RMN$

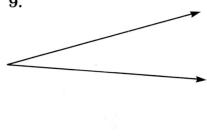

First estimate and then use a protractor to measure each angle.

7.

8.

9.

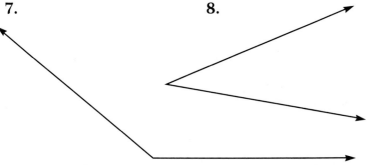

Draw an angle with each measure.

10. $75°$ 11. $105°$ 12. $15°$ 13. $167°$

Extend and Apply

14. A clock shows that it is 12:15. Which of the following is the best estimate of the angle formed by the hands of the clock?

   (A) $45°$ (B) $85°$ (C) $105°$ (D) $135°$

15. Use the figure for Exercises 1–3 to find the measure of $\angle EBD$.

16. Use the figure for Exercises 4–6 to find the measure of $\angle RMP$.

Mixed Review

Find the coordinates of each point when reflected over the x-axis.

17. $(3, 2)$ 18. $(-2, -4)$ 19. $(5, -1)$

Solve and check.

20. $x + 31 = 65$ 21. $40 = m - 2$ 22. $5n = 30$

Algebra Connection

Order of Operations and Solving Equations

When a problem involves more than one operation, mathematicians have agreed that all multiplications and divisions should be computed first, followed by any additions and subtractions.

Does your calculator follow this order of operations? Find each answer using your calculator.

1. $8 + 10 \times 4$ **2.** $20 - 4 \div 8$ **3.** $6 + 20 \div 5$

If you got 48, 19.5, and 10, your calculator automatically follows the order of operations. If you got 72, 2, and 5.2, your calculator does not follow the order of operations, and you will have to enter the numbers in the correct order to get the correct answers.

To solve equations that involve more than one operation, you *undo* the equation in the *reverse* order of the order of operations. This makes sense, as you are undoing the operations.

Sample   Solve and check.

$$9y + 6 = 51 \quad \text{We want the variable } y \text{ by itself.}$$
$$9y + 6 = 51 \quad \text{Subtracting 6 undoes adding 6.}$$
$$\underline{-6 \quad -6}$$
$$9y \quad\;\; = 45$$
$$\frac{9y}{9} = \frac{45}{9} \quad \text{Dividing by 9 undoes multiplying by 9.}$$
$$y = 5$$

Check: $9 \cdot 5 + 6 \overset{?}{=} 51 \quad \text{Replace } y \text{ with 5.}$
$$45 + 6 \overset{?}{=} 51$$
$$51 = 51 \quad \checkmark$$

Problems   Solve and check.

1. $5x - 9 = 26$ **2.** $2x - 12 = 14$ **3.** $3x - 7 = 41$

4. $6y + 4 = 28$ **5.** $4y + 23 = 31$ **6.** $7y + 19 = 75$

7. $9n - 5 = 31$ **8.** $8n + 14 = 86$ **9.** $11n - 3 = 19$

10. $42 = 2x + 6$ **11.** $31 = 4y - 5$ **12.** $53 = 8n + 5$

13. $89 = 8x + 1$ **14.** $37 = 5x + 7$ **15.** $16 = 7t + 9$

16. $45 = 3y - 15$ **17.** $8 = 9n - 37$ **18.** $12 = 6y - 6$

3-6 Adding Angle Measures

Objective: Apply the Angle Addition Postulate and angle bisectors in problems involving angle measures.

Explore

Work in groups of three.

1. Each person in the group should draw an angle labeled $\angle ABC$. There should be an acute angle, an obtuse angle, and a right angle.

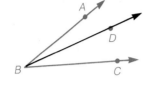

2. Choose a point D in the interior of $\angle ABC$ and draw $\overrightarrow{BD}$. Measure $\angle ABD$, $\angle DBC$, and $\angle ABC$.

Discuss In each case, how do the measures of the three angles compare?

The following postulate summarizes what you may have discovered in the Explore. It says that the measure of an angle is equal to the sum of the measures of the angles "inside" it.

Postulate 8 **The Angle Addition Postulate**

For any nonstraight angle $\angle ABC$, if D is a point in its interior, then $m\angle ABD + m\angle DBC = m\angle ABC$.

For any straight angle $\angle ABC$, if D is a point not on $\angle ABC$, then $m\angle ABD + m\angle DBC = m\angle ABC$.

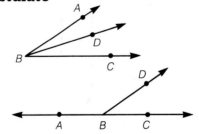

Examples Refer to the figure to complete these statements.

1. $m\angle AFB + m\angle BFC = \underline{m\angle AFC}$

2. $m\angle AFC + m\angle CFD = \underline{m\angle AFD}$

3. $m\angle BFD - m\angle CFD = \underline{m\angle BFC}$

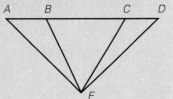

Try This... Refer to the figure to complete these statements.

a. $m\angle BFC + m\angle CFD =$ _____

b. $m\angle FCG + m\angle GCD =$ _____

c. $m\angle AFD - m\angle AFB =$ _____

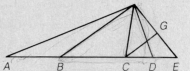

Postulate 8 can often be used to find the measure of an angle when the measures of other angles are known.

Examples **4.** Find $m\angle ABC$.

$m\angle ABC = 30° + 80° = 110°$

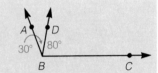

5. $\angle ABC$ is a straight angle. Find $m\angle ABD$.

$m\angle ABD = 180° - 45° = 135°$

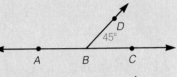

6. $m\angle ABC = 51°$. Find x.

$$m\angle ABD + m\angle DBC = m\angle ABC$$
$$2x + 37 = 51$$
$$2x = 14$$
$$x = 7$$

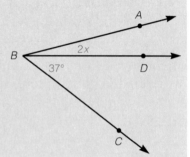

Try This... **d.** Find $m\angle PQR$.

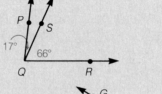

e. $\angle DEF$ is a straight angle. Find x.

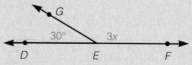

f. $m\angle TRS = 90°$. Find y.

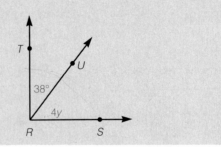

Just as every segment has a midpoint that bisects the segment, for every nonstraight angle there is a ray that bisects the angle.

Definition Given point D in the interior of $\angle ABC$, $\overrightarrow{BD}$ is the **bisector** of $\angle ABC$ whenever $m\angle ABD = m\angle DBC$ (or $m\angle ABD = m\angle DBC = \frac{1}{2}m\angle ABC$).

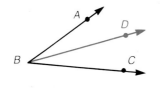

We also say $\overrightarrow{BD}$ bisects $\angle ABC$.

Examples In each figure, $\overrightarrow{BD}$ is the angle bisector of $\angle ABC$.

7. Find $m\angle ABC$.

$$m\angle ABC = 2m\angle DBC$$
$$= 2 \times 30°$$
$$= 60°$$

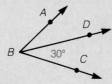

8. $m\angle ABC = 136°$. Find $m\angle DBC$ and $m\angle ABD$.

$$m\angle DBC = m\angle ABD = \frac{1}{2}m\angle ABC$$
$$= \frac{1}{2} \times 136°$$
$$= 68°$$

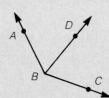

Try This... In each figure, $\overrightarrow{BD}$ is the angle bisector of $\angle ABC$.

g. Find $m\angle ABC$.

h. $m\angle ABC = 146°$. Find $m\angle DBC$ and $m\angle ABD$.

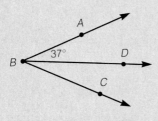

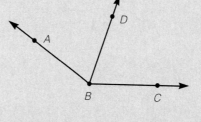

◈ i. $m\angle ABC = 176°$ and $m\angle DBC = 11x$. Find x.

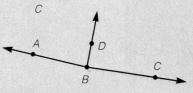

Discussion Is it possible to get two obtuse angles after bisecting an angle? What kind of angles would you get if you bisected a straight angle?

Exercises

Practice

Use the figures to complete these statements.

1. $m\angle AFC + m\angle CFE =$ _____
2. $m\angle ACB + m\angle ACF =$ _____
3. $m\angle EFC + m\angle CFA =$ _____
4. $m\angle FCB + m\angle DCF =$ _____
5. $m\angle DCB - m\angle BCA =$ _____
6. $m\angle ACE - m\angle FCE =$ _____

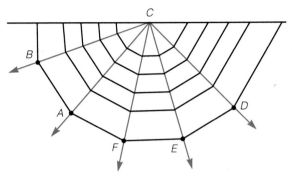

7. $m\angle GPH + m\angle HPK =$ _____
8. $m\angle OPH + m\angle HPK =$ _____
9. $m\angle NPL - m\angle NPM =$ _____
10. $m\angle OPK - m\angle OPH =$ _____

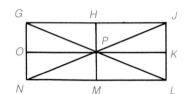

11. Find $m\angle PQR$.

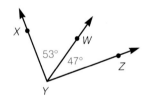

12. Find $m\angle DEF$.

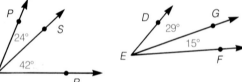

13. $\angle ABC$ is a straight angle. Find $m\angle ABD$.

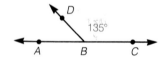

14. Find $m\angle XYZ$.

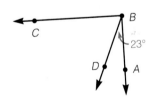

15. $\angle ABC$ is a right angle. Find $m\angle DBC$.

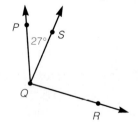

16. Find $m\angle DEF$.

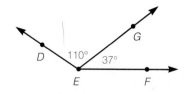

17. $m\angle PQR = 110°$. Find $m\angle SQR$.

18. $m\angle GHI = 52°$.
Find y.

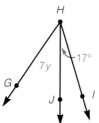

19. $m\angle DEF = 180°$.
Find t.

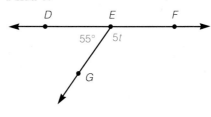

In each figure, $\overrightarrow{BD}$ is the angle bisector of $\angle ABC$.

20. Find $m\angle ABC$.

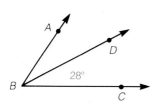

21. $m\angle ABC = 174°$. Find $m\angle DBC$ and $m\angle ABD$.

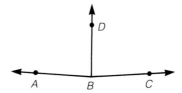

22. Find $m\angle ABC$.

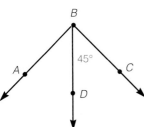

23. $m\angle ABC = 150°$.
Find $m\angle ABD$ and $m\angle DBC$.

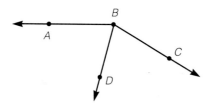

24. Find x.

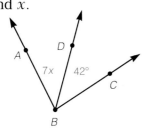

25. $\angle ABC$ is a straight angle.
Find w.

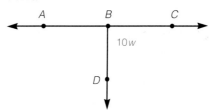

Extend and Apply

26. If $\overrightarrow{QS}$ is the angle bisector of $\angle PQR$ and $m\angle PQS = 24°$, then $m\angle SQR =$ ____ .

(A) 12°   (B) 24°   (C) 48°   (D) 90°

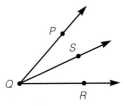

27. Two sides of a picture frame are glued together to form a corner. Each side is first cut at an angle of 45°. What is the measure of the angle formed by the corner of the frame?

28. Suppose you have an angle drawn on a sheet of paper. How could you fold the paper to find the bisector of the angle?

$\overrightarrow{XV}$ bisects $\angle TXW$. $\overrightarrow{XU}$ bisects $\angle TXV$.

29. If $m\angle TXU = 20°$, find $m\angle TXV$.

30. If $m\angle TXU = 20°$, find $m\angle TXW$.

31. If $m\angle TXW = 88°$, find $m\angle TXU$ and $m\angle UXV$.

32. If $m\angle VXW = 32°$, find $m\angle TXU$.

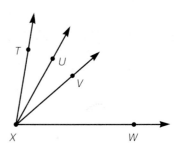

Use Mathematical Reasoning

33. If $m\angle NMR = 80°$, $\overrightarrow{MQ}$ bisects $\angle NMR$, and $m\angle PMQ = 12°$, find $m\angle NMP$.

34. If $m\angle NMR = 88°$, $m\angle QMR = 40°$, and $\overrightarrow{MP}$ bisects $\angle NMQ$, find $m\angle PMQ$.

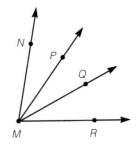

Mixed Review

Use the figure at the right for Exercises 35–39.

35. Identify the sides and vertex of $\angle 1$.

36. Name $\angle 3$ two other ways.

37. Identify the points in the interior of $\angle LMT$.

38. Identify the points in the exterior of $\angle PNT$.

39. Identify the points on $\angle 1$.

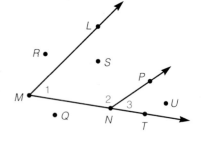

⟨⟩ Solve and check.

40. $3y - 4 = 20$ **41.** $9v + 5 = 23$ **42.** $6 = 12x - 42$

Problem Solving

When one ray is drawn in the interior of an angle, three different angles are formed. For example, when $\overrightarrow{BD}$ is drawn in $\angle ABC$, three angles are formed: $\angle ABD$; $\angle DBC$; and the original angle, $\angle ABC$.

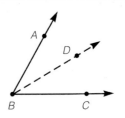

If two rays are drawn, six angles are formed. Drawing $\overrightarrow{BE}$ and $\overrightarrow{BF}$ in $\angle ABC$ forms $\angle ABE$, $\angle EBF$, $\angle FBC$, $\angle ABF$, $\angle EBC$, and $\angle ABC$.

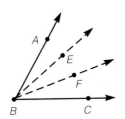

How many angles are formed when 9 rays are drawn in the interior of an angle? Make a table and look for a pattern to find the answer.

3-7 Greek Constructions — Bisecting and Copying Angles

Objective: Use a compass and a straightedge to bisect and copy angles.

▄ *Explore* ▄▄▄▄▄▄▄▄▄▄

Use paper folding or computer software to complete the following activity.

1. On a sheet of paper, draw a large acute angle. Label the angle *ABC*.

2. Fold the paper so that $\overrightarrow{BA}$ falls on top of $\overrightarrow{BC}$, and make a crease.

3. Label a point *D* on the crease.

4. Repeat the procedure using an obtuse angle and a right angle.

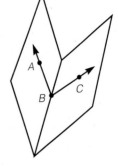

Discuss In each case, how do ∠*ABD* and ∠*CBD* compare? What is an appropriate name for $\overrightarrow{BD}$?

In the Explore you may have found angle bisectors by paper folding. We can also construct angle bisectors using only a compass and straightedge as the ancient Greeks did.

Example 1. Use a straightedge and a compass to bisect ∠*A*.

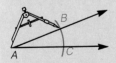

Step 1
Place the compass point on *A* and draw an arc, locating points *B* and *C*.

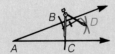

Step 2
Place the compass point on *B* and draw an arc. With the same compass setting, place the compass point on *C* and draw an arc, locating point *D*.

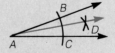

Step 3
Draw a ray, $\overrightarrow{AD}$, which is the angle bisector of ∠*A*.

Another basic construction developed by the ancient Greeks is copying an angle.

When we copy an angle, we construct an angle that is the same size as the one given.

Example **2.** Use a compass and a straightedge to copy ∠A.

Step 1
Place the compass point on A and draw an arc, locating points B and C.

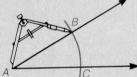

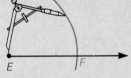

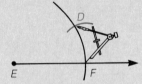

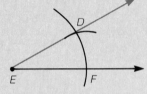

Step 2
Draw a ray with endpoint E. Place the compass point on E. With the same compass setting as in Step 1, draw an arc, locating point F.

Step 3
Place the compass point on B and set the opening so the pencil is at C. With this setting, place the compass point on F and draw an arc, locating point D.

Step 4
Draw a ray, $\vec{ED}$. ∠DEF has the same measure as ∠CAB.

Discussion Which do you think is more accurate: bisecting an angle with a protractor, or with a compass and a straightedge? Defend your answer.

Exercises

Practice

Copy or trace each angle. Then use a compass and a straightedge to bisect the angle.

1.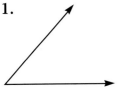

2.

3.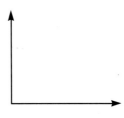

Use a compass and a straightedge to copy each angle.

4.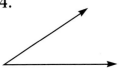

5.

6.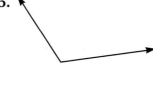

7. Draw an acute angle. Use a compass and a straightedge to bisect the angle.

8. Draw an obtuse angle. Use a compass and a straightedge to bisect the angle.

9. Draw an acute angle. Use a compass and a straightedge to copy the angle.

10. Draw an obtuse angle. Use a compass and a straightedge to copy the angle.

Extend and Apply

11. Draw a straight angle. Use a compass and a straightedge to bisect the angle.

12. Which of the following shows a correct construction to bisect ∠P?

13. While preparing scenery for a school play, Mei needed to draw the bisector of one angle of a large triangular prop. How could she use string and a piece of chalk to construct the angle bisector?

14. Draw a large triangle with three acute angles. Construct the bisector of each of the three angles.

15. Draw a large triangle with an obtuse angle. Construct the bisector of each of the three angles.

16. Draw a large triangle with a right angle. Construct the bisector of each of the three angles.

Use Mathematical Reasoning

17. Draw an acute angle, $\angle A$. Use the construction for copying an angle twice to create an angle with twice the measure of $\angle A$.

18. Draw an obtuse angle, $\angle B$. Use the construction for bisecting an angle twice to construct an angle with one-fourth the measure of $\angle B$.

The angles below have measures a, b, and c as shown. Use a compass and a straightedge to construct angles with these measures.

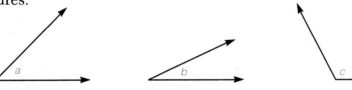

19. $a + b$ **20.** $b + c$ **21.** $c - b$ **22.** $a - b$ **23.** $2a$

24. $2b$ **25.** $\frac{1}{2}c$ **26.** $\frac{1}{2}a$ **27.** $2a - b$ **28.** $a + 2b$

Mixed Review

Use the figure at the right for Exercises 29–35.

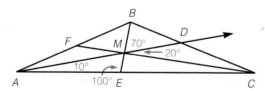

29. Complete the statement: $m\angle CME + m\angle EMA =$ _____.

30. Complete the statement: $m\angle BMD + m\angle DMC =$ _____.

31. Complete the statement: $m\angle AMB - m\angle FMB =$ _____.

32. Complete the statement: $m\angle FMD - m\angle BMD =$ _____.

33. Find $m\angle BMC$.

34. Find $m\angle MEC$.

35. $\overrightarrow{AD}$ bisects $\angle BAC$. Find $m\angle DAB$.

◈ Solve and check.

36. $20 = 7x - 50$ **37.** $6m + 39 = 63$ **38.** $5y - 9 = 16$

39. $6y - 9 = 63$ **40.** $11x - 14 = 41$ **41.** $65 = 8t + 41$

Chapter 3 Review

Use the number line below for Problems 1–3.

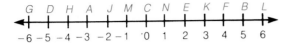

1. Name the coordinates of *A*, *B*, and *C*.
2. Name the points having coordinates −5 and 6.
3. Find the distances *DE*, *KJ*, and *GA*.

4. What is the endpoint of $\overrightarrow{YV}$?

5. Name all pairs of opposite rays shown.

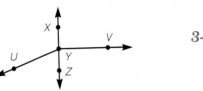

6. Find the coordinate of the midpoint of $\overline{AB}$.

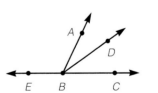

7. Find the coordinates of points *A*, *B*, and *C*.

8. Find the coordinates of the reflection of point *A* when reflected over the *x*-axis.
9. Find the coordinates of the translation of point *C* when translated upward 2 units.

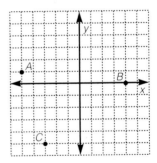

10. Name ∠*ABC* in another way.

11. Identify the vertex and sides of ∠*ABD*.
12. Name a point in the exterior of ∠*DBE*.

Read the measure of each angle.

13. ∠*KML*
14. ∠*KMN*
15. ∠*LMJ*

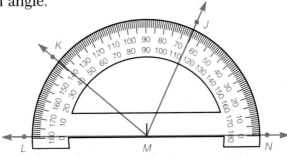

First estimate and then use a protractor to measure each angle.

16.

17.

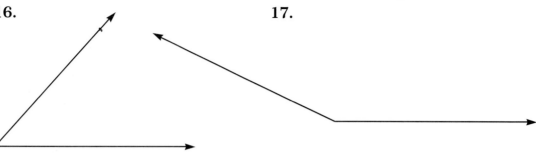

3-6

18. $m\angle DEF = 70°$ and $m\angle CEF = 22°$. Find $m\angle DEC$.

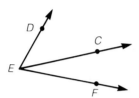

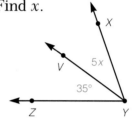 **19.** $\overrightarrow{YV}$ is the angle bisector of $\angle XYZ$. Find x.

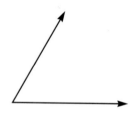

3-7

20. Draw an obtuse angle. Use a compass and a straightedge to bisect the angle.

21. Use a compass and a straightedge to copy the angle at the right.

Chapter 3 Test

Use the number line at the right for Problems 1–3.

1. Name the coordinates of C, E, and F.
2. Name the points having coordinates -3 and 0.
3. Find the distances AL, JK, and DJ.

4. What is the endpoint of $\overrightarrow{RP}$?
5. Name all pairs of opposite rays shown.

6. Find the coordinate of the midpoint of $\overline{AB}$.

7. Find the coordinates of points *P*, *Q*, and *R*.

8. Find the coordinates of the reflection of point *P* when reflected over the *x*-axis.

9. Find the coordinates of the translation of point *Q* when translated upward 5 units.

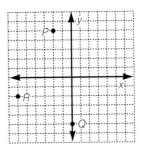

10. Name ∠*FJH* in another way.

11. Identify the vertex and the sides of ∠*EJH*.

12. Name a point in the interior of ∠*HJF*.

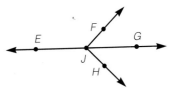

Read the measure of each angle.

13. ∠*NQP*

14. ∠*RQN*

15. ∠*MQR*

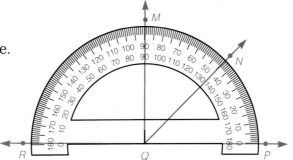

First estimate and then use a protractor to measure each angle.

16.

17.

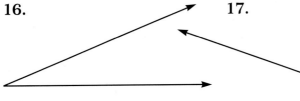

18. $m\angle ABC = 105°$ and $m\angle DBC = 32°$. Find $m\angle ABD$.

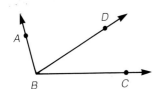

◇ **19.** $\overrightarrow{GE}$ is the angle bisector of ∠*DGF*. Find *y*.

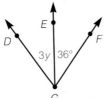

20. Draw a right angle. Use a compass and a straightedge to bisect the angle.

21. Use a compass and a straightedge to copy the angle at the right.

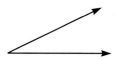

4 Angle Relationships

Lt. Col. Guion Bluford, an aerospace engineer, uses angle relationships to predict flight paths.

4-1 Congruent Angles and Segments

Objective: Identify congruent segments and angles.

In the photo, we see many segments that appear to be the same size. We call such segments **congruent.** In general, **congruent figures** are geometric figures with the exact same size and shape. They would fit together exactly.

The symbol "≅" is read "is congruent to," and the symbol "≇" is read "is *not* congruent to."

Definition **Congruent segments** are segments that have the same length.

We can use a ruler to show segments are congruent by their measures.

Example 1. Use a ruler to show that $\overline{PQ}$ and $\overline{RS}$ are congruent.

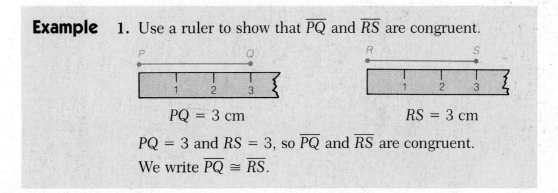

$PQ = 3$ cm $RS = 3$ cm

$PQ = 3$ and $RS = 3$, so $\overline{PQ}$ and $\overline{RS}$ are congruent. We write $\overline{PQ} \cong \overline{RS}$.

We can also use a compass to show that segments are congruent. Open the compass to fit one segment. Then place the compass on the other segment. If it fits exactly, the segments are congruent.

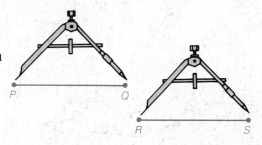

Example 2. Which segments are congruent? Use a ruler or compass.

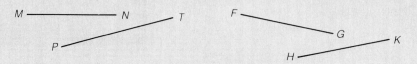

$$\overline{AB} \cong \overline{CD} \text{ and } \overline{PQ} \cong \overline{XY}$$

Try This... a. Which segments are congruent? Use a ruler or compass.

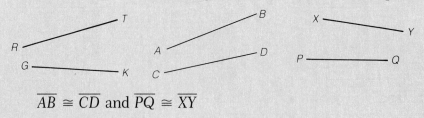

Just as we may have congruent segments that are the same size and shape, we may also have congruent *angles*.

Definition **Congruent angles** are angles that have the same measure.

To show that two angles are congruent, we can use a protractor.

Example 3. Use a protractor to show that $\angle P$ and $\angle Q$ are congruent.

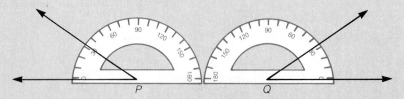

Because $m\angle P = m\angle Q = 34°$, $\angle P$ and $\angle Q$ are congruent. To say that $\angle P$ and $\angle Q$ are congruent, we write

$$\angle P \cong \angle Q$$

Example   **4.** Use a protractor to find a pair of congruent angles.

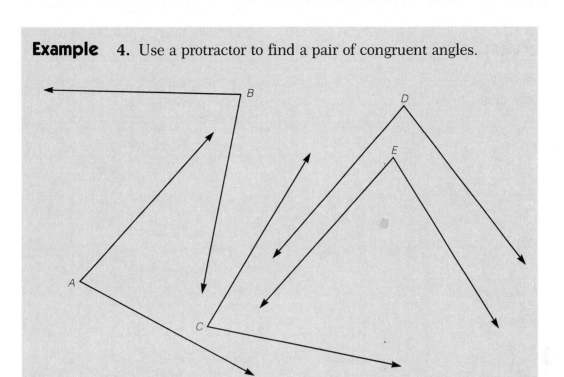

$\angle B \cong \angle D$ because $m\angle B = m\angle D$.

Try This...   **b.** Use a protractor to find another pair of congruent angles.

Angles and segments are sets of points, but their measures are numbers.

$\angle A \cong \angle B$ is a statement about two geometric figures. It says that the two figures have the same *size* and *shape*.

$m\angle A = m\angle B$ is a statement about the *numbers* that are assigned to the two angles. It says that the angle *measures* are the same.

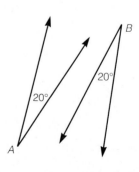

If two angles have the same measure, they are congruent. For the above angles, since $20° = 20°$, we know $\angle A \cong \angle B$.

Discussion   Erin wrote the following statement:
"$m\angle R \cong m\angle S$." Is her statement a sensible statement? Why or why not?

Exercises

Practice

Which segments are congruent? Use a ruler or compass.

1. 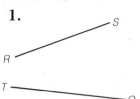  **2.**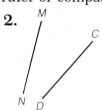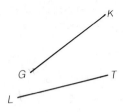

Which angles are congruent? Use a protractor.

3.

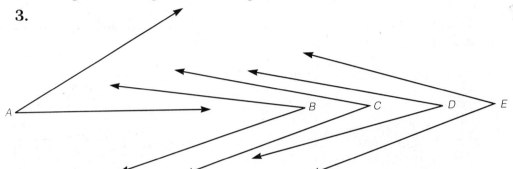

4.

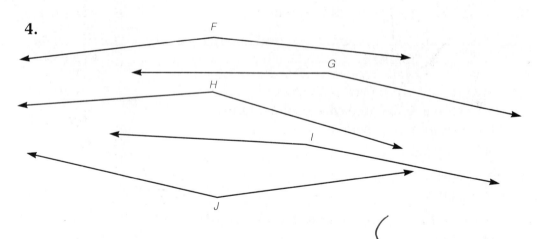

Extend and Apply

5. $\angle A \cong \angle C$ and $m\angle A = 48°$. What is $m\angle C$?

6. $\overline{AB} \cong \overline{XY}$ and $AB = 15$ cm. What is XY?

7. $RS = 9$ cm and $TV = 9$ cm. $\overline{RS}$ is congruent to what segment?

8. If $m\angle B = 54°$ and $m\angle G = 54°$, then $\angle B \cong$ _____ .

9. If $\angle R \cong \angle T$ and $\angle T \cong \angle S$, then $\angle R \cong$ _____ .

10. If $\overline{AB} \cong \overline{CD}$ and $\overline{CD} \cong \overline{HK}$, then $\overline{AB} \cong$ _____ .

11. Explain the difference between the statements "$m\angle ABC = m\angle STR$" and "$\angle ABC \cong \angle STR$."

12. If $\angle A \cong \angle B$ and $\angle B \not\cong \angle C$, then

(A) $\angle A \cong \angle C$ (B) $\angle B \cong \angle C$
(C) $\angle B \not\cong \angle A$ (D) $\angle C \not\cong \angle A$

Use the figure for Exercises 13–17.
Find an angle congruent to the
given angle, and find its measure.

13. $\angle DGE$

14. $\angle CGD$

15. $\angle CGF$

16. $\angle AGD$

17. Find two angles congruent
to $\angle BGC$. What are their
measures?

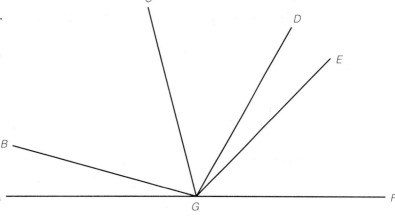

18. a. Measure to identify congruent segments
in the kite figure.

b. Identify *sticks* that have the same
lengths.

c. Are the answers to **a.** and **b.** the same?
Why or why not?

19. Identify congruent angles in the kite.

Use Mathematical Reasoning

20. Suppose $AB = 12$ in. for the kite shown.
Estimate the lengths of each stick of the kite,
and the length of the package that originally
contained the kite parts. Explain your
reasoning.

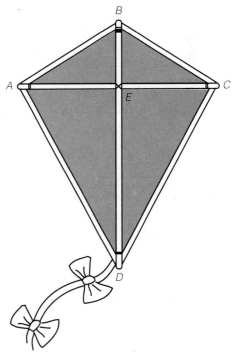

Mixed Review

21. True or false: The points of a triangle are
coplanar.

22. True or false: The points of a triangle are
collinear.

Which theorem or postulate explains the following?

23. If a bird flies in a straight line back and forth
between two buildings, it flies the same
distance each way.

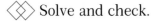 Solve and check.

24. $2x + 3 = 5$ **25.** $5a - 6 = 14$

26. $65 = 8z + 9$ **27.** $37 = 6u + 7$

Combining Like Terms and Solving Equations

In an algebraic expression like $7y-8x$, $7y$ and $-8x$ are **terms.**
Terms with the same variables are called **like terms.**

$7y + 3y - 8x$ The like terms are $7y$ and $3y$, since both have the variable y.

The distributive property allows us to combine like terms.

$7y + 3y = (7 + 3)y = 10y$

You can also do this mentally, as long as you are sure to add *only like terms*.

$8n + 3n = 11n$ $7y + 12y = 19y$
$22d - 13d = 9d$ $-6a + 8a = 2a$
$-9a + 4a = -5a$ $6x + 9x - 7y = 15x - 7y$ (15x and 7y are not like terms.)

Combine like terms before solving an equation.

Sample   Solve $3x + 2x - 1 = 14$ for x.

$3x + 2x - 1 = 14$
$5x - 1 = 14$ Combining like terms 3x and 2x
$\underline{ +1 = +1}$ Adding 1 undoes subtracting 1.
$5x = 15$
$x = 3$

$3x + 2x - 1 = 14$ Check.
$3(3) + 2(3) - 1 = 14$
$9 + 6 - 1 = 14$
$14 = 14$ ✓   The answer, 3, checks.

Problems

Solve and check.

1. $9n + 8n = 17$ **2.** $5k + 2k = 14$ **3.** $6b + 6b = 36$

4. $29a + 18a = 94$ **5.** $7y + 16y - 2 = 90$ **6.** $5r + 15r - 10 = 0$

7. $x + 3x - 15 = 13$ **8.** $11p + p - 40 = 8$

9. $21d - 18d + 24 = 36$ **10.** $15u - 8u - 6 = 50$

11. $19r - 18r + 99 = 101$ **12.** $17t - 17t + t = 3$

13. $-2a + 7a - 7 = 90$ **14.** $-e + 8e - 16 = 1$

15. $-10j + 11j + 12 = 13$ **16.** $-3f + 9f + 17 = 17$

4-2 Complementary and Supplementary Angles

Objective: Solve problems involving complementary and supplementary angles.

The house bracing shows a pair of **complementary angles.**

$$m\angle 1 + m\angle 2 = 90°$$

Definition   Two angles are **complementary** whenever the sum of their measures is 90°. Each angle is a **complement** of the other.

Examples **1.** Identify each pair of complementary angles.

∠1 and ∠2
(because 25° + 65° = 90°)
∠1 and ∠3
∠2 and ∠4
∠3 and ∠4

2. Find the measure of a complement of an angle of 29°.

$$90° - 29° = 61°.$$

Thus, the measure of the complement is 61°.

Try This... **a.** Identify each pair of complementary angles.

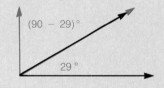

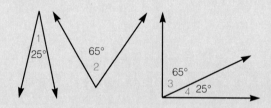

Find the measure of a complement of each angle.

b.   **c.**   **d.**

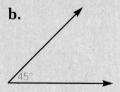

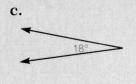

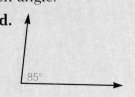

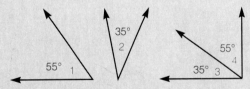

The skier and the skis show a pair of **supplementary angles.**

$$m\angle 1 + m\angle 2 = 180°$$

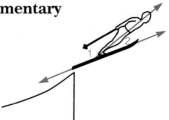

Definition Two angles are **supplementary** whenever the sum of their measures is 180°. Each angle is a **supplement** of the other.

Example 3. Identify each pair of supplementary angles.

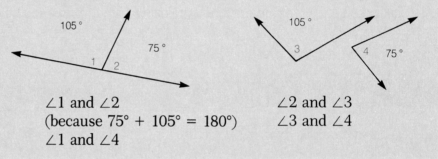

∠1 and ∠2
(because 75° + 105° = 180°)
∠1 and ∠4

∠2 and ∠3
∠3 and ∠4

Try This... e. Identify each pair of supplementary angles.

Example 4. Find the measure of a supplement of 112°.

180° − 112° = 68°.
Thus, the measure of
the supplement is 68°.   180° − 112°   112°

Try This... Find the measure of a supplement of each angle.

    f. 38°   **g.** 157°   **h.** 90°

1. Draw an acute angle, and copy this angle.
2. Draw the complements of both angles, and measure them.
3. Draw the supplements of both angles, and measure them.
4. What is true of the complements? the supplements?

The Explore suggests the following theorems:

Theorem 4.1 If two angles are congruent, then their supplements are congruent.

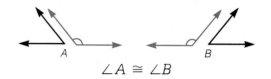

$\angle A \cong \angle B$

Theorem 4.2 If two angles are congruent, then their complements are congruent.

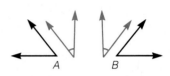

Example 5. $\angle C \cong \angle D$ and $m\angle C = 30°$. Find the measures of a complement and a supplement of $\angle D$.

We can find the complement of $\angle C$. $90 - 30 = 60$.
Since $\angle C \cong \angle D$, the complement of $\angle D$ measures $60°$.

Likewise, we can find the supplement of $\angle C$.
$180 - 30 = 150$. Therefore, the supplement of $\angle D$ measures $150°$.

Try This... i. $\angle J \cong \angle K$ and $m\angle J = 55°$. Find the measures of the complement and the supplement of $\angle K$.

If you have a store key on your calculator, $\boxed{\text{STO}}$, and a recall key, $\boxed{\text{RCL}}$, you can find measures of the complement and supplement of an angle quickly.

Example 6. Find the measures of the complement and the supplement of a $54°$ angle.

First store the number 54.   54 $\boxed{\text{STO}}$

For the complement,   90 − $\boxed{\text{RCL}}$ $\boxed{=}$ $\boxed{\qquad 36}$

then for the supplement,   180 − $\boxed{\text{RCL}}$ $\boxed{=}$ $\boxed{\qquad 126}$

The complement measures $36°$, the supplement $126°$.

Try This... j. Find the measures of the complement and the supplement of a 71° angle.

Discussion How many complements does an acute angle have? How many supplements does any angle have? Defend your answer.

Exercises

Practice

Identify each pair of complementary angles.

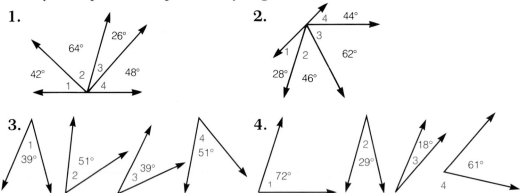

Find the measures of complements of angles with these measures.

5. 22°   **6.** 45°   **7.** 56°   **8.** 73°   **9.** 89°   **10.** $t°$

Identify each pair of supplementary angles.

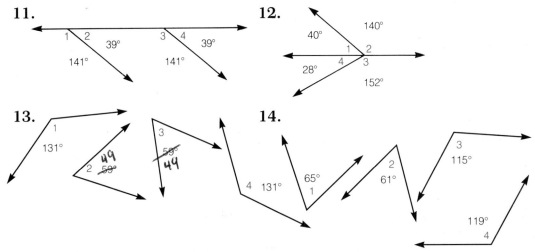

Find the measures of supplements of angles with these measures.

15. 119°   **16.** 129°   **17.** 143°   **18.** 57°   **19.** 74°   **20.** $s°$

21. $\angle T \cong \angle S$ and $m\angle T = 78°$. Find the measures of the complement and the supplement of $\angle S$.

22. $\angle L \cong \angle K$ and $m\angle K = 32°$. Find the measures of the complement and the supplement of $\angle L$.

23. If $\angle A \cong \angle B$ and $m\angle A = 63°$, then the measure of the complement of $\angle B = $ _____ .

24. If $\angle P \cong \angle Q$ and $m\angle Q = 108°$, then the measure of the supplement of $\angle P = $ _____ .

Extend and Apply

25. Two complementary angles have the same measure. What is the measure of each angle?

26. What type of angle is the supplement of any acute angle?

27. $m\angle AEC = m\angle BED$. What other angles have the same measure?

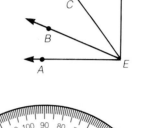

28. Examine the protractor shown. How does the outer scale relate to the inner scale? Why is there no inner reading at the 90° outer reading?

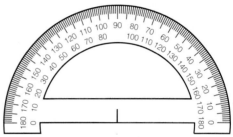

29. An angle's supplement has a complement. The angle must be

(A) acute or right   (B) right or obtuse
(C) right only   (D) obtuse only

30. Which angles are complementary?

31. Which angles are supplementary?

32. A single windshield wiper clears a 133° angle. What angle is not cleared by the wiper?

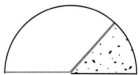

Use Mathematical Reasoning

33. Write a convincing argument to show that if an angle is congruent to a complement, then its measure must be 45°.

   Hints: a. What is true of the measures of complementary angles?
   b. What is true of the measures of congruent angles?
   c. If 2 times a number is 90, what is the number?

34. Write a convincing argument to show that if an angle is congruent to its supplement, then it must be a right angle.

35. Write a convincing argument to show that if two angles are complementary, then they are each an acute angle.

Mixed Review

36. Which segments are congruent? Use a compass.

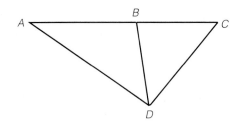

37. $\overline{ZT}$ is a segment on the number line. The coordinate of Z is 115, and the coordinate of T is 38. Find ZT and the coordinate of the midpoint of $\overline{ZT}$.

Find the translation of the point $(3, -2)$ when translated

38. 4 units left.　　　**39.** 7 units down.　　　**40.** 10 units up.

◇ Solve and check.

41. $5a + 3a = 24$　　　**42.** $6n - n + 2 = 7$　　　**43.** $17y + y - 2 = 34$

Discover

Fold a square piece of paper in half. Unfold the square and label the endpoints of the fold A and B.

Now fold the square in half the other way. Unfold the paper and label the endpoints of this fold C and D.

Label the intersection of $\overline{AB}$ and $\overline{CD}$, E.

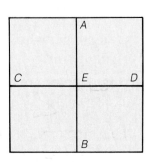

What is the measure of $\angle DEB$?
What is the measure of $\angle BEC$?
What can you conclude about $\angle DEB$ and $\angle BEC$?

Now fold the square along a diagonal. Unfold the paper and label the endpoints of this fold F and G.

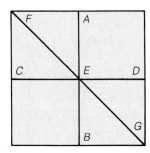

What is the measure of $\angle DEG$?
What is the measure of $\angle GEB$?
What can you conclude about $\angle DEG$ and $\angle GEB$?

What can you conclude about $\overline{EG}$?

Problem-Solving Strategies

Write an Equation

An important problem-solving strategy involves translating the problem to an equation. Then, by solving the equation and checking the answer in the original problem, we can often solve the problem.

◇◇◇ **Sample** The measure of an angle is 20° less than three times the measure of its complement. Find the measure of each angle.

Let x be the measure of the smaller angle. Then the measure of the complement is $3x - 20$.

The angles are complementary, so their sum is 90°.

Translate this to an equation and solve.

$$
\begin{aligned}
x + 3x - 20 &= 90 \\
4x - 20 &= 90 \quad \text{Combining like terms} \\
+20 &= +20 \quad \text{Adding 20 to both sides} \\
\hline
4x &= 110 \\
x &= 27.5 \quad \text{Dividing both sides by 4}
\end{aligned}
$$

The measure of one angle is 27.5°. The measure of its complement is $3(27.5) - 20 = 82.5 - 20 = 62.5°$.

Check to see if the answer makes sense. Are 27.5° and 62.5° complements? $27.5 + 62.5 = 90°$. Yes.

◇◇◇ **Problems**

1. The measure of an angle is twice the measure of its complement. Find the measures of the two angles.

2. The measure of an angle is 15° greater than the measure of its supplement. Find the measures of the two angles.

3. The measure of an angle is 60° less than twice the measure of its supplement. Find the measures of the two angles.

4. The measure of an angle is 88° less than the measure of its complement. Find the measures of the two angles.

5. The measure of an angle equals the measure of a supplement added to the measure of a complement. Find the measure of the angle.

4-3 Adjacent Angles and Linear Pairs

Objective: Solve problems involving adjacent angles and linear pairs.

The angles made by the ribs of the leaf are **adjacent angles.** Adjacent angles share a ray and a vertex.

Definition   Two angles are **adjacent** whenever
(1) they have a common side, (2) they have a common vertex, and (3) their interiors do not intersect.

In the figure, $\angle AEB$ and $\angle BEC$ are adjacent. $\angle AEB$ and $\angle CED$ are not adjacent because they do not have a common side. $\angle AEB$ and $\angle AEC$ are not adjacent because their interiors intersect.

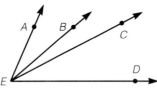

Examples   Tell whether $\angle 1$ and $\angle 2$ are adjacent.

1.

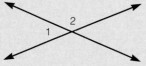

Yes, they are adjacent.

2.

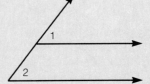

No. The interiors of $\angle 1$ and $\angle 2$ intersect.

Try This... Tell whether $\angle 1$ and $\angle 2$ are adjacent.

a.

b.

Adjacent angles whose sides form a straight line are called a
linear pair of angles, or just a *linear pair*.

Definition   Two angles form a **linear pair** whenever
  1. they are adjacent, and
  2. their noncommon sides are opposite rays.

Examples   Tell whether ∠1 and ∠2 form a linear pair.

3.

Yes.
(1) ∠1 and ∠2 are adjacent.
(2) Their noncommon sides
are opposite rays.

4.

No.
∠1 and ∠2 are not adjacent.

5.

No.
∠1 and ∠2 are adjacent, but
their noncommon sides are not
opposite rays.

6.

No.
∠1 and ∠2 are not adjacent.

Try This...   Tell whether ∠1 and ∠2 form a linear pair.

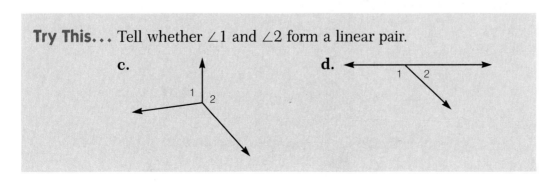

c.

d.

Explore

1. Draw several linear pairs of angles.
2. Measure the two angles of each linear pair.
3. What is the sum of the measures of each linear pair?

The following theorem summarizes an important relationship between angles of a linear pair:

Theorem 4.3   If two angles form a linear pair, then they are supplementary.

Example   7. ∠P and ∠Q form a linear pair. $m\angle P = 112°$. Find $m\angle Q$.

Since a linear pair of angles are supplementary,

$$m\angle Q = 180° - m\angle P$$
$$= 180° - 112°$$
$$= 68°$$

Try This... e. ∠S and ∠T form a linear pair. $m\angle T = 23°$. Find $m\angle S$.

Discussion   Why do you think the name *linear pair* is used to describe a pair of angles such as ∠1 and ∠2 in Example 3?

Exercises

Practice

Tell whether ∠1 and ∠2 are adjacent.

1.   2.   3.

4.   5.   6.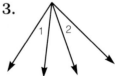

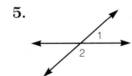

Tell whether $\angle 1$ and $\angle 2$ form a linear pair.

7.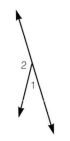

8.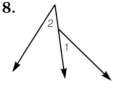

9.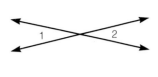

10.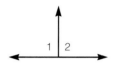

11.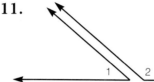

12.

13. $\angle R$ and $\angle S$ form a linear pair. $m\angle R = 132°$. Find $m\angle S$.

14. $\angle X$ and $\angle Y$ form a linear pair. $m\angle Y = 39°$. Find $m\angle X$.

Extend and Apply

15. Write *always, sometimes,* or *never* to make the statement true. One angle of a linear pair is _____ congruent to the supplement of the other angle.

16. Three lines are drawn through each pair of three noncollinear points. How many linear pairs are formed?

17. Are $\angle BFC$ and $\angle CFD$ adjacent angles?

18. Do $\angle BFC$ and $\angle CFD$ form a linear pair?

19. Are $\angle BFE$ and $\angle EFA$ adjacent angles?

20. Do $\angle BFE$ and $\angle EFA$ form a linear pair?

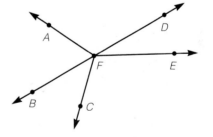

21. An automobile traveling on Arcana Avenue makes a 68° turn onto Chestnut Street. Later, the driver turns around on Chestnut and returns to Arcana to continue heading in the original direction. At what angle will the car turn onto Arcana?

(A) 68°   (B) 22°   (C) 112°   (D) 158°

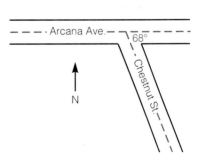

Use Mathematical Reasoning

22. Vandell said, "Every linear pair of angles are supplementary, but not every pair of supplementary angles form a linear pair." Do you agree or disagree? Explain.

23. Write a convincing argument to show that if two angles form a linear pair and have the same measure, then each is a right angle.

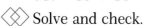 **24.** Two angles form a linear pair. One is 90° larger than the other. The larger angle is _____ times the size of the smaller angle.

Mixed Review

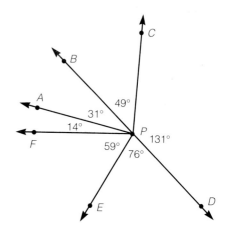

25. Name two pairs of complementary angles.

26. Name four pairs of supplementary angles.

27. What is $m\angle APD$?

28. What is $m\angle BPD$?

29. Name a pair of opposite rays.

30. Find a 153° angle.

Add or subtract.

31. $39 + (-14)$ **32.** $39 - (-14)$

33. $-39 + 14$ **34.** $-39 - 14$

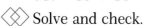 Solve and check.

35. $5x + 9 = 64$

36. $4a + 7a - 2 = 9$

37. $6t - 12 - 2t = 22$

Problem Solving

How many linear pairs can you find in each figure?

1.

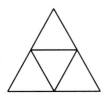

2.

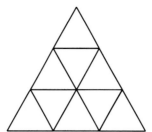

3.

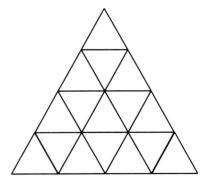

Algebra Connection

Solving Equations with Variables on Both Sides

Recall that when you solve an equation, the variable terms
must be combined. If there are variable terms on both
sides of an equation, you can "move" one of the terms
by subtracting it (from both sides).

Sample Solve $5x - 8 = 3x + 2$ for x.

$$\begin{array}{rl}
5x - 8 = & 3x + 2 \\
-3x & -3x \\
\hline
2x - 8 = & 2 \\
+8 & +8 \\
\hline
2x = & 10 \\
x = & 5
\end{array}$$

We want the variable x by itself.

Subtracting $3x$ from both sides

Adding 8 to both sides

We have solved for x

$$\begin{array}{l}
5x - 8 = 3x + 2 \\
5(5) - 8 = 3(5) + 2 \\
25 - 8 = 15 + 2 \\
17 = 17 \checkmark
\end{array}$$

We check by replacing x with 5.

The answer, 5, checks.

Problems

Solve and check.

1. $5x = 4x + 6$

2. $12y = 11y + 15$

3. $3y = y + 8$

4. $7p = 5p + 16$

5. $6t = 3t + 15$

6. $9y = 5y - 20$

7. $4r = 7r - 18$

8. $3q = 8q + 17$

9. $7t + 6 = 15t$

10. $19r - 4 = 28r - 13$

11. $2x + 5 = 3x - 7$

12. $5y - 1 = 2y + 20$

13. $5x - 9 = 3x + 9$

14. $4x - 17 = 2x + 5$

15. $13y - 5 = 6y + 23$

16. $9x + 16 = 3x + 25$

17. $4x + 18 = 8x + 14$

18. $5y - 16 = 11y - 30$

19. $4y + 3 = 2y + 7$

20. $5x - 2 = 4x + 1$

21. $7.6a - 12.1 = 0.8a + 1.5$

22. $\frac{1}{2}x - 2 = \frac{1}{4}x - 1$

23. $y + \frac{2}{3} = 3y + \frac{1}{3}$

24. $360y = 30 + 90y$

4-4 Vertical Angles

Objective: Use the Vertical Angle Theorem to find measures of angles.

▬ *Explore* ▬▬▬▬▬▬▬▬▬▬

1. Use a piece of waxed paper about 5″ × 5″. Make two creases that intersect to form angles as shown.

2. Label the angles ∠SPT, ∠RPQ, ∠SPR, and ∠QPT as shown.

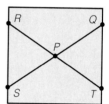

3. Fold the paper through vertex P so that $\overrightarrow{PS}$ falls on $\overrightarrow{PT}$. Does $\overrightarrow{PR}$ fall on $\overrightarrow{PQ}$?

4. Unfold the paper from Step 3. Now fold the paper through vertex P so that $\overrightarrow{PT}$ falls on $\overrightarrow{PQ}$. What happens to $\overrightarrow{PS}$ and $\overrightarrow{PR}$?

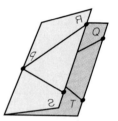

Discuss Recall that figures that fit together exactly are congruent. What can you say about ∠SPT, ∠RPQ, ∠SPR, and ∠QPT?

Pairs of opposite angles such as ∠SPT and ∠RPQ and the pair ∠SPR and ∠QPT are called *vertical angles*.

> **Definition** Two nonstraight angles are **vertical angles** whenever their sides form two pairs of opposite rays.

The Explore above illustrates that pairs of vertical angles are congruent. Note the many pairs of vertical angles in this photo.

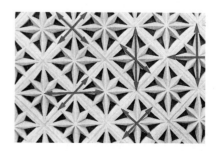

Theorem 4.4   The Vertical Angle Theorem

Vertical angles are congruent.

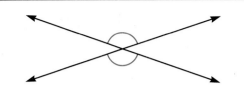

Examples **1.** In the power-line tower, $m\angle 1 = 124°$ and $m\angle 2 = 56°$. Find $m\angle 3$.

$\angle 1$ and $\angle 3$ are vertical angles, so $m\angle 1 = m\angle 3$. Thus, $m\angle 3 = 124°$.

2. Find $m\angle APC$, $m\angle APD$, and $m\angle CPB$.

$\angle APC$ and $\angle DPB$ are vertical angles. Thus, $m\angle APC = m\angle DPB$ and $m\angle APC = 38°$.

Because $\angle APD$ and $\angle DPB$ are supplementary, we know that $m\angle APD + m\angle DPB = 180°$. Thus,

$$m\angle APD = 180 - m\angle DPB$$
$$= 180 - 38$$
$$= 142°$$

It then follows that $m\angle CPB = 142°$.

Try This... **a.** In the fence, $m\angle 1 = 98°$. Find $m\angle 2$, $m\angle 3$, and $m\angle 4$.

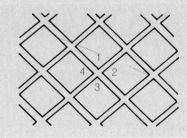

Example **3.** In the drawing, $m\angle 1 = 23°$ and $m\angle 3 = 34°$. Find $m\angle 2$, $m\angle 4$, $m\angle 5$, and $m\angle 6$.

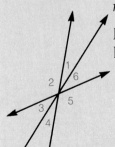

Because $\angle 1$ and $\angle 4$ are vertical angles, $m\angle 4 = 23°$. Likewise, $\angle 3$ and $\angle 6$ are vertical angles, so $m\angle 6 = 34°$.

$$m\angle 1 + m\angle 2 + m\angle 3 = 180$$
$$23 + m\angle 2 + 34 = 180 \quad \text{Substituting}$$
$$m\angle 2 + 57 = 180$$
$$m\angle 2 = 180 - 57$$
$$m\angle 2 = 123°$$

Because $\angle 2$ and $\angle 5$ are vertical angles, $m\angle 5 = 123°$.

Try This... **b.** In the drawing, $m\angle 2 = 41°$ and $m\angle 4 = 10°$.

Find $m\angle 1$, $m\angle 3$, $m\angle 5$, and $m\angle 6$.

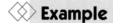

Example **4.** Find $m\angle ABC$ and $m\angle DBE$.

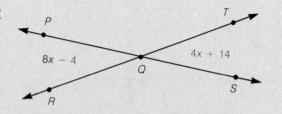

Because $\angle ABC$ and $\angle DBE$ are vertical angles, $m\angle ABC = m\angle DBE$. Thus,

$$6y - 10 = 3y + 5$$
$$6y - 3y = 5 + 10$$
$$3y = 15$$
$$y = 5$$

$m\angle ABC = 3y + 5 = 3(5) + 5 = 20$
So, $m\angle ABC = 20° = m\angle DBE$.

Try This... **c.** Find $m\angle PQR$ and $m\angle TQS$.

Discussion Ramon said that his friend convinced him that the Vertical Angle Theorem was true by using supplements of the same angle. Explain how his friend could have convinced Ramon.

Exercises

Practice

Find the measures of the indicated angles.

1. $m\angle 2$ and $m\angle 3$

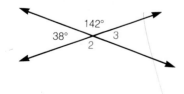

2. $m\angle 1$ and $m\angle 2$

3. $m\angle 2$, $m\angle 3$, and $m\angle 4$

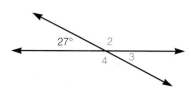

4. $m\angle 1$, $m\angle 2$, and $m\angle 4$

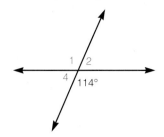

5. $m\angle 1$, $m\angle 2$, $m\angle 3$, and $m\angle 4$

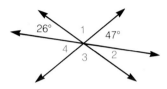

6. $m\angle 1$, $m\angle 2$, $m\angle 3$, and $m\angle 4$

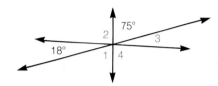

7. $m\angle 1$, $m\angle 3$, $m\angle 4$, $m\angle 5$, and $m\angle 6$

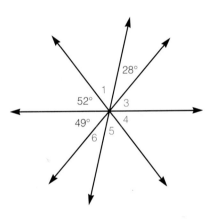

8. $m\angle 1$, $m\angle 2$, $m\angle 5$, $m\angle 7$, and $m\angle 8$

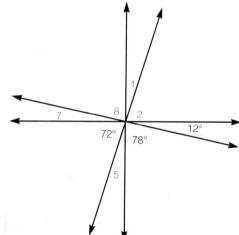

9. $m\angle PQR$ and $m\angle TQS$

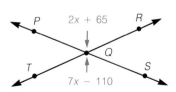

10. $m\angle ABC$ and $m\angle DBE$

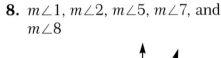

11. $m\angle TQS$ and $m\angle GQR$

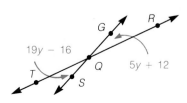

12. $m\angle VPS$ and $m\angle QPR$

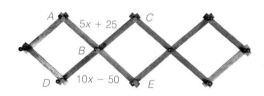

13. $m\angle ABC$ and $m\angle DBF$

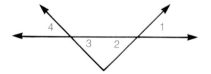

14. The photograph shows some vertical angles. Find $m\angle RST$ and $m\angle VSQ$.

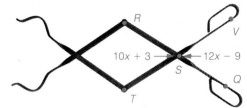

Extend and Apply

15. Suppose $\angle 2 \cong \angle 3$. What can you conclude about $\angle 4$ and $\angle 1$? Explain how you know.

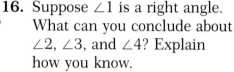

16. Suppose $\angle 1$ is a right angle. What can you conclude about $\angle 2$, $\angle 3$, and $\angle 4$? Explain how you know.

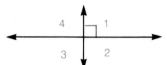

17. In the spinner at right, $m\angle BGC = 60°$ and $m\angle AGF = 60°$. What do we know of the six small angles shown?

(A) All six angles measure 60°.
(B) 4 angles measure 60°.
(C) 2 angles measure 60°.
(D) None of these.

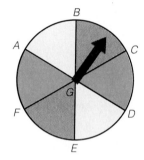

Use Mathematical Reasoning

Tell whether each of the following statements is always true, sometimes true, or never true.

18. Vertical angles are supplementary.

19. Vertical angles are congruent.

20. Vertical angles are complementary.

21. Vertical angles are adjacent angles.

22. Vertical angles form a linear pair.

23. Vertical angles are right angles.

24. Write a convincing argument that two angles that are both supplementary and vertical must be right angles.

25. Write a convincing argument that two angles that are both complementary and vertical must have a measure of 45°.

26. Pearl correctly solved a problem where she found that the angles formed by two intersecting lines were 87°, 93°, 87°, and 93°. The problem had shown two marked angles. One was $5x + 12°$, and the other was $7x - 12°$. Were the marked angles vertical or adjacent? Why?

27. Can you construct two intersecting lines so that the angles formed are 60°, 120°, 75°, and 105°?

Mixed Review

Write the converse of each statement.

28. If you're cool, then you wear 602 jeans.

29. If two angles are congruent, then they have the same measure.

30. Vertical angles have the same measure.

31. Two angles form a linear pair if they are adjacent and supplementary.

32. If there's a will, there's a way.

33. Name all linear pairs.

34. Find the missing measures of the angles.

 Solve and check.

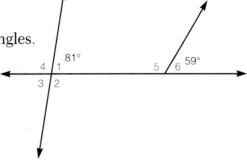

35. $6w - 7 = 5w + 7$

36. $7k - 7 = 5k + 7$

37. $17t - 7 = 15t + 7$

Enrichment

A billiard ball bounces off the sides of a billiard table at the same angle at which it arrives.

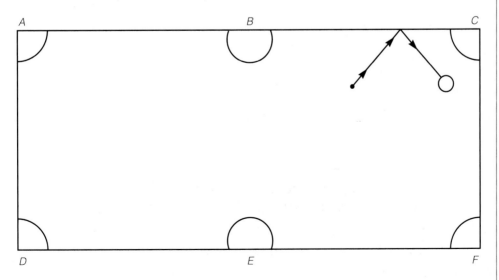

Into which pocket will the billiard ball fall?
(Use your protractor and straightedge to help you.)

Spirals

The Logarithmic Spiral

A curve that appears in nature is the *spiral*. Spirals may be found in the tusk of an elephant, the web of a spider, the shell of a chambered nautilus, and the head of a daisy. The stars of our galaxy, the Milky Way, are arranged in a shape called a *logarithmic spiral*.

1. Use your protractor to mark a point every 20° until it completes a circle. Join opposite points through the center as shown.

2. Use a file card or a piece of stiff cardboard with a square corner. Place the square corner along a segment, and from a point *A* near the center of the circle, draw a connecting segment to a point *B* on the next segment. $\angle A$ is a right angle.

3. Repeat Step 2 from point *B* to locate a point *C*.

4. Complete the spiral by repeating Step 2 until the spiral reaches the circle.

The Archimedean Spiral

Do Step 1 as outlined above.

Use your ruler. Choose one segment and place a point $\frac{1}{8}$ in. from the center. Move clockwise to the next segment and place a point $\frac{2}{8}$ in. (or $\frac{1}{4}$ in.) from the center. Continue adding $\frac{1}{8}$ in. to the previous measurement and placing points as you move clockwise until you reach the circle.

Connect the marks to form a spiral.

4-5 Right Angles and Perpendicular Lines

Objective: Identify perpendicular lines and construct them with straightedge and compass.

Explore

1. Fold a piece of paper in half.

2. Label the endpoints of the fold R and S.

3. Mark a point on the paper (not on $\overline{RS}$) and label it Q.

4. Now join the edges of your fold to create a new fold that passes through Q. Unfold the paper.

5. Try this again, creating a new fold and marking a new point. Are your results the same?

Discuss   What kind of angles are formed at the intersection of $\overline{RS}$ and the creases?

Definition   Two lines are **perpendicular** whenever they intersect to form a right angle.

To say that $\overleftrightarrow{AB}$ is perpendicular to $\overleftrightarrow{RS}$, we write

$$\overleftrightarrow{AB} \perp \overleftrightarrow{RS}, \text{ or } \ell \perp m$$

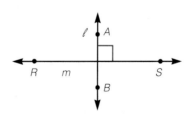

Example   1. Which pairs of lines are perpendicular? Use a protractor.

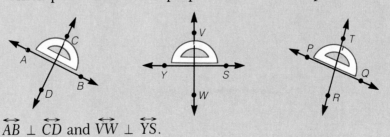

$$\overleftrightarrow{AB} \perp \overleftrightarrow{CD} \text{ and } \overleftrightarrow{VW} \perp \overleftrightarrow{YS}.$$

To say that $\overleftrightarrow{PQ}$ is not perpendicular to $\overleftrightarrow{TR}$, we write

$$\overleftrightarrow{PQ} \not\perp \overleftrightarrow{TR}$$

Try This... a. Which pairs of lines are perpendicular? Use a protractor.

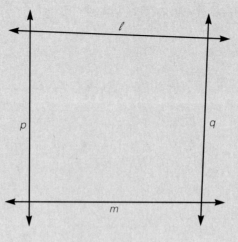

When two lines intersect and are perpendicular, one right angle is formed. What are the measures of the other three angles that are formed?

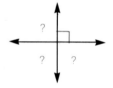

Theorem 4.5 If two lines form one right angle, then they form four right angles.

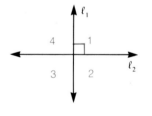

Example 2. Consider point P on line ℓ. Construct a line perpendicular to ℓ at P.

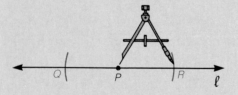

Step 1
Place the compass point on P. Using one opening, draw arcs that intersect ℓ at two points, Q and R.

Step 2
Open the compass to about the length of $\overline{QR}$. Place the point on Q and make an arc as shown. Place the point on R and make another arc so that the two arcs intersect at a point S.

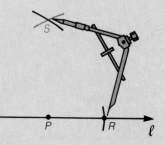

Step 3

Use a straightedge to join points S and P. $\overleftrightarrow{SP}$ is perpendicular to ℓ at P.

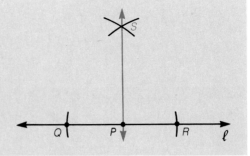

Since $QP = QR$, P is the midpoint of $\overline{QR}$, and thus bisects $\overline{QR}$. $\overleftrightarrow{SP}$ is called the **perpendicular bisector** of $\overline{QR}$.

> **Definition**   The line that is perpendicular to a segment at its midpoint is the **perpendicular bisector** of the segment.

Try This... **b.** Draw a line, $\overleftrightarrow{BC}$. Choose a point A on $\overleftrightarrow{BC}$ and construct a line perpendicular to $\overleftrightarrow{BC}$ at A.

We can use a compass and a straightedge to construct a perpendicular from a point not on the line.

Example   **3.** Consider point P not on line ℓ. Construct a line perpendicular to ℓ containing P.

Step 1

Place the compass point on P. Using one opening, draw arcs that intersect ℓ at two points, Q and R.

Step 2

Open the compass to about the length of $\overline{QR}$. Place the point on Q and make another arc as shown. Place the point on R and make another arc so that the two arcs intersect in a point T.

Step 3

Use a straightedge to join points T and P. $\overleftrightarrow{PT}$ is perpendicular to ℓ.

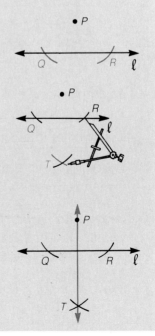

Try This... c. Draw a line, $\overleftrightarrow{QR}$. Choose a point A not on $\overleftrightarrow{QR}$. Construct a line containing A that is perpendicular to $\overleftrightarrow{QR}$.

Discussion Ariel said, "Given a line ℓ and a point P on the line, there may be more than one line perpendicular to ℓ at P." Do you agree? Why or why not?

Exercises

Practice

Which pairs of lines are perpendicular? Use a protractor.

1. **2.**

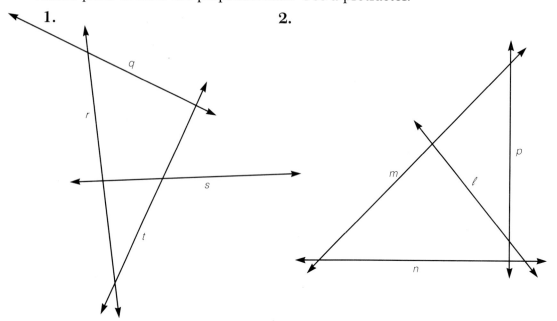

Use a compass and a straightedge for these Exercises.

3. Draw a line, $\overleftrightarrow{ST}$. Choose a point Q on $\overleftrightarrow{ST}$. Construct a line perpendicular to $\overleftrightarrow{ST}$ at Q.

4. Draw a line, ℓ. Choose points R and P on ℓ. Construct perpendiculars to ℓ at R and P.

5. Draw a line, $\overleftrightarrow{AB}$. Choose a point C not on $\overleftrightarrow{AB}$. Construct a line containing C that is perpendicular to $\overleftrightarrow{AB}$.

6. Draw a line, ℓ. Choose points X and Y not on ℓ. Construct lines containing X and Y that are perpendicular to ℓ.

Extend and Apply

Find the measure of each angle.

7.

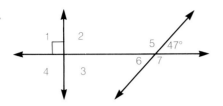

8.

Make drawings that illustrate the following.

9. $\overleftrightarrow{AB} \perp \overline{CD}$ **10.** $\overrightarrow{RS} \perp \overline{GK}$

11. $\overline{MN} \perp \overline{TV}$ **12.** $\overrightarrow{FL} \perp \overrightarrow{BD}$

13. Using a straightedge and compass, draw a segment and construct its perpendicular bisector.

A road that runs north to south is perpendicular to one that runs east to west. Use the compass at right to tell the direction of the road perpendicular to each road named below.

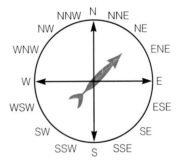

14. A road that runs east to west

15. A road that runs southeast to northwest

16. A road that runs north-northwest to south-southeast

17. A road that runs east-southeast to west-northwest

18. Two perpendicular lines form each of these, except _____ .

   (A) vertical angles
   (B) linear pairs
   (C) complementary angles
   (D) straight angles

Use Mathematical Reasoning

19. Suppose line ℓ is in a plane and E is any point on ℓ. Why is there only one line in the plane that is perpendicular to ℓ at E? (Hint: Consider the figure shown.)

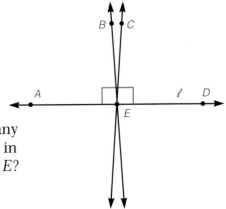

20. Suppose two lines intersect to form congruent adjacent angles. What is the measure of each angle? Explain why the lines are perpendicular.

21. Write a convincing argument for Theorem 4.5. Hint: Suppose $\ell_1 \perp \ell_2$ with $\angle 1$ a right angle.

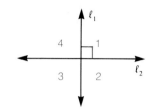

  a. Since $\angle 1$ is a right angle, what do you know about $\angle 3$?

  b. Since $\angle 4$ and $\angle 1$ form a linear pair, they are supplementary. What does this tell you about $\angle 4$?

  c. If $\angle 4$ is a right angle, what is true of $\angle 2$?

  d. If two lines intersect to form a right angle, what do you know about the other three angles?

Mixed Review

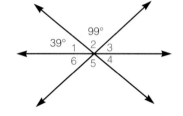

22. Which angles are congruent?

23. Find the missing angle measures.

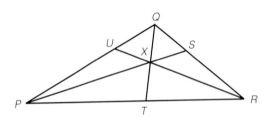

Use the figure above for Exercises 24–28.

24. $m\angle PUR + m\angle RUQ = m\angle$ _____

25. $m\angle PXR - m\angle TXR = m\angle$ _____

26. Which points are in the interior of $\angle QPR$?

27. Which points are in the exterior of $\angle QXU$?

28. $\angle QXS$ and $\angle PXT$ are _____ angles.

Use the following assumptions to reach a conclusion.

29. (i) All heptagons are polygons
  (ii) All 7-sided figures are heptagons

30. (i) You will go home if you click your heels together three times and say "there's no place like home."
  (ii) You click your heels together three times and say "there's no place like home."

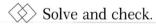 Solve and check.

  31. $3x + 4 = 5$ **32.** $5y - 120 = 180 - y$

In the photograph, as the minute hand of the clock moves from 12 to 1, it is traveling in a *clockwise* direction. Its reflection in the mirror would move in a *counterclockwise* direction. Reflection changes the direction or **orientation** of a figure. Rotation keeps the same orientation.

Samples   Determine whether the figures have the same or a reverse orientation.

1.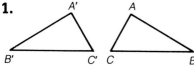

Reverse orientation

2.

Same orientation

3.

Reverse orientation

Problems

Determine whether the figures have the same or a reverse orientation.

1.

2.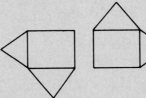

3.

4. Find the next picture in the pattern.

(A)   (B)   (C)   (D)

4-6 Reflections

Objective: Find the reflection of a figure over a line and recognize properties of reflected figures.

Explore

1. Draw △ABC on the left side of a sheet of paper.

2. Fold the paper in half so the triangle is outside. Use a compass point to punch a hole through the paper at points *A*, *B*, and *C*.

3. Open the paper and label the marks as *A′*, *B′*, and *C′*. Use a straightedge to draw △A′B′C′.

4. Draw $\overline{AA'}$, $\overline{BB'}$, and $\overline{CC'}$. Measure the angles of intersection of the fold line, ℓ, with each segment. Measure the distance of each point from the fold line.

Discuss Describe how the fold line is related to each of these segments.

Points *A* and *A′* are **reflections** of each other over ℓ, as are *B* and *B′*, and *C* and *C′*. △ABC and △A′B′C′ are also reflections of each other over ℓ. △A′B′C′ is the **image** of △ABC. ℓ is the **line of reflection.**

If you were to place a mirror along the fold line in the Explore above, you would see the image of △ABC in the mirror.

Examples Find the reflection image of each over line ℓ.

1. $\overline{BC}$

   F is the image of *B*, and *E* is the image of *C*. So $\overline{FE}$ is the image of $\overline{BC}$ over ℓ.

2. △BCD

   △FEG is the image of △BCD over ℓ.

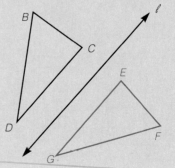

Try This... Find the reflection
image of each
over line *m*.

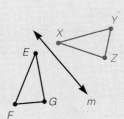

    a. $\overline{XY}$

    b. $\overline{FG}$

    c. $\triangle EFG$

Notice that as a point moves toward the reflection line, so
does its image. In fact, if the point is *on* the reflection line,
the point is its own image.

Definition   One point, *A*, is the **reflection image over line ℓ**
of another point, *A′*, if ℓ is the perpendicular bisector of $\overline{AA'}$.
A is the image of *A′*.

B is its own image.

A′ is the image of *A*.

We can use the definition given above to construct reflection
images.

Examples **3.** Construct the reflection image of the points *D, E, F, G,*
and *H* over line *m*.

Draw a perpendicular from *D* to
line *m*, and extend to *D′* so that *D*
and *D′* are **equidistant** (the same
distance) from line *m*.

In the same way, we can find *E′*,
F′, and *G′*. Point *H* is its own image
(reflection).

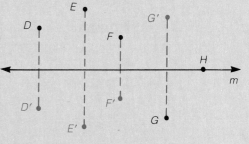

 4. Construct the reflection image of $\triangle ABC$
over line *m*.

Locate the image of each vertex
using the construction method
of Example 3. Then use a
straightedge to construct $\overline{A'B'}$, $\overline{B'C'}$,
and $\overline{A'C'}$.

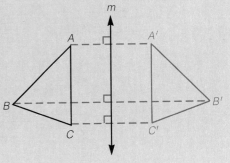

Try This...d. Copy the figure at the right, and construct the reflection image over line ℓ.

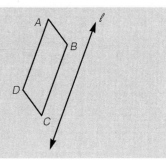

▬ *Explore* ▬

1. Measure all the sides and angles of figure $ABCD$, above, and of its image that you constructed in Try This d.
2. Which segments have the same length? Which angles have the same measure?
3. What appears to be the same about the figure and its image? What appears to be different about the figure and its image?

When any figure is reflected over a line, many aspects are unchanged. We say that these properties are *preserved* by the reflection.

Postulate 9   The reflection of a set of collinear points over a line is also a set of collinear points. (Reflections preserve collinearity.)

The following theorems state that line reflections preserve distance and angle measure:

Theorem 4.6   If A' and B' are the reflections of A and B over a line, then $AB = A'B'$.

Theorem 4.7   If $\angle A'B'C'$ is the reflection of $\angle ABC$ over a line, then $m\angle ABC = m\angle A'B'C'$.

Example   **5.** $\triangle RST$ is the reflection of $\triangle R'S'T'$ over line m. Name the congruent segments and angles.

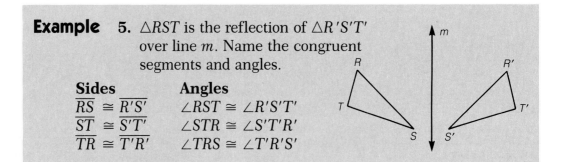

Sides	Angles
$\overline{RS} \cong \overline{R'S'}$	$\angle RST \cong \angle R'S'T'$
$\overline{ST} \cong \overline{S'T'}$	$\angle STR \cong \angle S'T'R'$
$\overline{TR} \cong \overline{T'R'}$	$\angle TRS \cong \angle T'R'S'$

Try This... e. △*ABC* is a reflection of △*A'B'C'* over line *m*. Name the congruent segments and angles.

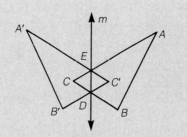

Discussion   Iban said, "When I raise my left hand in a mirror, the image raises its right hand. So a mirror must reverse left and right but not up and down." Do you agree?

Exercises

Practice

Find the reflection image of each over line *m* for each figure.

1. *T*

3. $\overline{TR}$

5. $\overline{RS}$

7. △*TRS*

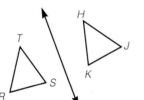

2. *A*

4. $\overline{AB}$

6. $\overline{AC}$

8. △*ABC*

Copy each line and each set of points or figure. Then construct the reflections of each over line *m*.

9.

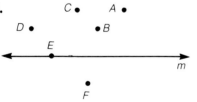

10.

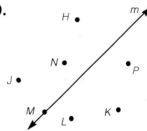

11.

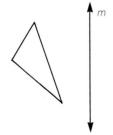

12.

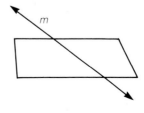

Use the figure for Exercises 13–17.
△ABC is the reflection of △DCB over line m.
Complete each statement.

13. $m\angle AEC = m\angle$ _____ .

14. If A, E, and B are collinear, then _____ are collinear.

15. The image of $\angle ACB$ is _____ .

16. Point _____ is its own image.

17. $CF =$ _____ .

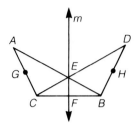

Extend and Apply

Copy each figure on dot paper and sketch the reflection image of each design over line m.

18.

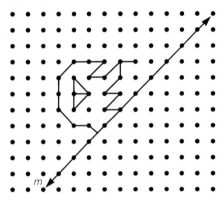

19.

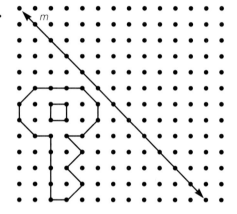

20. Which of the codes below could be a reflection image of GIVE IT A TRY over line ℓ?

(A) ԍIΛE Iʇ ∀ ⊥ᴚY

(B) YᴚT A TI ƎVIꓨ

(C) YRT A TI EVIG

(D) ꓨIΛƎ IT A TᴚY

Use Mathematical Reasoning

Determine whether each pair of figures is a reflection image. Copy the figure and sketch the line of reflection (if it exists).

21.

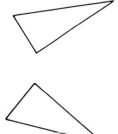

22.

23.

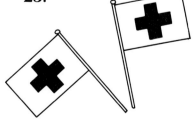

Classify the following angles as right, straight, obtuse, or acute.

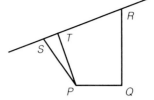

24. ∠*PQR*   **25.** ∠*PSR*   **26.** ∠*STR*

27. ∠*SPQ*   **28.** ∠*TRQ*   **29.** ∠*PTR*

30. Which postulate or theorem guarantees the statement, "Two highways that are perfectly straight will only cross once."

Solve and check.

31. $3x = 2x + 7$

32. $8t - 6 = 2t + 6$

33. $8a - 16 = 5a - 10$

Discover

This activity may be done with paper and pencil or with computer software.

1. Reflect point *A* over line ℓ as shown.

2. Place four points anywhere on line ℓ, and label them *B*, *C*, *D*, and *E*.

3. Use a straightedge to draw segments from each of the four points to *A* and to *A*′.

4. Measure each of the eight segments drawn in Step 3. Which segments are congruent?

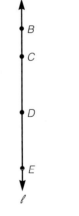

5. Draw $\overline{AA'}$.

Since *A*′ is the image of *A*, and line ℓ is the line of reflection, then line ℓ is also the perpendicular bisector of $\overline{AA'}$. Recall that a **locus** is a set of points satisfying a given condition. So *B*, *C*, *D*, and *E* are a locus of points that are on the perpendicular bisector of $\overline{AA'}$.

Use your discovery above to complete the Perpendicular Bisector Theorem. Fill in the blanks with one or more words.

Theorem 4.8 **The Perpendicular Bisector Theorem**
> The perpendicular bisector of a segment is the _____ of points that are _____ from the endpoints of the segment.

Chapter 4 Review

4-1   Which pairs of segments are congruent? Use a ruler or compass.

1.   **2.**

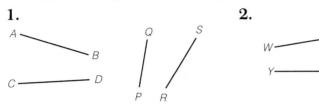

Which angles are congruent? Use a protractor.

3.

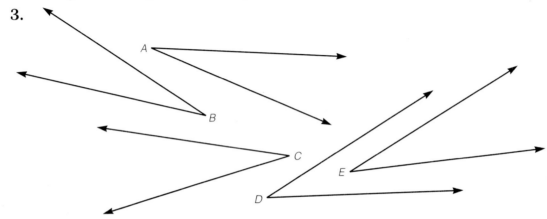

4-2   **4.** Identify each pair of complementary angles.

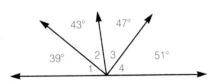

5. Identify each pair of supplementary angles.

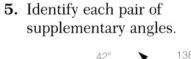

6. $\angle A \cong \angle C$ and $m\angle C = 72°$. What is the measure of $\angle A$, the measure of its complement, and the measure of its supplement?

4-3   **7.** Name two linear pairs of angles.

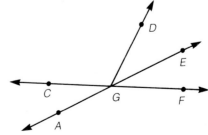

8. Find $m\angle 1$, $m\angle 2$, $m\angle 3$, and $m\angle 4$. ◇◇ **9.** Find $m\angle ACB$ and $m\angle ECD$.   **4-4**

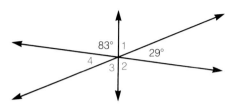

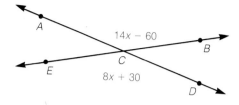

Which pairs of lines are perpendicular? Use a protractor.

10.   **11.**   **12.**

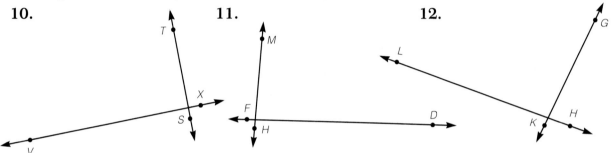

13. Use a compass and a straightedge.   **4-5**
Draw a line, $\overleftrightarrow{AB}$.
Choose a point D not on $\overleftrightarrow{AB}$.
Construct a line containing D that is
perpendicular to $\overleftrightarrow{AB}$.

Find the reflection image of each over line ℓ.   **4-6**

14. $\overline{BC}$ **15.** $\angle DBA$

16. $\angle BCD$ **17.** $\triangle ADC$

18. B **19.** $\overline{WZ}$

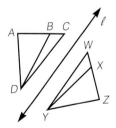

Copy each figure and find its reflection over line m.

20.   **21.**

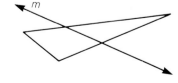

Chapter 4 Test

Which pairs of segments are congruent? Use a ruler or compass.

1.

2.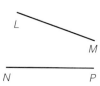

Which pairs of angles are congruent? Use a protractor.

3.

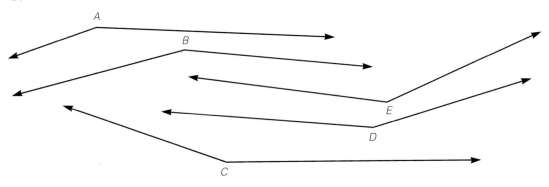

4. Identify each pair of complementary angles.

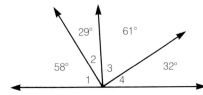

5. Identify each pair of supplementary angles.

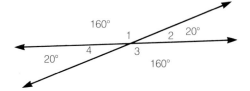

6. ∠B ≅ ∠Q and m∠B = 18°. What is the measure of ∠Q, the measure of its supplement, and the measure of its complement?

7. Name two linear pairs of angles.

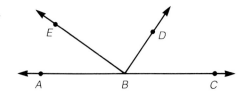

8. Find $m\angle 1$, $m\angle 2$, $m\angle 3$, and $m\angle 4$.

 9. Find $m\angle RMP$ and $m\angle SMN$.

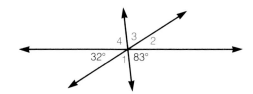

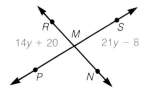

Which pairs of lines are perpendicular? Use a protractor.

10.

11.

12.

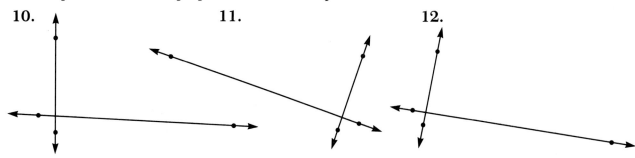

13. Use a compass and a straightedge.
Draw a line, $\overleftrightarrow{PQ}$.
Choose a point F on $\overleftrightarrow{PQ}$.
Construct a line perpendicular
to $\overleftrightarrow{PQ}$ at F.

Find the reflection image of each over line ℓ.

14. Q **15.** $\overline{SR}$

16. $\angle LNM$ **17.** $\angle QSR$

18. $\triangle NOL$ **19.** $\angle LNO$

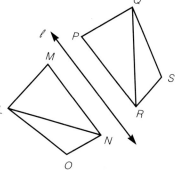

Copy each figure and find its reflection over line m.

20.

21.

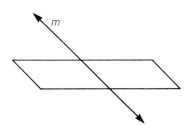

Cumulative Review Chapters 1–4

1-1 **1.** Study the figure made from cubes.
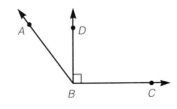
   a. How many cubes are in the figure?
   b. Draw the top view.
   c. Draw the front view.

1-2 Write using symbols.

   2.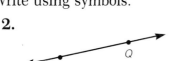

   3.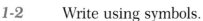

1-3 Classify each angle as acute, right, obtuse, or straight.

   4. ∠ABD   **5.** ∠ABC   **6.** ∠DBC

1-6 **7.** Suggest a possible counterexample for the statement "All apples are red."

2-1 **8.** Use the following assumptions. What can you conclude?
   (i) If a number is even, then it has 2 as a factor.
   (ii) The number 56 is even.

2-2 **9.** Draw three noncollinear points.

2-3 Complete each statement.

   10. Given any two points, there is exactly one _____ containing the two points.

   11. Given any three noncollinear points, there is exactly one _____ containing them.

2-5 **12.** Write the converse of "If the tomato is red, then it is ripe."

Use the number line for Exercises 13–16.

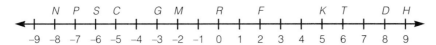

3-1 **13.** Name the coordinates of C, R, and T.

   14. Name the points having coordinates −8 and 5.

3-2 **15.** Find the distances PF, TG, and HS.

   16. Find the coordinate of the midpoint of $\overline{NT}$.

17. Find the coordinates of reflection of the point $(4, -5)$ when   **3-3**
    a. reflected over the x-axis, and   **b.** reflected over the y-axis.

18. Find the coordinates of translation of the point $(4, -5)$ when
    a. translated downward 4 units, and **b.** translated left 5 units.

19. Name in three ways the angle shown.   **3-4**

20. Name the vertex and sides of this angle.

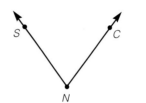

21. Use a protractor to draw an angle that measures $46°$.   **3-5**

22. $m\angle PQR = 172°$.   **3-6**
    Find $m\angle PQS$.

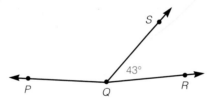

23. Find the measure of the complement and the supplement of $38°$.   **4-2**

24. Name two linear pairs of angles in the figure.   **4-3**

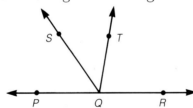

25. Find $m\angle 1$, $m\angle 2$, $m\angle 3$, and $m\angle 4$.   **4-4**

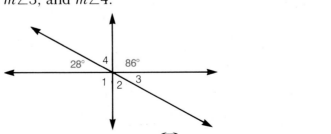

26. Use a compass and a straightedge. Draw $\overleftrightarrow{PQ}$. Choose a point R **4-5**
    on $\overleftrightarrow{PQ}$. Construct a line containing R that is perpendicular to $\overleftrightarrow{PQ}$.

5 Triangles and Congruence

Congruent triangles are prominent features of many towers. The rigid shape of the triangles gives strength to such structures.

5-1 Classifying Triangles

Objective: Classify triangles by sides and angles.

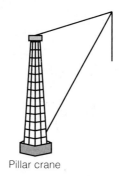
Pillar crane

Recall that a triangle consists of three segments determined by three noncollinear points. The segments are **sides** and the points the **vertices** (plural of **vertex**) of the triangle.

P, Q, and R are vertices.
$\overline{PQ}$, $\overline{QR}$, and $\overline{PR}$ are sides.

In $\triangle PQR$, $\overline{QR}$ and $\angle P$ are opposite each other. $\overline{QR}$ is called the *included side* of $\angle Q$ and $\angle R$. Likewise, $\angle P$ is called the *included angle* of $\overline{PQ}$ and $\overline{PR}$.

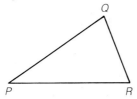

Examples Consider $\triangle ABC$.

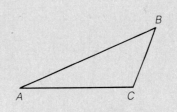

1. What angle is included between $\overline{AB}$ and $\overline{BC}$? $\angle B$ is included between $\overline{AB}$ and $\overline{BC}$.

2. What side is included between $\angle B$ and $\angle C$? $\overline{BC}$ is included between $\angle B$ and $\angle C$.

3. What angle is opposite $\overline{BC}$? $\angle A$ is opposite $\overline{BC}$.

4. What side is opposite $\angle C$? $\overline{AB}$ is opposite $\angle C$.

Try This... Consider $\triangle ABC$ above.

   a. What angle is included between $\overline{AB}$ and $\overline{AC}$?

   b. What side is included between $\angle A$ and $\angle C$?

   c. What angle is opposite $\overline{AB}$?

   d. What side is opposite $\angle B$?

Triangles may be classified by their angles.

Acute	Equiangular	Right	Obtuse
All angles acute	All angles congruent	One right angle	One obtuse angle

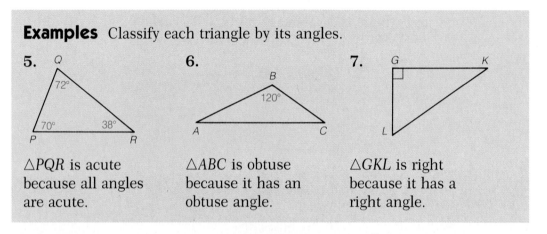

Examples Classify each triangle by its angles.

5.

Q
72°
70°   38°
P   R

△PQR is acute because all angles are acute.

6.

B
120°
A   C

△ABC is obtuse because it has an obtuse angle.

7.

G   K

L

△GKL is right because it has a right angle.

Try This... Classify each triangle by its angles.

e.

S
R   T

f.

E
110°
D   F

g.

X
Z   Y

Triangles may also be classified by their sides.

Equilateral	Isosceles	Scalene
All sides congruent	At least two sides congruent	No sides congruent

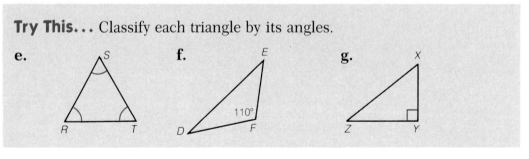

Examples Classify each triangle by its sides.

8.

△MNQ is isosceles because two sides are congruent.

9.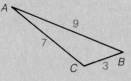

△ABC is scalene because no sides are congruent.

10.

△RST is equilateral because all sides are congruent.

Try This... Classify each triangle by its sides.

h.   **i.**   **j.**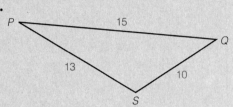

Discussion Do you think that you could draw an obtuse right triangle? Why or why not? Defend your answer.

Exercises

Practice

Refer to △QRS for Exercises 1–6.

1. What angle is included between $\overline{QR}$ and $\overline{QS}$?
2. What angle is included between $\overline{QR}$ and $\overline{RS}$?
3. What side is included between ∠Q and ∠S?
4. What side is included between ∠S and ∠R?
5. What side is opposite ∠Q?
6. What angle is opposite $\overline{QR}$?

Classify each triangle by its angles.

7. 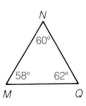 **8.**   **9.**   **10.**

11.

12.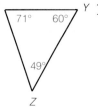

13.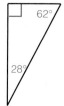

14.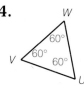

Classify each triangle by its sides.

15.

16.

17.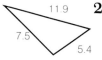

18.

19.

20.

21.

22.

Extend and Apply

23. Draw a triangle with side $\overline{MN}$ opposite angle Q.

24. Draw a triangle with $\angle P$ opposite side $\overline{RS}$.

25. Draw a triangle with $\angle T$ the included angle of $\overline{TQ}$ and $\overline{TS}$.

26. Without drawing $\triangle XYZ$, tell which angle is included between $\overline{XZ}$ and $\overline{YZ}$.

27. Without drawing $\triangle XYZ$, tell which side is included between $\angle Y$ and $\angle Z$.

28. Without drawing $\triangle XYZ$, tell which angle is opposite $\overline{XY}$.

29. Without drawing $\triangle XYZ$, tell which side is opposite $\angle Z$.

30. Write a convincing argument that an equilateral triangle is also isosceles.

31. A billiard rack is a triangle
with sides of length
15 in., 15 in., and 15 in.
The triangle is

(A) Isosceles only
(B) Isosceles and equilateral
(C) Right
(D) Scalene

Use Mathematical Reasoning

32. Draw an acute scalene triangle.

33. Draw a right scalene triangle.

34. Draw a right isosceles triangle.

35. Try to draw an obtuse right triangle.

36. The word *isosceles* contains the prefix *iso.* Look up some other words that have this prefix, such as *isobar, isometric,* and *isotope.* Why do you think that this prefix is used in the name for triangles that have at least two sides with the same length?

37. How many ways can the rack in Exercise 31 be placed so that the triangle of billiard balls is in the position shown? Explain. (Hint: The rack may also be turned over.)

Mixed Review

Use the figure below for Exercises 38–40.

38. Which point, X or Y, is a reflection of a point of the figure over ℓ?

39. Copy the figure and draw its image over ℓ.

40. Find the measure of each angle of the figure.

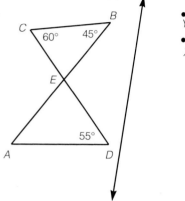

◇ Solve and check.

41. $4x + 12 = 180$

42. $4x - 12 = 180$

Enrichment

Create a "word web" by listing as many words as you can think of that are related to the idea of *triangle* and then connecting associated terms. The web is started below.

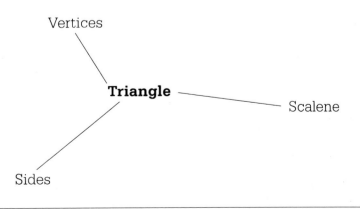

Using Equilateral Triangles

Many designs can be created using compass arcs and equilateral triangles.

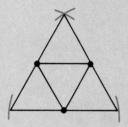 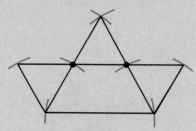

These designs are made up of equilateral triangles and compass arcs drawn from various vertices.

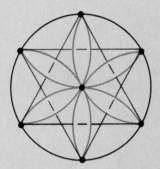

A six-leaved rose

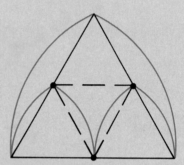

An arch from Gothic architecture

A trefoil design

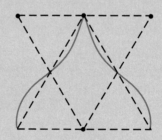

An ogee design

Problems

1. Find a logo that uses equilateral triangles on the cover of this book.

2. Make your own design based upon equilateral triangles and arcs.

5-2 Symmetry

Objective: Determine and draw lines of symmetry.

Symmetry is found everywhere you look. We find symmetry in nature, as in the kitten's face at the right. Many objects are designed to be symmetric. Business logos, such as that of the Addison-Wesley Publishing Company, are often symmetrical.

When you were younger, did you make valentine hearts by cutting a folded piece of paper?

 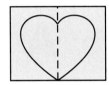

Cutting a design around the folded edge of a paper produces a **symmetrical** shape. The folded edge divides the shape into two identical parts. The folded edge is a **line of symmetry.**

Explore

1. Draw line ℓ and $\overline{AB}$ as shown.
2. Find the reflection of points A and B over ℓ, and draw $\triangle BAB'$.
3. Fold on line ℓ. Does $\triangle BAB'$ appear to be symmetrical?

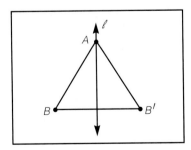

Definition   A figure has **reflectional symmetry** over line ℓ if the reflected image of each point on one side of ℓ dividing the figure is also a point of the figure. Line ℓ is a **line of symmetry.**

Examples Copy each figure. Determine whether it has a line of symmetry and draw the line if it exists.

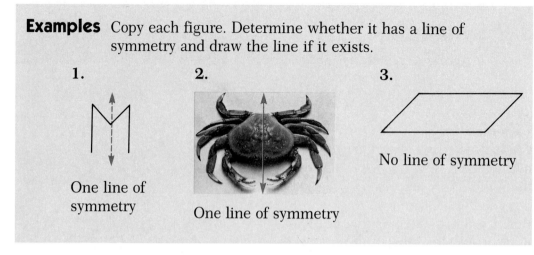

1.

One line of symmetry

2.

One line of symmetry

3.

No line of symmetry

Try This... Copy each figure. Determine whether it has a line of symmetry and draw the line if it exists.

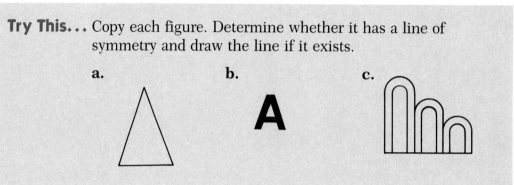

a.

b.

A

c.

Many shapes have more than one line of symmetry.

Examples Find all lines of symmetry for each letter or figure.

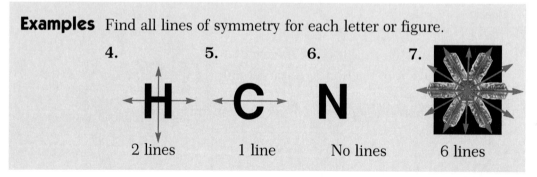

4.

H

2 lines

5.

C

1 line

6.

N

No lines

7.

6 lines

Try This... Copy each letter or figure. Determine how many lines of symmetry each one has and draw the lines.

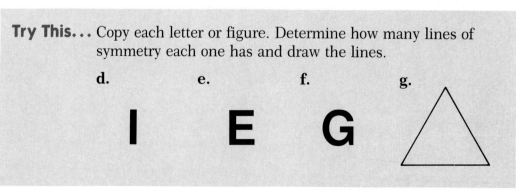

d.

I

e.

E

f.

G

g.

Exercises

Practice

Copy each letter or figure. Determine how many lines of symmetry each one has and draw the lines.

1. **D** 2. **F** 3. **Z** 4. **Y**

5.  6. 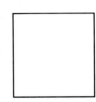 7.   8.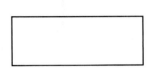

Extend and Apply

9. Find all the capital letters of the English alphabet that have lines of symmetry. Which letters have exactly one line of symmetry? Which letters have exactly two lines of symmetry? Do any letters have three or more lines of symmetry?

10. Copy the design on the right and shade additional squares so that the design has line symmetry.

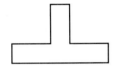

11. Which word has a vertical line of symmetry?
   (A) HIX   (B) OW   (C) OTTO   (D) BOB

Use Mathematical Reasoning

12. Make up a word that has both vertical and horizontal symmetry.

Mixed Review

Draw a triangle with the description given.

13. Isosceles   14. Right isosceles   15. Obtuse   16. Equilateral

Solve and check.

17. $4a - 7 = 3a$ 18. $5b - 2 = 28 - b$ 19. $5x - x = 37$

5-3 Congruent Triangles

Objective: Identify corresponding parts of congruent triangles.

Explore

1. Draw a large scalene triangle on a sheet of paper. Place a second sheet of paper under the first sheet. Cut out two triangles by cutting both sheets of paper as you cut out the drawn triangle. Be sure the paper does not slide.

2. Stack the triangles so that each vertex of the top triangle is on a vertex of the bottom triangle.

3. Label the vertices of the top triangle A, B, and C, and the corresponding vertices of the other triangle X, Y, and Z.

4. Compare $m\angle A$ with $m\angle X$, $m\angle B$ with $m\angle Y$, and $m\angle C$ with $m\angle Z$. Compare AB with XY, AC with XZ, and BC with YZ.

Discuss   What appears to be true of the matching angles and sides of congruent triangles?

Can you match the vertices of your triangles in a different way to show that they are congruent?

If all three vertices of a triangle can be matched, what can be said about corresponding sides and angles?

The Explore suggests that we define congruent triangles as follows:

Definition   **Two triangles are congruent** whenever their vertices can be matched so that the corresponding sides and the corresponding angles are congruent.

Example   1. Consider $\triangle ABC$ and $\triangle XYZ$. If we match A with X, B with Y, and C with Z, what are the corresponding sides? What are the corresponding angles?

Sides: *Angles:*
$\overline{AB} \leftrightarrow \overline{XY}$ $\angle A \leftrightarrow \angle X$
$\overline{BC} \leftrightarrow \overline{YZ}$ $\angle B \leftrightarrow \angle Y$
$\overline{AC} \leftrightarrow \overline{XZ}$ $\angle C \leftrightarrow \angle Z$

$\leftrightarrow$ means "corresponds to."

There are six corresponding parts of two congruent triangles: three sides and three angles. These are called **corresponding parts of congruent triangles.** Corresponding parts of congruent triangles are congruent.

We write $\triangle ABC \cong \triangle XYZ$ to say that $\triangle ABC$ and $\triangle XYZ$ are congruent. We agree that this symbol tells us the order in which the vertices are matched.

$\triangle ABC \cong \triangle XYZ$, so $\triangle ABC \cong \triangle XYZ$ means
$\angle A \cong \angle X$ and $\overline{AB} \cong \overline{XY}$
$\angle B \cong \angle Y$ $\overline{AC} \cong \overline{XZ}$
$\angle C \cong \angle Z$ $\overline{BC} \cong \overline{YZ}$

Example   2. Suppose $\triangle PQR \cong \triangle STV$. What are the congruent corresponding parts?

Angles: *Sides:*
$\angle P \cong \angle S$ $\overline{PQ} \cong \overline{ST}$
$\angle Q \cong \angle T$ $\overline{PR} \cong \overline{SV}$
$\angle R \cong \angle V$ $\overline{QR} \cong \overline{TV}$

Try This... a. Suppose $\triangle ABC \cong \triangle DEF$. What are the congruent corresponding parts?

Example   3. Name the corresponding parts of these congruent triangles.

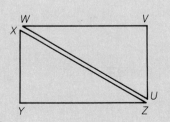

Angles: *Sides:*
$\angle X \cong \angle U$ $\overline{XY} \cong \overline{UV}$
$\angle Y \cong \angle V$ $\overline{YZ} \cong \overline{VW}$
$\angle Z \cong \angle W$ $\overline{ZX} \cong \overline{WU}$

Try This... **b.** Name the corresponding parts of these congruent triangles.

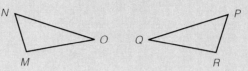

Discussion Suppose you know that △GHK ≅ △RST. Could you also say △HKG ≅ △STR? Could you say △KHG ≅ △RTS? Defend your answer.

Exercises

Practice

Name the corresponding parts of these congruent triangles.

1. △ABC ≅ △RST

2. △DEF ≅ △GHK

3. △XYZ ≅ △UVW

4. △MNQ ≅ △HJK

5. △ABC ≅ △ABC

6. △ABC ≅ △ACB

Name the corresponding parts of these congruent triangles.

7.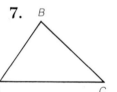

8.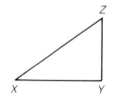

9.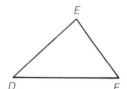

10.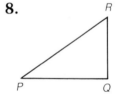

11.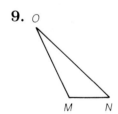

12.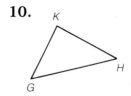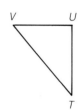

13.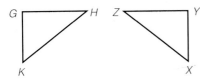

14.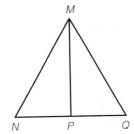

Extend and Apply

15. Here are corresponding parts of two congruent triangles. Find a matching of vertices that shows the congruence.

$$\angle T \cong \angle S \qquad \overline{PT} \cong \overline{RS}$$
$$\angle U \cong \angle Q \qquad \overline{PU} \cong \overline{RQ}$$

16. Write a convincing argument that a triangle is congruent to itself.

17. $\triangle ABC \cong \triangle JQM$. Which of the following is also a true statement?

(A) $\triangle BAC \cong \triangle QMJ$ (B) $\triangle MQJ \cong \triangle CBA$

(C) $\overline{CB} \cong \overline{JM}$ (D) $\triangle ABC \cong \triangle MQJ$

18. How many congruent triangle pairs can you find in the shuffleboard court?

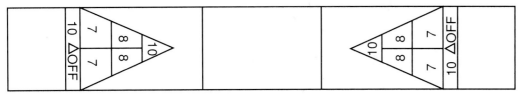

Use Mathematical Reasoning

19. Suppose $\triangle ABC \cong \triangle DEF$ and $\triangle DEF \cong \triangle GHK$. What can you say about $\triangle ABC$ and $\triangle GHK$? Explain your answer.

20. $\triangle ABC$ is both equilateral and equiangular. List all the possible matchings of vertices that show that $\triangle ABC$ is congruent to itself.

Mixed Review

How many lines of symmetry does each figure have?

21. §   **22.** Σ   **23.** ¥   **24.** Å   **25.** 🔲   **26.** ⇉

◇ Solve and check.

27. $15y - 45 = 135$ **28.** $15x - 135 = 45$

29. $20z - 45 = 5z + 135$ **30.** $135w - 45 = 15w - 45$

A 4-sided solid figure called a *tetrahedron* can be made by folding an equilateral triangle along the dashes shown in the pattern.

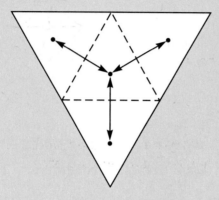

1. Suppose the pattern above is folded and the resulting tetrahedron is placed on one triangular *base*. Which cannot be a view of the tetrahedron?

(A) 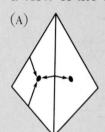   (B) 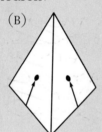   (C) 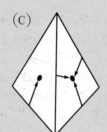   (D)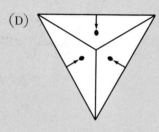

2. Suppose a tetrahedron-shaped box is manufactured and labeled so that the box is never turned on its side. Two views of it are shown below. What is wrong with the labeling?

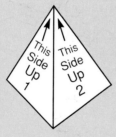   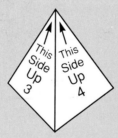

5-4 Reading and Making Drawings

Objective: Tell what information is given in a drawing and make drawings to show given information.

As we have seen, drawings can be used to give information about geometric figures. A jeweler might use a drawing as a guide for cutting a diamond. Drawings can be used to show that points are collinear, that angles are adjacent, and that lines, rays, and segments intersect.

Marks are added to drawings to show right angles and to indicate the congruent parts in a figure. Segments with an equal number of marks are congruent. Congruent angles are indicated in the same way.

Examples What information about the triangles is shown in the drawing?

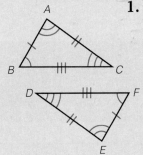

1. Segments with 1 mark are congruent $\rightarrow \overline{AB} \cong \overline{EF}$
Segments with 2 marks are congruent $\rightarrow \overline{AC} \cong \overline{DE}$
Segments with 3 marks are congruent $\rightarrow \overline{BC} \cong \overline{DF}$
 Angles with 1 mark are congruent $\rightarrow \angle B \cong \angle F$
 Angles with 2 marks are congruent $\rightarrow \angle A \cong \angle E$
 Angles with 3 marks are congruent $\rightarrow \angle C \cong \angle D$

2. $\overline{DB} \perp \overline{AC}$
$\overline{AD} \cong \overline{CD}$
$\overline{AB} \cong \overline{BC}$
$\angle A \cong \angle C$

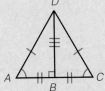

Also, since $\overline{BD}$ is a segment of both $\triangle ABD$ and $\triangle CBD$, we can state that $\overline{BD} \cong \overline{BD}$. This may or may not be marked in the drawing.

Try This... What information about the triangles is shown?

a.

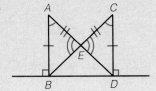

b.

It is also helpful to be able to make accurately marked drawings to show specific information.

Example   **3.** Make drawings to show the given information about $\triangle ABC$ and $\triangle RSP$.

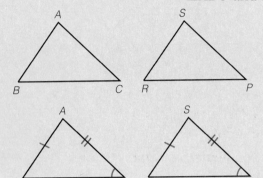

$\overline{AB} \cong \overline{SR}$, $\overline{AC} \cong \overline{SP}$, and $\angle C \cong \angle P$

Note the corresponding vertices. There are two triangles, $\triangle ABC$ and $\triangle SRP$. Sketch them.

Then mark the drawing accordingly.

Try This...   Make drawings to show the given information.

c. $\angle C \cong \angle F$, $\overline{CB} \cong \overline{FE}$, and $\angle B \cong \angle E$

d. $\overline{DE} \cong \overline{XY}$, $\overline{DF} \cong \overline{XZ}$, and $\overline{EF} \cong \overline{YZ}$

Discussion   Alaa said, "If segments of two triangles are marked with a slash, they are congruent." Jodi said, "You still need to measure the segments." With whom do you agree?

Exercises

Practice

What information about the triangles is shown in these drawings?

1.
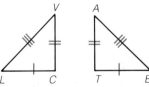

2.

3.
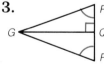

4.
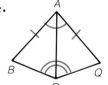

5.
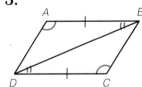

6.
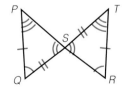

7.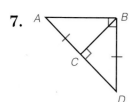

8.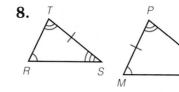

Make drawings to show the given information.

9. $\angle GED \cong \angle GEK$ and $\angle EGD \cong \angle EGK$.

10. $\angle DAC \cong \angle ACB$ and $\angle BAC \cong \angle DCA$.

11. B is the midpoint of $\overline{AD}$, and $\angle BAE \cong \angle BDC$.

12. $\overline{XT}$ is perpendicular to $\overline{MR}$ and bisects $\angle MXR$.

13. $\overline{AB} \cong \overline{CB}$ and $\angle BAE \cong \angle DCB$.

14. $\angle PTS \cong \angle QTS$, and $\overrightarrow{ST}$ is the bisector of $\angle PSQ$.

15. $\overline{FG}$ is perpendicular to $\overline{GH}$, $\overline{GI}$ is perpendicular to $\overline{FH}$, and $\angle FGI \cong \angle H$.

16. $\overline{AB}$ is perpendicular to $\overline{CD}$ at E, and $\overline{AC} \cong \overline{BD}$.

17. $\angle ABC$ and $\angle ADC$ are right angles, and $\overline{AB}$, $\overline{BC}$, $\overline{AD}$, and $\overline{DC}$ are congruent.

18. A is the midpoint of $\overline{BD}$, E is the midpoint of $\overline{CD}$, $\angle CBD \cong \angle EAD$, and $\angle C \cong \angle AED$.

Extend and Apply

Make drawings to show the given information.

19. $\triangle ABC$ and $\triangle ACD$ are equilateral triangles.

20. If $\overline{AR} \perp \overleftrightarrow{RQ}$, $\overline{BQ} \perp \overleftrightarrow{RQ}$, $\overline{AR} \cong \overline{BQ}$, and C is the midpoint of $\overline{RQ}$, then $\overline{AC} \cong \overline{BC}$.

21. $\angle ACB \cong \angle ECD$, and D and B are in the interior of $\angle ACE$.

22. $\angle DFE$ and $\angle DCE$ are right angles, $\angle FDE \cong \angle B$, D bisects $\overline{AC}$, F bisects $\overline{AB}$, and E is the midpoint of $\overline{BC}$.

23. Which of the following is true?

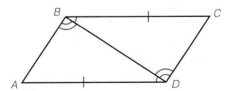

(A) $\overline{BD} \cong \overline{BD}$, $\angle ABD \cong \angle BDC$
(B) $\overline{AD} \cong \overline{BC}$, $\angle CBD \cong \angle BDA$
(C) Both (A) and (B)
(D) None of these

Use Mathematical Reasoning

24. Work with a partner. Make up information about two congruent triangles. Write down your information for your partner and have him or her make a drawing. Then check the drawing and reverse roles.

Mixed Review

25. Construct an equilateral triangle.

Solve and check.

26. $90x + 270 = 18x + 27x + 360$

Make a Drawing

Information can be used to construct accurate triangles, either with a compass and a straightedge or by using a computer.

Samples

Make a drawing to show the information given.

1. $\overline{AB} \cong \overline{SR}$, $\overline{AC} \cong \overline{SP}$, and $\overline{BC} \cong \overline{RP}$. Does $\triangle ABC$ appear to be congruent to $\triangle SRP$?

Draw two rays and mark off $\overline{AB} \cong \overline{SR}$.

Set a compass opening of AC (or SP) and draw arcs.

Set a compass opening of BC (or RP) and draw arcs.

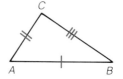
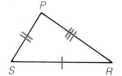

Label the intersections of the arcs C and P and draw triangles. Mark the congruent sides. $\triangle ABC$ and $\triangle SRP$ appear to be congruent.

2. $\angle R \cong \angle H$, $\overline{RS} \cong \overline{HG}$, and $\angle S \cong \angle G$. Does $\triangle RST$ appear to be congruent to $\triangle HGK$?

Two angles are mentioned so we draw them first.

Use the other information about the sides to complete the drawing.

Label the drawing accordingly. $\triangle RST$ and $\triangle HGK$ appear to be congruent.

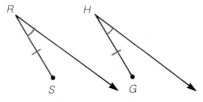
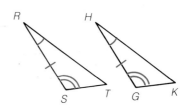

Problem

Make a drawing to show the information given.

$\angle C \cong \angle F$, $\overline{CB} \cong \overline{FE}$, and $\angle B \cong \angle E$.
Does $\triangle ABC$ appear to be congruent to $\triangle DEF$?

5-5 SAS, SSS, and ASA Congruence

Objective: Identify triangles that can be shown congruent by the SAS, SSS, or ASA Postulates.

Explore

Sometimes we can state that triangles are congruent without knowing that all six corresponding parts are congruent.

1. Draw △ABC with one 40° angle.

2. Draw △DEF with a 3 cm side.

3. Draw △GHI with a 50° angle and a side of length 4 cm.

4. Draw △JKL with $m\angle J = 50°$ and $m\angle K = 40°$.

5. Draw △MNO with $m\angle M = 40°$, $m\angle N = 50°$, and $MN = 3$ cm.

6. Draw △PQR with $m\angle P = 45°$, $PQ = 3$ cm, and $PR = 4$ cm.

7. Draw △STU with $m\angle S = 50°$, $m\angle T = 60°$, and $m\angle U = 70°$.

8. Draw △XYZ with $XY = 4$ cm, $YZ = 5$ cm, and $ZX = 6$ cm.

Discuss   Compare your triangles with those drawn by the members of other groups. Which triangles does *everyone* have congruent?

How many corresponding congruent parts are necessary to show that two triangles are congruent?

In Chapter 1 you copied a triangle by copying each of its sides. Obviously, a triangle and its copy are congruent. This suggests the following postulate:

Postulate 10   The SSS (Side-Side-Side) Postulate

If three sides of one triangle are congruent to three sides of another triangle, then the triangles are congruent.

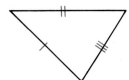 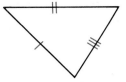

Examples Are the triangles congruent by the SSS Postulate?

1.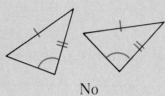

No

2.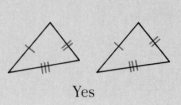

Yes

3.

No

4.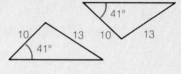

Yes

Try This... Are the triangles congruent by the SSS Postulate?

a.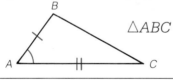

b.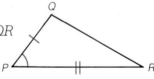

Postulate 11 The SAS (Side-Angle-Side) Postulate

Two triangles are congruent if two sides and the included angle of one triangle are congruent to two sides and the included angle of the other triangle.

$\triangle ABC \cong \triangle PQR$

Examples Are the triangles congruent by the SAS Postulate?

5.

No

6.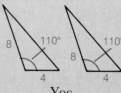

Yes

7.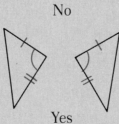

Yes

8.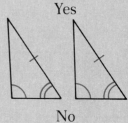

No

Try This... Are the triangles congruent by the SAS Postulate?

c.  **d.**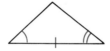

e. 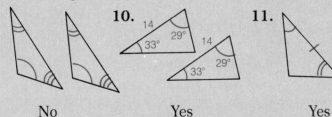 **f.**

We have shown triangles congruent by SAS and SSS. Here is a third way to show congruence.

Postulate 12 **The ASA (Angle-Side-Angle) Postulate**

If two angles and the included side of a triangle are congruent to two angles and the included side of another triangle, then the triangles are congruent.

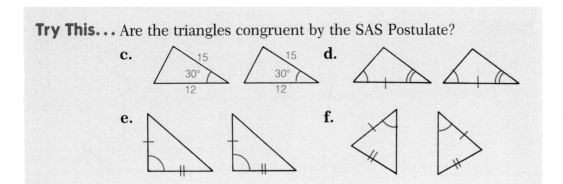

Examples Are the triangles congruent by the ASA Postulate?

9.

No

10. 14 33° 29° 14 29° 33°

Yes

11.

Yes

Try This... Are the triangles congruent by the ASA Postulate?

g. 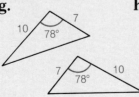 10 7 78° 7 78° 10

h.

i.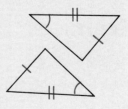

When showing that triangles are congruent, we need to decide whether to use SAS, SSS, or ASA.

Examples Which postulate (if any) should be used to show these pairs of triangles congruent?

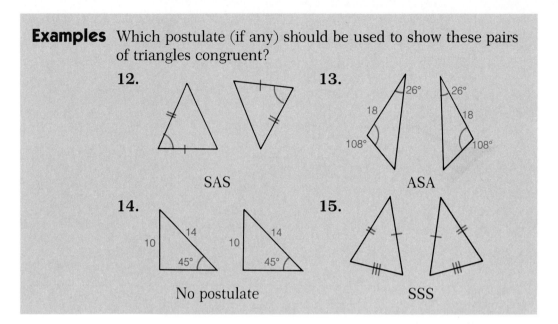

12.

SAS

13.

ASA

14.

No postulate

15.

SSS

Try This... Which postulate (if any) should be used to tell whether these pairs of triangles are congruent?

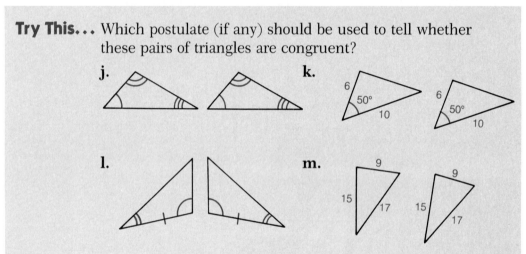

j.

k.

l.

m.

Discussion Describe the three methods of showing triangles congruent, and the special conditions that must exist for each.

Exercises

Practice

Which postulate (if any) should be used to show the triangles congruent?

1.

2.

3.

4.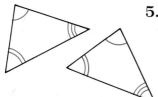

5.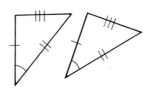

6.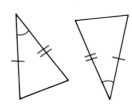

7.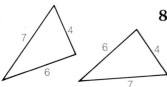

8.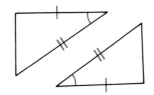

9.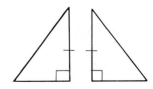

10.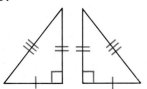

11.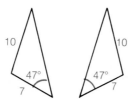

12.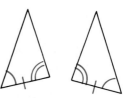

13.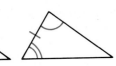

14.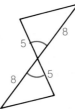

15.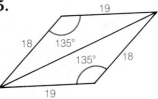

16.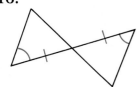

17.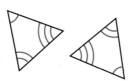

18.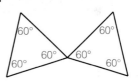

Extend and Apply

We want to know if $\triangle PQR \cong \triangle STV$.

19. We know $\overline{RQ} \cong \overline{VT}$. What else must we know to use ASA?

20. We know $\angle P \cong \angle S$. What else must we know to use SAS?

21. We know $\overline{PR} \cong \overline{SV}$ and $\overline{RQ} \cong \overline{VT}$. What else must we know to use SSS?

22. We know $\angle R \cong \angle V$ and $\angle Q \cong \angle T$. What else must we know to use ASA?

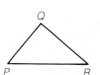

23. If $\overline{AB} \cong \overline{PQ}$, then

  (A) the triangles are congruent by ASA.
  (B) the triangles are congruent by SAS.
  (C) the triangles are already congruent by AAA.
  (D) the triangles are not necessarily congruent.

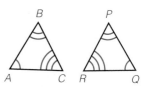

24. If four rods are connected as shown, the figure wobbles. Why does the addition of a fifth rod make the structure rigid?

25. How do you know that the faces of this pyramid are congruent triangles?

26. How can you find congruent triangles in the photograph if all the tubes are the same size?

27. *M* is the midpoint of both crosspieces. Explain why the triangles are congruent.

Use Mathematical Reasoning

28. Draw $\triangle ABC$ where $m\angle A = 35°$, $AB = 6$ cm, and $BC = 4$ cm. Try to draw another triangle, $\triangle DEF$, with the same measurements but not congruent to $\triangle ABC$. Are triangles necessarily congruent by SSA?

29. Draw $\triangle TOP$ where $m\angle T = 30°$, $m\angle O = 60°$, and $TP = 5$ cm. Try to draw another triangle, $\triangle QRS$, with the same measurements but not congruent to $\triangle TOP$. Are triangles necessarily congruent by SAA?

30. Draw $\triangle KMH$ where $m\angle K = 20°$, $m\angle H = 100°$, and $m\angle M = 60°$. Try to draw another triangle, $\triangle LNI$, with the same measurements but not congruent to $\triangle KMH$. Are triangles necessarily congruent by AAA?

Mixed Review

Determine how many lines of symmetry each figure has.

31. I   **32.** I   **33.** I   **34.** *H*   **35.** *H*   **36.** H

◈ Solve and check.

37. $2x = x + 100$ **38.** $100x = 98x + 2$

5-6 Showing Triangles Congruent

Objective: To show in a drawing why pairs of triangles are congruent.

There are many ways to make a convincing argument that two triangles are congruent. One way is to make a drawing, mark the congruent parts, and decide whether the triangles are congruent by SAS, SSS, or ASA.

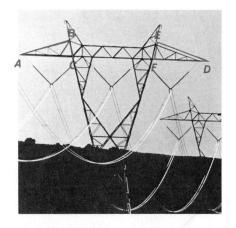

Examples **1.** In the photo above, $\angle A \cong \angle D$, $\overline{AC} \cong \overline{DF}$, and $\overline{AB} \cong \overline{DE}$. Make a drawing to show that $\triangle ABC \cong \triangle DEF$.

We make a drawing and mark the congruent parts.

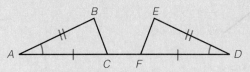

We can see that $\triangle ABC \cong \triangle DEF$ by SAS.

2. Make a drawing to show that if D is the midpoint of $\overline{QP}$, $\overline{EQ} \perp \overline{QP}$, $\overline{CP} \perp \overline{QP}$, and $\angle QDE \cong \angle PDC$, then $\triangle CPD \cong \triangle EQD$.

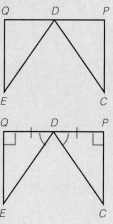

We mark the congruent parts.

We can see that $\triangle CPD \cong \triangle EQD$ by ASA.

Try This... a. Make a drawing to show that if $\overline{RS} \cong \overline{DE}$, $\overline{RT} \cong \overline{DF}$, and $\angle R \cong \angle D$, then $\triangle RST \cong \triangle DEF$.

Try This... b. Make a drawing to show that if $\overline{AB} \perp \overline{ED}$ and B is the midpoint of ED, then $\triangle ABD \cong \triangle ABE$.

Discussion In Example 2, how did you know that $\overline{QD} \cong \overline{DP}$? How did you know that $\angle Q \cong \angle P$?

Exercises

Practice

Make a drawing to show the given information, and fill in the blanks.

1. R is the midpoint of both $\overline{PT}$ and $\overline{QS}$.
   $\triangle PRQ \cong$ _____ by _____ .

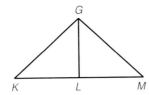

2. $\angle 1$ and $\angle 2$ are right angles, X is the midpoint of $\overline{AY}$, and $\overline{XB} \cong \overline{YZ}$.
   $\triangle ABX \cong$ _____ by _____ .

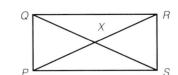

3. L is the midpoint of $\overline{KM}$ and $\overline{GL} \perp \overline{KM}$.
   $\triangle KLG \cong$ _____ by _____ .

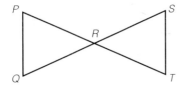

4. X is the midpoint of both $\overline{QS}$ and $\overline{RP}$, with $RQ = SP$.
   $\triangle RQX \cong$ _____ by _____ or _____ .

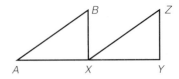

5. $\triangle AEB$ and $\triangle CDB$ are isosceles, with $\overline{AE} \cong \overline{AB} \cong \overline{CB} \cong \overline{CD}$. Also, B is the midpoint of $\overline{ED}$.
   _____ $\cong \triangle CDB$ by _____ .

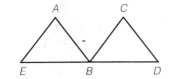

6. $\overline{AB} \perp \overline{BE}$ and $\overline{DE} \perp \overline{BE}$. $\overline{AB} \cong \overline{DE}$ and $\angle BAC \cong \angle EDC$.
   _____ $\cong \triangle DEC$ by _____ .

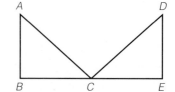

7. $\overline{GK} \cong \overline{ML}$ and $\angle GKM \cong \angle LMK$.

_____ $\cong \triangle LMK$ by _____.

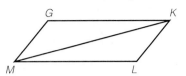

8. $\overline{SX} \cong \overline{RX}$ and $\overrightarrow{XT}$ bisects $\angle SXR$.

_____ $\cong \triangle RXT$ by _____.

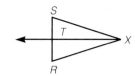

9. $\overline{FT} \cong \overline{FR}$ and $\overline{ST} \cong \overline{SR}$.

$\triangle FTS \cong$ _____ by _____.

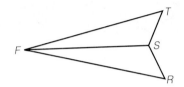

10. H is the midpoint of $\overline{QK}$, $\overline{QM} \cong \overline{KD}$, and $\overline{MH} \cong \overline{DH}$.

$\triangle QHM \cong$ _____ by _____.

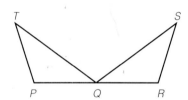

11. $\overline{TQ} \cong \overline{SQ}$, $\overline{TP} \cong \overline{SR}$, and Q is the midpoint of $\overline{PR}$.

_____ $\cong \triangle SQR$ by _____.

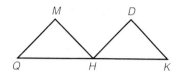

12. L is the midpoint of both $\overline{GN}$ and $\overline{KM}$ and $\overline{GK} \cong \overline{MN}$.

_____ $\cong \triangle NML$ by _____

or _____.

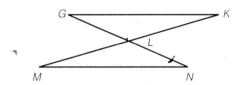

Extend and Apply

13. T is the midpoint of $\overline{RS}$ and $\angle AST$ is supplementary to $\angle 3$. Are $\triangle AST$ and $\triangle PRT$ congruent? If so, why?

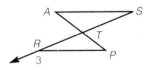

14. $\angle ATR \cong \angle STR$ and $\angle 1 \cong \angle 2$. Are $\triangle ART$ and $\triangle SRT$ congruent? If so, why?

15. Iona and Ben began at the same spot. Iona walked 1 mile south, then 2 miles east. Ben walked 1 mile east, then 2 miles south. Who was farther from the starting point?

(A) Iona

(B) Ben

(C) They were the same distance away.

(D) It cannot be determined from the information given.

16. On the flag of Jamaica, X marks the midpoint of $\overline{QS}$ and $\overline{RP}$. Which triangles are congruent, and why?

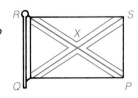

17. The photograph shows that $\overline{AB} \perp \overline{BC}$, $\overline{DE} \perp \overline{EF}$, $\overline{AB} \cong \overline{DE}$, and $\angle BAC \cong \angle EDF$. Find two congruent triangles. How do you know they are congruent?

Use Mathematical Reasoning

18. To brace a door, identical crossboards are spaced at equal distances perpendicular to the vertical boards on the door. Diagonal braces $\overline{AB}$ and $\overline{XZ}$ are also added. Find two congruent triangles. How do you know they are congruent?

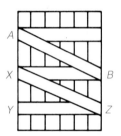

19. On the sailboat pennants, $\angle 1 \cong \angle 2$, $\overline{AB} \cong \overline{DE}$, and $\overline{EF} \cong \overline{BC}$. Explain why $\triangle ABC \cong \triangle DEF$.

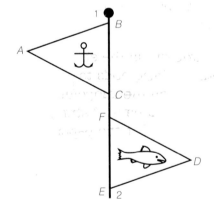

20. Suppose the sun is positioned behind a pyramid such that the indicated sides are congruent. Explain why the shadow is congruent to the face of the pyramid.

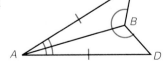

Mixed Review

Use the figure for Exercises 21–23.

21. What information is shown by the marks in the figure?

22. Name two pairs of adjacent angles in the figure.

23. Assume $\triangle ABC \cong \triangle ABD$. List all the corresponding parts.

24. Make a drawing of $\triangle DEF$ and $\triangle UVW$ and mark it to show that $\angle D$ is a right angle, $\angle U$ is a right angle, $\overline{DE} \cong \overline{UV}$, and $\overline{DF} \cong \overline{UW}$.

25. Assume $\triangle DEF \cong \triangle UVW$ in Exercise 24. List all the corresponding parts.

26. Draw $\overleftrightarrow{JK}$ intersecting $\overleftrightarrow{LM}$ at N. Indicate with marks that the vertical angles are congruent.

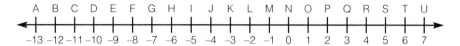

Give the letter of the point named by each coordinate to find the coded word.

27. $\dfrac{}{-5} \ \dfrac{}{5} \ \dfrac{}{1} \ \dfrac{}{5} \ \dfrac{}{-11} \ \dfrac{}{-9} \ \dfrac{}{-2} \ \dfrac{}{-9} \ \dfrac{}{5}$

28. $\dfrac{}{-11} \ \dfrac{}{1} \ \dfrac{}{4} \ \dfrac{}{4} \ \dfrac{}{-9} \ \dfrac{}{5} \ \dfrac{}{2} \ \dfrac{}{1} \ \dfrac{}{0} \ \dfrac{}{-10} \ \dfrac{}{-5} \ \dfrac{}{0} \ \dfrac{}{-7}$

◇ Solve and check.

29. $3y + 17 = y + 45$

$24 + 17 = 45$
$\underline{-17}$
28

Enrichment

A bricklayer often uses an instrument called a *level* to make a row of bricks horizontal. You can make a simple level based upon the principle of congruent triangles.

Cut out three congruent pieces of stiff cardboard in the shapes of rectangles. Tape the rectangles together, joining ends to form an equilateral triangle. Draw a line through the midpoint of the triangle's base. Tie a large paper clip to a piece of string. Attach the string to the triangle's top vertex so that the paper clip hangs just above the base.

What will happen if the level is resting on a non-horizontal surface?

What will happen if the level is resting on a level surface?

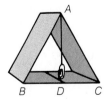

If the surface is horizontal, what can you conclude about $\triangle ABD$ and $\triangle ACD$? If the surface is not horizontal, can you make the same conclusion? Explain how you know.

5-7 Showing Corresponding Parts Congruent

Objective: To use congruent triangles to show corresponding parts congruent.

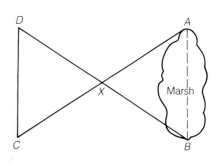

To find the distance AB across a marsh, a surveyor marks off distances AC and BD with X as the midpoint of $\overline{AC}$ and $\overline{BD}$. She observes that $\triangle ABX \cong \triangle CDX$, which means that $\overline{CD} \cong \overline{AB}$, because they are corresponding parts of congruent triangles. She then measures CD, finds it to be 125 feet, and concludes that $AB = 125$ feet.

Sometimes we can conclude that angles and segments are congruent by first showing that triangles are congruent.

Examples **1.** $\angle 1 \cong \angle 2$ and $\overline{AB} \cong \overline{AD}$. Explain why $\angle B \cong \angle D$.

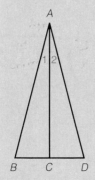

Step 1
We will first show that $\triangle ABC \cong \triangle ADC$. Mark the congruent parts.

$\overline{AB} \cong \overline{AD}$ and $\angle 1 \cong \angle 2$.
$\overline{AC} \cong \overline{AC}$, since it is the same segment for both triangles. We see that $\triangle ABC \cong \triangle ADC$ by SAS.

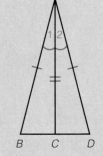

Step 2
$\angle B \cong \angle D$ because they are corresponding parts of congruent triangles.

2. $\angle 1 \cong \angle 4$ and $\overline{BC} \cong \overline{CD}$. Explain why $\overline{AB} \cong \overline{ED}$.

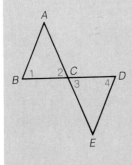

Step 1
We will first show that $\triangle ABC \cong \triangle EDC$. Mark the congruent parts.

$\angle 1 \cong \angle 4$ and $\overline{BC} \cong \overline{CD}$. Since $\angle 2$ and $\angle 3$ are vertical angles, $\angle 2 \cong \angle 3$. We see that $\triangle ABC \cong \triangle EDC$ by ASA.

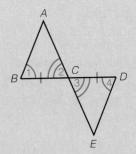

Step 2
$\overline{AB} \cong \overline{ED}$ because they are corresponding parts of congruent triangles.

Try This... **a.** $\overline{RS} \cong \overline{RT}$ and $\overline{SQ} \cong \overline{QT}$. Explain why $\angle S \cong \angle T$.

b. Explain how the surveyor found AB in the marsh problem on p. 178.

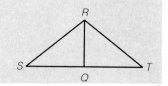

Example **3.** $\overrightarrow{QS}$ bisects $\angle PQR$, and $\overline{RQ} \cong \overline{PQ}$. Explain why $\angle R \cong \angle P$.

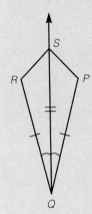

Step 1
We will first show that $\triangle QRS \cong \triangle QPS$.
Mark the congruent parts on a drawing.

$\overline{RQ} \cong \overline{PQ}$
Since $\overrightarrow{QS}$ bisects $\angle PQR$, we know $\angle RQS \cong \angle PQS$.
$\overline{QS} \cong \overline{QS}$, since it is the same segment in both triangles.
We see that $\triangle QRS \cong \triangle QPS$ by SAS.

Step 2
$\angle R \cong \angle P$ because they are corresponding parts of congruent triangles.

Try This... **c.** T is the midpoint of $\overline{AG}$, $\overline{AB} \cong \overline{FG}$, and $\overline{BT} \cong \overline{FT}$. Explain why $\angle B \cong \angle F$.

d. Explain how you could use congruent triangles to find the distance PQ across the lake.

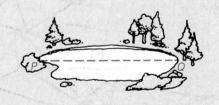

According to legend, an officer of Napoleon used the following method to determine the width of a river.

He looked across the river at point Q by sighting under the visor of his cap. He then turned and sighted up the river to point S. Pacing the distance SB, he announced that SB was the river's width.

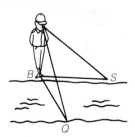

Discussion Explain the method Napoleon's officer used. How accurate do you think his result was? Why?

Exercises

Practice

Explain each conclusion. Mark a drawing for each exercise.

1. $\overline{AB} \cong \overline{BC}$ and $\overline{EB} \cong \overline{DB}$. Explain why $\overline{AE} \cong \overline{CD}$.

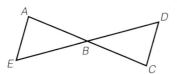

2. $\angle R \cong \angle T$, $\angle W \cong \angle V$, and $\overline{RW} \cong \overline{TV}$. Explain why $\angle RSW \cong \angle TSV$.

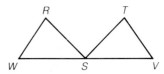

3. $\overline{GK} \perp \overline{LJ}$, $\overline{HK} \cong \overline{KJ}$, and $\overline{GK} \cong \overline{LK}$. Explain why $\angle G \cong \angle L$.

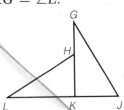

4. $\overline{AB} \cong \overline{DC}$ and $\angle BAC \cong \angle DCA$. Explain why $\overline{AD} \cong \overline{CB}$.

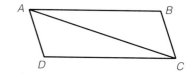

5. $\overline{AR} \cong \overline{AK}$ and $\overline{RT} \cong \overline{KT}$. Explain why $\angle RAT \cong \angle KAT$.

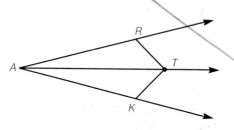

6. $\angle M \cong \angle S$, $\angle NPM \cong \angle TWS$, and $\overline{MP} \cong \overline{SW}$. Explain why $\angle T \cong \angle N$.

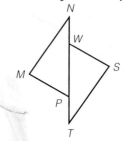

7. $\overline{PQ} \cong \overline{ST}$, $\angle P \cong \angle S$, and $\angle Q \cong \angle T$. Explain why $\overline{PR} \cong \overline{SV}$.

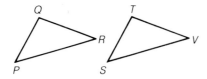

8. $\overline{PU} \cong \overline{PT}$ and $\angle UPS \cong \angle TPS$. Explain why $\overline{US} \cong \overline{TS}$.

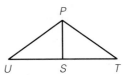

Extend and Apply

9. The indicated sides of the kite are congruent. Explain why $\angle 1 \cong \angle 2$.

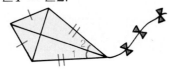

10. On the arrowhead, the indicated angles are congruent. Explain why $\angle R \cong \angle M$.

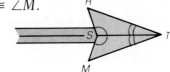

11. On a pair of pinking shears, the indicated angles and sides are congruent. Explain why P is the midpoint of $\overline{GR}$.

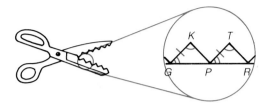

12. On this national flag, the indicated segments and angles are congruent. Explain why P is the midpoint of $\overline{EF}$.

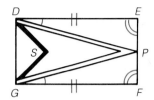

13. $\angle K \cong \angle T$, $\angle G \cong \angle R$, and $\overline{GK} \cong \overline{RT}$. Explain why P is the midpoint of $\overline{GR}$.

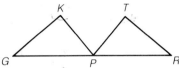

14. $\overline{AR} \cong \overline{AK}$ and $\overline{RT} \cong \overline{KT}$. Explain why $\overrightarrow{AT}$ bisects $\angle RAK$.

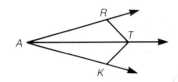

15. $\angle CAB \cong \angle DAB$ and $\angle CBA \cong \angle DBA$. Explain why $\overline{CB} \cong \overline{DB}$.

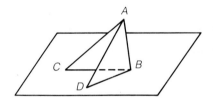

16. $\overline{RH} \cong \overline{CM}$, $\overline{TH} \cong \overline{LM}$, and $\angle 1$ is supplementary to $\angle 3$. Which is not necessarily true?

(A) $\overline{RT} \cong \overline{CL}$
(B) $\triangle MCL \cong \triangle HRT$ by SAS
(C) $\overline{CL} \cong \overline{RH}$
(D) $\angle 1 \cong \angle 2$

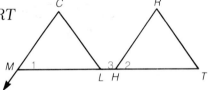

Use Mathematical Reasoning

17. $\angle N \cong \angle G$ and $\overline{NK} \cong \overline{GK}$. Explain why $\triangle NKH \cong \triangle GKL$.

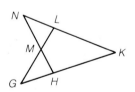

18. $\angle D \cong \angle K$, $\overline{DF} \cong \overline{KE}$, and $\angle DFS \cong \angle KES$. Explain why $\angle DSF \cong \angle KSE$.

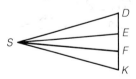

Mixed Review

19. $\angle QPR$ and $\angle RPS$ form a linear pair. $m\angle QPR = 60°$. Find $m\angle RPS$.

20. $\angle XYZ$ and $\angle O$ are supplementary angles. $m\angle XYZ = 60°$. Find $m\angle O$.

21. $\angle W$ and $\angle C$ are vertical angles. $m\angle C = 60°$. Find $m\angle W$.

22. $\angle KPQ$, $\angle PKQ$, and $\angle PQK$ are congruent angles. What kind of triangle is $\triangle KPQ$?

Solve and check.

23. $16t - 8 = 8t + 16$

24. $21q - 7 = 7q + 21$

5-8 The AAS and HL Postulates

Objective: Identify pairs of triangles that are congruent by either AAS or HL.

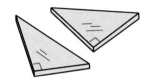

Two students cut out right-triangular pieces of glass for an art project. The students found that the sides opposite the right angles were of equal length and that a second pair of sides were of equal length. Could the students conclude that the triangles were congruent?

We already know that SAS, ASA, and SSS allow us to conclude that two triangles are congruent. In this lesson we will look at two more ways of showing triangles to be congruent.

Postulate 13   The AAS (Angle-Angle-Side) Postulate

If two angles and a non-included side of one triangle are congruent to two angles and the corresponding non-included side of another triangle, then the triangles are congruent.

Examples Which postulate (SAS, SSS, ASA, or AAS) can be used to show each pair of triangles congruent?

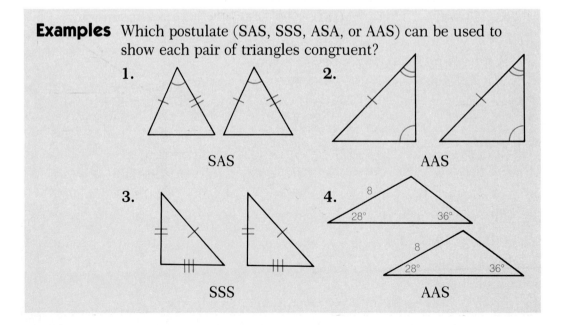

1. SAS

2. AAS

3. SSS

4. AAS

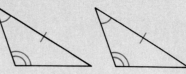
Explore

1. Draw a horizontal segment $\overline{AB}$ at least 3 inches long.
2. Draw a 2-inch segment $\overline{AC}$ at a 45° angle from A.
3. Open the compass so that the compass point is at C and the pencil point is just across $\overline{AB}$.
4. Draw an arc that intersects $\overline{AB}$ in as many points as possible.
5. At how many points does the arc intersect $\overline{AB}$?
6. Label the intersection point(s) D, E, and so on. Are the triangles formed by A, C, and these points congruent (is $\triangle ACD \cong \triangle ACE$)?
7. Is SSA sufficient to show triangles congruent?
8. Repeat Steps 1–4, this time drawing a 90° angle. At how many points does the arc intersect $\overline{AB}$? Is SSA sufficient to show triangles congruent when the angle is 90°?

Is SSA sufficient to state that two triangles are congruent? It turns out that triangles are *not necessarily* congruent if SSA holds, *unless* they are right triangles. This is stated in the following postulate.

In a right triangle, the **hypotenuse** is the side opposite the right angle. The other two sides are the **legs.**

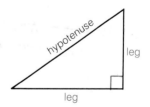

Postulate 14 The HL (Hypotenuse-Leg) Postulate

If the hypotenuse and a leg of one right triangle are congruent to the hypotenuse and a leg of another right triangle, then the two right triangles are congruent.

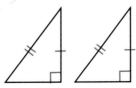

Examples Which postulate (SAS, SSS, ASA, AAS, or HL), if any, could be used to show each pair of triangles congruent?

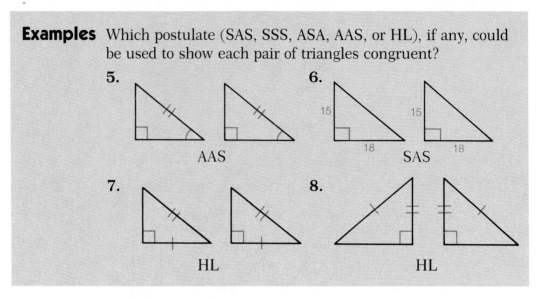

5.

AAS

6.

15   15

18   SAS   18

7.

HL

8.

HL

Try This... Which postulate (SAS, SSS, ASA, AAS, or HL), if any, could be used to show each pair of triangles congruent?

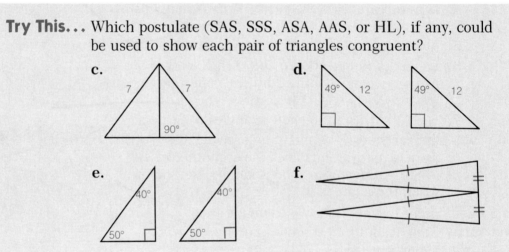

c.

7   7

90°

d.

49°   12   49°   12

e.

40°   40°

50°   50°

f.

The AAS and HL postulates may also be used to show that corresponding parts of two triangles are congruent.

Example 9. M is the midpoint of $\overline{AB}$, and $\angle C \cong \angle D$. Mark a diagram and show $\triangle ACM \cong \triangle BDM$ to explain why $\overline{AC} \cong \overline{BD}$.

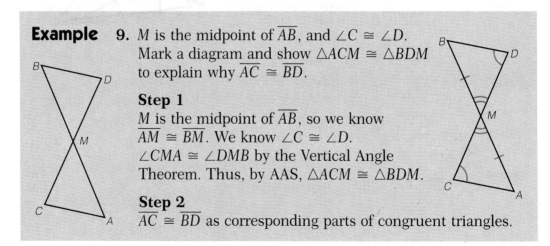

Step 1
M is the midpoint of $\overline{AB}$, so we know $\overline{AM} \cong \overline{BM}$. We know $\angle C \cong \angle D$. $\angle CMA \cong \angle DMB$ by the Vertical Angle Theorem. Thus, by AAS, $\triangle ACM \cong \triangle BDM$.

Step 2
$\overline{AC} \cong \overline{BD}$ as corresponding parts of congruent triangles.

Try This... g. $\overline{AL}$ is the bisector of $\angle GAK$, and $\angle G \cong \angle K$. Mark a diagram and show $\triangle GAL \cong \triangle KAL$ to explain why $\overline{GL} \cong \overline{KL}$.

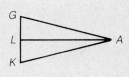

Example 10. $\overline{QR} \perp \overline{PS}$ and $\overline{PQ} \cong \overline{SQ}$. Explain why $\angle PQR \cong \angle SQR$.

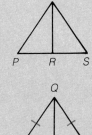

Step 1
Because $\overline{QR} \perp \overline{PS}$, both $\triangle PRQ$ and $\triangle SRQ$ are right triangles. We know $\overline{PQ} \cong \overline{SQ}$. We know $\overline{QR} \cong \overline{QR}$. By HL, $\triangle PRQ \cong \triangle SRQ$.

Step 2
Thus, $\angle PQR \cong \angle SQR$ as corresponding parts of congruent triangles.

Try This... h. $\overline{MN} \perp \overline{RT}$ at P, $\overline{RM} \cong \overline{TN}$, and $\overline{RP} \cong \overline{TP}$. Explain why $\overline{MP} \cong \overline{NP}$.

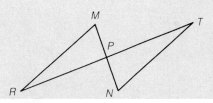

Discussion   Why is the HL Postulate the same as an SSA Postulate for right triangles? If both legs of a right triangle are congruent to both legs of another right triangle, are the triangles congruent? How do you know?

Exercises

Practice

Which postulate (SAS, SSS, ASA, AAS, or HL) can be used to show each pair of triangles congruent?

1.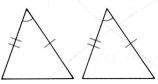

2.

3.

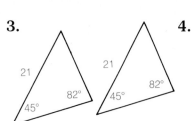

4.

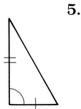

5.

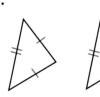

6.

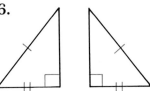

7.

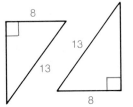

8.

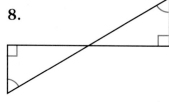

9.

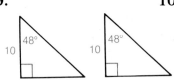

10.

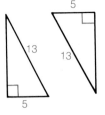

11.

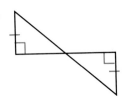

Extend and Apply

12. $\overline{PQ} \cong \overline{TS}$ and $\angle Q \cong \angle S$. Explain why $\triangle PQR \cong \triangle TSR$.

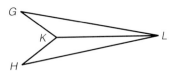

13. $\angle G \cong \angle H$ and $\angle GLK \cong \angle HLK$. Explain why $\overline{GL} \cong \overline{HL}$.

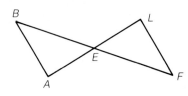

14. C is the midpoint of $\overline{AE}$, $\overline{BA} \perp \overline{AE}$, $\overline{DE} \perp \overline{AE}$, and $\overline{BC} \cong \overline{DC}$. Explain why $\triangle BAC \cong \triangle DEC$.

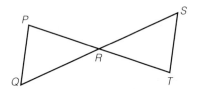

15. $\angle A$ and $\angle L$ are right angles, and E is the midpoint of $\overline{BF}$ and $\overline{AL}$. Explain why $\angle B \cong \angle F$.

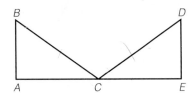

16. Y is the midpoint of $\overline{RS}$, and $\angle T \cong \angle Q$. Explain why $\overline{TS} \cong \overline{QR}$.

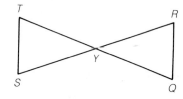

17. $\overline{BD} \cong \overline{CD}$ and $\angle A \cong \angle F$. Explain why $\triangle ABC \cong \triangle FCB$.

18. $\overline{AB} \cong \overline{DE}$, $\angle A \cong \angle F$, and $\angle C \cong \angle D$.
△*ABC* and △*DEF* are _____ .
  (A) congruent by HL
  (B) congruent by AAS
  (C) congruent by ASA
  (D) not necessarily congruent

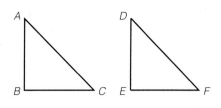

19. A cliff makes a 90° angle with the ground. You note, using a protractor, that the top of the cliff and the ground make a 45° angle. If you are at *C*, 100 feet from the base of the cliff, how high is the cliff? How do you know?

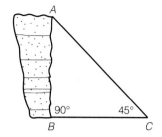

Use Mathematical Reasoning

20. $\overline{JT} \cong \overline{RT}$, $\overline{TK} \perp \overline{DJ}$, $\overline{TS} \perp \overline{DR}$, and $\overline{TK} \cong \overline{TS}$. Explain why △*DJR* is isosceles.

21. $\overline{YM} \perp \overline{MS}$, $\overline{XS} \perp \overline{MS}$, and $\overline{MY} \cong \overline{SX}$. Explain why $\overline{MS}$ and $\overline{YX}$ bisect each other.

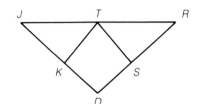

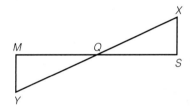

Mixed Review

Use the figure for Exercises 22–26.

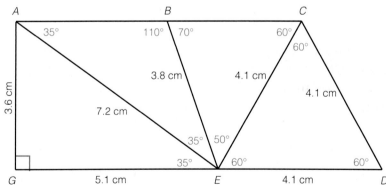

22. $\angle BAG$ is a right angle. Find $m\angle GAE$.

23. Classify each triangle according to its angles.

24. Classify each triangle according to its sides.

25. Which angle is included between $\overline{AE}$ and $\overline{AB}$?

26. Which side is included between $\angle CBE$ and $\angle CEB$?

Solve and check.

27. $2x + 35 = x + 70$

28. $6y + 35 = y + 70$

Number Properties

When you add two numbers, does the order of addition matter? Does 2180 + 1232 have the same answer as 1232 + 2180?

2180 + 1232 = 3412
1232 + 2180 = 3412

Both have the same answer. Is this always true for any two numbers? for positive numbers? for negative numbers? for fractions?

Try many different types of numbers, using a calculator if you wish.

After many trials, we can use *inductive reasoning* to conclude that numbers can be added in any order. We say that **addition is commutative.**

$a + b = b + a$ for all real numbers

Is subtraction commutative? Does 20 − 17 = 17 − 20?

20 − 17 = 3
17 − 20 = 17 + (−20) = −3

We have found one *counterexample*. Thus, subtraction is *not* commutative.

Problems

1. Is multiplication commutative? (Is $a \times b = b \times a$ always true?)

2. Is division commutative? (Is $a \div b = b \div a$ always true?)

3. **Addition is associative** if, when three numbers are added, it does not matter which two numbers are added first. Is addition *associative*? (Is $(a + b) + c = a + (b + c)$ always true for any three numbers?)

4. Is subtraction *associative*? (Is $(a − b) − c = a − (b − c)$ always true?)

5. Is multiplication *associative*? (Is $(a \times b) \times c = a \times (b \times c)$ always true?)

6. Is division *associative*? (Is $(a \div b) \div c = a \div (b \div c)$ always true?)

7. The **identity element for addition** is the number that when added to a equals a. What number is the identity element for addition?

8. The **identity element for multiplication** is the number that when multiplied by a equals a. What number is the identity element for multiplication?

Chapter 5 Review

Classify each triangle by its angles.

1.

2.

3.

Classify each triangle by its sides.

4.

5.

6.

7. Name the corresponding parts if $\triangle GKL \cong \triangle PTL$.

Determine how many lines of symmetry each figure has.

8. X   **9.** V   **10.** X

11. List the information shown in this drawing.

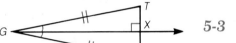

Make a drawing to show the given information.

12. X is the midpoint of $\overline{BD}$, $\overline{AB} \cong \overline{AD}$, and $\angle AXB \cong \angle AXD$.

Which postulate (if any) can be used to show these pairs of triangles congruent?

13.

14.

15.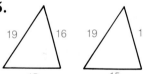

16.

17.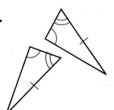

18.

19.

20.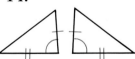

21.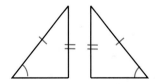

Make a drawing to show the given information, and fill in the blanks.

5-6   **22.** $\overline{MS} \cong \overline{LS}$, $\overline{ES} \cong \overline{BS}$.
$\triangle LBS = $ _____ by _____ .

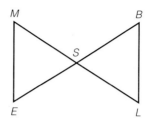

5-7   **23.** P is the midpoint of $\overline{RS}$, and $\angle R \cong \angle S$. Explain why $\angle T \cong \angle V$.

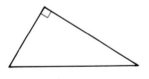

24. $\overline{TF} \perp \overline{SM}$, and F is the midpoint of $\overline{SM}$. Explain why $\angle S \cong \angle M$.

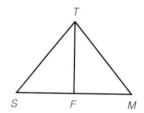

Chapter 5 Test

Classify each triangle by its angles.

1.

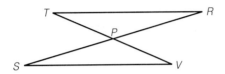

2.

3.

Classify each triangle by its sides.

4.

5.

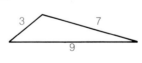

6.

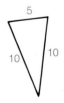

7. Name the corresponding parts if $\triangle YAH \cong \triangle KEN$.

Determine how many lines of symmetry each figure has.

8.

9.

10.

11. List the information shown in this drawing.

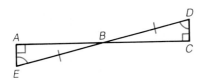

Make a drawing to show the given information about △ADM and △FMB.

12. $\overline{AB}$ and $\overline{DF}$ intersect at M, $\overline{AM} \cong \overline{MB}$, and $\angle FAM \cong \angle DBM$.

Which postulate (if any) can be used to show these pairs of triangles congruent?

13.

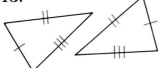

14.

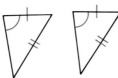

15.

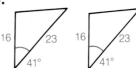

16.

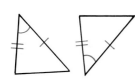

17.

18.

19.

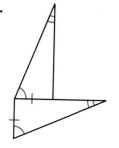

20.

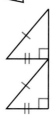

21.

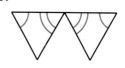

22. If $\overrightarrow{LF}$ bisects $\angle CLE$, and $\overline{CL} \cong \overline{EL}$, then △_____ $\cong$ △ELF by _____.

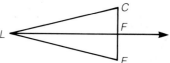

23. M is the midpoint of both $\overline{DK}$ and $\overline{RP}$. Explain why $\angle R \cong \angle P$.

24. $\angle 1 \cong \angle 2$ and $\angle 3 \cong \angle 4$. Explain why $\angle S \cong \angle F$.

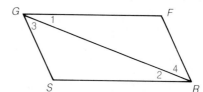

6 Triangle Relationships

The sails of these yachts are adjusted to make the best use of the wind. Their shapes reflect many of the properties of triangles.

6-1 Isosceles Triangles

Objective: Use the Isosceles Triangle Theorem to solve problems.

Explore

1. Draw △ABC so that AB = BC = 5 cm.

2. Cut out △ABC and fold it as shown to compare to ∠A and ∠C.

3. Draw △PQR so that ∠P ≅ ∠R.

4. Cut out △PQR and fold it as shown to compare $\overline{PQ}$ and $\overline{RQ}$.

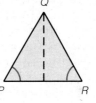

Discuss What is true of ∠A and ∠C? What is true of $\overline{PQ}$ and $\overline{RQ}$?

We know that isosceles triangles have at least two congruent sides, called **legs.** The third side is called the **base.** The angles opposite the congruent sides are called the **base angles.** The angle opposite the base is called the **vertex angle.**

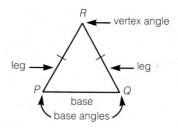

The Explore suggests that whenever two sides of a triangle are congruent, the angles opposite them are congruent. Conversely, it seems that whenever two angles of a triangle are congruent, the sides opposite them are congruent.

Theorem 6.1 **The Isosceles Triangle Theorem**

Two sides of a triangle are congruent whenever the angles opposite them are congruent.

Two angles of a triangle are congruent whenever the sides opposite them are congruent.

Examples **1.** Find $m\angle S$.

Because $\overline{TS} \cong \overline{TR}$, we know $m\angle S = m\angle R$. Thus, $m\angle S = 65°$.

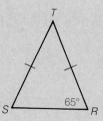

2. In $\triangle GHK$, $GK = 27$. Find GH.

Because $\angle H \cong \angle K$, we know $\overline{GH} \cong \overline{GK}$. Thus, because $GK = 27$, we know $GH = 27$.

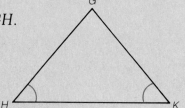

In Example 1, we used the fact that if two sides of a triangle are congruent, then the angles opposite them are congruent. In Example 2, we used the converse.

Try This... **a.** Find $m\angle A$.

b. In $\triangle PQR$, $PR = 16$. Find PQ.

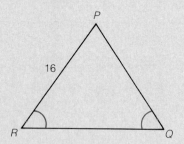

◇◇ **Examples** **3.** Find x.

Because $MN = QN$, we know $m\angle M = m\angle Q$. Thus,

$3x = 51$

$x = 17$ Solving for x

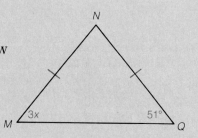

4. Find NS and TS.

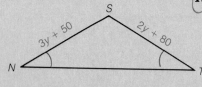

Because $\angle N \cong \angle T$, we know $NS = TS$. Thus,

$3y + 50 = 2y + 80$

$\quad 3y = 2y + 30$ Solving for y

$\quad\quad y = 30$

$NS = 3y + 50 = 3(30) + 50 = 140$

$TS = NS = 140$

Try This... c. Find x.

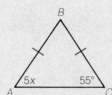

d. Find y.

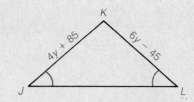

Discussion Daryl said that if a triangle is equilateral, then it is also equiangular and vice versa. Use the Isosceles Triangle Theorem to give a convincing argument that this is true.

Exercises

Practice

1. Find $m\angle Q$.

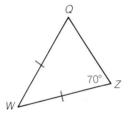

2. Find $m\angle S$.

3. In $\triangle ABC$, $AB = 62$. Find AC.

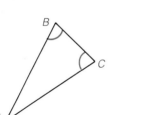

4. In $\triangle DEF$, $DE = 78$. Find DF.

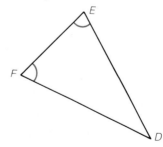

In Exercises 5–12, find the indicated value.

5. Find x.

6. Find x.

7. Find ST.

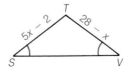

8. Find DF.

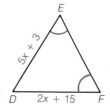

9. Find x.

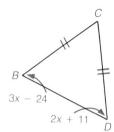

C

B

$3x - 24$

$2x + 11$ D

10. Find x.

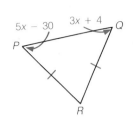

$5x - 30$ $3x + 4$ Q

P

R

11. Find FE.

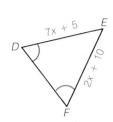

$7x + 5$ E

D

$2x + 10$

F

12. Find LM.

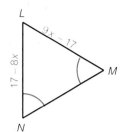

L

$9x - 17$

$17 - 8x$

M

N

Extend and Apply

13. In $\triangle GRP$, $m\angle 1 = 140°$. Find $m\angle 3$.

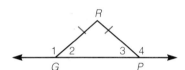

R

1 2 3 4

G P

14. In $\triangle VST$, $m\angle 1 = 49°$. Find $m\angle 4$.

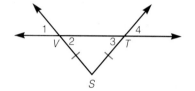

1 4

V 2 3 T

S

15. In $\triangle ABC$, $m\angle 1 = 45°$, $m\angle 2 = 135°$, and $AC = 8$. Find AB.

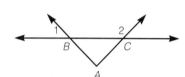

1 2

B C

A

16. In $\triangle ABC$, $m\angle 1 = 60°$, $m\angle 2 = 60°$, and $AB = 12$. Find BC and AC.

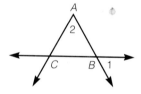

A

2

C B 1

17. In the building at the right, the overhang makes a 30° angle with the walls. Find the measures of $\angle 1$, $\angle 2$, and $\angle 3$.

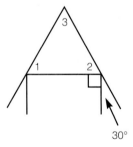

3

1 2

$30°$

18. Which of the following is $m\angle 1$?

(A) 20°   (B) 30°   (C) 60°   (D) 150°

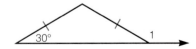

$30°$ 1

Use Mathematical Reasoning

19. Draw several isosceles triangles. Draw any lines of symmetry in each. Does every isosceles triangle have a line of symmetry? If so, what line is it?

20. An isosceles triangle has three lines of symmetry. What else can you say about the triangle? Draw a picture to illustrate.

21. A surveyor stands at point B and sites a tree at point A across the river. She then turns 90° and walks along $\overrightarrow{BD}$ until she reaches a point C so that $m\angle ACB = 45°$. How can she find the distance across the river from B to A?

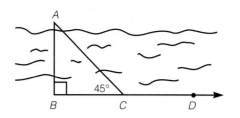

Mixed Review

For Exercises 22–26, copy the figure at right, mark the information given, and tell which postulate or theorem shows that $\triangle ADC \cong \triangle BDC$.

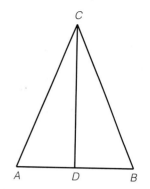

22. $\overline{CD}$ is the perpendicular bisector of $\overline{AB}$.

23. $m\angle ACD = 23°$, $m\angle A = 67°$, $m\angle BCD = 23°$, $m\angle B = 67°$.

24. $\overline{CD}$ is perpendicular to $\overline{AB}$, and $\overline{CD}$ bisects $\angle ACB$.

$D = 5$, $AC = 13$, $CB = 13$.

lar to $\overline{AB}$, and $\overline{AC} \cong \overline{BC}$.

ng parts of $\triangle ACD$ and $\triangle BCD$.

ach statement true.

29. -7 ____ -5 **30.** 0 ____ -1

32. -9 ____ -10 **33.** 6 ____ -6

t

ngruent cardboard strips at one end with a Punch a hole at the other end of each strip. ald each be the same distance from the end of a string securely through one hole and then ng through the second hole. Pull the string tight.

f triangle does CAT

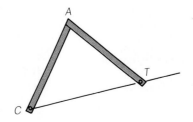

ong the string away from C, the string tight. What kind of triangle does CAT form now?

Does changing the distance between C and T affect the type of triangle formed? Explain how you know.

Use a Flowchart

Sometimes a flowchart can be used to help explain a conclusion. The samples below show how a flowchart organizes the ideas of a logical argument.

Samples **1.** $\overline{LM} \cong \overline{LN}$. P is the midpoint of $\overline{MN}$. Explain why $\triangle LMP \cong \triangle LNP$.

2. $\angle 1 \cong \angle 4$. Explain why $\triangle PQR$ is isosceles.

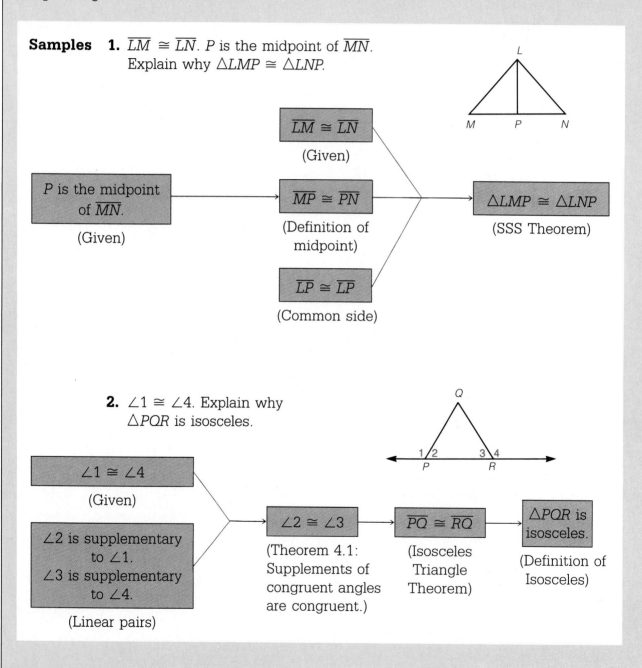

Problems

Use a flowchart to explain each conclusion.

1. $\overline{FD} \cong \overline{ED}$.
Explain why $\angle 3 \cong \angle 2$.

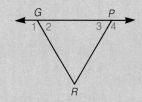

2. $\overline{GR} \cong \overline{PR}$.
Explain why $\angle 1 \cong \angle 4$.

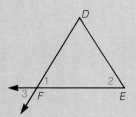

3. $\overline{MN} \cong \overline{MQ}$.
Explain why $\angle 2 \cong \angle 3$.

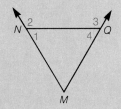

4. $\angle 1 \cong \angle 4$.
Explain why $\triangle KPT$ is isosceles.

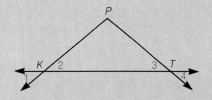

5. $\overline{RS} \cong \overline{WS}$ and $\overline{RT} \cong \overline{VW}$.
Explain why $\triangle STV$ is isosceles.

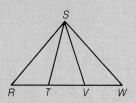

6. $\overline{AD}$ bisects $\angle BAC$ and $\overline{BA} \cong \overline{CA}$.
Explain why $\triangle BAD \cong \triangle CAD$.

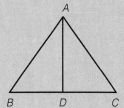

7. E is the midpoint of $\overline{AD}$ and $\overline{BC}$.
Explain why $\triangle AEB \cong \triangle DEC$.

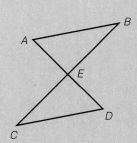

6-2 Exterior Angles of a Triangle

Objective: Identify exterior and remote interior angles of a triangle and use the Exterior Angle Theorem to solve problems.

Explore

Computer software may be used for this Explore.

1. Draw a triangle, △*ABC*, and extend side $\overline{BC}$ as shown in the figure.

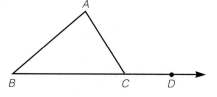

2. Measure ∠*ACD*. Then measure ∠*ABC* and ∠*BAC*.

3. Compare the measure of ∠*ACD* with the measure of each of the other two angles. In each case, which is larger?

Discuss Do you think the result in Step 3 holds true regardless of the side of the triangle extended? Do you think it holds true for all triangles?

In this lesson, we will begin to study some of the inequalities associated with triangles. First, some definitions are needed.

An angle that forms a linear pair with an angle of a triangle is called an **exterior angle** of the triangle. The angles of the triangle that are *not* adjacent to a specific exterior angle are called **remote interior angles.**

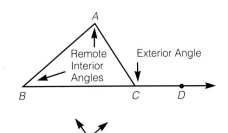

There are two exterior angles at each vertex of a triangle. They are the numbered angles in the figure at the right.

There are two remote interior angles associated with each pair of exterior angles. ∠9 and ∠10 are remote interior angles for ∠7 and ∠8.

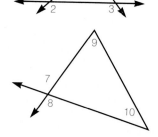

Example 1. Name the remote interior angles for each of the exterior angles shown for △BCD.

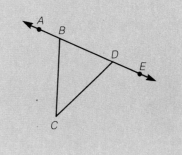

∠BCD and ∠CDB are remote interior angles for ∠ABC.

∠BCD and ∠CBD are remote interior angles for ∠CDE.

Try This... **a.** Name the remote interior angles for each of the exterior angles shown for △FGH.

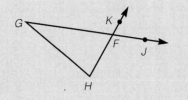

You may have discovered the following theorem in the Explore:

Theorem 6.2   The Exterior Angle Theorem

The measure of an exterior angle of a triangle is greater than the measure of either of its remote interior angles.

Example 2. Consider the triangle formed by a nautical identification flag. What is the relationship between ∠ACD and ∠ABC? Explain how you know.

$m\angle ACD > m\angle ABC$ because ∠ACD is an exterior angle of △ABC and ∠ABC is one of its remote interior angles.

Try This... **b.** What is the relationship between $m\angle SRT$ and $m\angle RTQ$ in this triangular shelf brace?

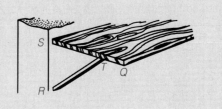

Exercises

Practice

For the exterior angles given, name the remote interior angles.

1. ∠ACD   **2.** ∠QLK   **3.** ∠VTR; ∠MQT   **4.** ∠NMG; ∠LMS

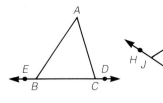 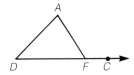 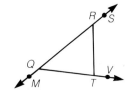 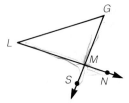

What is the relationship between the measures of the two indicated angles?

5. ∠AFC and ∠D   **6.** ∠TVR and ∠S   **7.** ∠K and ∠KSH

 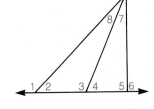 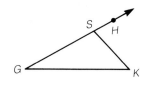

Extend and Apply

Insert <, >, or = to make a true statement about the figure.

8. m∠10 _____ m∠12

9. m∠9 _____ m∠13

10. m∠11 _____ m∠9

11. m∠13 _____ m∠12

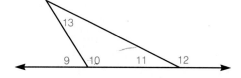

12. m∠8 _____ m∠4

13. m∠1 _____ m∠3

14. m∠3 _____ m∠7

15. m∠1 _____ m∠5

16. Which of the following could be the measure of ∠A?

(A) 120°   (B) 150°   (C) 110°   (D) 180°

Complete each statement with *always*, *sometimes*, or *never*.

17. If an exterior angle of △ABC is a right angle, then △ABC is _____ a right triangle.

18. If an exterior angle of △PQR is an acute angle, then it is _____ possible for △PQR to be a right triangle.

19. If an exterior angle of △DEF is an obtuse angle, then △DEF will _____ be a right triangle.

20. If an exterior angle of △RST is an acute angle, then △RST will _____ be an obtuse triangle.

21. Which is greater, $m\angle DCE$ or $m\angle ABD$? How do you know?

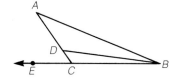

 Mixed Review

22. Find x.

23. Find *TV*.

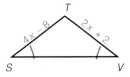

24. Find x.

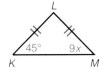

25. Find *DE*.

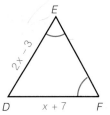

Visualization

Four different views of the same cube are shown below. Determine which figures must lie on opposite faces of the cube.

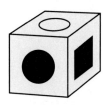
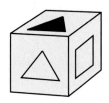
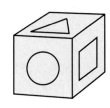
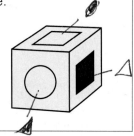

6-3 The Opposite Parts Theorem

Objective: Use the Opposite Parts Theorem and its converse to solve problems.

▪ Explore ▪

Computer software may be used for this activity.

1. Measure the sides of each of these triangles.

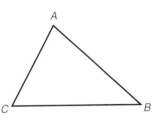
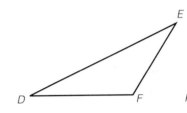

2. For each triangle, list the sides from largest to smallest.

3. Measure the angles of each triangle.

4. For each triangle, list the angles from largest to smallest.

Discuss   What relationship do you see between the sides and the angles of each triangle?

We can relate the measure of each angle of a triangle to the measure of the side opposite the angle.

Theorem 6.3   The Opposite Parts Theorem
 In any $\triangle ABC$, if $CA > CB$,
 then $m\angle B > m\angle A$.

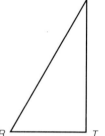

Example   **1.** In $\triangle PQR$, $PR > PQ$. What is true of $\angle Q$ and $\angle R$?

$$m\angle Q > m\angle R$$

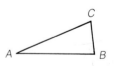

Try This... **a.** In $\triangle RST$, $RS > RT$. What is true of $\angle T$ and $\angle S$?

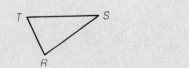

Example 2. Jordan (*J*), Cohagen (*C*), and Sand Springs (*S*) are three towns in Garfield County, Montana. They form the vertices of △*JCS*. List the angles of △*JCS* from smallest to largest.

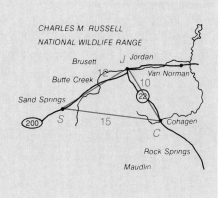

∠*S*, ∠*C*, ∠*J*

Try This... b. In △*ABC*, *AB* = 9, *BC* = 18, and *AC* = 24. List the angles from smallest to largest.

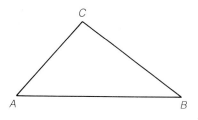

The converse of the Opposite Parts Theorem is also true. This is stated as Theorem 6.4.

Theorem 6.4 In any △*ABC*, if $m\angle C > m\angle B$, then $AB > AC$.

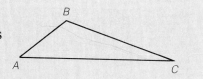

Example 3. In △*GKH*,

$$m\angle G = 55°,$$
$$m\angle H = 61°,$$
$$\text{and } m\angle K = 64°.$$

List the sides from shortest to longest.
$\overline{KH}, \overline{GK}, \overline{GH}$

Try This... c. Three stars determine $\triangle LMN$, with $m\angle L = 45°$, $m\angle M = 55°$, and $m\angle N = 80°$. List the sides from shortest to longest.

Discussion How can you use the Opposite Parts Theorem to give a convincing argument that no two angles of a scalene triangle are congruent?

Exercises

Practice

1. In $\triangle GHK$, $GH > GK$. What is true of $\angle K$ and $\angle H$?

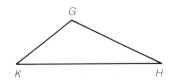

2. On the road sign, $KJ > KH$. What is true of $\angle J$ and $\angle H$?

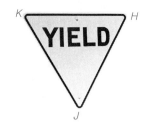

List the angles from smallest to largest.

3.

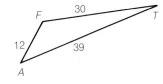

4.

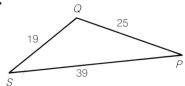

5.

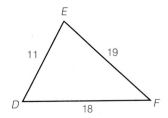

6.

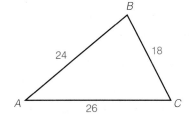

List the sides from shortest to longest.

7.

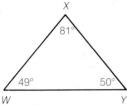

8.

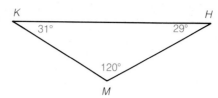

Extend and Apply

9. Which of the following is true?

(A) $ML < LK$ (B) $MK < KL$ (C) $ML > MK$ (D) $MK > ML$

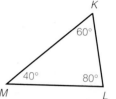

10. $m\angle 2 > m\angle 1$. How does $m\angle 2$ compare with $m\angle 3$? How does AC compare with BC?

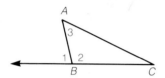

11. $m\angle G > m\angle N$. How does NK compare with MK? (Hint: Consider $\angle KMN$ as an exterior angle.)

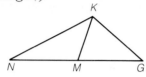

12. Two forest rangers each spot some smoke from their observation towers. The first ranger determines that $m\angle CAB = 82°$. The second ranger determines that $m\angle CBA = 67°$. Which ranger is closest to the smoke area? How do you know?

Use Mathematical Reasoning

13. $\overline{DE} \cong \overline{DF}$, and B is between E and S and also between D and F, as shown. Compare ES and FS. What theorems did you use to make the comparison?

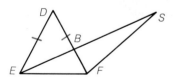

14. $\overline{PQ} \cong \overline{PR}$, and R is between Q and S, as shown. Compare PS and PQ. What theorems did you use to make the comparison?

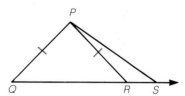

Mixed Review

Write the converse of each sentence.

15. If $\overline{AB} \cong \overline{CD}$, then $AB = CD$.

16. If a triangle is equiangular, it is equilateral.

◇ Solve and check.

17. $3x + 6 = 4x - 1$ **18.** $4x + 3x - 2 = 6x + 3$ **19.** $5x + 1 = 6x$

6-4 The Triangle Inequality

Objective: Use the Triangle Inequality to solve problems.

Explore

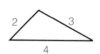

1. Cut several straws or sticks into lengths of 1, 2, 3, 4, 5, and 6 inches.

2. Choose any three pieces and see if they fit together to form a triangle. For example, the 1-in., 2-in., and 6-in. pieces do not fit; but the 2-in., 3-in., and 4-in. pieces do.

3. Repeat Step 2 several times. As you experiment, keep a record of your results and look for a pattern.

Discuss When will three segments form a triangle? Use your results to make a conjecture.

The Wong family is driving from Kingsville to Perryville. The map shows that going from Kingsville to Perryville by way of Essex is longer than it is to go directly from Kingsville to Perryville.

The Triangle Inequality helps to explain the above situation.

Theorem 6.5 **The Triangle Inequality**

The sum of the lengths of any two sides of a triangle is greater than the length of the third side.

$$AB + BC > AC$$
$$AC + CB > AB$$
$$AB + AC > BC$$

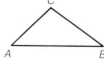

The Triangle Inequality has many important applications. Some of these are explored in the Examples and exercises of this lesson.

Example 1. Consider △*PQR* in the bridge shown at the right. List three inequalities.

$PQ + QR > PR$
$QP + PR > QR$
$PR + RQ > PQ$

Try This... **a.** Consider △*GKH* in the photo of the bridge. List three inequalities.

Examples Can these numbers be the lengths of the sides of a triangle?

2. 5, 7, 9
   Yes, because $5 + 7 > 9$, $5 + 9 > 7$, and $9 + 7 > 5$.

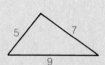

3. 7, 2, 3
   No, because $2 + 3 < 7$.

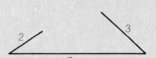

4. 3, 6, 3
   No, because $3 + 3 = 6$.

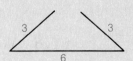

Try This... Can these numbers be the lengths of the sides of a triangle?

b. 5, 6, 9   **c.** 2, 2, 4   **d.** 7, 2, 11

Discussion A triangle has sides of length 6 and 8. What can you say about the possible lengths for the third side?

Exercises

Practice

List three inequalities for each triangle.

1.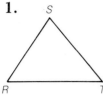

2.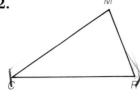

3.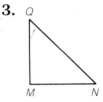

4.

Can these numbers be the lengths of the sides of a triangle?

5. 8, 9, 10   **6.** 1, 1, 2   **7.** 6.2, 9.3, 8.6

8. 2.4, 5.2, 7.6   **9.** 9.4, 0.8, 10.1   **10.** 0.5, 8.3, 1.03

11. $3\frac{1}{2}$, $5\frac{1}{8}$, 6   **12.** $7\frac{1}{4}$, $9\frac{1}{2}$, $2\frac{3}{4}$ **13.** $8\frac{5}{8}$, 3, $6\frac{1}{4}$

Extend and Apply

In Exercises 14–19, the lengths of two sides of a triangle are given. What can you say about the possible lengths for the third side?

14. 8, 5   **15.** 9, 2   **16.** 10, 10

17. 4, 13   **18.** 27, 39   **19.** 0.5, 0.3

20. Which of the following could be the length of $\overline{XZ}$?

(A) 12   (B) 4   (C) 14   (D) 21

21. Ralph has a pet rabbit and wants to build a pen for it. He has three pieces of wood, 3 ft, 7 ft, and 8 ft long. Can he build a closed pen with these pieces of wood?

22. Mr. Diaz is going to build a chicken coop and a barn so that they are situated as shown at the right. He wants both buildings to be at least 200 ft from his house and no less than 125 ft from each other. His land area will allow for a maximum of 650 ft around the triangle. Which of the five layouts below would satisfy all of the conditions above?

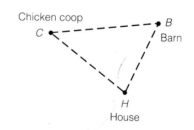

	1	2	3	4	5
CH	250	225	320	300	200
BH	250	200	200	250	350
CB	100	150	125	125	125

23. For years Tina has driven the 29 miles from Winston to Milton and then the 49 miles from Milton to Troy. One day she was in a hurry and decided to take the direct route from Winston to Troy. What is the range in mileage from Winston to Troy.

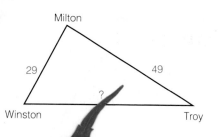

Use Mathematical Reasoning

24. Suppose a 12-inch ___k is marked at one-inch interval___ ___hich two intervals can the ___ broken so that the three pie___ ___gether to form a triangle?

25. Co___pare the diffe___ ___ the lengths ___ ___ sides of a tr___gle with th___ ___ of the third

Mixed Rev___

Use the fig___ ___ the r___ ___ xercise

26. If $HJ >$ ___ ___at ___ $\angle HK$ ___ ?

27. If $m\angle M$___ ___ $\angle$___ ___s tr___ and ML?

28. If $JK = 7$ ___ ___ ___hat ___ of $\angle J$ and $\angle H$?

29. If $m\angle MKL = $ ___ a___ ___ $n\angle$___ $44°$, ___ of the sides of ___ $1K$

30. If $m\angle HKJ > m\angle$___ ___nd ___ $\angle H$ ___ $\angle J$, list the sides of $\triangle HJK$ fro___ ___ho___est ___ longest.

31. In $\triangle MKL$, name the remote inte___ exterior ___ ___

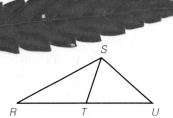

32. Name the remote interio___ ___ngles of exterior angle $\angle STU$.

◇ **33.** Find AB.   ◇ **34.** ___nd x.

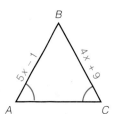

Inequalities on a Number Line

Allena draws a right triangle. What are the possible measurements for ∠B if ∠A can be any measure?

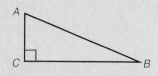

Many problems, like the one above, have multiple solutions. Often there are too many solutions to list. So we use a *number line* to make a picture of the solutions. Here are some examples of inequalities and the number lines that depict them.

All numbers greater than three:

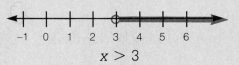

$x > 3$

All numbers between -2 and 4:

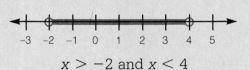

$x > -2$ and $x < 4$

For the problem stated above, since there are 180° in the angles of a triangle and we know one of the angles is 90°, we have 90° left for the other two angles. We can make an organized list and look for a pattern.

m∠A	10°	20°	35°	50°	80°	86°	89.5°
m∠B	80°	70°	55°	40°	10°	4°	0.5°

In this chart we see that $m\angle B$ can range from 80° to 0.5°. Looking for a pattern, we see that as long as $m\angle B$ is less than 90° and $m\angle B$ is greater than 0°, we will have a triangle. We can write $m\angle B > 0$ and $m\angle B < 90$, and can picture this on a number line as shown.

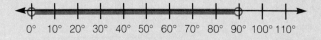

Notice that we use an *open* circle to indicate that $m\angle B$ cannot be either 0° or 90°. If we wanted to include 0° and 90°, we would use a solid dot.

Problems

Graph each inequality on a number line.

1. $x > 4$ **2.** $y < 1$ **3.** $m\angle A \leq 6$ **4.** $z \geq 2$ **5.** $x > 2$ and $x < 5$

6-5 Inequalities in Two Triangles

Objective: Use the Hinge Theorem and its converse to reach conclusions about pairs of triangles.

Explore

1. Draw a 6-in.-long segment $\overline{AB}$ on a piece of paper.

2. Put one end of a pencil on point A, attach a rubber band to the top of the pencil (point C), and hold the other end of the rubber band on point B.

3. Keeping one end of the pencil on A, move the pencil like a hinge so that you increase the measure of $\angle CAB$. Does the length of the pencil change? Does the length of segment $\overline{AB}$ change?

4. As $\angle CAB$ gets larger, what happens to the length of the rubber band?

5. Look at $\triangle PQR$ and $\triangle DEF$. Compare $m\angle P$ and $m\angle D$. Compare QR and EF.

Discuss   Make a conjecture from what you have observed above.

The inequalities we have used so far have involved just one triangle at a time. We will now look at a theorem that tells us about inequalities in two triangles.

Theorem 6.6   The Hinge Theorem

In $\triangle ABC$ and $\triangle DEF$, if $AB = DE$ and $AC = DF$, and $m\angle A < m\angle D$, then $BC < EF$.

Example   **1.** In $\triangle ABC$ and $\triangle GHK$, $AB = GH$, $AC = GK$, and $m\angle A < m\angle G$. Use the Hinge Theorem to reach a conclusion.

By the Hinge Theorem, $BC < HK$.

Try This... a. In △DEF and △RST, DE = RS, DF = RT, and $m \angle D < m \angle R$. Use the Hinge Theorem to reach a conclusion.

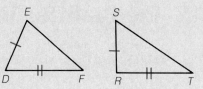

Example **2.** Consider △CLF and △HJK. Use the Hinge Theorem to reach a conclusion.

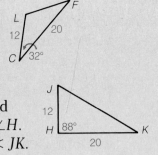

Because CL = HJ = 12, we know $\overline{CL} \cong \overline{HJ}$. Also, CF = HK = 20, so $\overline{CF} \cong \overline{HK}$. Because $m \angle C = 32°$ and $m \angle H = 88°$, we know $m \angle C < m \angle H$. Thus, by the Hinge Theorem, LF < JK.

Try This... b. Consider △XYZ and △MNK. Use the Hinge Theorem to reach a conclusion.

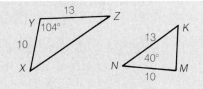

The following converse of the Hinge Theorem is also true.

Theorem 6.7 In △ABC and △DEF, if AB = DE, AC = DF, and BC < EF, then $m \angle A < m \angle D$.

Example **3.** Consider △ABC and △GHK. Use the converse of the Hinge Theorem to reach a conclusion.

AB = GH = 10
AC = GK = 15
BC = 17 and HK = 23, so BC < HK. Thus, by the converse of the Hinge Theorem, we know that $m \angle A < m \angle G$.

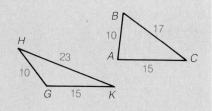

Try This... c. Consider $\triangle PQR$ and $\triangle TUV$. Use the converse of the Hinge Theorem to reach a conclusion.

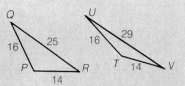

Discussion Why do you think the Hinge Theorem is also known as the SAS Inequality Theorem?

Exercises

Practice

Use the Hinge Theorem to reach a conclusion.

1. In $\triangle GHK$ and $\triangle DEF$, $\overline{GH} \cong \overline{DE}$, $\overline{GK} \cong \overline{DF}$, and $m\angle G < m\angle D$.

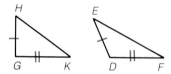

2. In $\triangle PRS$ and $\triangle FQT$, $\overline{PR} \cong \overline{FQ}$, $\overline{PS} \cong \overline{FT}$, and $m\angle P < m\angle F$.

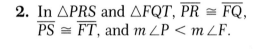

3. $\triangle TPS$ and $\triangle RFP$

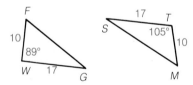

4. $\triangle WFG$ and $\triangle TMS$

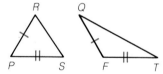

Use the converse of the Hinge Theorem to reach a conclusion.

5.

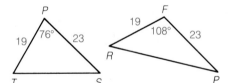

6.

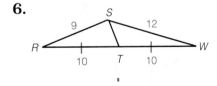

7.

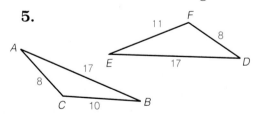

8.

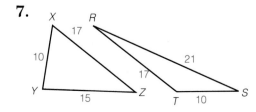

Extend and Apply

Insert $<$, $>$, or $=$ to make a true statement about the figure.

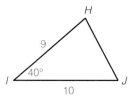

9. If $m \angle 1 = 30°$ and $m \angle 2 = 43°$,
   then AC _____ CD.

10. If $\angle 1 \cong \angle 2$, then AC _____ CD.

11. If $AC = 14$ and $CD = 20$,
    then $m \angle 1$ _____ $m \angle 2$.

12. If $\overline{AC} \cong \overline{CD}$, then $m \angle 1$ _____ $m \angle 2$.

13. If $m \angle 1 = 29°$ and $m \angle 2 = 26°$,
    then AC _____ CD.

14. Which of the following
    inequalities must be
    true?
    (A) $EF < HI$ (B) $EG < HJ$
    (C) $EG > HJ$ (D) $HJ > IJ$

15. What does the Hinge Theorem say
    about two circles drawn with a
    compass at different settings?

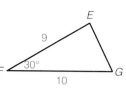

Mixed Review

List three inequalities for each triangle.

16. 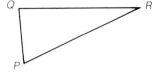   17.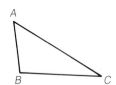

Can these numbers be the lengths of the sides of a triangle?

18. 3, 7, 9
19. 1.5, 9, 11
20. 1, 2, 5
21. 2.5, 3, 3.5
22. 11, 4, 7
23. 8, 9, 10

Find the coordinates of each point when
reflected over the x-axis.

24. A 25. B
26. C 27. D
28. E 29. F

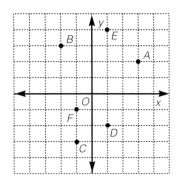

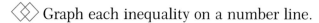 Graph each inequality on a number line.

30. $x > -3$ 31. $x \leq 0$ 32. $x > 2$ and $x < 4$

6-6 Concurrent Lines in Triangles

Objective: Construct the incenter and circumcenter of a triangle.

Explore

Do the following activity by paper folding or with computer software.

1. Draw an acute triangle, an obtuse triangle, and a right triangle.

2. If you are using paper, cut out the acute triangle and fold together one pair of sides. Make a crease to show the angle bisector. If you are using software, draw the bisector of one angle of the acute triangle.

3. Repeat Step 2 to find the bisectors of the other two angles.

4. Repeat this process with the obtuse triangle and the right triangle.

Discuss What do you observe about the three angle bisectors in each triangle? Do you think this is true for every triangle?

Certain lines associated with triangles always intersect in a single point. There is a special name for such lines.

Definition Lines are **concurrent** if they intersect in a single point. The point of intersection is called the **point of concurrency.** Segments and rays are concurrent if they are contained in concurrent lines.

point of concurrency

The following theorem summarizes what you may have discovered in the Explore:

Theorem 6.8 The angle bisectors of a triangle are concurrent in a point that is equidistant from the sides of a triangle.

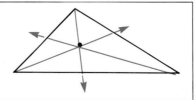

> **Definition** The point of concurrency of the angle bisectors of a triangle is called the **incenter.**

Theorem 6.8 says that the distance from the incenter to each side of the triangle is the same. When we talk about the distance from a point (such as the incenter) to a line (such as the side of a triangle) we mean the length of the perpendicular segment from the point to the line.

Example 1. Construct the incenter of △PQR.

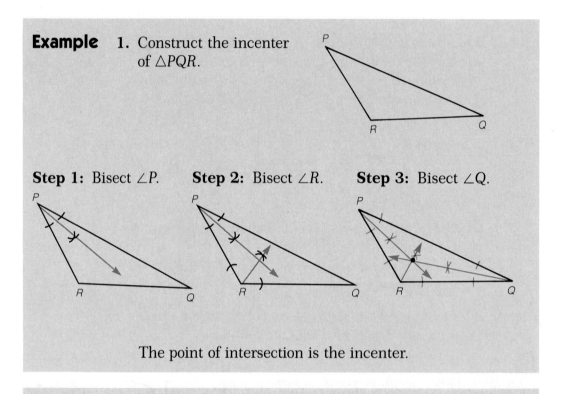

Step 1: Bisect ∠P.   **Step 2:** Bisect ∠R.   **Step 3:** Bisect ∠Q.

The point of intersection is the incenter.

Try This... **a.** Draw a triangle and construct its incenter.

We will now look at another point of concurrency related to triangles.

> **Theorem 6.9** The perpendicular bisectors of the sides of a triangle are concurrent in a point that is equidistant from the vertices of the triangle.

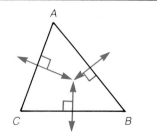

Definition The point of concurrency of the perpendicular bisectors of a triangle is called the **circumcenter.**

Example 2. Construct the circumcenter of $\triangle PQR$.

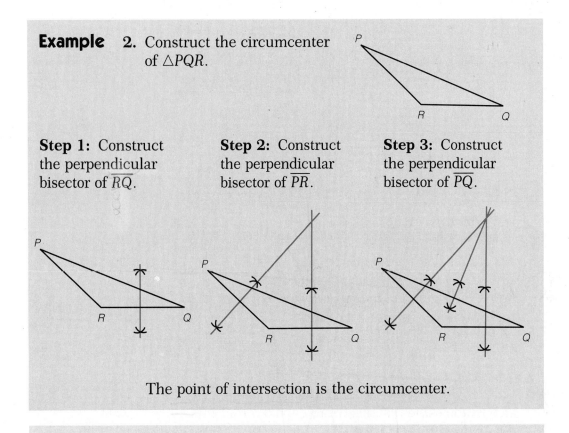

Step 1: Construct the perpendicular bisector of $\overline{RQ}$.

Step 2: Construct the perpendicular bisector of $\overline{PR}$.

Step 3: Construct the perpendicular bisector of $\overline{PQ}$.

The point of intersection is the circumcenter.

Try This... **b.** Draw an acute triangle and construct its circumcenter.

Discussion Consider the four angle bisectors of a rectangle. Are they always concurrent? Are the perpendicular bisectors of the sides of a rectangle always concurrent?

Exercises

Practice

Draw the indicated triangle and construct its incenter.

1. An obtuse triangle
2. A right isosceles triangle
3. An acute triangle
4. An obtuse scalene triangle
5. An equilateral triangle
6. An acute isosceles triangle

Draw the indicated triangle and construct its circumcenter.

7. An acute triangle

8. An obtuse isosceles triangle

9. An equilateral triangle

10. A right isosceles triangle

11. A right triangle

12. An obtuse triangle

13. What do we call point *P*?

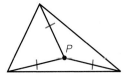

14. What do we call point *Q*?

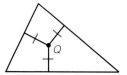

Extend and Apply

15. A ranch is bordered by three highways. The ranch owners want to build their home the same distance from each highway. How can they do this?

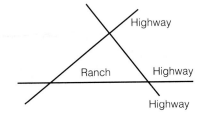

16. The residents of Avon, Brockton, and Cartersville want to build a tri-city airport the same distance from each city. How can they do this?

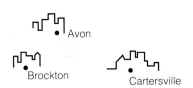

17. Point *X* is the circumcenter of △*RST*. Which of the following is *RX*?

(A) 4   (B) 8   (C) 9   (D) 20

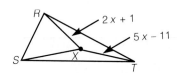

Use Mathematical Reasoning

Draw several figures to help you complete each sentence with *always*, *sometimes*, or *never*.

18. The circumcenter of a triangle is _____ in the interior of the triangle.

19. The incenter of a triangle is _____ in the exterior of the triangle.

20. The incenter of a right triangle is _____ in the interior of the triangle.

21. The circumcenter of a right triangle is _____ on the hypotenuse of the triangle.

Mixed Review

Use the Hinge Theorem to reach a conclusion.

22. In $\triangle GKH$ and $\triangle FQT$, $\overline{GH} \cong \overline{FQ}$, $\overline{GK} \cong \overline{FT}$, and $m\angle G < m\angle F$.

23. In $\triangle EDF$ and $\triangle RPS$, $\overline{ED} \cong \overline{RP}$, $\overline{DF} \cong \overline{PS}$, and $m\angle D > m\angle P$.

For each segment, find its length and the coordinate of its midpoint.

24.

25.

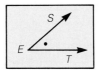

◇◇ Solve and check.

26. $3y - 4 = 2y + 4$ **27.** $x + 14 = 5x - 26$ **28.** $3w + 1 = 5w - 3$

Discover

Draw the same angle, $\angle SET$, on a series of index cards. On each card construct a point that is equidistant from the sides of the angle as follows. First construct an arc across the angle. Then, with the same compass opening, draw two intersecting arcs as shown. Draw a dot at the point of intersection. Use different compass openings on each card.

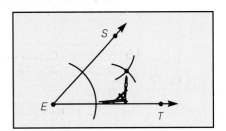

When you have created a set of such cards, hold one end of the stack firmly in one hand. Flip through the other end with your thumb. What do you see?

Sample cards

Use your discovery to complete the Angle Bisector Theorem. Fill in the blanks with one or more words.

Theorem 6.10   The Angle Bisector Theorem
The bisector of an angle is the _____ of points that are _____ from the sides of the angle.

The Centroid

Physics is the branch of science dealing with the properties of matter and energy. Here is one situation a physicist might consider.

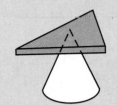

A large, thin triangular piece of wood is perfectly balanced on a cone beneath it. How can we find the point of the triangle at which it will balance?

The point at which a figure balances is called its **centroid.** For a triangle, the centroid is the point of concurrency of the triangle's medians.

Sample   Find the centroid of this triangle.

First use a compass and a straightedge to find the midpoint of each side of the triangle.

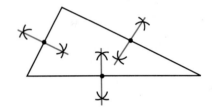

Use these midpoints to draw the medians.

The point of concurrency is the centroid.

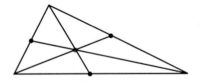

Problems

Draw each of the following triangles and then construct its centroid.

1. An obtuse triangle   **2.** A right triangle

3. An equilateral triangle   **4.** A scalene triangle

Cut out of cardboard each of the following shapes. Try to balance each on the tip of a pencil. Describe a method of finding the centroid for each shape.

5. Square   **6.** Rectangle

7. Circle   **8.** Diamond

Chapter 6 Review

1. In △ABC, AC = 14. Find AB.

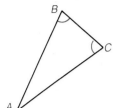

     6-1

For the exterior angles given, name the remote interior angles.

6-2

3. ∠ABE

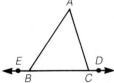

4. ∠LJH

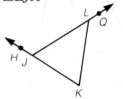

5. What is the relationship between ∠HTJ and ∠V?

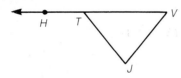

6. In △ABC, AB > AC. What is true of ∠C and ∠B?

6-3

7. In △MES, ME = 21, MS = 8, and SE = 16. List the angles from smallest to largest.

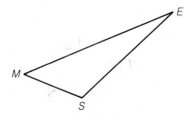

8. In △RST, m∠R = 119°, m∠S = 31°, and m∠T = 30°. List the sides from shortest to longest.

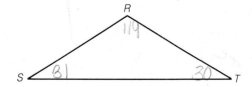

9. List three inequalities for △JLK.

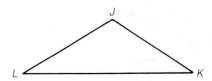

6-4

Can these numbers be the lengths of the sides of a triangle?

10. 5, 9, 11

11. 2.3, 2.5, 4.8

6-5

12. Use the Hinge Theorem to reach a conclusion about △TFS and △PRM.

13. Use the converse of the Hinge Theorem to reach a conclusion about △WAV and △PLG.

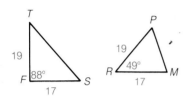

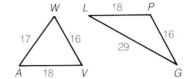

6-6

14. Draw an obtuse triangle and construct its incenter.

15. Draw a right triangle and construct its circumcenter.

Chapter 6 Test

1. In △DEF, DF = 31. Find DE.

⟨⟨⟩⟩ **2.** Find DP.

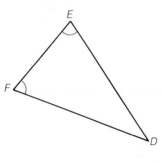

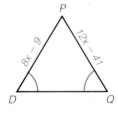

For the exterior angles given, name the remote interior angles.

3. ∠SRT

4. ∠SCB

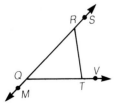

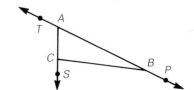

5. What is the relationship between ∠S and ∠STV?

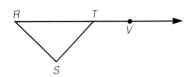

6. In △PQR, PR > PQ. What is true of ∠Q and ∠R?

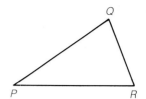

7. In △ARK, AR = 31, AK = 26, and RK = 9. List the angles from smallest to largest.

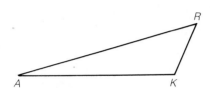

8. In △EPT, m∠P = 111°, m∠E = 33°, and m∠T = 36°. List the sides from shortest to longest.

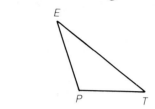

9. List three inequalities for △MRQ.

Can these numbers be the lengths of the sides of a triangle?

10. 9, 3, 5

11. 2.5, 2.5, 4

12. Use the Hinge Theorem to reach a conclusion about △GRS and △FCT.

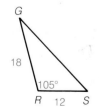

13. Use the converse of the Hinge Theorem to reach a conclusion about △STN and △CLF.

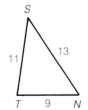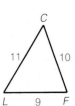

14. Draw an equilateral triangle and construct its incenter.

15. Draw a scalene triangle and construct its circumcenter.

7

Parallel Lines

Farmers usually plant fields in parallel rows to allow for easier harvesting.
Because of perspective, these rows appear to meet in the distance.

7-1 Parallel and Skew Lines

Objective: Construct parallel lines and distinguish between parallel and skew lines.

When laying railroad tracks, workers must be careful that the tracks are always the same distance apart. Standard gauge tracks, for example, are $56\frac{1}{2}$ inches apart. Because they never intersect, we say that the tracks are *parallel*.

Definition   Two lines are **parallel** whenever they are in the same plane and do not intersect. We write $\ell \parallel m$ to say ℓ is parallel to m.

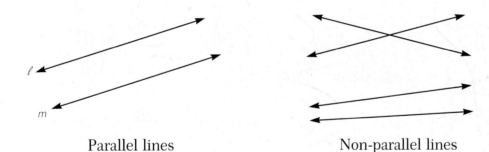

Parallel lines   Non-parallel lines

We can also talk about parallel segments and parallel rays. Two segments or rays are *parallel* whenever the lines containing them are parallel.

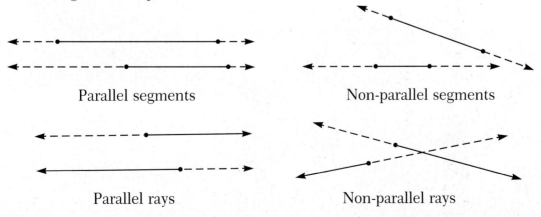

Parallel segments   Non-parallel segments

Parallel rays   Non-parallel rays

Our intuition tells us that lines in a plane will not intersect if they are always the same distance apart. We can use this idea to construct parallel lines.

Example 1. Construct a line that contains P and is parallel to ℓ.

Drop a perpendicular from P to line ℓ. This determines point Q.

Choose a point R on ℓ, and construct a perpendicular to ℓ through R.

Open the compass to PQ, then put the point on R and mark off S so that $PQ = SR$.

Draw line $\overleftrightarrow{PS}$. $\overleftrightarrow{PS} \parallel \ell$.

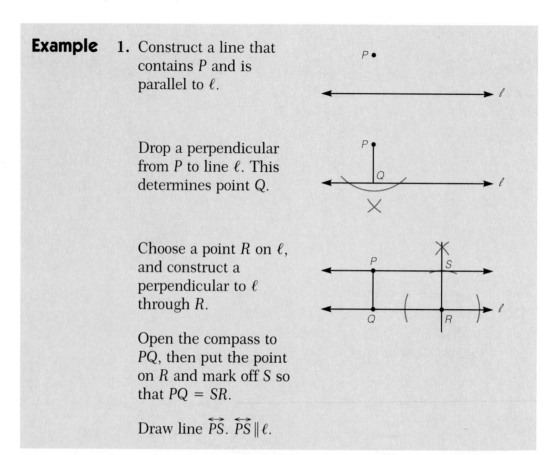

Try This... a. Draw a line ℓ and a point P not on ℓ. Use the method above to construct a line that contains P and is parallel to ℓ.

The main highways in this interchange are an example of lines that do not lie in the same plane. We call them *skew lines*.

Definition Two lines that do not lie in the same plane are called **skew lines.** Two segments contained in two skew lines are called **skew segments.**

Note that skew lines are noncoplanar and do not intersect. If skew lines did intersect, they would be coplanar—that is, they would lie in the same plane. Examples 2 and 3 illustrate parallel and skew segments.

Examples 2. Name the segments of the cube that are parallel to $\overline{AB}$.

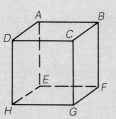

The segments $\overline{DC}$, $\overline{HG}$, and $\overline{EF}$ are parallel to $\overline{AB}$.

3. Name the segments of the cube that are skew to $\overline{GF}$.

The segments $\overline{DC}$, $\overline{AB}$, $\overline{AE}$, and $\overline{DH}$ are skew to $\overline{GF}$.

Try This... b. Name the segments of the above cube that are parallel to $\overline{DH}$.

c. Name the segments of the above cube that are skew to $\overline{HG}$.

Discussion Eileen said, "If line ℓ is parallel to line m, and line m is parallel to line n, then line ℓ must be parallel to line n." Do you agree or disagree?

Exercises

Practice

Use the cube shown at the right for Exercises 1–8.

Name the segments in the cube that are parallel to the given segment.

1. $\overline{PQ}$ 2. $\overline{JK}$

3. $\overline{PS}$ 4. $\overline{LK}$

Name the segments in the cube that are skew to the given segment.

5. $\overline{PS}$ 6. $\overline{PQ}$

7. $\overline{JK}$ 8. $\overline{SL}$

Copy each figure. Then use a compass and a straightedge to construct a line that contains the point and is parallel to the line.

9. •

10.

11.

Extend and Apply

Use the pyramid at the right for Exercises 12–17.

Name all segments that are skew to the given segment.

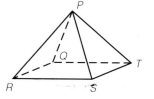

12. $\overline{PR}$ **13.** $\overline{RQ}$

14. $\overline{PT}$ **15.** $\overline{TS}$

16. Which of the following segments is skew to $\overline{RS}$?
  (A) $\overline{QT}$ (B) $\overline{PR}$ (C) $\overline{PT}$ (D) $\overline{PS}$

17. Name all pairs of parallel segments.

Tell whether the lines in these pictures are illustrations of parallel or skew lines.

18.    **19.**

Tell whether each of the following is an example of intersecting, parallel, or skew lines.

20. Rabbit-ear TV antenna

21. Goal lines of a football field

22. Foul lines of a baseball diamond

23. Venetian blinds

24. Window frame

25. Steel girders in a skyscraper

Tell whether each statement is true or false. If it is false, tell why.

26. Skew lines are always noncoplanar.

27. Lines that are not parallel always intersect.

28. Two lines that do not intersect are always parallel.

29. Two lines in the same plane that do not intersect are parallel.

Use Mathematical Reasoning

Complete each statement with *always*, *sometimes*, or *never*.

30. Two intersecting lines are _____ skew.

31. Two lines are _____ parallel, skew, or intersecting.

32. Two skew lines are _____ coplanar.

33. Two lines that do not intersect are _____ parallel.

Mixed Review

List three inequalities for each triangle.

34.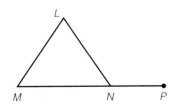

35.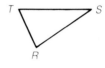

Can these numbers be the lengths of the sides of a triangle?

36. 6, 8, 10   **37.** 5, 5, 9   **38.** 17, 8, 7

Use the Hinge Theorem to reach a conclusion about each pair of triangles.

39. △TPS and △RFP

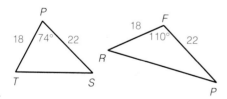

40. △WFG and △TMS

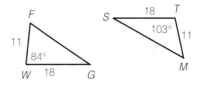

What is the relationship between the measures of the two indicated angles?

41. ∠LNP and ∠L

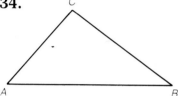

42. ∠RSV and ∠V

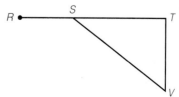

◇ Solve and check.

43. $3x + 7 = 4x + 2$ **44.** $4y - 6 = y$ **45.** $6w - 5 = 5w + 1$

Perspective Drawing

You have probably noticed that objects seem to get smaller as they move farther away. Railroad tracks appear to get closer together as they approach the horizon. These are two examples of the effects of perspective. Artists use perspective in paintings to create the illusion of depth.

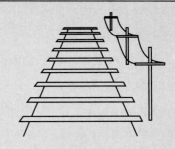

When we draw three-dimensional objects on a flat surface in such a way that they still look three-dimensional, we call it a *perspective drawing*. The following sample shows how to make your own perspective drawing.

Sample 1. Make a perspective drawing of a rectangular box. Computer software may be used.

Draw a rectangle *ABCD* for the front of the box. Then draw a *horizon line* parallel to the top (or bottom) of the rectangle.

Select a point on the horizon line. This is called the *vanishing point*. Connect this point to each of the vertices *A*, *B*, *C*, and *D*. We call these the vanishing lines.

To show the visible back edge, draw a segment $\overline{EF}$ parallel to the horizon line. Drawing lines through *E* and *F* parallel to $\overline{BA}$ and $\overline{CD}$ determines the hidden vertices *G* and *H*.

Hidden edges are usually drawn with dashed lines.

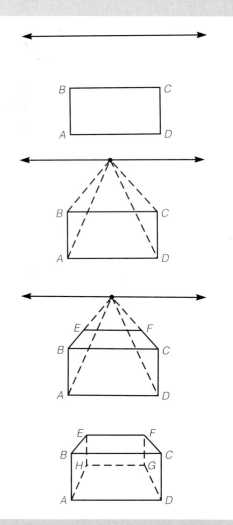

The position of the horizon line and the vanishing point determines the position of the viewer. In the previous sample, the viewer was above and directly in front of the box. Here are some more illustrations.

Samples Where is the viewer of each drawing?

2.

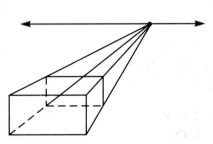

The viewer is above and to the right of the box.

3.

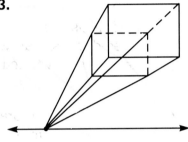

The viewer is below and to the left of the box.

Problems

Copy the given horizon line, vanishing point, and front side of a figure. Complete a perspective drawing. Computer software may be used.

1.

2.

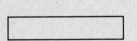

Where is the viewer of each drawing?

3.

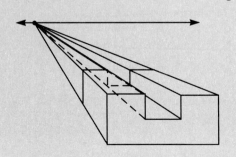

4.

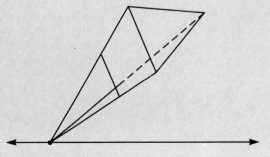

7-2 Transversals and Angles

Objective: Identify corresponding, alternate interior, and alternate exterior angles associated with lines and transversals.

Carpenters sometimes nail a board across several uprights to reinforce a wall or partition until it is in place. Such a board is like a *transversal*.

Definition   A **transversal** is a line that intersects two or more coplanar lines in different points.

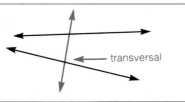

transversal

When a transversal intersects a pair of lines, eight angles are formed. Certain pairs of these angles have special names.

Interior angles:
∠3, ∠4, ∠5, and ∠6

interior angles on the same side of the transversal

Corresponding angles:
∠2 and ∠6
∠3 and ∠7
∠1 and ∠5
∠4 and ∠8

corresponding angles

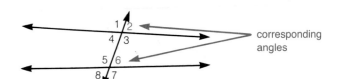

Alternate interior angles:
∠4 and ∠6
∠3 and ∠5

alternate interior angles

Alternate exterior angles:
∠1 and ∠7
∠2 and ∠8

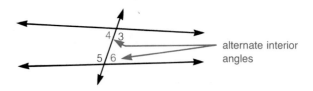

alternate exterior angles

Examples Identify each pair of angles as corresponding angles, interior angles, alternate interior angles, or alternate exterior angles.

1. $\angle AGH$ and $\angle CFG$
   Corresponding angles

2. $\angle BGF$ and $\angle GFC$
   Alternate interior angles

3. $\angle AGF$ and $\angle CFG$
   Interior angles

4. $\angle HGB$ and $\angle CFE$
   Alternate exterior angles

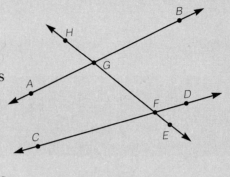

Try This... Identify each pair of angles as corresponding angles, interior angles, alternate interior angles, or alternate exterior angles.

a. $\angle DEB$ and $\angle EBC$

b. $\angle FEB$ and $\angle CBE$

c. $\angle HED$ and $\angle EBA$

d. $\angle DEH$ and $\angle CBG$

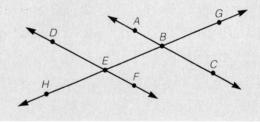

Sometimes a line or a segment is a transversal for more than one pair of lines.

Examples Identify each pair of angles and the pair of lines and transversal associated with them.

5. $\angle 1$ and $\angle 4$

These are alternate interior angles associated with lines m and n and transversal ℓ.

6. $\angle 5$ and $\angle 6$

These are corresponding angles associated with lines p and q and transversal n.

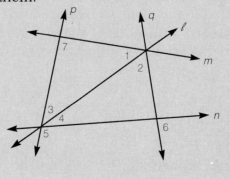

Discussion Francisco looked at the figure for Examples 5 and 6 and said, "Every line shown is a transversal." Do you agree or disagree?

Exercises

Practice

Identify each pair of angles as interior angles, corresponding angles, alternate interior angles, or alternate exterior angles.

1. ∠1 and ∠5   **2.** ∠3 and ∠5

3. ∠8 and ∠3   **4.** ∠2 and ∠5

5. ∠4 and ∠8   **6.** ∠2 and ∠8

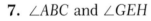

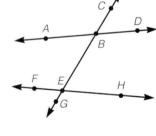

7. ∠ABC and ∠GEH   **8.** ∠DBE and ∠BEH

9. ∠DBE and ∠FEB   **10.** ∠ABC and ∠FEB

11. ∠CBD and ∠FEG   **12.** ∠ABE and ∠HEB

For each pair of angles, tell whether they are interior, corresponding, alternate interior, or alternate exterior angles, and identify the pair of lines and transversal associated with them.

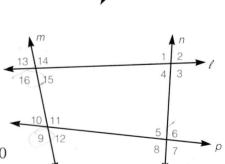

13. ∠13 and ∠3   **14.** ∠13 and ∠10

15. ∠3 and ∠5   **16.** ∠9 and ∠6

17. ∠15 and ∠11   **18.** ∠16 and ∠4

19. ∠1 and ∠7   **20.** ∠14 and ∠1

21. ∠16 and ∠11   **22.** ∠12 and ∠7

Extend and Apply

Identify the transversal associated with each pair of angles.

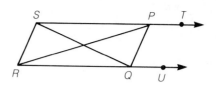

23. $\angle QPS$ and $\angle RSP$ are interior angles.

24. $\angle RPQ$ and $\angle PRS$ are alternate interior angles.

25. $\angle RQS$ and $\angle PSQ$ are alternate interior angles.

26. $\angle SPQ$ and $\angle RQP$ are interior angles.

27. $\angle TPQ$ and $\angle PSR$ are corresponding angles.

28. $\angle PQU$ and $\angle SRQ$ are corresponding angles.

29. Which of the following pairs of angles are alternate interior angles?

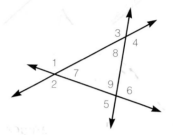

(A) $\angle 1$ and $\angle 4$ (B) $\angle 3$ and $\angle 9$
(C) $\angle 5$ and $\angle 7$ (D) $\angle 3$ and $\angle 2$

30. The roof of a house is designed as shown at the right. The crossbeam forms a transversal. What type of angles are $\angle 1$ and $\angle 2$?

Mixed Review

Use the figure at the right for Exercises 31–34.

Name the segments that are parallel to the given segment.

31. $\overline{TU}$ **32.** $\overline{SW}$

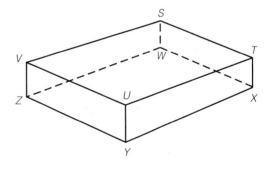

Name the segments that are skew to the given segment.

33. $\overline{ZW}$ **34.** $\overline{TU}$

35. Copy the figure. Then use a compass and a straightedge to construct a line that contains the point and is parallel to the line.

36. Is subtraction commutative?
(Is $a - b = b - a$ always true?)

37. Is addition associative?
(Is $(a + b) + c = a + (b + c)$ always true?)

Solving Equations Involving Parentheses

Laurel, Brooke, and Ingri ordered three chickenburgers, three cartons of milk, and three orders of potato skins. Chickenburgers cost $1.25, milk costs $0.55, and potato skins cost $0.80. What was their total bill?

Laurel first added up one order. $1.25 + $0.55 + $0.80 = $2.60
Then she multiplied this answer by 3. $3 \times $2.60 = $7.80

Brooke used her calculator to multiply the price of each item by 3 and then added. She also got a total of $7.80. $3 \times 1.25 + 3 \times 0.55 + 3 \times 0.80 = 7.80$

This illustrates a basic property of numbers called the *distributive property*. In general, $a(b + c) = ab + ac$.

Samples   Use the distributive property to remove the parentheses.

1. $3(2 + x)$

$$3(2 + x) = 3 \cdot 2 + 3x$$
$$= 6 + 3x$$

2. $8(2 - 3y)$

$$8(2 - 3y) = 8 \cdot 2 - 8 \cdot 3y$$
$$= 16 - 24y$$

If there are parentheses in an equation, you can use the distributive property to remove the parentheses before solving.

Sample   **3.** Solve $5(a + 3) = 25$.

$$5(a + 3) = 25$$
$5a + 15 = 25$ Use the distributive property to remove the parentheses.
$5a = 10$ Subtract 15 from both sides.
$a = 2$ Divide both sides by 5.

The solution is 2. Replacing a with 2 will show that the solution checks.

Problems

Use the distributive property to remove the parentheses.

1. $3(x + 8)$ **2.** $8(7 - 2y)$ **3.** $10(3a - 2)$

Solve and check.

4. $2(x + 1) = 10$ **5.** $5(y - 3) = 30$ **6.** $4(10 - 2x) = 16$

7. $7(2a + 4) = 42$ **8.** $9(6 - 2y) = 18$ **9.** $93 = 3(5x + 6)$

7-3 When Are Lines Parallel?

Objective: Identify conditions that enable you to conclude that two lines are parallel.

Explore

Complete the following activity with a compass and a straightedge or with computer software.

1. Draw a line $\overleftrightarrow{AB}$, choose a point P not on the line, and then draw a line containing P and some point Q on the line.

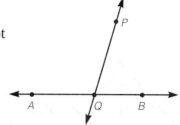

2. With vertex at P, copy $\angle PQB$, as shown, so that $m\angle RPQ = m\angle PQB$.

3. Draw line $\overleftrightarrow{RP}$.

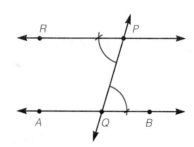

Discuss How are lines $\overleftrightarrow{RP}$ and $\overleftrightarrow{AB}$ related? If a transversal intersects a pair of lines and forms congruent alternate interior angles with the lines, how are the lines related?

When a transversal intersects two lines, we can look at the related angles that are formed and determine if the lines are parallel. The following theorem summarizes what you may have discovered in the Explore.

Theorem 7.1 If two lines and a transversal form congruent alternate interior angles, then the lines are parallel.

If $\angle 1 \cong \angle 2$, then $\ell \parallel m$.

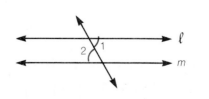

The next three theorems all follow from Theorem 7.1. They give additional methods for determining whether two lines are parallel.

Theorem 7.2 If two lines and a transversal form congruent corresponding angles, then the lines are parallel.

If $\angle 1 \cong \angle 2$, then $\ell \parallel m$.

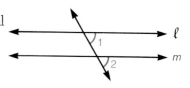

Theorem 7.3 If two lines and a transversal form congruent alternate exterior angles, then the lines are parallel.

If $\angle 1 \cong \angle 2$, then $\ell \parallel m$.

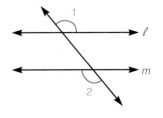

Theorem 7.4 If two lines and a transversal form supplementary interior angles on the same side of the transversal, then the lines are parallel.

If $m\angle 1 + m\angle 2 = 180$, then $\ell \parallel m$.

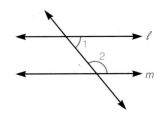

The following Examples illustrate how we can use the above theorems to show that lines are parallel.

Example 1. Consider lines ℓ, m, n, and p and transversal t. Which pairs of lines are parallel? Explain how you know.

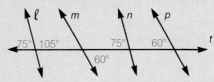

$\ell \parallel n$ because corresponding angles are congruent. (Also, $\ell \parallel n$ because interior angles on the same side of the transversal are supplementary.)

$m \parallel p$ because alternate interior angles are congruent.

Examples 2. Vertical boards in walls are often called studs. The studs shown at the right are all perpendicular to the floor. Why are the studs parallel?

The floor is like a transversal and each of the studs forms a 90° angle with it. The studs are parallel because corresponding angles are congruent.

Also, the studs are parallel because interior angles on the same side of the transversal are supplementary.

3. △ABC and △BCD are equilateral triangles. Which segments in the figure are parallel? Explain how you know.

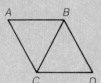

In this example, $\overline{BC}$ is a transversal for each pair of sides. Since equilateral triangles are also equiangular, $m\angle A = m\angle ABC = m\angle BCA = m\angle DCB = m\angle CBD = m\angle D = 60°$.

Because alternate interior angles $\angle ABC$ and $\angle BCD$ are congruent, $\overline{AB} \parallel \overline{CD}$.
Because alternate interior angles $\angle ACB$ and $\angle CBD$ are congruent, $\overline{AC} \parallel \overline{BD}$.

Try This... a. Which lines are parallel? Explain how you know.

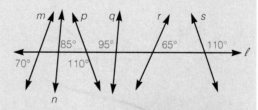

b. Why is the top of the window frame parallel to the bottom?

Discussion Two lines in a plane are perpendicular to a transversal. Give a convincing argument that the two lines are parallel. Can you think of more than one argument?

Exercises

Practice

Identify the parallel lines. In each case, explain how you know the lines are parallel.

1.

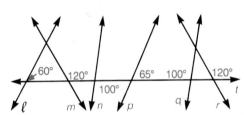

2.

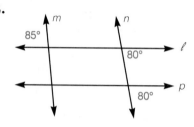

Tell why $\overleftrightarrow{AB} \parallel \overleftrightarrow{CD}$.

3.

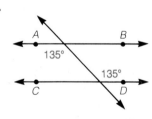

4.

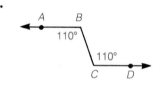

5.

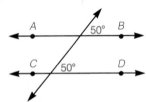

6.

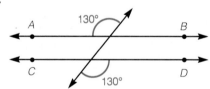

7.

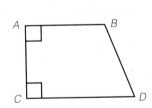

8.

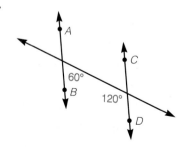

Extend and Apply

9. When Vonetta builds a picnic table, she always makes sure that the marked angles measure 40°. How can Vonetta be sure that this makes the table level?

10. Which of the following does *not* guarantee that $m \parallel n$?

(A) $\angle 1 \cong \angle 5$ (B) $\angle 5 \cong \angle 4$

(C) $\angle 2 \cong \angle 7$ (D) $\angle 6 \cong \angle 7$

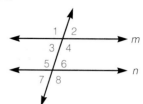

Use Mathematical Reasoning

11. $\triangle ABC \cong \triangle DCB$. Which segments are parallel? Explain how you know.

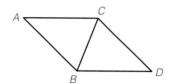

Mixed Review

Identify each pair of angles as interior angles, corresponding angles, alternate interior angles, or alternate exterior angles.

12. $\angle 1$ and $\angle 5$ **13.** $\angle 2$ and $\angle 8$

14. $\angle 8$ and $\angle 3$ **15.** $\angle 7$ and $\angle 1$

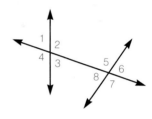

◇ Solve and check.

16. $3(y + 8) = 30$ **17.** $6(4x + 1) = 102$ **18.** $3(4 - x) = 9$

19. $7(2 + 7w) = 112$ **20.** $3(10 - 3x) = 12$ **21.** $27 = 3(x + 4)$

Enrichment

Periscopes are often used to look over objects. Periscopes are based on the fact that when light reflects off of a mirror, the angle of incidence has the same measure as the angle of reflection.

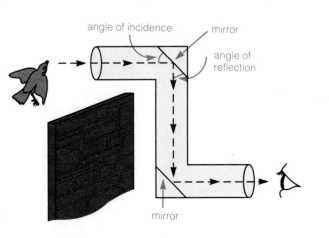

1. If the light is to make two right angles, what must be the measure of the angle of incidence and the angle of reflection?

2. If the light makes two right angles, explain why the light entering the periscope is parallel to the light leaving the periscope.

7-4 The Parallel Postulate

Objective: Identify congruent pairs of angles associated with parallel lines and transversals.

Jerome wants to put up a closet shelf five feet from the floor. He uses a level because he knows that the shelf will be parallel to the ground if the air bubble in the level is centered.

It seems reasonable for Jerome to assume that there is only one way for the shelf to be parallel at a given height from the floor. Jerome is assuming that there is *only one* line that contains *P* and is parallel to the floor.

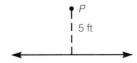

Although this seems obvious, many mathematicians tried to prove this, but none succeeded. Today we know that it is not possible to prove that parallels are unique using only the postulates and theorems we have so far. Thus, we must add this statement as a postulate.

Postulate 15   The Parallel Postulate
Given a line ℓ and a point *P* not on ℓ, there is only one line that contains *P* and is parallel to ℓ.

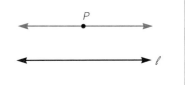

In the previous lesson, we found that if alternate interior angles are congruent, then lines are parallel. Now that we have assumed that parallels are unique, we can state the converse.

You will recognize the following theorems as converses of those in Lesson 7-3.

Theorem 7.5 If a transversal intersects two parallel lines, then the alternate interior angles are congruent.

If $\ell \parallel m$, then $\angle 1 \cong \angle 2$.

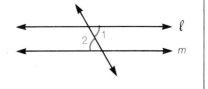

Theorem 7.6 If a transversal intersects two parallel lines, then the corresponding angles are congruent.

If $\ell \parallel m$, then $\angle 1 \cong \angle 2$.

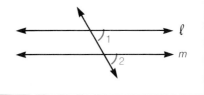

Theorem 7.7 If a transversal intersects two parallel lines, then the alternate exterior angles are congruent.

If $\ell \parallel m$, then $\angle 1 \cong \angle 2$.

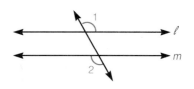

Theorem 7.8 If a transversal intersects two parallel lines, then the interior angles on the same side of the transversal are supplementary.

If $\ell \parallel m$, then $m\angle 1 + m\angle 2 = 180°$.

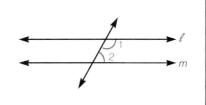

Example 1. If $\ell \parallel m$ and $m\angle 1 = 40°$, what are the measures of the other angles?

$m\angle 7 = 40°$, since alternate interior angles are congruent.
$m\angle 5 = 40°$, since corresponding angles are congruent.
$m\angle 8 = 140°$, since interior angles on the same side of a transversal are supplementary.

By similar reasoning, we have the measures of the other angles as shown.

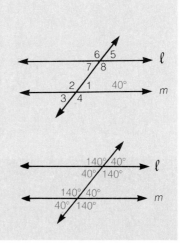

Examples 2. $\overline{PT} \| \overline{SR}$. Which pairs of angles are congruent?

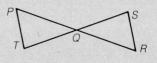

Since alternate interior angles are congruent, $\angle TPQ \cong \angle SRQ$, and $\angle PTQ \cong \angle RSQ$.

Also, $\angle PQT \cong \angle SQR$ because they are vertical angles.

3. $\overline{DE} \| \overline{BC}$. Which pairs of angles are congruent? Which pairs of angles are supplementary?

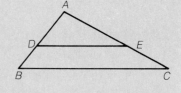

Since corresponding angles are congruent, $\angle ADE \cong \angle ABC$, and $\angle AED \cong \angle ACB$.

Since they are interior angles on the same side of a transversal, $\angle EDB$ and $\angle DBC$ are supplementary. Similarly, $\angle DEC$ and $\angle ECB$ are supplementary. As linear pairs, $\angle ADE$ and $\angle BDE$ are supplementary and $\angle AED$ and $\angle CED$ are supplementary.

Try This... **a.** If $p \| q$, and $m\angle 1 = 130°$, what are the measures of the other angles?

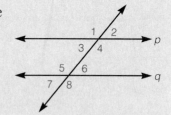

b. $\overleftrightarrow{AB} \| \overleftrightarrow{CD}$. Which pairs of angles are congruent? Which pairs of angles are supplementary?

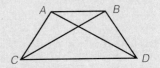

c. $\overrightarrow{YZ} \| \overrightarrow{VU}$. Which pairs of angles are congruent?

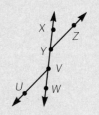

Discussion   Two parallel lines are cut by a transversal, and you know the measure of one of the eight angles formed. Is it always possible to find the measures of the other angles? If so, how?

Exercises

Practice

In each figure, $\overline{AB} \parallel \overline{CD}$. Find the measure of each of the numbered angles.

1.

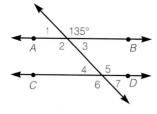

2.

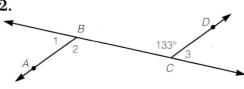

In each figure, $\overline{AB} \parallel \overline{CD}$. Which pairs of angles are congruent? Which pairs of angles, if any, are supplementary?

3.

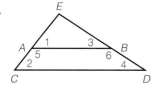

4.

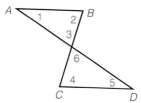

5.

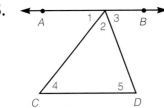

6.

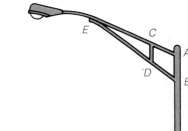

7.

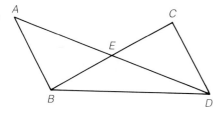

8.

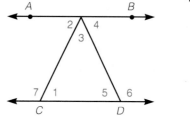

Extend and Apply

9. If $p \parallel q$, which of the following is $m\angle 1$?

(A) 70° (B) 50° (C) 60° (D) 110°

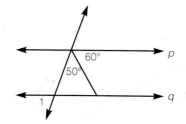

10. Xavier wants the cross legs of his workbench to make a 45° angle with the floor. At what angle should he adjust the legs?

11. How are the indicated angles on the stairs related? Explain how you know.

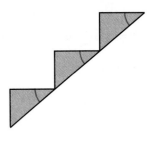

Use Mathematical Reasoning

12. $\angle A$, $\angle B$, and $\angle C$ are right angles. Is $\angle D$ a right angle? Explain how you know.

13. $\overline{AB} \cong \overline{AC}$, and $\overleftrightarrow{DE} \parallel \overleftrightarrow{AB}$. Is $\triangle CDE$ isosceles? Explain how you know.

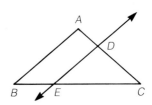

Mixed Review

Can these numbers be the lengths of the sides of a triangle?

14. 5, 2, 3   **15.** 16, 12, 14   **16.** 7, 5, 13

Solve and check.

17. $6(3y + 1) = 60$ **18.** $3(6 - y) = 15$ **19.** $9 = 3(15 - 2y)$

Enrichment

On a clean sheet of paper, draw a line ℓ and a point A not on the line. How can you fold the paper to produce a line that contains point A and is parallel to ℓ?

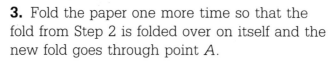

1. Fold the paper along line ℓ.

2. Fold the paper over again so that the fold from Step 1 is folded over on itself and the new fold goes through point A.

3. Fold the paper one more time so that the fold from Step 2 is folded over on itself and the new fold goes through point A.

4. Open the paper and draw a line m on the crease from Step 3. What do you notice?

5. Why does this series of folds produce the required line?

7-5 The Angles of a Triangle

Objective: Use the Angle Sum Theorem to solve problems.

Explore

1. Draw a large triangle on a sheet of paper and cut it out.

2. Draw a segment on the triangle that is parallel to the longest side and that intersects the other two sides at their midpoints.

3. Fold in the three corners as shown. What do you notice?

4. Repeat this activity with a different triangle.

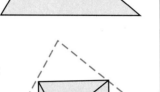

Discuss What does the activity show regarding the sum of the measures of the angles of a triangle? Do you get the same result for acute, obtuse, and right triangles?

In Chapter 1 you tore off the corners of a triangle to conclude that the sum of its angles' measures was 180°. However, we were not able to list this theorem until now because, deductively, it follows from the Parallel Postulate.

Theorem 7.9 The Angle Sum Theorem

The sum of the measures of the angles of a triangle is 180°.

$m\angle A + m\angle B + m\angle C = 180°$

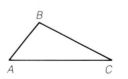

Example **1.** Find $m\angle C$.

$$m\angle A + m\angle B + m\angle C = 180$$
$$80 + 30 + m\angle C = 180$$
$$110 + m\angle C = 180$$
$$m\angle C = 70°$$

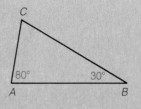

Alternatively, we can use a calculator.

180 $\boxed{-}$ 80 $\boxed{-}$ 30 $\boxed{=}$ $\boxed{70}$

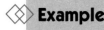

Example 2. Find the measure of each angle.

$$m\angle A + m\angle B + m\angle C = 180$$
$$x + 2x + 3x = 180$$
$$6x = 180$$
$$x = 30$$

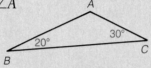

$m\angle A = 30°$, $m\angle B = 60°$, and $m\angle C = 90°$.

Try This... **a.** Find $m\angle A$

b. Find the measure of each angle.

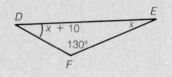

There are several theorems that follow from the Angle Sum Theorem.

> **Theorem 7.10**   If two angles of one triangle are congruent to two angles of another triangle, then the third angles are congruent.

Example 3. $\angle 1 \cong \angle 2$. How are $\angle A$ and $\angle C$ related? Explain how you know.

We know that $\angle 1 \cong \angle 2$. We know that $\angle CBE \cong \angle ABD$ by the Vertical Angle Theorem. Thus, by Theorem 7.10, $\angle A \cong \angle C$.

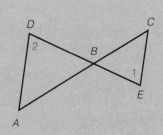

Try This... **c.** How are $\angle 1$ and $\angle 2$ related? Explain how you know.

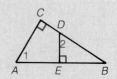

Here is another theorem that follows from the Angle Sum Theorem.

Theorem 7.11 The acute angles of a right triangle are complementary.

Example 4. Find $m\angle 1$ and $m\angle 2$.

By Theorem 7.11,
$m\angle A + m\angle 1 = 90°$. Thus,
$m\angle 1 = 30°$.

By Theorem 7.11,
$m\angle C + m\angle 2 = 90°$. Thus,
$m\angle 2 = 60°$.

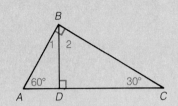

Try This... d. $\triangle ABC$ is a right triangle, $\overline{BC} \cong \overline{BD}$, and $m\angle BDC = 55°$.

Find the measures of all the angles in the figure.

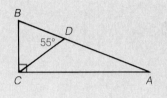

The next theorem is a stronger version of the Exterior Angle Theorem from Chapter 6.

Theorem 7.12 The measure of an exterior angle of a triangle is the sum of the measures of the two remote interior angles.

$m\angle 1 = m\angle 2 + m\angle 3$

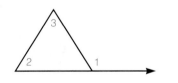

Example 5. Find $m\angle A$ and $m\angle ACB$ in $\triangle ABC$.

By Theorem 7.12,
$m\angle A + m\angle B = m\angle ACD$.
$m\angle A + 75 = 146$
$m\angle A = 71°$

Thus, $m\angle A = 71°$ and, since
$\angle ACB$ and $\angle ACD$ are supplementary,
$m\angle ACB = 180 - 146 = 34°$.

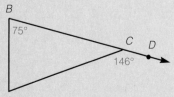

Try This... e. Find $m\angle A$ and $m\angle ACB$ in $\triangle ABC$.

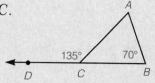

Discussion Draw a triangle. Draw the line through one vertex parallel to the opposite side. Use the fact that parallel lines have congruent alternate interior angles to give a convincing argument for the Angle Sum Theorem.

Exercises

 Practice

Find $m\angle A$ in each figure.

1.

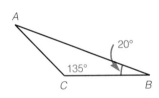

2.

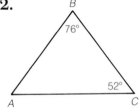

3.

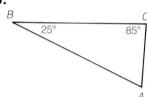

4.

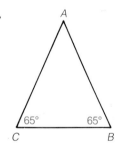

5.

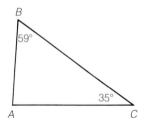

6.

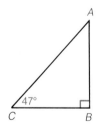

7. Find $m\angle E$ and $m\angle F$.

8. Find $m\angle A$, $m\angle C$, and $m\angle CBD$.

9. Find $m\angle B$ and $m\angle C$.

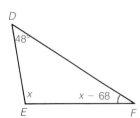

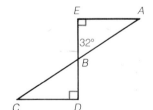

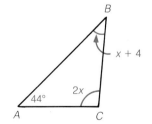

10. Find $m\angle 1$ and $m\angle 2$.

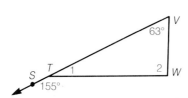

11. Find $m\angle 1$ and $m\angle 2$.

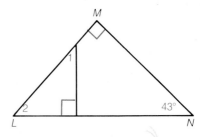

12. $\angle A \cong \angle B$. How are $\angle 1$ and $\angle 2$ related? Explain how you know.

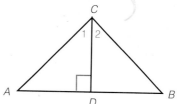

Extend and Apply

13. Which of the following is $m\angle 1$?

   (A) $40°$ (B) $120°$ (C) $100°$ (D) $160°$

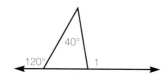

Find the measure of each angle.

14. $\overline{AB} \parallel \overline{DE}$ **15.** $\angle 1 \cong \angle 2$ **16.** $m\angle 1 = m\angle 2 = 62°$

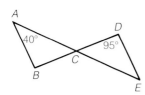 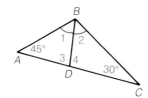 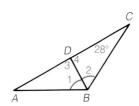

⬦ Solve for x and find the measure of each angle.

17.

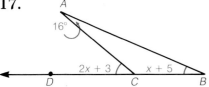

18.

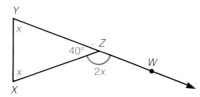

19. $\overline{PR} \cong \overline{RQ}$. Find the measures of $\angle 1$, $\angle 2$, and $\angle 3$.

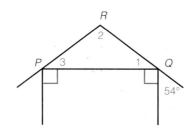

20. Find the measure of $\angle ABC$.

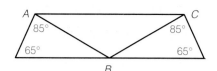

Use Mathematical Reasoning

21. $\triangle ABC$ and $\triangle DEF$ are equilateral. Is $\angle DEF \cong \angle C$? Explain how you know.

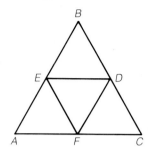

Mixed Review

Tell why $\overleftrightarrow{AB} \parallel \overleftrightarrow{CD}$.

22.

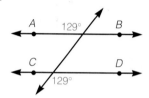

23.

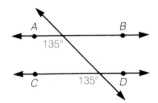

24.

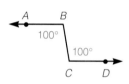

25.

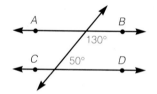

◇ Solve and check.

26. $2x + 7 = 3x - 3$ **27.** $4y - 1 = 5 + 3y$ **28.** $8z + 2 = 5z + 5$

Problem Solving

Use logical reasoning and the following list of clues to determine the measures of $\angle A$, $\angle B$, and $\angle C$.

1. All of the angles' measures are divisible by 10.

2. The measure of $\angle A$ is greater than the measure of $\angle B$.

3. The measure of $\angle B$ is greater than the measure of $\angle C$.

4. The measure of $\angle C$ is greater than 40°.

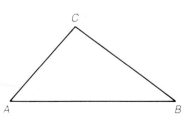

7-6 Translations

Objective: Find the translation image of a figure and recognize properties of translated figures.

Artist M. C. Escher often created his art by taking a figure and sliding it in the plane. In the print at the right, any single bird can be shifted to any other.

A slide moves all points the same distance in the same direction. Recall that the formal mathematical name for a slide is a *translation*.

Definition Given distance XY and direction $\overrightarrow{XY}$ in a plane, the **translation image** of any point P is a point P' such that $PP' = XY$, and $\overrightarrow{PP'}$ is in the same direction as $\overrightarrow{XY}$.

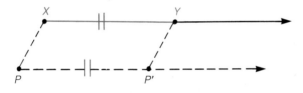

In the above figure, notice that rays $\overrightarrow{XY}$ and $\overrightarrow{PP'}$ are parallel.

Examples Name the translation image of each figure under the translation $\overrightarrow{XY}$.

1. $\overline{RT}$
   U is the image of R, and W is the image of T. So $\overline{UW}$ is the translation image of $\overline{RT}$.

2. $\triangle RTS$
   $\triangle UWV$ is the translation image of $\triangle RTS$.

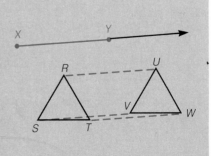

Try This... Name the translation image of each figure under the translation $\overrightarrow{MN}$.

    a. $\overline{CD}$

    b. Point B

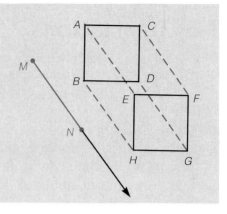

We can use the above definition and the Parallel Postulate to draw the translation image of a figure.

Example   **3.** Draw the image of $\triangle DEF$ under the translation $\overrightarrow{XY}$.

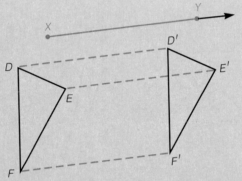

Draw lines through D, E, and F parallel to $\overrightarrow{XY}$.

Choose points D', E', and F' so that $XY = DD' = EE' = FF'$, and $\overrightarrow{DD'}$, $\overrightarrow{EE'}$, and $\overrightarrow{FF'}$ are in the same direction as $\overrightarrow{XY}$.

Connecting the image points, we find that $\triangle D'E'F'$ is the translation image of $\triangle DEF$.

Try This... **c.** Copy the figure below and draw the image of $\triangle KLM$ under the translation $\overrightarrow{XY}$.

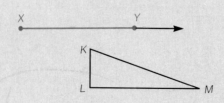

1. Measure all the sides and angles of $\triangle DEF$ in Example 3. Measure all the corresponding sides and angles of $\triangle D'E'F'$.
2. Which segments have the same length? Which angles have the same measure?
3. What appears to be the same about a figure and its translation image?

The following theorem summarizes facts that you may have discovered in the Explore:

Theorem 7.13   Translations preserve collinearity of points, distance, and angle measure.

We can use Theorem 7.13 to identify figures that are congruent under a translation.

Examples   *LMNP* is the image of *RSTU* under the translation $\overrightarrow{YZ}$.

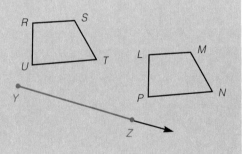

4. Name a segment congruent to $\overline{ST}$.
$\overline{MN}$ is congruent to $\overline{ST}$.
5. Name an angle congruent to $\angle P$.
$\angle U$ is congruent to $\angle P$.

Try This... $\triangle DEF$ is the image of $\triangle ABC$ under the translation $\overrightarrow{XY}$.

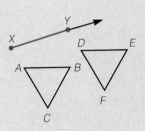

d. Name an angle congruent to $\angle B$.
e. Name a segment congruent to $\overline{DF}$.

Discussion   Give a convincing argument that the translation image of a square is another square congruent to the original.

Exercises

Practice

Name the image of each figure under the translation $\overrightarrow{UV}$.

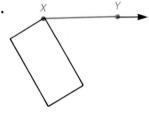

1. Point A **2.** $\overline{AC}$

3. $\overline{DE}$ **4.** Point D

5. $\triangle ECD$ **6.** $\triangle ACB$

Copy each figure and draw its image under the translation $\overrightarrow{XY}$.

7.

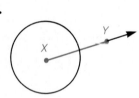

8.

9.

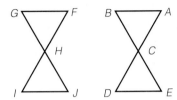

10.

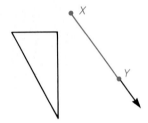

11.

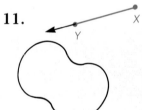

12.

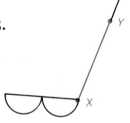

$TUVW$ is the image of $PQRS$ under the translation $\overrightarrow{MN}$.

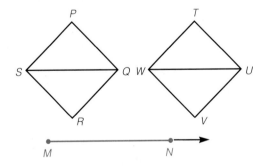

13. Name a segment congruent to $\overline{PQ}$.

14. Name a segment congruent to $\overline{SQ}$.

15. Name an angle congruent to $\angle PSR$.

16. Name an angle congruent to $\angle T$.

17. Name an angle congruent to $\angle TUW$.

18. △*DEF* is the translation image of △*ABC*, with *D* the image of *A*, *E* the image of *B*, and *F* the image of *C*. Which of the following must be true?

(A) △*DEF* ≅ △*ABC* (B) △*DEF* is equilateral.

(C) $\overline{AB} \cong \overline{EF}$ (D) ∠*A* ≅ ∠*F*

Which illustrations are examples of translations?

19. Car on track

20. Door closing

21. What two transformations are illustrated by locking a door bolt as shown in the photos? Does the order of the transformations matter?

Sketch the image of each figure under the translation $\overrightarrow{XY}$.

22.

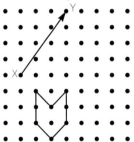

23.

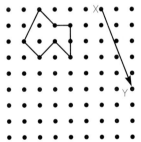

Use Mathematical Reasoning

24. Lines ℓ and *m* are parallel. Copy the figure and sketch the reflection image of △*ABC* over line ℓ. Label the image △*DEF*. Sketch the reflection image of △*DEF* over line *m*. Label this image △*GHI*. How do △*ABC* and △*GHI* appear to be related? Consider the reflection of other figures over lines ℓ and *m* and form a generalization.

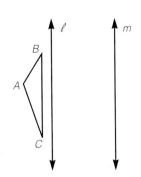

25. Recall that reflections reverse orientation. Do translations preserve or reverse orientation? Draw several figures to help you reach a conclusion.

Mixed Review

In Exercises 26–29, find the required angle measures.

26. Find $m\angle P$, $m\angle 1$, and $m\angle 2$.

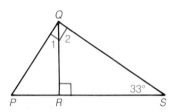

27. $\angle A \cong \angle D$ and $\angle B \cong \angle E$. Find $m\angle F$.

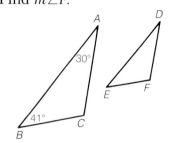

28. Find $m\angle 1$ and $m\angle 2$.

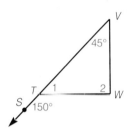

29. $\angle M$ is a right angle. Find $m\angle 1$ and $m\angle 2$.

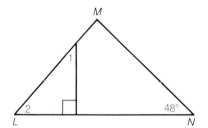

◇ Solve and check.

30. $3x + 8 + 4x = 2x + 23$

31. $6x - 4 - 2x = x - 1$

Visualization

The three-dimensional solids shown below are made up of smaller blocks. Assume there are no blocks missing from the back. How many blocks cannot be seen from *any* viewpoint?

1.

2.

Chapter 7 Review

Use the cube at the right for Problems 1 and 2.

7-1

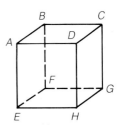

1. Name the segments in the cube that are parallel to $\overrightarrow{FG}$.

2. Name the segments in the cube that are skew to $\overline{AD}$.

3. Copy the figure and use a compass and a straightedge to construct a line that contains A and is parallel to $\overrightarrow{BC}$.

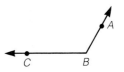

Use the figure at the right for Problems 4–6.

7-2

Identify each pair of angles as interior angles, corresponding angles, alternate interior angles, or alternate exterior angles. Identify the pair of lines and transversal associated with them.

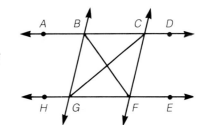

4. $\angle ABG$ and $\angle BGH$

5. $\angle CGF$ and $\angle GCB$

6. $\angle GBC$ and $\angle FCD$

7. Identify the parallel lines. Explain how you know.

7-3

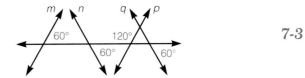

8. Tell why $\overrightarrow{AB} \parallel \overrightarrow{CD}$.

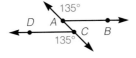

In each figure, $\overline{AB} \parallel \overline{CD}$.

7-4

9. Find $m\angle EAB$ and $m\angle EBA$.

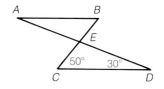

10. Find $m\angle BED$ and $m\angle ECD$. Which pairs of angles are supplementary?

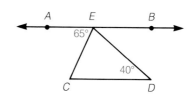

7-5 Find the measure of each angle.

11.

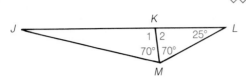

◇◇ **12.**

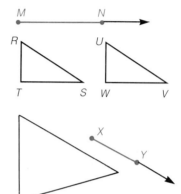

7-6 △*UVW* is the image of △*RST* under translation $\overrightarrow{MN}$.

13. Name the image of $\overline{RS}$.

14. Name an angle congruent to ∠*W*.

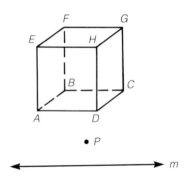

15. Copy the figure and draw its image under the translation $\overrightarrow{XY}$.

Chapter 7 Test

Use the cube at the right for Problems 1 and 2.

1. Name the segments in the cube that are parallel to $\overline{HD}$.

2. Name the segments in the cube that are skew to $\overline{DC}$.

3. Copy the figure and use a compass and a straightedge to construct a line that contains *P* and is parallel to line *m*.

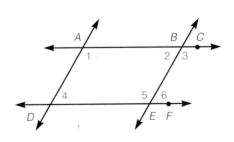

Use the figure at the right for Problems 4–6.

Identify each pair of angles as interior angles, corresponding angles, alternate interior angles, or alternate exterior angles. Identify the pair of lines and transversal associated with them.

4. ∠2 and ∠5

5. ∠6 and ∠2

6. ∠4 and ∠6

7. Identify the parallel lines. Explain how you know.

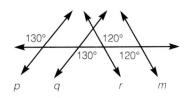

8. Tell why $\overrightarrow{AB} \parallel \overrightarrow{CD}$.

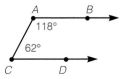

In each figure, $\overline{AB} \parallel \overline{CD}$.

9. Find $m\angle EDC$ and $m\angle ECD$.

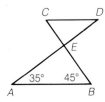

10. Find $m\angle CDE$. Which pairs of angles are supplementary?

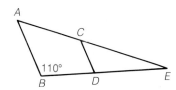

Find the measure of each angle.

11.

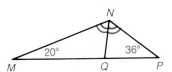

12.

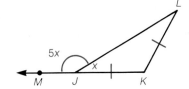

$\triangle ABC$ is the image of $\triangle DEF$ under translation $\overrightarrow{MN}$.

13. Name the image of $\overline{DF}$.

14. Name a segment congruent to $\overline{CB}$.

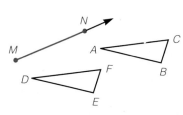

15. Copy the figure and draw its image under the translation $\overrightarrow{XY}$.

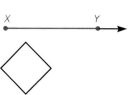

8 Quadrilaterals

Since they contain only right angles, rectangles and squares are the quadrilaterals most commonly used in architecture.

8-1 Classifying Quadrilaterals

Objective: Identify the parts of a quadrilateral and identify special types of quadrilaterals.

■ Explore ■

1. Draw several 4-sided figures.

2. For each figure, measure the angles. Then find the sum of their measures.

3. Tear off the corners and assemble them around a point.

Discuss What is true about the sum of the measures of the angles of a 4-sided figure?

Definition A **quadrilateral** consists of four coplanar segments that intersect only at their endpoints. Each endpoint belongs to exactly two segments.

Segments $\overline{AB}$, $\overline{BC}$, $\overline{CD}$, and $\overline{DA}$ are the **sides**, points A, B, C, and D are the **vertices**, and $\angle A$, $\angle B$, $\angle C$, and $\angle D$ are the **angles** of quadrilateral $ABCD$. Sides and angles may be **consecutive** ($\overline{AB}$ and $\overline{BC}$, $\angle A$ and $\angle B$) or **opposite** ($\overline{AB}$ and $\overline{DC}$, $\angle D$ and $\angle B$).

A **diagonal** of a quadrilateral is a segment that joins two opposite vertices. Segments $\overline{AC}$ and $\overline{BD}$ are diagonals.

Some quadrilaterals are "caved in." They are **concave quadrilaterals.** All others are **convex quadrilaterals.** What may be true about a diagonal of a concave quadrilateral?

In this chapter we will discuss convex quadrilaterals only.

In the Explore you used inductive reasoning to develop the following theorem:

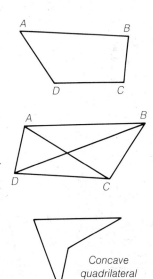

Concave
quadrilateral

Theorem 8.1 The sum of the measures of the angles of a quadrilateral is 360°.

We can use Theorem 8.1 to find the missing angle in a quadrilateral.

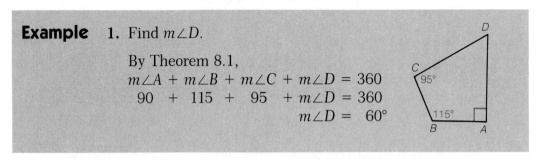

Example   1. Find $m\angle D$.

By Theorem 8.1,
$$m\angle A + m\angle B + m\angle C + m\angle D = 360$$
$$90 + 115 + 95 + m\angle D = 360$$
$$m\angle D = 60°$$

Try This...a. Find $m\angle F$.   **b.** Find $m\angle G$.

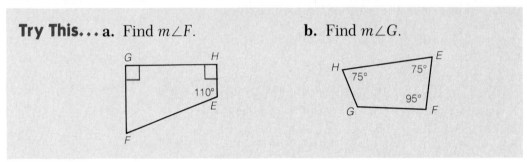

Some special types of quadrilaterals are defined in the table.

Type of Quadrilateral	Definition	Picture
Parallelogram	Two pairs of parallel sides	
Rectangle	A parallelogram with four right angles	
Rhombus	A parallelogram with four congruent sides	
Square	A rectangle with four congruent sides or A rhombus with four right angles	
Trapezoid	Exactly one pair of parallel sides	
Kite	Exactly two pairs of congruent consecutive sides	

Although a quadrilateral may have several classifications, there is one term that best classifies it.

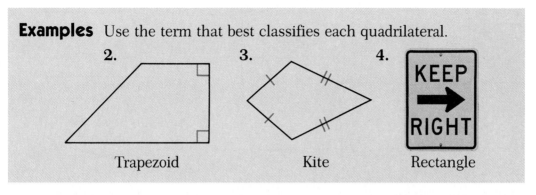
Try This... Use the term that best classifies each quadrilateral.

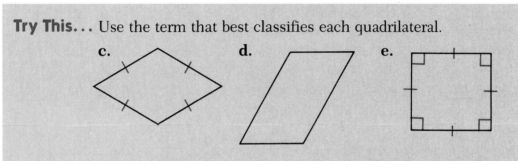

c.   **d.**   **e.**

Discussion Why are rectangles, rhombuses, and squares special types of parallelograms? How are they alike? How are they different?

Exercises

Practice

Find the measure of the missing angle.

1.

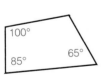

100°
85°   65°

2.

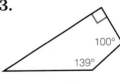

120°

3.

100°
139°

4.

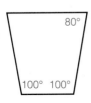

80°
100° 100°

5.

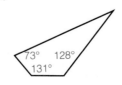

73° 128°
131°

6.

Use the term that best classifies each quadrilateral.

7.

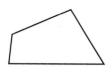

8.

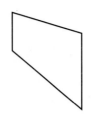

9.

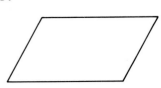

10.

11.

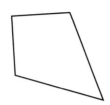

12.

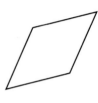

13.

14.

15.

16.

17.

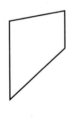

18.

Extend and Apply

For Exercises 19–25, complete each statement with *always*, *sometimes*, or *never* to make each statement true.

19. A parallelogram is _____ a rhombus.

20. A square is _____ a parallelogram.

21. A trapezoid is _____ a parallelogram.

22. A kite is _____ a quadrilateral.

23. A rectangle is _____ a rhombus.

24. A rhombus is _____ a square.

25. A square is _____ a trapezoid.

Try to draw a quadrilateral with the following properties.
Then classify it.

26. Exactly two right angles

27. Exactly one pair of parallel sides

28. Exactly three right angles

29. Exactly one right angle

30. Four right angles and four congruent sides

31. Exactly one pair of sides parallel and the other two sides congruent

32. Two pairs of parallel sides and no congruent sides

33. Four congruent sides and no right angles

34. The quadrilateral shown is made up of congruent equilateral triangles. What is the best classification for the quadrilateral?

(A) Kite   (B) Parallelogram   (C) Square   (D) Rhombus

Use Mathematical Reasoning

35. When you cut off both the top and the bottom of a square box, place the box on its side, and push against the box, what happens to the square shape?

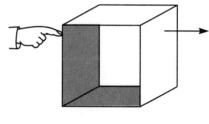

36. When you cut off both the top and the bottom of a rectangular box, place the box on its side, and push against the box, what happens to the rectangular shape?

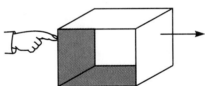

Mixed Review

Use the figure for Exercises 37–46.

37. Which theorem or postulate shows that $\overleftrightarrow{AD} \parallel \overleftrightarrow{FJ}$?

38. Given that $\overleftrightarrow{AD} \parallel \overleftrightarrow{FJ}$, which theorem or postulate shows that $m\angle 10 = m\angle 3$?

Find the measure of each angle.

39. $\angle 1$

40. $\angle 5$

41. $\angle 7$

42. $\angle 10$

43. $\angle 9$

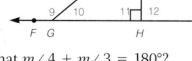

44. Which theorem or postulate shows that $m\angle 4 + m\angle 3 = 180°$?

45. Which theorem or postulate shows that $m\angle 4 = m\angle 2 + m\angle 5$?

46. Which theorem or postulate shows that if $\overline{GC}$ and $\overline{BH}$ bisect each other, then $\triangle BCE \cong \triangle HGE$?

47. Determine whether the following is true or false: If two sides and any angle of $\triangle ABC$ are congruent to two sides and any angle of $\triangle DEF$, the triangles are congruent.

◇ Solve and check.

48. $3x + x + 1 = 4$ **49.** $5y + 9y - 2 = 6$ **50.** $5z - 45 = 4z + 45$

Enrichment

Draw as many different types of quadrilaterals as you can. You may wish to work in a group so that each of you may draw two types of quadrilaterals on one sheet of paper.

Include an *isosceles* trapezoid with congruent angles as shown,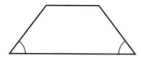

and a non-isosceles trapezoid.

Cut out each quadrilateral and fold it to find all lines of symmetry.

Problems

Which types of quadrilaterals have:
1. exactly 1 line of symmetry?
2. exactly 2 lines of symmetry?
3. exactly 3 lines of symmetry?
4. exactly 4 lines of symmetry?
5. more than 4 lines of symmetry?
6. no lines of symmetry?
7. 2 lines of symmetry and no right angles?
8. 2 lines of symmetry and four right angles?
9. no lines of symmetry and opposite angles congruent?
10. no lines of symmetry and one pair of parallel sides?

Lines of symmetry always bisect the sides or angles of a quadrilateral.
11. Which type(s) of quadrilaterals have lines of symmetry that are side bisectors?
12. Which type(s) of quadrilaterals have lines of symmetry that are angle bisectors?
13. Which type(s) of quadrilaterals have lines of symmetry that are side bisectors and lines of symmetry that are angle bisectors?

Draw a Venn Diagram

To show classification, we can use a diagram with circles called a **Venn diagram.** A Venn diagram uses circles to show classification of things within a larger set. Here are some examples:

Samples

1. Within the set *Animals,* shown by the rectangle, are *Cats* and *Dogs.* Since no cats are dogs and no dogs are cats, the *Cats* circle is completely separate from the *Dogs* circle.

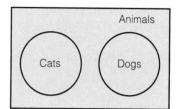

2. Within the set *Canadians* are *French-speaking Canadians* and *English-speaking Canadians.* Since many Canadians speak French and English, the circles overlap. Where the circles overlap represents Canadians who speak both French and English.

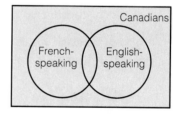

Problems

1. Complete and label the Venn diagram for the set *Quadrilaterals* so that each type of quadrilateral is represented.

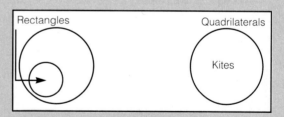

2. Why is there no circle for the set *Squares* in Problem 1?

3. Which of these Venn diagrams shows the correct shading for people who are swimmers and hikers but not skateboarders?

(A)

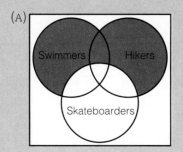

(B)

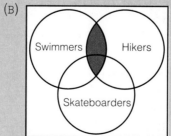

(C)

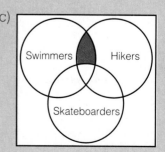

8-2 Properties of Parallelograms

Objective: Identify and apply the properties of parallelograms to solve problems.

Explore

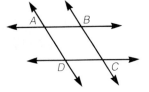

1. Draw two pairs of parallel lines to form ▱ (parallelogram) *ABCD*.

2. Compare the lengths of opposite sides.

3. Compare the measures of opposite angles.

4. Compare the measures of consecutive angles.

5. Draw diagonal $\overline{AC}$ and compare △*ADC* and △*CBA*.

6. Draw diagonal $\overline{BD}$, intersecting $\overline{AC}$ at point *E*. Compare *DE* and *EB*.

Discuss What properties of parallelograms did you discover?

You may have found some of the following properties:

Theorem 8.2 A diagonal of a parallelogram determines two congruent triangles.

△*ADC* ≅ △*CBA*

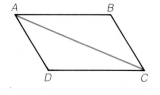

Because corresponding parts of congruent triangles are also congruent, we have the following:

Theorem 8.3 The opposite angles of a parallelogram are congruent.

Theorem 8.4 The opposite sides of a parallelogram are congruent.

Examples 1. $m\angle A = 120°$. Find the measures of the other angles of □ ABCD.

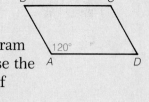

Since opposite angles of a parallelogram are congruent, $m\angle C = 120°$. Because the sum of the measures of the angles of a quadrilateral is 360°,

$$m\angle A + m\angle B + m\angle C + m\angle D = 360°.$$

Because $m\angle B = m\angle D$, we have

$$120 + 120 + 2m\angle B = 360$$
$$2m\angle B = 360 - 240 = 120$$
$$m\angle B = 60$$

Thus, $m\angle D = m\angle B = 60°$.

2. $m\angle A = 2x$, and $m\angle B = x - 15$. Find the measures of the angles of □ABCD.

The sum of the angle measures is 360°, so

$$m\angle A + m\angle B + m\angle C + m\angle D = 360.$$

Since opposite angles of a parallelogram are congruent, $m\angle C = m\angle A$, and $m\angle D = m\angle B$. So,

$$2m\angle A + 2m\angle B = 360.$$

Substituting, we solve for x:

$$2(2x) + 2(x - 15) = 360$$
$$4x + 2x - 30 = 360$$
$$6x = 390$$
$$x = 65°$$

Thus, $m\angle A = m\angle C = 2 \times 65 = 130°$, and $m\angle B = m\angle D = 65 - 15 = 50°$.

Try This... Find the measure of each angle.

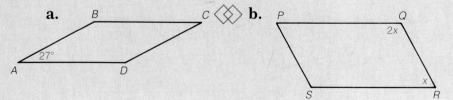

a.   **b.**

Definition   The **perimeter** of a quadrilateral is the sum of the lengths of its four sides. We often refer to the perimeter of a quadrilateral as the distance around it.

Examples 3. Find the perimeter.

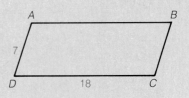

Because the opposite sides of a parallelogram are congruent, $AB = 18$ and $BC = 7$. Thus, the perimeter is $2(7) + 2(18) = 50$.

◆◆ **4.** The perimeter of ▱$MNOP$ is 144. One side is twice as long as an adjacent side. Find the length of each side of ▱$MNOP$.

Let $MP = x$. Then $MN = 2x$. Because the perimeter is 144, we have

$$x + 2x + x + 2x = 144$$
$$6x = 144$$
$$x = 24$$

Thus, $MP = NO = 24$, and $MN = PO = 48$.

Try This... c. Find the perimeter.

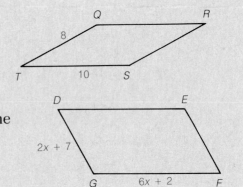

◆◆ **d.** The perimeter of ▱$DEFG$ is 68. Find the length of each side of ▱$DEFG$.

Recall that when a transversal intersects two parallel lines, the interior angles on the same side of the transversal are supplementary (Theorem 7.8). From this theorem we get the following:

Theorem 8.5 Consecutive angles of a parallelogram are supplementary.

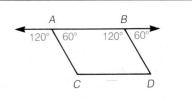

Example 5. $m\angle A = 32°$. Find $m\angle B$ and $m\angle D$.

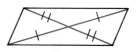

Since consecutive angles of a parallelogram are supplementary, $m\angle A + m\angle B = 180°$. Hence, $32 + m\angle B = 180°$
$$m\angle B = 148°$$

Similarly, $m\angle D = 148°$

Using the properties of parallel lines and the ASA Postulate, we obtain the following theorem:

Theorem 8.6   The diagonals of a parallelogram bisect each other.

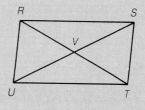

Example 6. $RV = 12$, and $VS = 13$. Find the length of each diagonal.

Since the diagonals of a parallelogram bisect each other, V is the midpoint of $\overline{RT}$ and $\overline{US}$. Thus, $RT = 2(12) = 24$, and $US = 2(13) = 26$.

Try This... e. Find $m\angle D$, $m\angle E$, and $m\angle F$.

f. $AC = 14$, and $BD = 8$. Find AE, BE, CE, and DE.

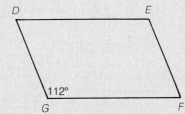

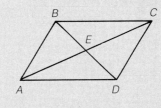

Discussion   Divide into groups. Each group take turns stating a "property" of a parallelogram (which may be true or false). The other group should decide if the "property" holds true for all parallelograms.

Exercises

Practice

Find the measures of the angles of each parallelogram.

1.

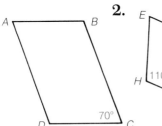

2.

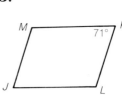

3.

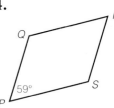

4.

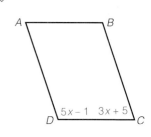

5.

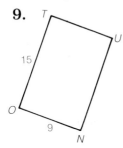

6.

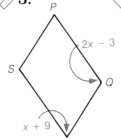

7.

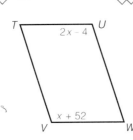

8.

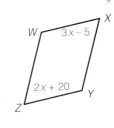

Find the perimeter of each parallelogram.

9.

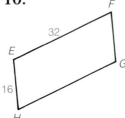

10.

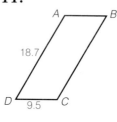

11.

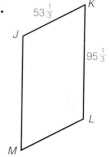

12.

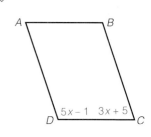

Find the lengths of the sides of each parallelogram.

13.

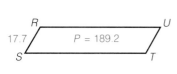

14.

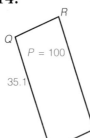

15.

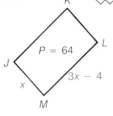

16.

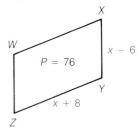

17. $AB = 14$, and $BD = 19$. Find the length of each diagonal.

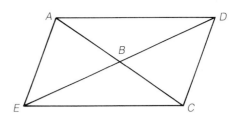

18. $EJ = 23$, and $GJ = 13$. Find the length of each diagonal.

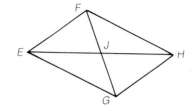

Extend and Apply

19. The perimeter of $\square ABCD$ is 60. $AD = 12$, $DB = 16$, and $EC = 11$. Find AB, EB, and AC.

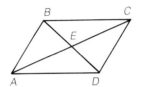

20. Can two parallelograms have congruent sides but different angle measures? Support your answer with drawings.

21. Name all pairs of congruent triangles in $\square ABCD$ above. Explain how you know.

22. Which is not necessarily true of $\square ABCD$?

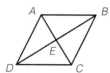

    (A) $AE = EC$ (B) $m\angle ABC = m\angle ADC$
    (C) $AD = BC$ (D) $m\angle DEC = m\angle BEC$

23. A builder claims that the top board of a partition is parallel to the bottom board, and that the vertical studs are also parallel. Does this mean the structure is rigid and cannot move? Why?

Use Mathematical Reasoning

24. Name three pairs of parallel segments. Explain how you know.

25. Name three pairs of congruent triangles. Explain how you know.

26. Name three pairs of congruent segments (other than the sides). Explain how you know.

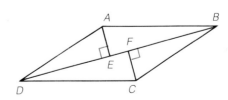

$ABCD$ is a parallelogram.

Mixed Review

For Exercises 27–31, determine whether each statement is true or false.

27. If two lines do not intersect, then they are parallel.

28. The acute angles of a right triangle are congruent.

29. Every quadrilateral has 4 diagonals.

30. The sum of the measures of the angles of a quadrilateral is 380°.

31. In a plane, if line ℓ intersects line m, and line m and line n are parallel, then line ℓ and line n are parallel.

32. In $\triangle HIJ$ and $\triangle LMN$, if $IJ = MN$, $HJ = LN$, and $HI > LM$, then what can you conclude?

33. In $\triangle PQR$ and $\triangle STU$, if $QR = TU$, $PR = SU$, and $PQ = ST$, then what can you conclude?

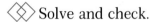 Solve and check.

34. $2(x + 20) = 100$ **35.** $3(r - 10) = 180$ **36.** $2(z - 1) = 27$

Visualization

Try to solve these problems *without* using paper.

1. A piece of paper is folded once horizontally and once vertically.

One hole is punched completely through the folded paper.

Then the paper is unfolded and all of the holes are connected by segments. What shape will occur?

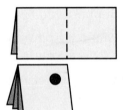

2. A square piece of paper is folded diagonally and then diagonally again.

One hole is punched through the folded paper.

Then the paper is unfolded and all of the holes are connected by segments. What shape will occur?

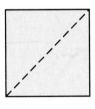

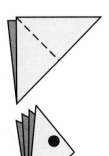

Computer Graphics

More and more often, people are using personal computers to create art that is both illustrative and technically accurate. Many computer "paint" programs have options for creating quadrilaterals and other common shapes quickly and for transforming figures. These programs can also show reflections of your drawings over lines of symmetry.

Here are some shapes and patterns that can be drawn by computer software:

You can also create your own patterns or drawings.

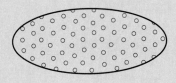  

When lines of *brush symmetry* are selected in a paint program and a figure is drawn, images over each line of symmetry are drawn by the computer.

Figure drawn: *Brush symmetry lines chosen:*   *Result:*

Anything drawn in one region will be mirrored in the other regions.

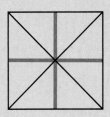

Only one shape has been drawn by the computer user. Determine the symmetry.

1. F

2.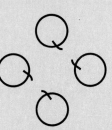

3.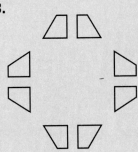

8-3 Determining Parallelograms

Objective: Tell whether a given quadrilateral is a parallelogram.

Explore

Use Geo D-Stix, construction toys, straws, or a ruler and a protractor to build or draw the following quadrilaterals:

1. *ABCD* with sides $\overline{AB}$ and $\overline{CD}$ one length, and $\overline{BC}$ and $\overline{DA}$ a different length.

2. *EFGH* with $\overline{EF} \parallel \overline{GH}$, and $\overline{EF}$ and $\overline{GH}$ the same length.

3. *MNOP* with $m\angle M = m\angle O$, and $m\angle N = m\angle P$.

4. Draw two segments that bisect each other, then connect the endpoints of these segments to form a quadrilateral.

Discuss Which of the quadrilaterals above are parallelograms? What are the conditions that determine a parallelogram?

We know from the previous lesson that if quadrilateral *ABCD* is a parallelogram, then

opposite angles are congruent.

opposite sides are congruent.

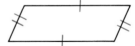

diagonals bisect each other.

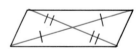

a diagonal determines two congruent triangles.

As you discovered above, the converses of several of these statements are also true.

Theorem 8.7 If the opposite angles of a quadrilateral are congruent, then the quadrilateral is a parallelogram. If ∠A ≅ ∠C, and ∠B ≅ ∠D, then *ABCD* is a parallelogram.

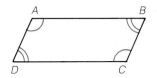

Theorem 8.8 If the opposite sides of a quadrilateral are congruent, then the quadrilateral is a parallelogram. If $\overline{AB} ≅ \overline{DC}$, and $\overline{AD} ≅ \overline{BC}$, then *ABCD* is a parallelogram.

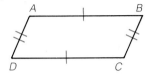

Theorem 8.9 If the diagonals of a quadrilateral bisect each other, then the quadrilateral is a parallelogram. If $\overline{AE} ≅ \overline{EC}$, and $\overline{BE} ≅ \overline{ED}$, then *ABCD* is a parallelogram.

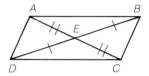

Theorem 8.10 If two sides of a quadrilateral are both parallel and congruent, then the quadrilateral is a parallelogram. If $\overline{AB} \parallel \overline{DC}$, and $\overline{AB} ≅ \overline{DC}$, then *ABCD* is a parallelogram.

Examples Is the figure a parallelogram? Why or why not?

1.

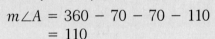

$m∠A = 360 - 70 - 70 - 110$
$\quad = 110$

Since opposite angles are congruent, *ABCD* is a parallelogram by Theorem 8.7.

2.

Because ∠D and ∠C are supplementary angles, $\overline{AD} \parallel \overline{BC}$ by Theorem 7.4. Also, $AD = BC$. Thus, by Theorem 8.10, *ABCD* is a parallelogram.

3.

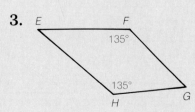

EFGH is not necessarily a parallelogram, because all we know is that one pair of opposite angles are congruent.

Example 4. Diagonal $\overline{AC}$ determines two congruent triangles.

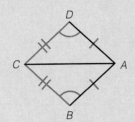

Although $\triangle ADC \cong \triangle ABC$, this quadrilateral is not necessarily a parallelogram. It is a kite.

Thus, the converse of Theorem 8.2 is not true. If a diagonal determines two congruent triangles, the quadrilateral is not necessarily a parallelogram.

Try This... Is the figure a parallelogram? Why or why not?

a.

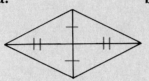

b.

c.

Discussion Which condition is sufficient to determine that quadrilateral $ABCD$ is a parallelogram: $\overline{AC}$ bisects $\overline{BD}$; $\angle A$ and $\angle B$ are supplementary, and $\angle C$ and $\angle D$ are supplementary; or the diagonals are congruent?

Exercises

Practice

Is the quadrilateral a parallelogram? Why or why not?

1.

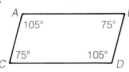

2.

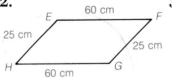

3.

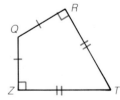

4.

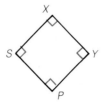

5.

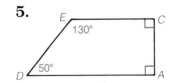

6.

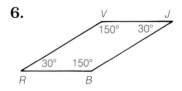

7.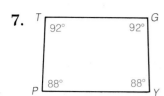

8.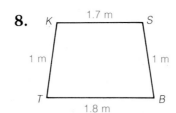

9.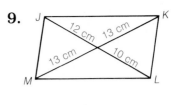

10.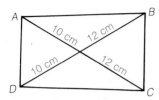

11.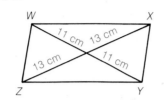

12.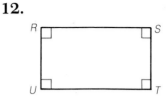

13.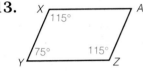

14.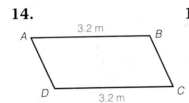

15.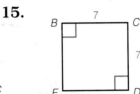

Extend and Apply

16. a. Draw a parallelogram with diagonals measuring 5 cm and 7 cm.

b. Draw a quadrilateral with diagonals measuring 5 cm and 7 cm that is *not* a parallelogram.

17. a. Draw a parallelogram with one pair of opposite angles measuring 45°.

b. Draw a quadrilateral with one pair of opposite angles measuring 45° that is *not* a parallelogram.

18. Parallel rulers are often used in navigation. The rulers pivot at points A, B, C, and D. Explain how the parallel ruler works.

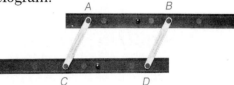

19. For quadrilateral $ACDB$ to be a parallelogram, what must be true?

(A) $\triangle ABC \cong \triangle DBC$ (B) $\triangle ABC \cong \triangle DCB$

(C) $\triangle ABC \cong \triangle BCD$ (D) $\triangle ABC \cong \triangle BDC$

Use Mathematical Reasoning

20. Is the statement "If $\triangle ZWX \cong \triangle YXW$, then $WXYZ$ is a parallelogram" true? If not, can you find a counterexample?

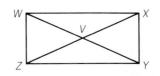

21. Explain how the design of these pliers guarantees that the jaws will always be parallel.

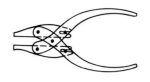

22. *ABCD* is a parallelogram, $\overline{AD} \cong \overline{DE}$, and $\overline{BC} \cong \overline{BF}$. What can you conclude about *AFCE*? Explain how you know.

23. Draw a parallelogram *ABCD*. Let *E* and *F* be the midpoints of $\overline{AB}$ and $\overline{CD}$. What can you conclude about *AEFD*? Explain how you know.

24. Draw □*ABCD* and diagonal $\overline{BD}$. Bisect ∠*A* and ∠*C*, with the bisectors meeting $\overline{DB}$ at points *E* and *F*, respectively. Draw $\overline{AF}$ and $\overline{CE}$. What can you conclude about quadrilateral *AFCE*? Explain how you know.

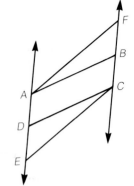

Mixed Review

Fill in the blanks.

25. A quadrilateral with exactly 2 pairs of congruent consecutive sides is a _____ .

26. A triangle with exactly 2 congruent sides is a(n) _____ triangle.

27. Two lines that are noncoplanar are called _____ .

28. Find all angle measures in the figure at right.

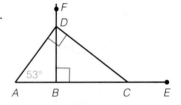

◇ Graph on a number line.

29. $4 > x > \frac{3}{2}$

Enrichment

Copy this Chinese tangram puzzle and cut it into seven pieces as shown. Now arrange the seven pieces to form a rectangle. See if you can form a parallelogram and a trapezoid, too.

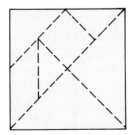

8-4 Special Parallelograms

Objective: Identify properties of rhombuses, rectangles, and squares, and identify rhombuses, rectangles, and squares by their properties.

Explore

1. Draw several rhombuses and several rectangles on a piece of paper. Then draw their diagonals.

2. Compare the lengths of each pair of diagonals.

3. Measure the angles where the diagonals intersect.

4. Draw several pairs of *perpendicular* segments that bisect each other. Connect the endpoints and describe the quadrilaterals formed.

5. Draw several pairs of *congruent* segments that bisect each other. Connect the endpoints and describe the quadrilaterals formed.

Discuss What special properties did you observe about the diagonals of a rhombus?
What special properties did you observe about the diagonals of a rectangle?

Rhombuses, rectangles, and squares are special kinds of parallelograms. Thus they have two pairs of parallel sides, along with the other properties shown in the figures.

Here are some additional properties involving diagonals:

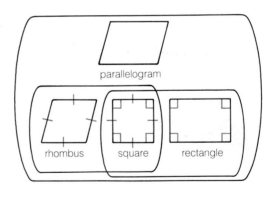

Theorem 8.11 A parallelogram is a rhombus if its diagonals are perpendicular.

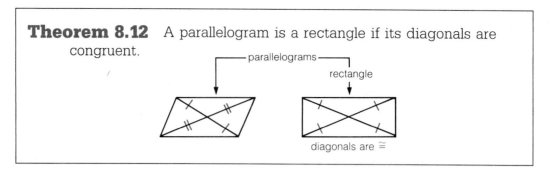

Theorem 8.12 A parallelogram is a rectangle if its diagonals are congruent.

A square is **both** a rhombus and a rectangle. Hence, its diagonals bisect each other and are perpendicular and congruent.

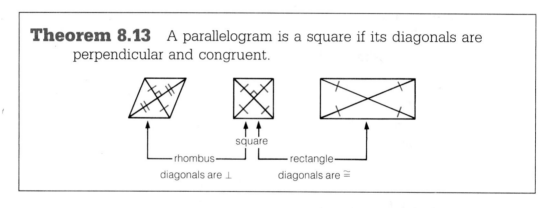

Theorem 8.13 A parallelogram is a square if its diagonals are perpendicular and congruent.

A quadrilateral can often be determined by its diagonals alone.

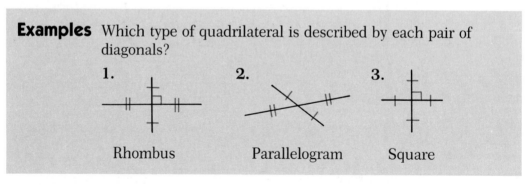

Examples Which type of quadrilateral is described by each pair of diagonals?

1.   2.   3.

Rhombus   Parallelogram   Square

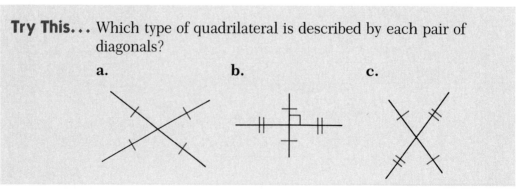

Try This... Which type of quadrilateral is described by each pair of diagonals?

a.   b.   c.

Examples Consider rhombus *ABCD*.

4. Which triangle is congruent to
   △*AMB*? Why? Since the diagonals
   bisect each other,

   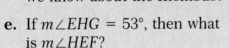

   $\overline{AM} \cong \overline{MC}$, and $\overline{DM} \cong \overline{MB}$.

   $\angle 1 \cong \angle 2 \cong \angle 3 \cong \angle 4$
   (right angles).

   Hence, by SAS,
   △*AMB* ≅ △*CMD* ≅ △*CMB* ≅ △*AMD*.
   Because the sides of a rhombus are congruent,
   the triangles are also congruent by SSS.

5. If *AM* + *DM* = 17, then what is *AC* + *BD*?
   Since the diagonals bisect each other,

   $$2(AM + DM) = AC + BD$$
   $$= 2 \cdot 17 = 34.$$

6. If $\overline{AC} \cong \overline{BD}$, then what do we know about the
   rhombus? If the diagonals are congruent, then
   by Theorem 8.13, the rhombus is a square.

Try This... Consider rhombus *EFGH*.

d. If $m\angle HEF = 90°$, then what do
   we know about the rhombus?

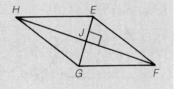

e. If $m\angle EHG = 53°$, then what
   is $m\angle HEF$?

f. Name all of the right triangles.

Examples Consider rectangle *RSTU*.

7. What type of triangle is △*RVS*?
   $\overline{RT} \cong \overline{SU}$, and *V* is the midpoint
   of $\overline{RT}$ and $\overline{SU}$. Thus, $\overline{RV} \cong \overline{VS}$,
   and △*RVS* is isosceles.

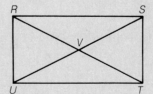

8. If *RV* = 4, then what is *SU*?
   If *RV* = 4, then *RT* = *SU* = 8.

9. If $\overline{RS} \cong \overline{ST}$, then what do we know about the rectangle?
   If $\overline{RS} \cong \overline{ST}$, then all four sides are congruent, and *RSTU*
   is a square.

Try This... Consider rectangle *JKLM*.

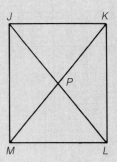

g. What type of triangle is △*JML*?

h. If *m*∠*JPK* = 90°, then what do we know about the rectangle?

i. If *MP* + *PL* = 14, then what is *JL*?

Discussion How are the properties of congruence and perpendicularity used to classify special quadrilaterals?

Exercises

Practice

Which type of quadrilateral is described by each pair of diagonals?

1.

2.

3.

4.

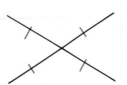

5.

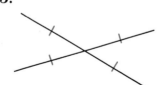

6.

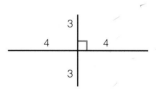

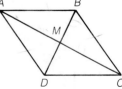

Use rhombus *ABCD* for Exercises 7–14.

7. If *BM* = 6, then what is *DM*?

8. If *AC* = 13, then what is *AM*?

9. If *AD* = 10, then what is *BC*?

10. If *BC* = 12, then what is *DC*?

11. What triangle is congruent to △*ADC*?

12. What triangle is congruent to △*ADB*?

13. If the rhombus is a square, then what is *m*∠*DAB*?

14. If the rhombus is a square, then what segment has length *AC*?

Use rectangle *PQRS* for Exercises 15–22.

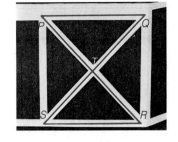

15. If *PR* = 12, then what is *SQ*?

16. If *SQ* = 13, then what is *PR*?

17. Name three other triangles congruent to △*PSR*.

18. What triangle is congruent to △*PQT*?

19. △*PTS*, △*PTQ*, △*QTR*, and △*STR* are what type of triangles?

20. Name four right triangles.

21. If the rectangle is a square, then what is *m*∠*PTQ*?

22. If the rectangle is a square, then how are $\overline{PS}$ and $\overline{PQ}$ related?

Extend and Apply

23. Name the properties of a square that are not properties of every parallelogram.

24. Name the properties of a rectangle that are not properties of every parallelogram.

25. Name the properties of a rhombus that are not properties of every parallelogram.

26. Ayana had a picture frame constructed that was supposed to be rectangular. Ayana checked that it was a rectangle by using only a tape measure. How did she check?

27. Which of these is false?

(A) *ADEB* is a square.

(B) *ADGC* is a rectangle.

(C) *BEHF* is a rhombus.

(D) *CFHG* is a parallelogram.

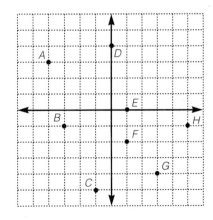

Use Mathematical Reasoning

28. Draw several rhombuses. For each, find the midpoint of each side. Connect the midpoints of consecutive sides to form a quadrilateral. What kind of quadrilateral did you get? Use inductive reasoning.

29. Draw ▱*ABCD* with *AB* > *AD*. Bisect ∠*A* and ∠*D*. Extend the bisectors to intersect $\overline{DC}$ and $\overline{AB}$ at *E* and *F*, respectively. What can you conclude about *AFED*? Explain how you know.

Mixed Review

What postulate (if any) shows the two triangles congruent?

30. $\overline{AB} \cong \overline{DE}$, $\angle A \cong \angle E$; $\triangle ABC \cong \triangle$_____ by _____.

31. $\overline{AB} \cong \overline{DE}$, $\overline{AC} \cong \overline{CE}$; $\triangle ABC \cong \triangle$_____ by _____.

32. $\overline{AC} \cong \overline{CD}$, $\angle A \cong \angle D$; $\triangle ABC \cong \triangle$_____ by _____.

33. $\overline{BC} \cong \overline{CD}$, $\angle B \cong \angle D$; $\triangle ABC \cong \triangle$_____ by _____.

34. $\overline{AE} \perp \overline{BD}$, $\overline{AB} \cong \overline{DE}$, $\overline{BC} \cong \overline{CE}$; $\triangle ABC \cong \triangle$_____ by _____.

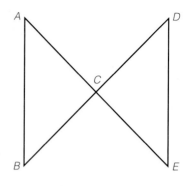

◇ Solve and check.

35. $18x - 27 = 15x$ **36.** $18x - 27 = 15$

Enrichment

Follow the directions below to construct a rhombus.

Step 1

Draw a segment, $\overline{AD}$. Use a compass setting of AD to draw all the arcs. Draw arc ①.

Step 2

Choose any point B on arc ① and draw arc ②.

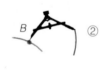

Step 3

Draw arc ③ from point D. Point C is the intersection of arcs ② and ③.

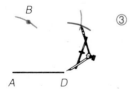

Step 4

Draw rhombus $ABCD$.

8-5 Trapezoids

Objective: Identify the parts and properties of trapezoids, and solve problems involving trapezoids.

One shape that is used for stability in everyday life is a special kind of trapezoid like the one shown in the vaulting horse at right.

Recall that a trapezoid is a quadrilateral that has exactly one pair of parallel sides. The **bases** of a trapezoid are the pair of parallel sides. The **legs** of a trapezoid are the other two sides. The **median** of a trapezoid is the segment determined by the midpoints of the legs.

Examples Name the bases, legs, and median of trapezoid $ABCD$. The points between the vertices are the midpoints of the sides.

1.

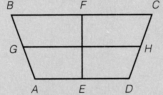

Bases: $\overline{AB}$ and $\overline{CD}$
Legs: $\overline{AD}$ and $\overline{BC}$
Median: $\overline{EF}$

2.

Bases: $\overline{BC}$ and $\overline{AD}$
Legs: $\overline{AB}$ and $\overline{CD}$
Median: $\overline{GH}$

3.

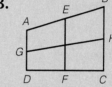

Bases: $\overline{AD}$ and $\overline{BC}$
Legs: $\overline{AB}$ and $\overline{DC}$
Median: $\overline{EF}$

Try This... Name the bases, legs, and median of trapezoid $ABCD$. The points between the vertices are the midpoints of the sides.

a.

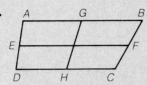

b.

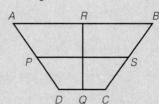

c.

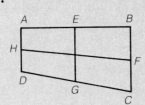

d.

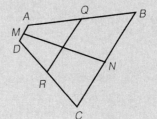

1. Draw several large trapezoids and their medians.
2. For each trapezoid, measure the median and the two bases. Compare the length of the median to the sum of the lengths of the bases.
3. Measure all the angles in the trapezoid. How do the angle measures tell you which segments are parallel?

Theorem 8.14 A median of a trapezoid is parallel to the bases and is half as long as the sum of the lengths of the two bases.

$\overline{MN} \parallel \overline{AB} \parallel \overline{DC}$
$MN = \frac{1}{2}(AB + DC)$

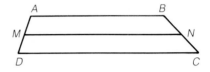

The length of the median of a trapezoid is sometimes called the "average width" of the trapezoid.

Examples 4. Find MN.

$MN = \frac{1}{2}(6 + 12) = 9$

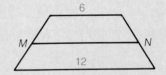

5. Find RS and $m\angle STU$.

$\frac{1}{2}(PQ + RS) = TU$
$\frac{1}{2}(18 + RS) = 22$
$18 + RS = 44$
$RS = 26$

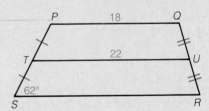

Because $\overline{TU} \parallel \overline{SR}$, and interior angles $\angle S$ and $\angle STU$ are supplementary, $m\angle STU = 180 - 62 = 118°$.

◇ 6. Find the lengths of the bases and the median.

$FC = \frac{1}{2}(ED + AB)$
$x + 2 = \frac{1}{2}(x + 2x + 3)$
$2(x + 2) = 3x + 3$
$2x + 4 = 3x + 3$
$1 = x$

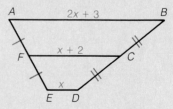

Substituting, we get $AB = 2(1) + 3 = 5$, $ED = 1$, and $FC = 1 + 2 = 3$. The lengths of the bases are 5 and 1, and the length of the median is 3.

Try This... **e.** Find *MN*.

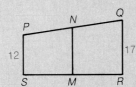

12, 17

f. Find *DE* and $m\angle GFE$.

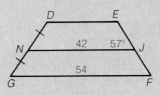

42, 57°, 54

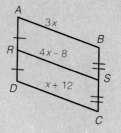

g. Find the lengths of the bases and the median.

3x, 4x − 8, x + 12

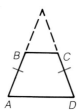

If we "cut off" part of an isosceles triangle parallel to its base, we would have an *isosceles trapezoid.*

Definition An **isosceles trapezoid** is a trapezoid with two congruent, non-parallel sides.

The following theorems describe properties of isosceles trapezoids:

Theorem 8.15 In an isosceles trapezoid, the angles of each pair of base angles are congruent.

Theorem 8.16 The diagonals of an isosceles trapezoid are congruent.

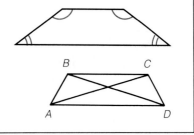

Examples Find the angle measures in each isosceles trapezoid.

7.

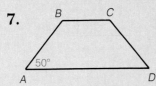

$m\angle D = 50°$. So, $m\angle B = m\angle C = 130°$, because they are supplementary to $\angle A$.

8.

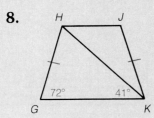

$m\angle HKJ = 72 - 41 = 31°$.
$m\angle J = 180 - 72 = 108°$.
$m\angle GHK = 180 - 72 - 41 = 67°$.
$m\angle JHK = m\angle J - 67$
$\qquad = 108 - 67 = 41°$.

Try This... Find the angle measures in each isosceles trapezoid.

h.

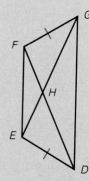

i.

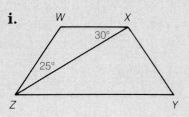

Examples

9. Name all pairs of congruent triangles. Which postulate explains the congruence?

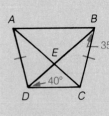

$\overline{ED} \cong \overline{FG}$, $\angle DEF \cong \angle GFE$ by Theorem 8.15, $\overline{DF} \cong \overline{EG}$ by Theorem 8.16. Since $\overline{EF} \cong \overline{EF}$. $\triangle DEF \cong \triangle GFE$ by SAS or SSS.

Similarly, $\triangle DEG \cong \triangle GFD$ by SAS or SSS.

$\angle EHD$ and $\angle FHG$ are vertical angles. Since $\angle EDF \cong \angle FGE$ as corresponding parts of congruent triangles, $\angle EDH \cong \angle FGH$. So, $\triangle EDH \cong \triangle FGH$ by AAS.

10. Find the missing angle measures.

$m\angle ACD = m\angle BDC = 40°$, so
$m\angle DEC = 180 - 2(40) = 100°$.
$m\angle AEB = m\angle DEC = 100°$,
$m\angle AED = m\angle BEC = 180 - 100 = 80°$.
$m\angle DAC = m\angle DBC = 35°$.
$m\angle ACB = m\angle BDA = 180 - 80 - 35 = 65°$.
$m\angle ADC = 65 + 40 = 105°$, so
$m\angle DAB = 180 - 105 = 75°$.
$m\angle CAB = m\angle DBA = 75 - 35 = 40°$.

Try This... j. Name all pairs of congruent triangles.

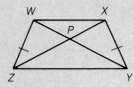

Which postulate explains the congruence?

k. Find the missing angle measures.

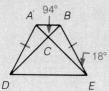

Examine the trapezoid shown.

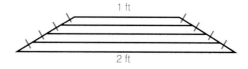

Exercises

Practice

Name the bases, legs, and median of each trapezoid. The points between vertices are midpoints.

1.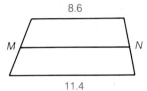

2.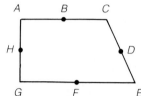

3.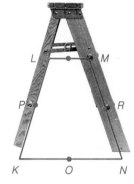

Find the indicated lengths and angle measure.

4. Find *MN*.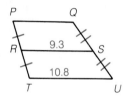

5. Find *PQ*.

6. Find *AB* and *m∠ACD*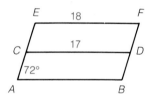

Find the lengths of the bases and the median.

7.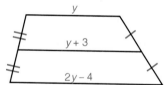
3x – 1
x + 3
x

8.
y
y + 3
2y – 4

9.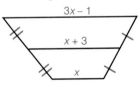
2x – 3
x + 4
x

10.
y + 10
2y – 1
y

Find the missing angle measures in these isosceles trapezoids. Name all pairs of congruent triangles in Exercises 15 and 16.

11.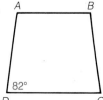
A   B
82°
D   C

12.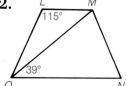
L   M
115°
39°
O   N

13.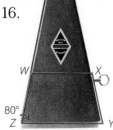
W   X
80°
Z   Y

14.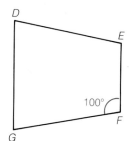

$100°$

15.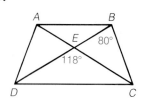

$80°$
$118°$

16.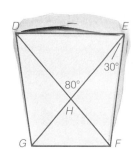

$30°$
$80°$

Extend and Apply

17. Which is not a property of an isosceles trapezoid?
  (A) The base angles are congruent.
  (B) The bases are parallel.
  (C) The diagonals bisect each other.
  (D) The legs are congruent.

18. The top brace on the irrigator is 1 ft, 9 in. wide. The bottom brace is 16 ft, 3 in. wide. How wide is the middle brace?

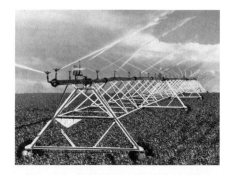

19. The bases of these tennis racket frames are 9 inches and 14 inches in length. What is the average width of the frame?

20. The angles at the top of the frame are each 115°. What are the measures of the angles at the bottom of the frame?

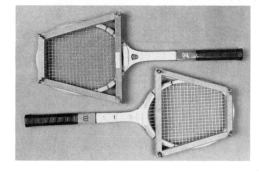

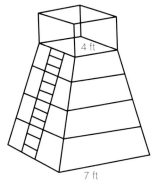

21. Coach Hall wants a coaching tower built for spring practice. The top platform is 4 ft by 4 ft and the base is 7 ft by 7 ft. What will be the lengths of the other three equally spaced braces on each side?

4 ft
7 ft

22. Jamal wants to put up a horizontal brace parallel to the base of his A-frame house. If he does not have a level, how can he use a measuring tape to be sure the brace is parallel to the base?

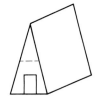

Use Mathematical Reasoning

23. Draw any isosceles trapezoid and measure the opposite angles. What can you conclude about these angles? Explain how you know.

24. Draw several trapezoids, their medians, and their diagonals. Where does the median intersect each of the diagonals? Use inductive reasoning.

Mixed Review

Describe each pair of angles as adjacent, vertical, interior, corresponding, alternate interior, or alternate exterior angles.

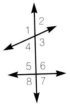

25. ∠1 and ∠2   **26.** ∠1 and ∠3   **27.** ∠1 and ∠4

28. ∠1 and ∠5   **29.** ∠1 and ∠7   **30.** ∠4 and ∠6

Solve and check.

31. $5x + 15 = 95$ **32.** $5(x + 15) = 95$

33. $5x - 15 = 95$ **34.** $5(x - 15) = 95$

Discover

1. Draw a triangle. Label it $\triangle ABC$.

2. Use a ruler to find the midpoints of $\overline{AC}$ and $\overline{AB}$. Label them E and D, respectively. Draw $\overline{ED}$.

3. Measure all of the angles in the figure.

4. Measure $\overline{ED}$ and $\overline{CB}$.

5. Repeat for other triangles.

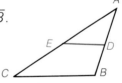

The segment $\overline{ED}$ is called a **midsegment** of the triangle. What observations did you make about the midsegment $\overline{ED}$?

Suppose one base of a trapezoid could be shortened until it had no length at all. What shape would be formed?
What would the length of its median be?

How is the midsegment of a triangle related to the median of a trapezoid?

Fill in the blanks to make the theorem below true.

Theorem 8.17   A midsegment connecting any two sides of a triangle is _____ to the third side and _____ as long as the third side.

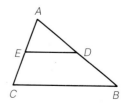

Collecting and Displaying Data

The five-member student council at La Entrada High School wanted to determine whether students prefer Hawks, Cardinals, Razorbacks, Bears, or Sharks as the school nickname.

First, each council member asked 20 different students for their preferences, recording the results in a table.

Nickname	Tally	Frequency
Hawks	||||	4
Cardinals	⫻ |||	8
Razorbacks	|||	3
Bears	||||	4
Sharks	|	1

After all five council members completed their surveys, the results were: Hawks, 21; Cardinals, 45; Razorbacks, 15; Bears, 11, and Sharks, 8.

To present this data to the student body, they prepared a bar graph by following these steps.

1. *Determine the greatest value for the data and the graph.* The greatest value is 45, so 50 is a good choice for the highest value of the graph.

2. *Select a scale to fit the data.* They decide that 5 is a good choice for the scale. They will need $\frac{50}{5}$, or 10, markings above 0.

3. *Draw and label the horizontal and vertical scales.*

4. *Draw and shade the bars of the graph.*

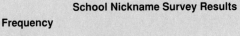
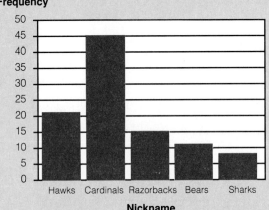

One of the shapes shown is the "golden rectangle," which you will learn about in Chapter 9. It is believed that people find the golden rectangle to be the most attractive.

(A)   (B)    (C)    (D)

Survey as many people as you can, to see which shape they prefer. Prepare a bar graph displaying your results.

Chapter 8 Review

Use the figure for Problems 1–3.

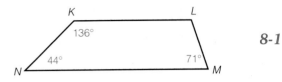

8-1

1. For quadrilateral *KLMN*, name one pair of opposite angles and one pair of consecutive angles.

2. Use the term that best classifies quadrilateral *KLMN*.

3. Find $m \angle L$.

4. The perimeter of $\square$ *DEFG* is 62. Find the measures of the angles and the lengths of the sides.

5. In $\square$ *JKLM*, *JN* = 2.3, and *KN* = 3. Find the lengths of the diagonals, $\overline{LJ}$ and $\overline{KM}$.

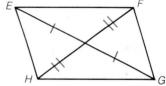

8-2

Is the figure a parallelogram? If so, why?

8-3

6.

7.

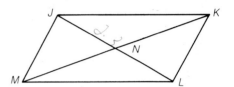

8.

9.

Which type of quadrilateral is described by each pair of diagonals?

8-4

10.

11.

Use the figure for Problems 12–13.

12. What triangles in rhombus *ABCD* are congruent to △*AEB*?

13. If *BE* = 4.5, then what is *DB* in *ABCD*?

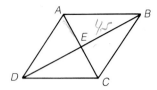

Use the figure for Problems 14–15.

14. If *MK* = 17 in rectangle *JKLM*, then what is *JN*?

15. Name two triangles in rectangle *JKLM* that are congruent to △*MKJ*.

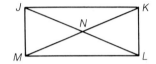

8-5 ◈ 16. Find the lengths of the bases and the median in trapezoid *ABCD*.

17. Find the missing angle measures in isosceles trapezoid *QRST*.

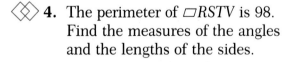

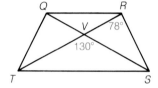

Chapter 8 Test

Use the figure for Problems 1–3.

1. For quadrilateral *ABCD*, name one pair of opposite sides and one pair of consecutive angles.

2. Use the term that best classifies quadrilateral *ABCD*.

3. Find *m∠D*.

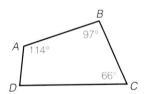

◈ 4. The perimeter of ▱*RSTV* is 98. Find the measures of the angles and the lengths of the sides.

5. In ▱*DEFG*, *GH* = 3.5, and *HF* = 4.8. Find the lengths of the diagonals, $\overline{DF}$ and $\overline{GE}$.

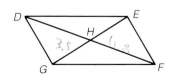

Is the figure a parallelogram? If so, why?

6.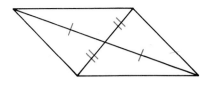

7.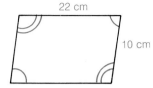

8.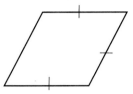

9.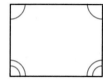

Which type of quadrilateral is described by each pair of diagonals?

10.

11.

Use the figure for Problems 12–13.

12. Name two triangles in rhombus *GHJK* that are congruent to △*GLK*.

13. In rhombus *GHJK*, *GH* = 5. What is *HJ*?

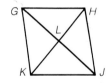

Use the figure for Problems 14–15.

14. In rectangle *ABCD*, if *AE* = 6.3, then what is *DE*?

15. If $m\angle AEB = 90°$, then what type of rectangle is *ABCD*?

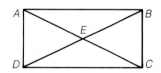

16. Find the lengths of the bases and the median in trapezoid *JKLM*.

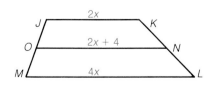

17. Find the missing angle measures in isosceles trapezoid *ABCD*.

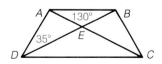

Cumulative Review Chapters 5–8

5-1   Classify each triangle by its sides.

1.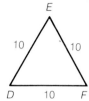

2.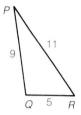

3.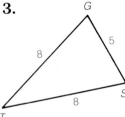

5-3   **4.** Name the corresponding parts of these congruent triangles.

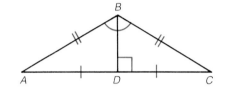

5-4   **5.** List four pieces of information shown in this drawing.

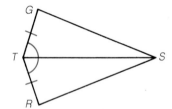

5-5   **6.** Which postulate should be used to show $\triangle STG \cong \triangle STR$?

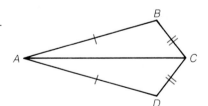

5-6   **7.** Complete the sentence.
$\triangle ABC \cong \triangle$_____ by the _____ Postulate.

5-7   **8.** Explain why $\angle D \cong \angle B$.

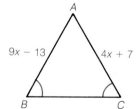

6-1   ◈ **9.** Find x.

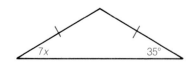

◈ **10.** Find AB.

11. List the angles in $\triangle WFP$ from largest to smallest.

6-3

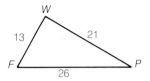

12. Can 4, 8, and 13 be the lengths of the sides of a triangle?

6-4

Identify each pair of angles as corresponding angles, interior angles, or alternate interior angles.

7-2

13. $\angle 1$ and $\angle 6$

14. $\angle 4$ and $\angle 5$

15. $\angle 3$ and $\angle 5$

16. Consider the figure in Exercises 13–15 with $\ell \parallel m$. Identify all congruent angles.

7-4

17. Find $m\angle C$ and $m\angle BAC$ in $\triangle ABC$.

7-5

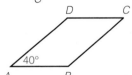

18. Find the measures of the angles of parallelogram $ABCD$.

8-2

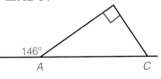

Tell why each figure is a parallelogram.

8-3

19.

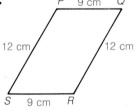

20.

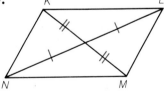

21. Which type of quadrilateral is described by this pair of diagonals?

8-4

22. Find the lengths of the bases and median in trapezoid $RVGT$.

8-5

9

Similarity and Scale Change

Architects build scale models of projects before the actual construction begins.
The scale of many architectural models is approximately 1 : 100.

9-1 Ratio

Objective: Find the ratio of one quantity to another.

Davis High School's orchestra includes seven violins and four cellos.

We say that the *ratio* of violins to cellos is 7 to 4. This ratio can be written as

7 to 4, 7:4, or $\frac{7}{4}$.

Definition   A comparison of two numbers by division is called a **ratio.** The ratio of a to b is the quotient, $\frac{a}{b}$, which may also be written as $a:b$.

Notice that for the ratio of a to b, a is the numerator and b is the denominator. Also, we often write ratios in *lowest terms*, which means that the only common factor of a and b is 1.

For example, in the measuring cup at the right, the ratio of oil to vinegar is 100 to 50, or in lowest terms, 2 to 1. In other words, there is twice as much oil as there is vinegar.

$$\frac{100}{50} = \frac{2}{1}, \text{ or } 2:1$$

The order in which the numbers of a ratio are written is important. The ratio of oil to vinegar, 2 to 1, is different from the ratio of vinegar to oil, 1 to 2.

The following Examples will give you practice in writing ratios and expressing them in lowest terms.

Examples Davis High School's Glee Club has 54 members. Their voice parts are shown at the right. Find these ratios and express them in lowest terms.

Sopranos	Altos
18	15
Tenors	**Basses**
12	9

1. Sopranos to altos
   The ratio is $\frac{18}{15}$, or $\frac{6}{5}$.
   We can also write the ratio as 6:5 or 6 to 5.

2. Tenors to basses
   The ratio is $\frac{12}{9}$, or $\frac{4}{3}$.

3. Basses to altos
   The ratio is $\frac{9}{15}$, or $\frac{3}{5}$.

4. Which two voice parts have a ratio of 2:1?
   The ratio of sopranos to basses is 18:9, or 2:1.

Try This... a. Find the ratio of altos to tenors.

b. Find the ratio of sopranos to tenors.

c. Find the ratio of tenors to sopranos.

d. Which two voice parts have a ratio of $\frac{5}{6}$?

When writing ratios, it is important that the numbers being compared are expressed in the same units.

Example 5. Find the ratio of 2 feet to 30 inches, and express the ratio in lowest terms.

Convert feet to inches: 2 feet = 24 inches.
The ratio of 2 feet to 30 inches is $\frac{24}{30}$, or $\frac{4}{5}$.

Try This... e. Find the ratio of 14 days to 4 weeks, and express the ratio in lowest terms.

Discussion When finding the ratio of 3 yards to 6 feet, Eric converted 3 yards to 9 feet. Sonia first converted 6 feet to 2 yards. Will they get the same ratio? Why?

Exercises

Practice

Express each ratio in lowest terms.

1. 21 to 3   **2.** 16:4   **3.** $\dfrac{4}{10}$

4. 80 to 30   **5.** 8:2   **6.** 18 to 3

7. 10 to 90   **8.** $\dfrac{15}{10}$ **9.** $\dfrac{8}{32}$

10. 100:25   **11.** $\dfrac{70}{80}$ **12.** 64:36

Write each ratio in lowest terms.

13. 75 meters to 10 meters   **14.** 28 years to 7 years

15. 18 kg to 76 kg   **16.** 2 weeks to 42 days

17. 80 cm to 2 m   **18.** 19 liters to 38 liters

19. 12 girls to 18 boys   **20.** 9 months to 3 years

21. 9 minutes to 2 hours   **22.** 45 seconds to 3 minutes

23. 28 cm to 50 cm   **24.** 54 km to 9 km

25. 96 apples to 168 apples   **26.** 125 peanuts to 50 peanuts

27. 700 seniors to 900 juniors   **28.** 2 pounds to 16 ounces

29. 30 lbs to 2 lbs   **30.** 10 yards to 15 feet

31. 100 m to 1 km   **32.** 20 hours on duty to 16 hours off duty

Extend and Apply

33. Which of the following is the ratio of 3 feet to 8 inches?

(A) $\dfrac{9}{2}$ (B) $\dfrac{3}{8}$ (C) $\dfrac{4}{1}$ (D) $\dfrac{1}{32}$

34. Find the ratio of △ to ○.

35. Find the ratio of ⬡ to ⬖.

36. Find the ratio of the perimeter of ▱*ABCD* to the perimeter of ▱*EFGH*.

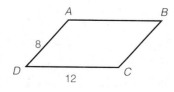

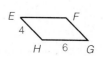

Omar won 14 of his 20 tennis matches. Use this information in Exercises 37–39.

37. What is the ratio of Omar's wins to losses?

38. What is the ratio of Omar's losses to wins?

39. What is the ratio of Omar's wins to the total number of matches he played?

Use Mathematical Reasoning

Ms. Kapisak said that for every 3 girls in her class, there are 2 boys. Use this information in Exercises 40–45.

40. What is the ratio of girls to boys in Ms. Kapisak's class?

41. What is the ratio of boys to girls in this class?

42. What is the least number of students that Ms. Kapisak could have in her class?

43. If the class has 20 students, how many are girls?

44. If the class has 30 students, how many are boys?

45. Explain why Ms. Kapisak could not have 23 students in her class.

46. A lawn fertilizer is composed of 22 parts nitrogen, 6 parts phosphorus, and 4 parts potassium. We can write the ratio of these chemicals as 22:6:4. Express this ratio in lowest terms.

Mixed Review

Find the indicated lengths and angle measure for each trapezoid.

47. Find *MN*.

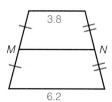

48. Find *AB*.

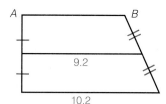

49. Find $m\angle S$.

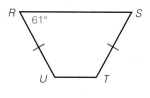

Find the lengths of the bases and median of each trapezoid.

50.

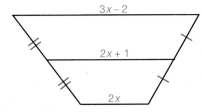

51.

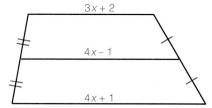

The Golden Ratio

The photo at the right shows a cross section of a chambered nautilus. Its spiral is related to a special figure known as the **golden rectangle.**

The golden rectangle is one in which the ratio of the length to the width is the **golden ratio,** $\frac{1 + \sqrt{5}}{2}$ to 1. The golden ratio is approximately 1.6 to 1.

Through the ages, people have found objects created with this ratio pleasing to the eye. Thus, many common items are made with this ratio in mind. Notice that the most common dimensions of picture frames — 3 × 5, 4 × 6, 5 × 8 — are all close to the golden ratio.

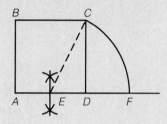

Here is one way to construct a golden rectangle.

1. Draw a square and label it *ABCD*.

2. Find the midpoint *E* of $\overline{AD}$. Place the point of a compass on *E* and, with setting *EC*, draw an arc intersecting $\overleftrightarrow{AD}$ at *F*. Construct a perpendicular at *F*, intersecting $\overleftrightarrow{BC}$ at *G*.

3. Quadrilateral *ABGF* is a golden rectangle. In fact, *FDCG* is also a golden rectangle.

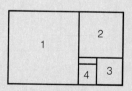

At this point, the following steps can be used to create a spiral like that of the chambered nautilus.

1. Divide *DCGF* into a square (on top) and another golden rectangle (on the bottom).

2. Continue this process, dividing the bottom rectangle as shown at the right.

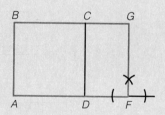

3. Place your compass point on the corners of the squares and, with the sides of each square as a compass setting, draw a circular arc.

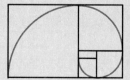

Problems

1. Construct a golden rectangle.

2. Construct a spiral as shown above.

9-2 Proportion and Scale

Objective: Solve problems involving proportions and scale.

Diana wants to design a patio that is twice as long as it is wide. The ratio of the length to the width of patio A is $\frac{24}{12}$. The ratio for patio B is $\frac{32}{16}$.

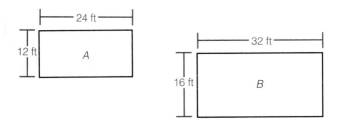

Both ratios can be expressed in lowest terms as $\frac{2}{1}$, so the ratios are equal. The equation $\frac{24}{12} = \frac{32}{16}$ is called a *proportion*.

Definition   A **proportion** is a statement that two ratios are equal. We say that the two ratios are **proportional.**

Explore

1. Use the numbers 1, 2, 3, and 6 to write as many different proportions as you can.

2. For the proportion $\frac{a}{b} = \frac{c}{d}$, the products ad and bc are sometimes called **cross products.** Find the cross products for the proportions in Step 1.

3. What do you notice about the cross products? Do you think this result is true for any proportion? Test your conjecture on some proportions with different numbers.

Theorem 9.1   If $\frac{a}{b} = \frac{c}{d}$, then $ad = bc$.
If $ad = bc$ with $b \neq 0$ and $d \neq 0$, then $\frac{a}{b} = \frac{c}{d}$.

Theorem 9.1 gives us a convenient way to check whether two ratios form a proportion.

The following Examples illustrate the use of Theorem 9.1.

Examples Tell whether each pair of ratios forms a proportion.

1. $\dfrac{7}{8}$ and $\dfrac{21}{24}$

   $\dfrac{7}{8} \stackrel{?}{=} \dfrac{21}{24}$

   This is a proportion because $7 \times 24 = 8 \times 21 = 168$.

2. $\dfrac{4}{5}$ and $\dfrac{8}{12}$

   $\dfrac{4}{5} \stackrel{?}{=} \dfrac{8}{12}$

   This is not a proportion because $4 \times 12 \neq 5 \times 8$.

Try This... Tell whether each pair of ratios forms a proportion.

a. $\dfrac{4}{5}$ and $\dfrac{20}{25}$ b. $\dfrac{75}{25}$ and $\dfrac{3}{2}$

c. $\dfrac{7}{8}$ and $\dfrac{6}{7}$ d. $\dfrac{15}{4}$ and $\dfrac{75}{20}$

We can use the cross product property of Theorem 9.1 to solve proportions.

Example 3. Solve the following proportion.

$$\dfrac{x}{72} = \dfrac{9}{12}$$

$x \cdot 12 = 72 \cdot 9$ Find the cross products.

$12x = 648$

$\dfrac{12x}{12} = \dfrac{648}{12}$ Divide both sides by 12.

$x = 54$

Try This... Solve each proportion.

e. $\dfrac{x}{8} = \dfrac{6}{4}$ f. $\dfrac{20}{7} = \dfrac{80}{x}$

We will see that solving proportions is useful in understanding maps and scale drawings.

One inch on the map at the right represents 8 miles on the ground. We say that the **scale** of the map is 1 inch = 8 miles.

Distances on the map and distances on the ground are proportional. This means that the ratio of two distances on the map will be *equal* to the ratio of two corresponding distances on the ground.

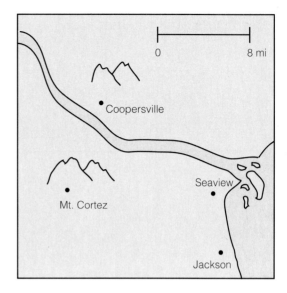

Examples **4.** How many miles are represented by 3.5 inches on the above map?

$$\frac{1 \text{ in.}}{3.5 \text{ in.}} = \frac{8 \text{ miles}}{x \text{ miles}}$$

$$1 \cdot x = 3.5 \cdot 8$$

$$x = 28$$

Thus, 3.5 inches on the map represent 28 miles.

5. Estimate the number of miles between Coopersville and Jackson.

The towns are about two inches apart on the map. This corresponds to about 16 miles.

Try This... g. How many miles are represented by 2.25 inches on the map?

h. How many inches on the map would be needed to represent 42 miles?

i. Estimate the number of miles between Mt. Cortez and Seaview.

Discussion   Can different proportions be used to solve the map problem in Example 4? How do the solutions compare to the original solution?

Exercises

Practice

Tell whether each pair of ratios forms a proportion.

1. $\dfrac{2}{7}$ and $\dfrac{3}{10}$

2. $\dfrac{9}{5}$ and $\dfrac{7}{4}$

3. $\dfrac{14}{8}$ and $\dfrac{42}{24}$

4. $\dfrac{3}{4}$ and $\dfrac{9}{12}$

5. $\dfrac{13}{3}$ and $\dfrac{39}{9}$

6. $\dfrac{10}{3}$ and $\dfrac{15}{8}$

7. $\dfrac{18}{4}$ and $\dfrac{9}{2}$

8. $\dfrac{5}{1}$ and $\dfrac{15}{3}$

Solve each proportion for x or y.

9. $\dfrac{x}{5} = \dfrac{3}{15}$

10. $\dfrac{y}{7} = \dfrac{3}{21}$

11. $\dfrac{32}{y} = \dfrac{8}{9}$

12. $\dfrac{45}{x} = \dfrac{15}{16}$

13. $\dfrac{16}{y} = \dfrac{1}{4}$

14. $\dfrac{25}{75} = \dfrac{1}{x}$

15. $\dfrac{5}{x} = \dfrac{10}{20}$

16. $\dfrac{5}{8} = \dfrac{10}{y}$

17. $\dfrac{12}{9} = \dfrac{x}{7}$

18. $\dfrac{16}{15} = \dfrac{y}{20}$

19. $\dfrac{24}{x} = \dfrac{4}{7}$

20. $\dfrac{18}{x} = \dfrac{3}{5}$

21. $\dfrac{125}{y} = \dfrac{35}{7}$

22. $\dfrac{100}{80} = \dfrac{25}{y}$

23. $\dfrac{11}{x} = \dfrac{2}{7.1}$

24. $\dfrac{10}{x} = \dfrac{5}{1.5}$

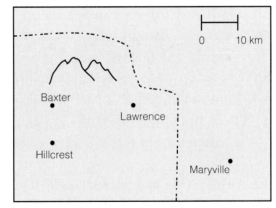

The scale of the map at the right is 1 cm = 10 km.

25. How many kilometers are represented by 8 cm?

26. How many kilometers are represented by 6.8 cm?

27. How many centimeters on the map would be needed to represent 27 kilometers?

Estimate the number of kilometers between each pair of towns on the above map.

28. Baxter and Hillcrest

29. Lawrence and Maryville

30. Baxter and Lawrence

Extend and Apply

31. On Karla's globe, the scale is 1 inch = approximately 660 miles. Karla determines that the distance from Los Angeles to Honolulu is $3\frac{3}{4}$ inches on the globe. What is the actual distance from Los Angeles to Honolulu?

32. The floor plan for the Hsus' living room has a scale of 1 inch = 4 feet. The actual dimensions of the living room are 13 feet by 23 feet. What are the dimensions of the floor plan?

33. The scale of a model airplane is $\frac{1}{4}$ in. = 1 ft. The wingspan of the airplane is 50 ft. Which of the following is the wingspan of the model?

(A) $12\frac{1}{2}$ ft   (B) 200 in.   (C) $12\frac{1}{2}$ in.   (D) 200 ft

Use Mathematical Reasoning

This map is drawn to scale. However, much of it is unreadable, including the scale and most of the distances. Using the information shown, first estimate the distances asked for in Exercises 34–38. Then use a ruler to check your estimates.

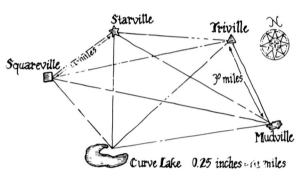

34. How far is it from Mudville to Starville?

35. How far is it from Squareville to Curve Lake?

36. How far is it from Starville to Triville?

37. How far is it from Starville to Squareville?

38. How far is it from Curve Lake to Mudville?

39. The measures of two complementary angles have a ratio of $\frac{2}{3}$. Find the measure of each angle.

40. The measures of two supplementary angles have a ratio of 4:5. Find the measure of each angle.

41. The ratio of the measures of two consecutive angles of a parallelogram is 13:7. Find the measure of each angle.

42. △ABC is an isosceles triangle. The ratio of y to x is 5 to 2. Find the measures of the angles of the triangle.

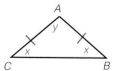

Mixed Review

Express each ratio in lowest terms.

43. 35 to 20   **44.** 11:77   **45.** $\dfrac{8}{20}$

Write each ratio in lowest terms.

46. 50 cm to 1 m   **47.** 1 day to 1 week   **48.** 2 ft to 2 in.

◇ Solve and check.

49. $6 = 3(y - 7)$ **50.** $4(x + 5) = 40$ **51.** $12(6 - z) = 24$

Figures with the same shape, regardless of their sizes, are called **similar**. To decide if one figure is similar to another, ask yourself whether one appears to be an enlargement or a reduction of the other.

Samples Which figure on the right appears to be similar to the figure on the left?

1.  (A) (B) (C)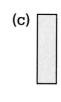

Figure B is similar, since it is a reduction of the figure on the left.

2. (A) (B) (C)

Figure C is similar, since it is an enlargement of the figure on the left.

Problems

Which figure on the right appears to be similar to the figure on the left?

1. 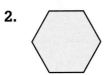 (A) 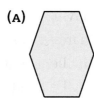 (B) (C)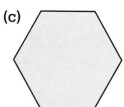

2. 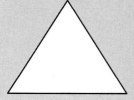 (A)  (B) (C)

9-3 Similarity and Size Transformations

Objective: Determine whether figures are similar, and use scale factors to draw similar figures.

Architects create "scaled down" models of buildings, like the one at the right. It is $\frac{1}{100}$ the size of the actual building. It has the *same shape* as the actual building, but it is not the same size.

Recall from the previous Visualization activity that figures with the same shape, regardless of their sizes, are said to be *similar*. The following definition gives us one way of deciding when two figures are similar:

Definition   Two figures are **similar** if their vertices can be matched so that corresponding sides are proportional and corresponding angles are congruent.

The ratio of any two corresponding sides gives the **scale factor** that transforms one figure into its image.

$ABCD$ is similar to $A'B'C'D'$, since
$\angle A \cong \angle A'$, $\angle B \cong \angle B'$, $\angle C \cong \angle C'$,
$\angle D \cong \angle D'$, and
$$\frac{A'B'}{AB} = \frac{B'C'}{BC} = \frac{C'D'}{CD} = \frac{D'A'}{DA} = 2.$$

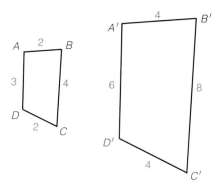

The scale factor that transforms $ABCD$ into $A'B'C'D'$ is 2.

To say that $ABCD$ and $A'B'C'D'$ are similar, we write $ABCD \sim A'B'C'D'$.

Similar figures can be thought of as **size transformations** of each other. Notice that the scale factor tells us whether the image figure is an enlargement or a reduction of the original figure. A scale factor greater than 1 means the image is an enlargement; a scale factor less than 1 means the image is a reduction.

Throughout this lesson, angles that appear to be congruent
may be assumed to be so.

Examples Determine whether the figures are similar. If so, what is the
scale factor that transforms the figure on the left to the
figure on the right?

1.

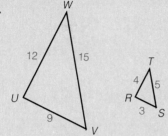

$\triangle UVW \sim \triangle RST$, since
$\angle U \cong \angle R$, $\angle V \cong \angle S$,
$\angle W \cong \angle T$, and
$$\frac{RS}{UV} = \frac{ST}{VW} = \frac{RT}{UW} = \frac{1}{3}.$$
The scale factor is $\frac{1}{3}$.

2.

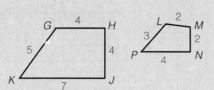

$GHJK$ is not similar to
$LMNP$, since
$$\frac{LM}{GH} \neq \frac{NP}{JK}.$$

Try This... Determine whether the figures are similar. If so, what is the
scale factor that transforms the figure on the left to the
figure on the right?

a.

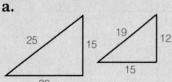

b.

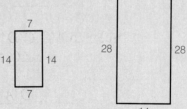

c.

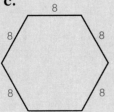

d.

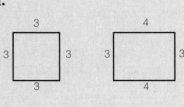

The next Example shows how grid paper can be used to
transform a figure given the scale factor of the
transformation.

Example 3. Use grid paper to draw a figure similar to *ABCD*, with a scale factor of 2.

Each side of *ABCD* is 3 units long. Since the scale factor is greater than 1, the required figure will be an enlargement. Each of its sides will be 2 times as long as that of *ABCD*. We draw a square with sides 6 units long.

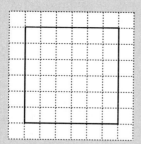

Try This... e. Use grid paper to draw a figure similar to *EFGH*, with a scale factor of $\frac{1}{3}$.

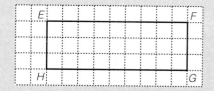

Discussion What can you say about two similar figures if the scale factor is 1?

Exercises

Practice

Determine whether the figures are similar. If so, what is the scale factor that transforms the figure on the left to the figure on the right?

(4

1.

```
    5
 ┌─────┐
5│     │5
 └─────┘
    5
```

```
        15
 ┌──────────────┐
 │              │
15│              │15
 │              │
 └──────────────┘
        15
```

2.

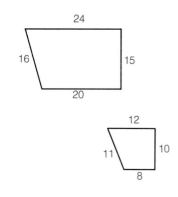

Determine whether the figures are similar. If so, what is the scale factor that transforms the figure on the left to the figure on the right?

3.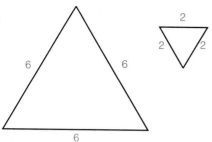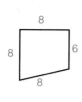

4.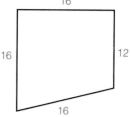

Use grid paper to draw a figure that is similar to the given figure, with the indicated scale factor.

5. Scale factor of 2

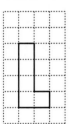

6. Scale factor of $\frac{1}{2}$

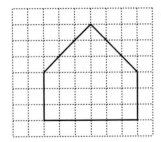

7. Scale factor of 4

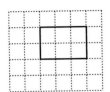

8. Scale factor of $\frac{1}{3}$

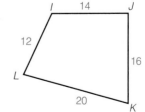

Extend and Apply

9. *EFGH ~ IJKL*. Which of the following scale factors transforms *EFGH* to its image?

(A) 2   (B) 4
(C) $\frac{1}{2}$ (D) $\frac{1}{3}$

10. A photograph is 10 cm long and 8 cm wide. A photographer enlarges the photo with a scale factor of 2.5. Find the dimensions of the enlarged photo.

Use Mathematical Reasoning

11. Suppose $\triangle ABC \sim \triangle A'B'C'$ and the scale factor that transforms $\triangle ABC$ to $\triangle A'B'C'$ is 4. What is the scale factor that transforms $\triangle A'B'C'$ to $\triangle ABC$?

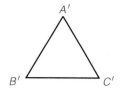

Mixed Review

Tell whether each pair of ratios forms a proportion.

12. $\dfrac{9}{12}$ and $\dfrac{12}{16}$

13. $\dfrac{7}{10}$ and $\dfrac{20}{14}$

14. $\dfrac{6}{9}$ and $\dfrac{14}{21}$

15. $\dfrac{7}{2}$ and $\dfrac{8}{3}$

16. $\dfrac{5}{3}$ and $\dfrac{12}{7}$

17. $\dfrac{12}{4}$ and $\dfrac{18}{6}$

Solve each proportion for x.

18. $\dfrac{5}{x} = \dfrac{15}{12}$

19. $\dfrac{6}{9} = \dfrac{x}{15}$

20. $\dfrac{4}{5} = \dfrac{16}{x}$

21. $\dfrac{x}{8} = \dfrac{25}{10}$

Find each absolute value.

22. $|-7|$ 23. $|2.6|$ 24. $|0|$ 25. $|-3.8|$

Enrichment

A pantograph is an instrument used to reduce or enlarge a drawing to scale. You can construct your own pantograph by cutting four strips of heavy cardboard and attaching them with paper fasteners as shown. $\overline{AT}$ is parallel to $\overline{BU}$, $\overline{SB}$ is parallel to $\overline{UT}$, and $\overline{UC}$ is longer than $\overline{TU}$. Place a paper fastener through A and securely tape the ends of the paper fastener to an even surface. Place a pointer through B and a pencil through C. As you trace a figure with the pointer, the pencil enlarges the figure to scale. To reduce the figure, simply interchange the pointer and the pencil.

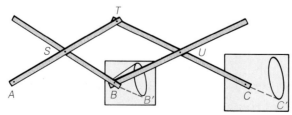

9-4 Similar Triangles

Objective: Find lengths of sides of similar triangles.

Explore

The following activity may be done using computer software.

1. Draw three rays with a common endpoint as shown. Select one point on each ray and connect the points to form $\triangle ABC$.

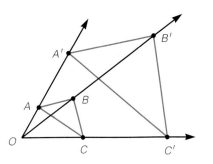

2. Now locate points A', B', and C' on the rays so that $3OA = OA'$, $3OB = OB'$, and $3OC = OC'$. Connect A', B', and C' to form $\triangle A'B'C'$.

3. Measure and compare corresponding angles of both triangles.

4. Measure and compare corresponding sides of both triangles. Are corresponding sides proportional?

Discuss What can you conclude about the triangles? What is the scale factor of the above transformation?

We will now consider similarity as it applies to triangles. Two triangles are *similar* whenever their vertices can be matched so that corresponding angles are congruent and the lengths of corresponding sides are proportional.

As we did for polygons, we write $\triangle ABC \sim \triangle DEF$ to say that $\triangle ABC$ and $\triangle DEF$ are similar. We agree that this symbol also tells us the way in which the vertices are matched.

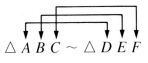

$\triangle\ A\ B\ C\ \sim\ \triangle\ D\ E\ F$

Thus, $\triangle ABC \sim \triangle DEF$ means

$\angle A \cong \angle D$
$\angle B \cong \angle E$ and $\dfrac{AB}{DE} = \dfrac{AC}{DF} = \dfrac{BC}{EF}$
$\angle C \cong \angle F$

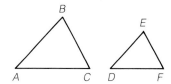

Let us agree to talk of "proportional sides" instead of "sides whose lengths are proportional." We understand that it is lengths that are proportional, not the sets of points.

Example   **1.** Suppose △PQR ~ △GHK. Which angles are congruent? Which sides are proportional?

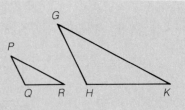

Angles:

∠P ≅ ∠G

∠Q ≅ ∠H

∠R ≅ ∠K

Sides:

$$\frac{PQ}{GH} = \frac{PR}{GK} = \frac{QR}{HK}$$

Try This... a. Suppose △JKL ~ △ABC. Which angles are congruent? Which sides are proportional?

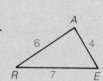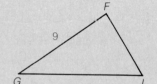

We can use porportions and the fact that triangles are similar to find missing lengths.

Example   **2.** △RAE ~ △GFL. Find FL and GL.

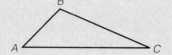

Because △RAE ~ △GFL, the corresponding sides are proportional. Thus,

$$\frac{6}{9} = \frac{4}{FL} \quad \text{and} \quad \frac{6}{9} = \frac{7}{GL}$$

We use the cross product property to solve the proportions.

6(FL) = 9 · 4	6(GL) = 9 · 7
6(FL) = 36	6(GL) = 63
FL = 6	GL = 10½

Try This... b. △WNE ~ △CBT. Find BT and CT.

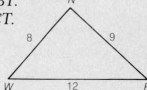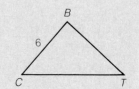

◇ **Example** 3. $\triangle JKH \sim \triangle WTR$. Find KH and TR.

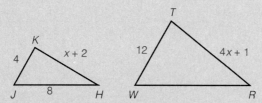

In $\triangle JKH$ and $\triangle WTR$,

$$\frac{JK}{WT} = \frac{KH}{TR}$$

$$\frac{4}{12} = \frac{x + 2}{4x + 1}$$

$$4(4x + 1) = 12(x + 2)$$

$$16x + 4 = 12x + 24$$

$$4x = 20$$

$$x = 5$$

Since $x = 5$, $KH = 7$ and $TR = 21$.

◇ **Try This...** c. $\triangle LUK \sim \triangle ZEN$.
Find UK and EN.

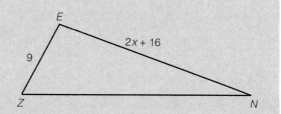

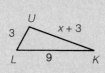

Discussion Can a different proportion be used to solve the problem in Example 3? If so, does it give the same answer?

Exercises

Practice

In Exercises 1–3, each pair of triangles are similar. Which angles are congruent? Which sides are proportional?

1.

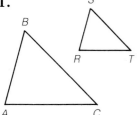

2.

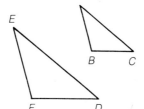

3.

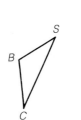

In Exercises 4–6, each pair of triangles are similar. Which angles are congruent? Which sides are proportional?

4.

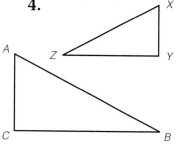

5.

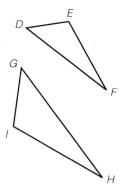

6.

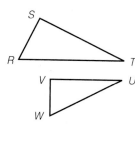

Find the missing lengths.

7. △ABC ~ △PQR   **8.** △MAC ~ △GET   **9.** △RST ~ △WYZ

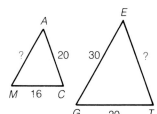
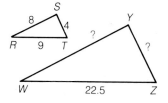

10. △JKL ~ △DEF   **11.** △ABC ~ △XYZ   **12.** △DEF ~ △RST

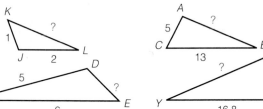
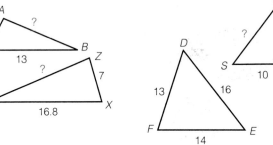

13. △DEF ~ △KGH ◇◇ **14.** △REG ~ △CUP. ◇◇ **15.** △MNP ~ △SQV.
                       Find RE and CU.   Find NP and QV.

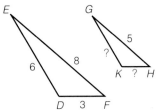
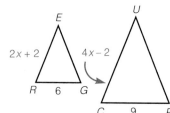
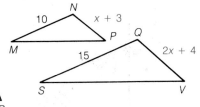

16. $\triangle RST \sim \triangle UVW$. Which of the following is VW?

(A) 4   (B) $\dfrac{5}{2}$ (C) $\dfrac{8}{5}$ (D) $\dfrac{5}{8}$

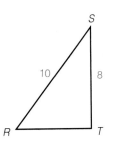

17. Bob drew a triangle with sides of length 4 cm, 6 cm, and 8 cm. When he photocopied it, the copier enlarged the triangle so that its shortest side was 10 cm. What were the lengths of the other two sides in the enlargement?

Use Mathematical Reasoning

Complete each statement with *always*, *sometimes*, or *never*.

18. Congruent triangles are _____ similar.

19. Similar triangles are _____ congruent.

20. If $\triangle BCD \sim \triangle EFG$, then $\angle B$ is _____ congruent to $\angle E$.

21. An equiangular triangle and a right triangle are _____ similar.

Mixed Review

Determine whether the figures are similar. If so, what is the scale factor that transforms the figure on the left to the figure on the right?

22.

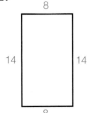

23.

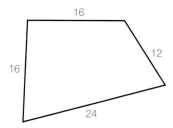

◈ Solve and check.

24. $5y + 6 = 8y - 3$ **25.** $6z - 9 = 1 + z$ **26.** $64 - 5x = 2x - 6$

Problem Solving

How many similar triangles are there in the figure at the right? (Hint: Look for a pattern.)

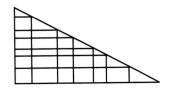

9-5 AA Similarity

Objective: Use the AA Similarity Theorem to identify pairs of similar triangles and to solve problems.

Explore

Work in pairs.

1. Each student should use a protractor and a straightedge to draw a triangle having a 50° angle and a 70° angle.

2. Label the triangle as shown, and measure $\overline{AB}$, $\overline{BC}$, and $\overline{AC}$.

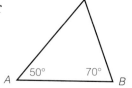

3. Calculate the ratio of the length of your side labeled $\overline{AB}$ to the length of your partner's side labeled $\overline{AB}$. Do the same for the sides labeled $\overline{BC}$ and $\overline{AC}$. What do you notice?

Discuss What can you conclude about two triangles that have congruent angles?

In the Explore, you may have discovered that if two triangles have congruent angles, then their corresponding sides are proportional. This means the triangles are similar. We state this as a postulate.

Postulate 16 **The AAA Similarity Postulate**
For any two triangles, if the corresponding angles are congruent, then the triangles are similar.

$\triangle ABC \sim \triangle PQR$

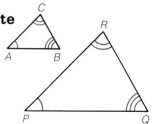

The theorem that follows from this postulate makes it even easier to determine when two triangles are similar.

Theorem 9.2 **The AA Similarity Theorem**
For any two triangles, if two pairs of corresponding angles are congruent, then the triangles are similar.

$\triangle PQR \sim \triangle ABC$

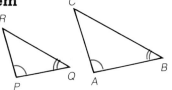

Examples Which pairs of triangles are similar by the AA Similarity Theorem?

1.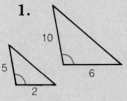

Not similar by
the AA Similarity
Theorem

2.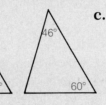

Similar

3.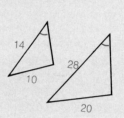

Similar

Try This... Which pairs of triangles are similar by the AA Similarity Theorem?

a.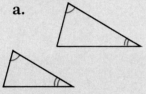

b.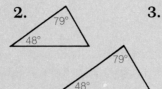

c.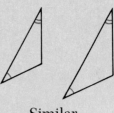

The AA Similarity Theorem can help us find the lengths of sides of triangles.

Example **4.** In $\triangle RST$, $\overline{GH}$ is parallel to $\overline{RS}$. Find ST.

$\overline{GH}$ is parallel to $\overline{RS}$, so $\angle SRT \cong \angle GHT$, since they are corresponding angles. Also, $\angle T \cong \angle T$. Thus, $\triangle RST \sim \triangle HGT$ by the AA Similarity Theorem. Hence,

$$\frac{RS}{HG} = \frac{ST}{GT}$$

$$\frac{18}{12} = \frac{ST}{26}$$

$$12 \cdot ST = 468$$

$$ST = 39$$

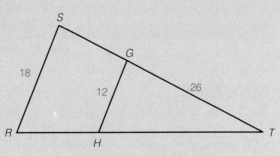

Try This... **d.** In $\triangle TAR$ and $\triangle MSR$, $\overline{TA} \parallel \overline{SM}$, $TA = 21$, $SM = 14$, and $TR = 18$. Find MR.

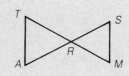

Example   **5.** A rod 3 m tall casts a shadow 5 m long. At the same time, the shadow of a tower is 110 m long. How tall is the tower?

The rod and the tower form right angles with the ground. The sun's rays form congruent angles with the rod and the tower. Thus, $\triangle ABC \sim \triangle PQR$, since $\angle PRQ \cong \angle ACB$, and $\angle RPQ \cong \angle CAB$ (AA Theorem). Hence,

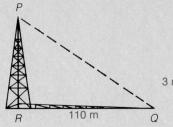

$$\frac{PR}{AC} = \frac{QR}{BC}$$
$$\frac{PR}{3} = \frac{110}{5}$$
$$5 \cdot PR = 3 \cdot 110$$
$$5 \cdot PR = 330$$
$$PR = 66$$

The tower is 66 m tall.

Try This...   **e.** The shadow of a tree is 25 m long. At the same time, a person who is 180 cm tall casts a shadow 240 cm long. How tall is the tree?

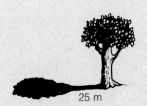

Discussion   Assume that the AAA Postulate is true. Use the fact that the angle measures of a triangle add up to 180° to give a convincing argument that the AA Similarity Theorem is true.

Exercises

Practice

Which pairs of triangles are similar by the AA Similarity Theorem?

1.

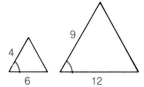

2.

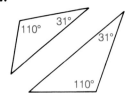

3.

Which pairs of triangles are similar by the AA Similarity Theorem?

4.

5.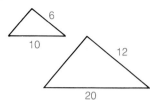

6.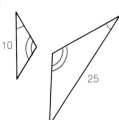

Find the missing lengths.

7.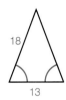

8.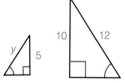

9.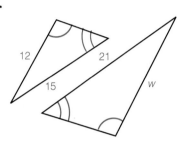

10.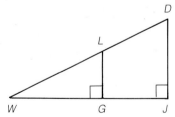

11. In $\triangle CLF$, $\overline{EM} \parallel \overline{CF}$, $LC = 12$, $LF = 16$, $CF = 22$, and $CE = 3$. Find EM.

12. In $\triangle WJD$, $\overline{LG} \perp \overline{WJ}$, $\overline{DJ} \perp \overline{WJ}$, $LG = 4$, $WJ = 20$, and $DJ = 15$. Find GW.

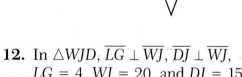

13. The shadow of a building is 72 feet long. At the same time, John, who is 5 feet tall, casts a shadow 6 feet long. How tall is the building?

14. At 4 p.m., a 100-foot pole casts a shadow 140 feet long. A tree near the pole casts a shadow 21 feet long. How tall is the tree?

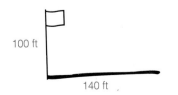

Extend and Apply

15. At the same time of day, a woman who is 66 in. tall casts an 88-in. shadow while her daughter casts a 44-in. shadow. Which of the following is the height of the woman's daughter?

66 in.

88 in.   44 in.

(A) 22 in.   (B) 40 in.   (C) 33 in.   (D) 36 in.

16. Consider two similar triangles. One triangle has sides whose lengths are 12, 18, and 23. The shortest side of the second triangle has length 8. Find the lengths of the other two sides of the second triangle.

Use Mathematical Reasoning

17. A surveyor found the distance AB across a river as follows. He paced off 10 ft for $\overline{AD}$ and 20 ft for $\overline{AC}$. Then he used a transit to sight from C to B and determined $m\angle ACB$ as shown. Using the same angle at D, he determined point E and found EA to be 13 ft. What is the distance from A to B?

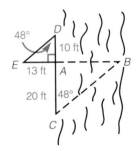

Mixed Review

Solve each proportion for y.

18. $\dfrac{3}{y} = \dfrac{9}{33}$ **19.** $\dfrac{y}{18} = \dfrac{16}{24}$ ◈ **20.** $\dfrac{6}{7} = \dfrac{x + 2}{14}$

Enrichment

You can use a mirror and similar triangles to find the height of a tree. Place a small mirror on the ground so that you can see the top of the tree reflected in the mirror. Measure the distance from the tree to the mirror and the distance from the mirror to yourself. Then measure the height of your eye level above the ground.

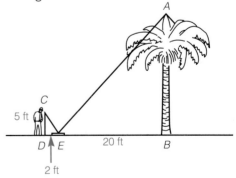

A law of physics tells us that $\angle CED \cong \angle AEB$, and we know the right angles $\angle CDE$ and $\angle ABE$ are congruent. Thus, $\triangle CDE \sim \triangle ABE$.

1. Find the height of the tree in the figure.

2. Try this method yourself with a tree or a flagpole.

9-6 SAS and SSS Similarity

Objective: Use the SAS and SSS Similarity Theorems to identify similar triangles and to solve problems involving similar triangles.

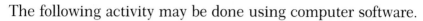

Explore

The following activity may be done using computer software.

1. Draw $\triangle ABC$ with $AB = 3$ cm, $m\angle A = 65°$, and $AC = 2$ cm. Draw $\triangle PQR$ with $PQ = 9$ cm, $m\angle P = 65°$, and $PR = 6$ cm.

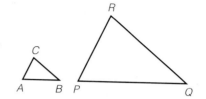

2. Measure and identify the corresponding angles.

3. Find the ratio $\frac{BC}{QR}$ and compare this to the other ratios of corresponding sides.

4. On another sheet of paper, draw $\triangle DEF$ with $DE = 3$ cm, $EF = 4$ cm, and $DF = 6$ cm. Draw $\triangle XYZ$ with $XY = 4.5$ cm, $YZ = 6$ cm, and $XZ = 9$ cm.

5. Measure and identify the corresponding angles.

6. Are the sides proportional?

Discuss Describe two new ways to identify similar triangles.

In addition to the AA Similarity Theorem, there are two other ways we can determine if triangles are similar. You may have discovered these methods in the Explore. They are summarized in the following theorems:

Theorem 9.3 **The SAS Similarity Theorem**

For any two triangles, if one pair of corresponding angles is congruent and the sides that include these angles are proportional, then the triangles are similar.

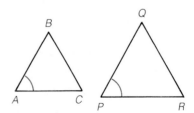

If $\dfrac{AB}{PQ} = \dfrac{AC}{PR}$, and $\angle A \cong \angle P$, then $\triangle ABC \sim \triangle PQR$.

Theorem 9.3 is useful in determining if triangles are similar when only one pair of angles are known to be congruent.

Theorem 9.4   The SSS Similarity Theorem

For any two triangles, if all
three pairs of corresponding
sides are proportional, then the
triangles are similar.

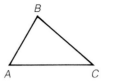

If $\dfrac{AB}{PQ} = \dfrac{AC}{PR} = \dfrac{BC}{QR}$,

then $\triangle ABC \sim \triangle PQR$.

The following Examples illustrate which theorem can be used
to show two triangles are similar.

Examples   Which theorem (AA, SAS, or SSS) can be used to show that
the triangles are similar?

1.

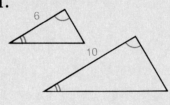

AA

2.

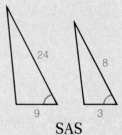

SAS

3.

SAS

4.

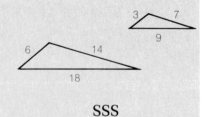

SSS

Try This...   Which theorem (AA, SAS, or SSS) can be used to show that
the triangles are similar?

a.

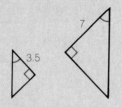

b.

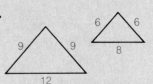

c.

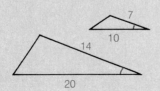

d.

Example **5.** A section of a roof is designed as shown. Find x.

Since $\angle A \cong \angle A$, and $\dfrac{AD}{AB} = \dfrac{AE}{AC}$,

$\triangle BAC \sim \triangle DAE$ by SAS. Thus,

$\dfrac{DE}{BC} = \dfrac{AD}{AB}$, so

$\dfrac{x}{4} = \dfrac{15}{5}$.

$5x = 4 \cdot 15$

$x = 12$

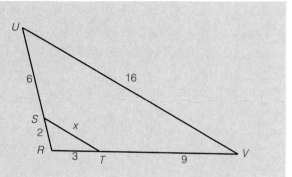

Try This... e. A section of a bridge is constructed as shown. Find x.

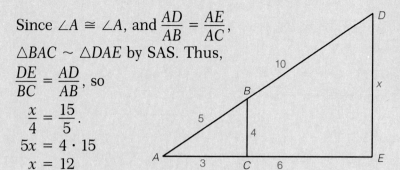

Discussion What is the difference between the SAS condition for congruent triangles and the SAS condition for similar triangles?

Exercises

Practice

Which theorem (AA, SAS, or SSS) can be used to show that the triangles are similar?

1.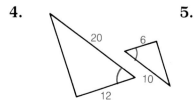

2.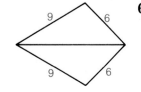

3.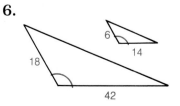

4.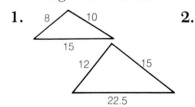

5.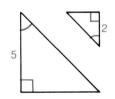

6.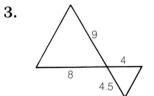

Which theorem (AA, SAS, or SSS) can be used to show that the triangles are similar?

7.

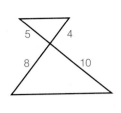

8.

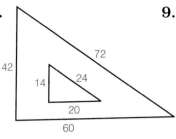

9.

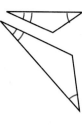

10. A support structure is designed as shown. Find t.

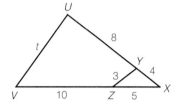

11. A section of a roof is designed as shown. Find y.

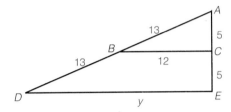

Extend and Apply

12. Which of the following is x?

   (A) 20   (B) 15   (C) 30   (D) 25

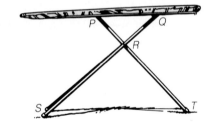

13. An adjustable ironing board is constructed as shown so that $PR = QR$, and $SR = TR$. How can you use the SAS Similarity Theorem to show that two triangles are similar? Explain why the board will always remain parallel to the floor.

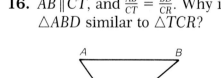

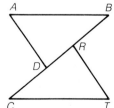

14. $\triangle ABC$ has sides of length 12, 16, and 18. $\triangle DEF$ has sides of length 9, 8, and 6. Explain why the triangles are similar. What is the scale factor that transforms $\triangle ABC$ to $\triangle DEF$?

Use Mathematical Reasoning

15. $MH = \frac{1}{4}MG, ML = \frac{1}{4}MK, HL = \frac{1}{4}GK$. Why is $\triangle KGM$ similar to $\triangle LHM$?

16. $\overline{AB} \parallel \overline{CT}$, and $\frac{AB}{CT} = \frac{BD}{CR}$. Why is $\triangle ABD$ similar to $\triangle TCR$?

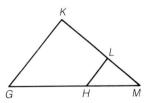

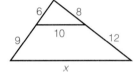

17. The fronts of two lots along First Street measure 120 ft and 150 ft. The lot lines on the sides are perpendicular to the back, on Main Street, as shown. Find the length of each lot along Main Street.

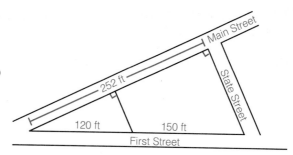

Mixed Review

18. A pole 4 m tall casts a shadow 7 m long. At the same time, the shadow of a tower is 133 m long. How tall is the tower?

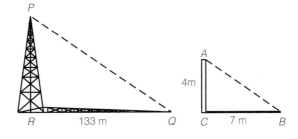

19. A 6-foot-tall man casts a shadow 7 feet long. At the same time, an 18-foot tree casts a shadow. How long is the tree's shadow?

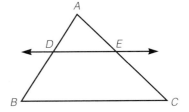

Solve and check.

20. $15 = 5(y - 8)$ **21.** $8(8z - 4) = 32$ **22.** $4(7 - 2x) = 4$

Discover

1. Draw a large triangle, $\triangle ABC$.

2. Using the sides of a straightedge, draw a line $\overleftrightarrow{DE}$ parallel to $\overline{BC}$ as shown.

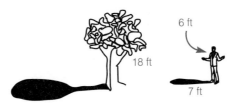

3. Since $\overleftrightarrow{DE}$ is parallel to $\overline{BC}$, what is true of $\angle ADE$ and $\angle ABC$? What is true of $\angle AED$ and $\angle ACB$? Measure the angles to check.

4. What does this say about $\triangle ADE$ and $\triangle ABC$?

5. Write a proportion that follows from your conclusion.

Use your discovery to complete the following theorem.

Theorem 9.5   If a line is _____ to one side of a triangle and intersects the other sides at any point except a vertex, then a triangle _____ to the given triangle is formed.

Radicals

A skateboard ramp is designed as shown at the right. The plans show that $\dfrac{AB}{CD} = \dfrac{CD}{EF}$. How long should the brace $\overline{CD}$ be?

To solve this problem, you might set up the following proportion.

$$\frac{1}{x} = \frac{x}{2}$$

$x \cdot x = 1 \cdot 2$ Cross products are equal.

$x^2 = 2$ $x \cdot x = x^2$

The answer to this problem is the positive number that multiplied by itself is 2. We write this number with a **radical** symbol as $\sqrt{2}$, the **square root** of 2.

This number is irrational—that is, it cannot be written as a repeating or terminating decimal. However, a calculator can be used as follows to find an approximation:

2 $\boxed{\sqrt{x}}$ $\boxed{\boxed{1.4142136}}$

Other square roots can be simplified easily. For example, $\sqrt{16} = 4$ because $4 \cdot 4 = 16$. Also, square roots can be estimated as follows: $\sqrt{18}$ must be between $\sqrt{16}$ and $\sqrt{25}$, so $\sqrt{18}$ is between 4 and 5.

Samples   **1.** Use a calculator to find the value of $5\sqrt{6}$.

5 $\boxed{\times}$ 6 $\boxed{\sqrt{x}}$ $\boxed{=}$ $\boxed{\boxed{12.247449}}$

2. Between which two whole numbers is $\sqrt{88}$?
$\sqrt{88}$ is between $\sqrt{81}$ and $\sqrt{100}$, so $\sqrt{88}$ is between 9 and 10.

Problems

Find the value of each radical to the nearest hundredth. Use a calculator when necessary.

1. $\sqrt{7}$ **2.** $3\sqrt{61}$ **3.** $\sqrt{25}$ **4.** $5\sqrt{19}$ **5.** $\sqrt{100}$

Between which two whole numbers is each square root?

6. $\sqrt{5}$ **7.** $\sqrt{11}$ **8.** $\sqrt{48}$ **9.** $\sqrt{72}$ **10.** $\sqrt{108}$

Chapter 9 Review

Express each ratio in lowest terms.

9-1

1. 33 to 11 **2.** 15:20 **3.** $\dfrac{4}{6}$

Write each ratio in lowest terms.

4. 2 ft to 9 in. **5.** 4 months to 1 year

Tell whether each pair of ratios forms a proportion.

9-2

6. $\dfrac{4}{7}$ and $\dfrac{12}{14}$ **7.** $\dfrac{6}{8}$ and $\dfrac{9}{12}$

Solve each proportion for x.

8. $\dfrac{9}{x} = \dfrac{6}{10}$ **9.** $\dfrac{4}{7} = \dfrac{x}{35}$

The scale of the map at the right is 1 cm = 10 km.

10. How many kilometers are represented by 5.7 cm?

11. How many centimeters on the map would be needed to represent 41 kilometers?

12. Estimate the number of kilometers between Greenville and Pleasant Valley.

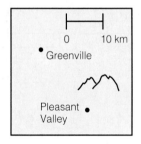

13. Determine whether *ABCD* is similar to *EFGH*. If so, what is the scale factor that transforms *ABCD* to *EFGH*?

9-3

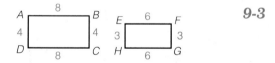

14. Use grid paper to draw a figure similar to the given figure, with a scale factor of 2.

15. △*ZEK* ~ △*GTR*. Which angles are congruent? Which sides are proportional?

9-4

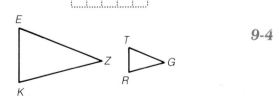

16. △*KAJ* ~ △*SYV*. Find *JA* and *VY*.

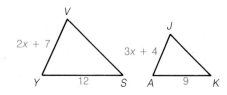

9-5   **17.** Find x.

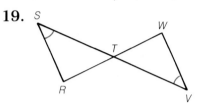

18. In $\triangle ABC$, $\overline{DE} \parallel \overline{BC}$,
$AB = 18$, $AC = 20$,
$BC = 16$, and $BD = 4$.
Find DE.

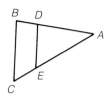

Which theorem (AA, SAS, or SSS) can be used to show the triangles are similar?

9-6   **19.**

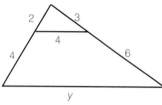

20.

21. Find y.

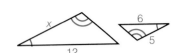

Chapter 9 Test

Express each ratio in lowest terms.

1. 12:18   **2.** 30 to 16   **3.** $\dfrac{14}{35}$

Write each ratio in lowest terms.

4. 2 days to 4 weeks   **5.** 4 feet to 2 yards

Tell whether each pair of ratios forms a proportion.

6. $\dfrac{8}{20}$ and $\dfrac{6}{15}$ **7.** $\dfrac{5}{8}$ and $\dfrac{9}{16}$

Solve each proportion for x.

8. $\dfrac{9}{24} = \dfrac{x}{16}$ **9.** $\dfrac{x}{5} = \dfrac{10}{100}$

The scale of the map at the right is 1 cm = 20 km.

10. How many kilometers are represented by
8 cm?

11. How many centimeters on the map would be
needed to represent 70 kilometers?

12. Estimate the number of kilometers between Dover
and Litton.

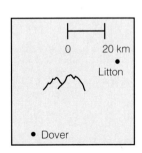

13. Determine whether *LMNP* is similar to *QRST*. If so, what is the scale factor that transforms *LMNP* to *QRST*?

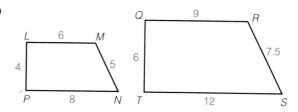

14. Use grid paper to draw a figure similar to the given figure, with a scale factor of $\frac{1}{3}$.

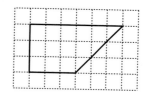

15. △*AQL* ~ △*EHK*. What angles are congruent? Which sides are proportional?

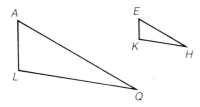

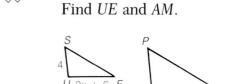 **16.** △*SUE* ~ △*PAM*. Find *UE* and *AM*.

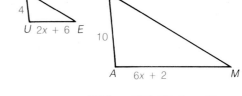

17. Find *x*.

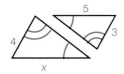

18. In △*TES*, $\overline{TE} \perp \overline{ES}$, $\overline{GM} \perp \overline{ES}$, *GM* = 6, *ST* = 24, and *TE* = 14. Find *GS*.

Which theorem (AA, SAS, or SSS) can be used to show the triangles are similar?

19.

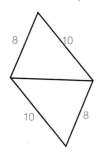

20.

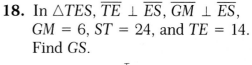

21. Find *z*.

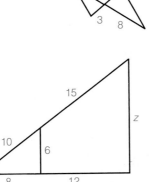

10 Using Similar Triangles

Graphic designers use standard right triangles to create similar right triangles that have the same shape but may have a different size.

10-1 Similar Right Triangles

Objective: Identify similar triangles formed by the altitude to the hypotenuse of a right triangle, and solve problems using the geometric mean.

Explore

In the figure at the right, $m\angle P = 50°$.

1. Find the measures of all angles in the figure.

2. How many different triangles are there in the figure? Name them.

3. Are any of these triangles similar? How do you know that they are similar?

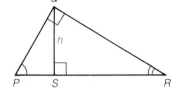

4. Name the triangles that are similar to $\triangle PQR$. Remember to write the correspondences correctly.

Discuss What are the proportional sides of each pair of triangles?

$\triangle PQR$ in the Explore is a right triangle with the right angle at Q. $\overline{QS}$ is an **altitude**, the length of which is the *height* of $\triangle PQR$. You may have discovered the following theorem.

Theorem 10.1 In a right triangle, the altitude to the hypotenuse forms two triangles, with both triangles similar to the right triangle and each one similar to the other.

Example **1.** Name the similar triangles in $\triangle GHK$.

The right angles are congruent. The common angles are congruent. Thus, $\triangle GHK \sim \triangle GTH \sim \triangle HTK$.

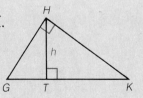

Try This... a. Name the similar triangles in $\triangle PFS$.

Recall from Chapter 9 that similar triangles have congruent angles, and that the sides of similar triangles are proportional.

Example 2. Complete the proportion. $\dfrac{u}{h} = \dfrac{h}{?}$

We know that $\triangle PSQ \sim \triangle QSR$. u and h are lengths of sides of $\triangle PSQ$. h is the length of a side of $\triangle QSR$, and v is the length of the other side. Therefore, $\dfrac{u}{h} = \dfrac{h}{v}$

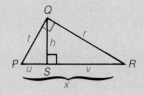

Try This... b. In the figure above, we also know that $\triangle PQR \sim \triangle PSQ$.

Complete the proportion. $\dfrac{x}{t} = \dfrac{?}{u}$

c. In the figure above, we know that $\triangle PQR \sim \triangle QSR$.

Complete the proportion. $\dfrac{?}{r} = \dfrac{r}{v}$

The *geometric mean* of two numbers is derived from a proportion involving the two numbers.

Definition The **geometric mean** of positive numbers a and b is the positive number x whenever

$$\frac{a}{x} = \frac{x}{b}$$

Example 3. Find the geometric mean of 3 and 5.

$$\frac{3}{x} = \frac{x}{5}$$
$$x^2 = 15 \quad \text{The cross products are equal}$$
$$x = \sqrt{15}$$

The geometric mean of 3 and 5 is $\sqrt{15}$.

We can use a calculator to find the approximate value of this geometric mean.

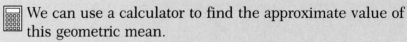

$\sqrt{15} \approx 3.9$, to the nearest tenth. $\quad \approx$ means "is approximately equal to."

In Example 2 we found the proportion $\frac{u}{h} = \frac{h}{v}$ for $\triangle PSQ$ and $\triangle QSR$. Notice that h is the geometric mean in this proportion. The proportion suggests the following theorem.

Theorem 10.2   In a right triangle, the length of the altitude to the hypotenuse is the geometric mean of the lengths of the segments on the hypotenuse.

$$\frac{a}{h} = \frac{h}{b}$$

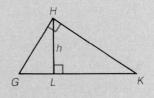

Example   **4.** $GL = 9$, and $LK = 25$. Find h.

h is the geometric mean between GL and LK.

$$\frac{GL}{h} = \frac{h}{LK}$$

$\dfrac{9}{h} = \dfrac{h}{25}$ Substituting

$h^2 = 9 \times 25$
$h^2 = 225$
$h = \sqrt{225}$
$h = 15$

This answer is reasonable since $\dfrac{9}{15} = \dfrac{15}{25}$.

Try This... f. Use $\triangle GHK$. $GL = 4$, and $LK = 9$.
Find HL.

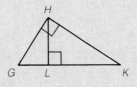

Discussion   Had you heard of a *mean* before? When you find the mean of two numbers by averaging them, we call this the *arithmetic mean*. Do you see any relationship between an arithmetic mean and a geometric mean?

Exercises

Practice

Name the similar triangles.

1.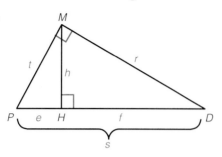

2.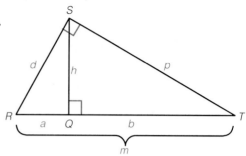

Complete the following proportions for the triangle in Exercise 1.

3. $\dfrac{?}{t} = \dfrac{t}{e}$

4. $\dfrac{e}{h} = \dfrac{?}{f}$

5. $\dfrac{s}{?} = \dfrac{r}{f}$

Complete the following proportions for the triangle in Exercise 2.

6. $\dfrac{a}{h} = \dfrac{h}{?}$

7. $\dfrac{m}{d} = \dfrac{d}{?}$

8. $\dfrac{m}{p} = \dfrac{p}{?}$

 For each pair of numbers, find the geometric mean to the nearest tenth.

9. 4 and 25

10. 1 and 16

11. 9 and 49

12. 36 and 25

13. 13 and 15

14. 26 and 30

15. 0.3 and 0.12

16. 0.4 and 0.2

17. 0.3 and 1.2

Use $\triangle TFR$ for Exercises 18–23.

 Find the missing length to the nearest tenth.

18. $RQ = 4$, $QF = 25$, $TQ = ?$

19. $RQ = 4$, $QF = 36$, $TQ = ?$

20. $RQ = 6$, $QF = 15$, $TQ = ?$

21. $RQ = 11$, $QF = 17$, $TQ = ?$

22. $FQ = 3$, $QR = 19$, $TQ = ?$

23. $FQ = 5$, $QR = 21$, $TQ = ?$

Extend and Apply

24. In which proportion is q the geometric mean?

(A) $\dfrac{x}{t} = \dfrac{t}{q}$ (B) $\dfrac{p}{q} = \dfrac{q}{r}$ (C) $\dfrac{q}{p} = \dfrac{r}{p}$ (D) $\dfrac{p}{q} = \dfrac{r}{s}$

 25. What is the geometric mean of $\dfrac{3}{4}$ and $\dfrac{7}{8}$, to the nearest hundredth?

 Find all answers to the nearest tenth.

26. Altitude HL is 30 feet, and GL is 13 feet. Find LK.

27. GL is 24 inches, and LK is 100 inches. Find HL.

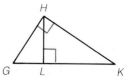

28. How far is it across the lake?

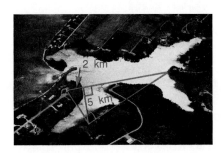

2 km

5 km

29. How far is it across the quicksand?

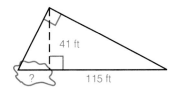

41 ft

115 ft

?

Use Mathematical Reasoning

30. Another theorem states that if the altitude is drawn to the hypotenuse of a right triangle, then the length of each leg is the geometric mean of the length of the hypotenuse and the length of the segment of the hypotenuse that is adjacent to that leg. Use this theorem to state two proportions in the figure at the right.

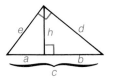

e   h   d

a   b

c

 Determine a proportion and use it to find the missing length.

31. $FQ = 5$, $FR = 20$, $FT = ?$

32. $RQ = 4$, $RT = 7$, $RF = ?$

33. $FT = 11$, $FR = 23$, $FQ = ?$

34. $QF = 10$, $TF = 20$, $RF = ?$

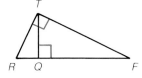

T

R   Q   F

Mixed Review

Which theorem (AA, SAS, or SSS) can be used to show the triangles are similar?

35.

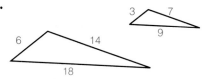

3   7

9

6   14

18

36.

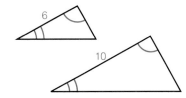

6

10

Write each ratio in lowest terms.

37. 2 hours to 12 minutes

38. 90 days to 12 weeks

39. 3 decades to 5 years

40. 9 muffins to 2 dozen muffins

 Find the value of each radical to the nearest hundredth.

41. $\sqrt{49}$ **42.** $\sqrt{55}$ **43.** $\sqrt{100}$ **44.** $\sqrt{102}$

Solve and check.

45. $27x - 3 = 78 + 18x$

46. $16y + 4 = 4(y + 16)$

10-2 The Pythagorean Theorem

Objective: Solve problems using the Pythagorean Theorem.

Explore

You may use computer software for the following activity.

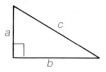

1. Draw three right triangles, each a different size. Label the legs of your triangles a and b, and label each hypotenuse c, as shown at the right.

2. Measure the lengths of the legs of each triangle, and calculate the sum of the square of the legs.

$$a^2 + b^2$$

3. Measure the hypotenuse of each triangle and calculate the square, c^2.

4. Compare the sum of the square of the legs, $a^2 + b^2$, with the square of the hypotenuse, c^2.

Discuss   What relationship do you find between the sum of the square of the legs, $a^2 + b^2$, and the square of the hypotenuse, c^2?

The Explore suggests one of the most important theorems in all of geometry, the **Pythagorean Theorem.** This theorem is named after the Greek mathematician Pythagoras, who lived about 550 B.C. and was one of the first to prove this theorem. It is believed, however, that the ancient Egyptians used the relationship as early as 2000 B.C. There is evidence that this theorem was known to Babylonian and ancient Chinese mathematicians as well.

Theorem 10.3   **The Pythagorean Theorem**

In a right triangle, the sum of the squares of the lengths of the two legs is equal to the square of the length of the hypotenuse.

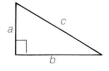

$$a^2 + b^2 = c^2$$

The diagram at the right gives a geometric illustration of the Pythagorean Theorem. Note that the square opposite the leg labeled a has an area of 4^2, or 16; the square opposite the leg labeled b has an area of 3^2, or 9; the area of the square opposite the hypotenuse, c, is 5^2, or 25.

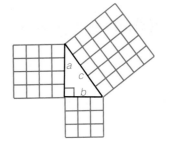

$$9 + 16 = 25$$

We can use the Pythagorean Theorem to find the lengths of sides of right triangles.

Examples 1. Find b.

$$a^2 + b^2 = c^2$$
$$6^2 + b^2 = 10^2$$
$$36 + b^2 = 100$$
$$b^2 = 64$$
$$b = \sqrt{64} = 8$$

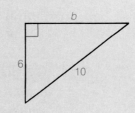

2. Find c.

$$a^2 + b^2 = c^2$$
$$7^2 + 24^2 = c^2$$
$$49 + 576 = c^2$$
$$625 = c^2$$
$$\sqrt{625} = c$$
$$25 = c$$

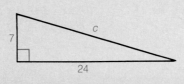

3. Find a.

$$a^2 + b^2 = c^2$$
$$a^2 + 7^2 = 9^2$$
$$a^2 + 49 = 81$$
$$a^2 = 32$$
$$a = \sqrt{32} \approx 5.6568542$$

$a \approx 5.66$, rounded to the nearest hundredth.

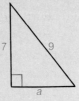

Try This... Find a in each triangle. Round to the nearest hundredth.

a.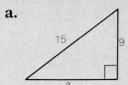

b.

The Pythagorean theorem has many applications in everyday life.

Example 4. On a little league baseball diamond, the distance between bases is 60 feet. About how far is it from home plate to second base?

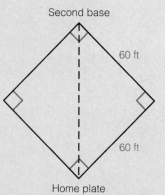

Second base

60 ft

60 ft

Home plate

$$a^2 + b^2 = c^2$$
$$60^2 + 60^2 = c^2$$
$$3600 + 3600 = c^2$$
$$7200 = c^2$$
$$\sqrt{7200} = c$$
$$c = \sqrt{7200} \approx 84.85281$$

The distance from home plate to second base is about 84.9 feet.

Try This... c. Kingsville is 7 miles due south of Belair. Chase is 12 miles due east of Kingsville. To the nearest tenth of a mile, what is the distance from Belair to Chase?

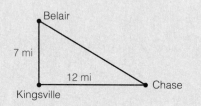

Belair

7 mi

12 mi

Kingsville

Chase

Discussion Does the Pythagorean Theorem work for triangles other than right triangles? Why or why not?

Exercises

Practice

 Use the information given to find the missing length to the nearest hundredth.

1. $b = 40$, $c = 50$, $a = ?$ **2.** $a = 24$, $c = 40$, $b = ?$

3. $a = 8$, $c = 17$, $b = ?$ **4.** $b = 16$, $c = 34$, $a = ?$

5. $a = 36$, $b = 15$, $c = ?$ **6.** $a = 16$, $b = 30$, $c = ?$

7. $b = 5$, $c = 9$, $a = ?$ **8.** $a = 8$, $c = 19$, $b = ?$

9. $a = 10$, $b = 7$, $c = ?$ **10.** $a = 5$, $b = 3$, $c = ?$

11. $b = 12$, $c = 18$, $a = ?$ **12.** $a = 9$, $b = 16$, $c = ?$

13. $a = 17$, $c = 36$, $b = ?$ **14.** $b = 11$, $c = 32$, $a = ?$

15. $a = 19$, $b = 42$, $c = ?$ **16.** $a = 46$, $c = 99$, $b = ?$

For Exercises 17–22, round answers to the nearest tenth.

17. *ABCD* is a rectangle. Find *BD*.

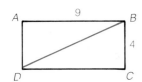

18. *GHKL* is a square. Find *GK*.

19. The distance between bases on a major league baseball diamond is 90 ft. About how far is it from home plate to second base?

20. The size of a television set is advertised in terms of the length, in inches, of its diagonal. How should this TV set be advertised?

21. A 10-foot ladder is placed three feet from the bottom of a wall. How far up the wall will the top of the ladder touch?

Extend and Apply

22. Hampstead is 52 miles from Millers. Millers is 18 miles due south of Tyler. Hampstead is due east of Tyler. About how far is Hampstead from Tyler? (Hint: Draw a diagram.)

23. In the triangle at the right, $a^2 = 25$ and $b^2 = 36$. What is the value of c^2?

(A) 11 (B) 10 (C) 61 (D) 7.8

24. A rectangular field is 70 meters by 155 meters. About how much shorter is it to walk diagonally from one corner to another than it is to walk along the edge (around the corners)?

25. An outfielder catches a ball on the first-base line, about 40 feet beyond first base. About how far would he have to throw the ball to third base?

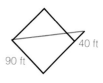

26. A person travels 4 miles east and 7 mi south, then 3 mi east and 1 mi north, and finally, 2 mi west. How far is the person from the original position?

27. Determine the tree's height before it fell.

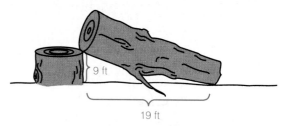

9 ft

19 ft

Use Mathematical Reasoning

28. Find the length of the sides of rhombus *PQRS*. $PR = 18$, $SQ = 26$

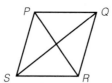

29. Find *PQ*.

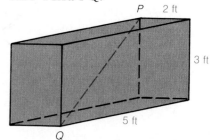

P   2 ft

3 ft

5 ft

Q

30. A doorway is three feet wide and 6 ft 8 in. high. What is the highest piece of paneling that can be slid through the door?

Mixed Review

31. Find the geometric mean of 12 and 24.

32. Find the geometric mean of 30 and 40.

33. Name the similar triangles in $\triangle SRP$.

34. $SM = 12$ cm and $MP = 15$ cm. Find *MR*.

35. $PM = 8$ ft and $MS = 6$ ft. Find *RM*.

◇◇ Solve and check.

36. $3(2x - 7) = 5$ **37.** $3x + 7x - 12 = 0$ **38.** $2(4x - 9) = 7x$

Visualization

Which figure could not be constructed with the shapes shown at right? (Assume an unlimited supply of each shape.)

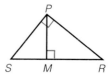

(A)   (B)   (C)   (D)   (E)

Probability of a Simple Event

What is the likelihood that a tossed coin will come up heads? There is an equally likely chance that it will come up heads or tails. Heads is one of two possible *outcomes.* The likelihood, or **probability**, of a head is $\frac{1}{2}$, or 0.5. We can abbreviate this with the notation $P(\text{head}) = 0.5$.

If an event can occur m ways out of a possible n equally likely ways, the probability of that event (E) is $P(E) = \frac{m}{n}$.

The probability of an event is always a fraction or decimal between, and including, 0 and 1.

Samples

1. What is the probability that an angle of a given square measures 90°?

   There are 4 angles of a square and each measures 90°, so the probability is $\frac{4}{4} = 1$.

2. What is the probability that when a six-sided die is rolled, the result is greater than 4?

   There are 6 possible outcomes: 1, 2, 3, 4, 5, and 6. Two of these are greater than 4, so the probability is $\frac{2}{6} = \frac{1}{3} \approx 0.33$.

3. What is $P(\;$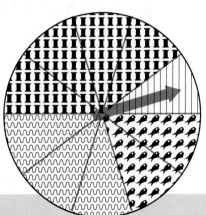$\;)$ for the spinner?

   Although there are 4 possible outcomes, they are not equally likely. However, each of the 10 36° segments is equally likely. So, $P(\;$ ✦ $\;)$ is $\frac{2}{10} = \frac{1}{5} = 0.2$.

Problems

1. What is $P(\;$▌▌$\;)$?

2. What is $P(not\;$ ∿∿ $\;)$?

3. What is the probability that when the battery of a digital watch stops, the hour is 1, 2, 3, 4, or 5?

10-3 The Converse of the Pythagorean Theorem

Objective: Determine whether a triangle is a right triangle from the lengths of its sides.

Carpenters often work with right triangles. It is important that these angles be very precise, and it is not always easy to measure them with a protractor.

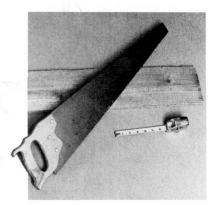

We can use the converse of the Pythagorean Theorem to determine whether a triangle is a right triangle. Recall that we find the converse of a conditional sentence by interchanging the hypothesis and the conclusion.

Theorem 10.4   The Converse of the Pythagorean Theorem

If the lengths of the sides of a triangle are a, b, and c, and $a^2 + b^2 = c^2$, then the triangle is a right triangle with right angle opposite the longest side, whose length is c.

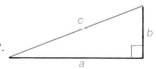

Example   **1.** Determine whether the triangle shown is a right triangle with the measurements given.

If $4^2 + 9^2 = 10^2$, then $\triangle ABC$ is a right triangle.

$$4^2 + 9^2 \overset{?}{=} 10^2$$
$$16 + 81 \overset{?}{=} 100$$
$$97 \overset{?}{=} 100$$

Since $97 \neq 100$, $\triangle ABC$ is not a right triangle.

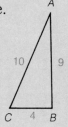

Try This... Determine whether each triangle is a right triangle.

a.

b.

Example 2. Determine whether the measurements 12, 16, and 20 can be the lengths of the sides of a right triangle.

First draw a diagram. Label the longest side, or the hypotenuse, 20, and label the legs 12 and 16 as shown.

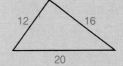

If $12^2 + 16^2 = 20^2$, then we have a right triangle.

$12^2 + 16^2 \stackrel{?}{=} 20^2$
$144 + 256 \stackrel{?}{=} 400$
$400 = 400$ ✓

Yes, 12, 16, and 20 can be lengths of the sides of a right triangle.

Try This... c. Determine whether the measurements 20, 25, and 32 can be the lengths of the sides of a right triangle.

Example 3. Determine whether the two pieces of wood are perpendicular if $\overline{AB}$ measures 11 in.

$8^2 + 6^2 \stackrel{?}{=} 11^2$
$64 + 36 \stackrel{?}{=} 121$
$100 \neq 121$

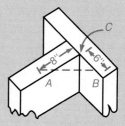

This means that $\triangle ABC$ is *not* a right triangle. So, the two pieces of wood are *not* perpendicular.

Try This... d. Determine whether the two pieces of wood shown in Example 3 are perpendicular if $\overline{AB}$ measures 10 in.

Discussion Egyptian surveyors carried loops of ropes with 12 equally spaced knots, which they used to determine right angles. The surveyors were called "rope stretchers." How could they have used the ropes to determine right angles?

Exercises

Determine whether each is a right triangle with the measurements given.

1.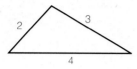

2.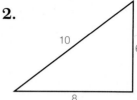

Determine whether these measurements can be the lengths of the sides of a right triangle.

3. $d = 7, e = 5, f = 9$

4. $d = 5, e = 2, f = 8$

5. $d = 40, e = 2, f = 8$

6. $d = 16, e = 2, f = 11$

7. $d = 48, e = 20, f = 52$

8. $d = 30, e = 16, f = 34$

9. $d = 1.2, e = 0.5, f = 1.3$

10. $d = 1.8, e = 1.2, f = 1.9$

11. $d = 106, e = 90, f = 50$

12. $d = 25, e = 24, f = 8$

13. $d = 12, e = 13, f = 9$

14. $d = 24, e = 40, f = 32$

15. $d = 48, e = 50, f = 14$

16. $d = 8, e = 15, f = 16$

17. Chi-Kai is installing a basketball net in his backyard. To make sure the pole is perpendicular to the ground, he marked point *A* 4 ft up on the pole, and point *B* 3 ft away on the ground. He finds distance *AB* to be 5 ft. Is the pole perpendicular to the ground?

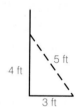

18. Tamara uses stakes to mark the corners for the foundation of a garage that will be 40 ft × 40 ft. To make sure that the corners are square, she measures *AC*, and finds it to be 50 ft. Are the corners square?

Extend and Apply

19. What is wrong with the diagram at the right?

20. What measure must $\overline{AB}$ have in order for ∠*C* to be a right angle?

   (A) 40   (B) 1600   (C) 21.2   (D) 56

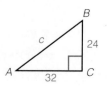

21. Is quadrilateral *ABCD* a rhombus with the given dimensions? Why or why not?

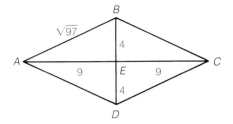

22. Is quadrilateral *GTRM* a square with the given dimensions? Why or why not?

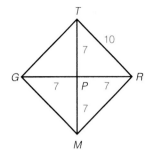

Use Mathematical Reasoning

23. Jessie works as a rec leader. In the spring she needs to lay out the softball field. After marking the four bases 60 ft apart, she checked the distance from home plate to second base to make sure the field was square. Explain her method.

24. Suppose you begin to lay a plywood floor on the foundation of a small shed that you are building. You find, much to your disappointment, that one of the corners of the plywood does not fit into the supposedly square corner of the foundation. Explain how you could use your tape measure to determine who made the mistake — you or the sawmill that made the plywood.

Mixed Review

Solve each proportion.

25. $\dfrac{s}{4} = \dfrac{15}{20}$ **26.** $\dfrac{7}{a} = \dfrac{21}{24}$ **27.** $\dfrac{x}{9} = \dfrac{4}{x}$ **28.** $\dfrac{4}{r} = \dfrac{r}{16}$

29. Draw an obtuse right triangle, if possible.

Add or subtract.

30. $3 - 6 - (-5)$ **31.** $4 + (-5) - 7$ **32.** $-15 + 7 - (-8)$

33. Find the ratio of © to Ω.

Ω   Ω
  ©   ©   Ω
©   Ω Ω   ©   © ΩΩ
©©   ΩΩ ©Ω©   ©

34. Find the ratio of ♠ to ♥.

♥   ♥ ♥   ♠
♠   ♥ ♥   ♠♥
♥ ♥ ♠ ♠   ♠ ♥
        ♥

35. △*ABC* ~ △*DEF*. *AC* = 18, *BC* = 27, and *DF* = 27. What is *EF*?

Between which two whole numbers is each square root?

36. $\sqrt{10}$ **37.** $\sqrt{70}$ **38.** $\sqrt{177}$

10-4 The Isosceles Right Triangle

Objective: Find lengths of the sides of isosceles right triangles.

Explore

1. Draw an isosceles right triangle of any size.

2. Measure the sides of the triangle as accurately as possible.

3. Measure the angles of the triangle.

4. Find the ratio of the hypotenuse to one of the sides, $\frac{h}{a}$.

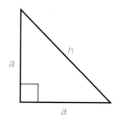

Discuss Compare your results with those of your classmates. Are the results similar for all isosceles right triangles regardless of size? Use your calculator to find $\sqrt{2}$, and compare this result with the ratio $\frac{h}{a}$.

The ratio that you found in the Explore suggests that you can find the hypotenuse of an isosceles right triangle by multiplying the length of a leg by a factor of $\sqrt{2}$.

Theorem 10.5 The Isosceles Right Triangle Theorem

For any isosceles right triangle, the hypotenuse is $\sqrt{2}$ times as long as either leg.

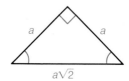

Example 1. Find the missing lengths to the nearest tenth.

Since △RGW is isosceles, $RG = 9$.

By the Isosceles Right Triangle Theorem, $RW = 9\sqrt{2}$. We can use a calculator to find an approximate value of RW.

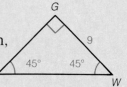

$9 \boxed{\times} 2 \boxed{\sqrt{x}} \boxed{=} \quad \boxed{12.72792}$

$RW \approx 12.7$, to the nearest tenth.

Example 2. Find the missing lengths to the nearest hundredth.

By the Isosceles Right Triangle Theorem, $GB = GT\sqrt{2}$. Thus,
$$7\sqrt{2} = GT\sqrt{2}$$
$$7 = GT$$
Since $\triangle GTB$ is isosceles, $BT = 7$.

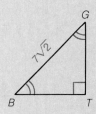

 Try This... Find the missing lengths in each triangle to the nearest tenth.

a.

b.

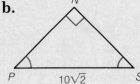

 Example 3. Find CE to the nearest tenth.

By the Isosceles Right Triangle Theorem, $CW = CE\sqrt{2}$. Thus,
$$19 = CE\sqrt{2}$$
$$\frac{19}{\sqrt{2}} = CE$$
$19 \boxed{\div} 2 \boxed{\sqrt{x}} = \boxed{13.43502}$
$CE \approx 13.4$

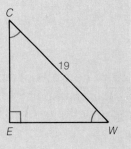

 Try This... **c.** Find GK to the nearest tenth.

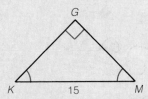

Discussion The *Isosceles Right Triangle Theorem* is sometimes called the *45°-45°-90° Triangle Theorem*. Explain why this is also an appropriate title.

Exercises

Practice

Find the lengths of the other two sides of △MRS to the nearest tenth.

1. $r = 2$

2. $s = 7$

3. $r = 4.5$

4. $r = 8.4$

5. $m = 14\sqrt{2}$

6. $m = 16\sqrt{2}$

7. $m = 3.5\sqrt{2}$

8. $m = 8.2\sqrt{2}$

9. $m = 24$

10. $m = 10$

11. $m = 3.8$

12. $m = 6.1$

13. $s = 22.6$

14. $r = 100.98$

15. $r = \dfrac{3}{5}$

16. $s = \dfrac{6}{7}$

17. $m = 142$

18. $m = 300$

Extend and Apply

19. What is ℓ in terms of s for the square?

(A) $\dfrac{s}{\sqrt{2}}$

(B) s

(C) s^2

(D) $s\sqrt{2}$

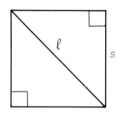

20. The walls of many buildings are supported by bracing that connects uprights. The bracing is put in at a 45° angle as shown below. Find the approximate length needed to brace the 8-ft uprights shown.

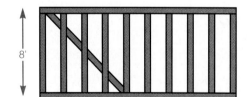

21. This toy is a series of isosceles right triangles. If $AB = 1$, find AC, BC, CD, CE, BD, DE, EF, and DF.

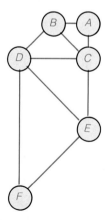

Use Mathematical Reasoning

22. Explain how a carpenter can lay out a 45° cut on the edge of a board using only a ruler.

23. Find AB in the diagram of an automobile valve.

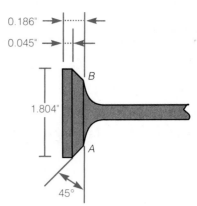

0.186"
0.045"
B
1.804"
A
45°

Mixed Review

 Find the missing lengths to the nearest tenth.

24. $a = 6$, $b = 8$, $c = ?$

25. $a = 23$, $c = 33$, $b = ?$

26. $b = 84$, $c = 91$, $a = ?$

c b a

Problem Solving

The numbers 3, 4, and 5 make up an unusual "triple" of whole numbers that can be the lengths of sides of a right triangle. These numbers make up a **Pythagorean triple.**

Multiples of 3, 4, 5—such as 6, 8, 10, and 9, 12, 15—are also Pythagorean triples. Why?

Are there others that are not multiples of 3, 4, and 5? Use your calculator to fill in the following table.

n	Shorter leg	Longer leg	Hypotenuse
1	2(1) + 1 = 3	2(1)(1 + 1) =	
2	2(2) + 1 =	2(2)(2 + 1) =	
3			
4			
5			

1. Given any whole number value of n, what is the length of the shorter leg? What is the length of the longer leg?

2. Choose any whole number for n. Will the above method always provide a Pythagorean triple?

3. How does the length of the hypotenuse compare with the length of the longer leg?

10-5 The 30°-60° Right Triangle

Objective: Find the lengths of the sides of 30°-60° right triangles.

Explore

You may use computer software for this Explore.

1. Draw an equilateral triangle of any size.

2. Measure the sides and the altitude of the triangle as accurately as possible.

3. Find the ratios $\frac{h}{a}$ and $\frac{b}{a}$.

4. Measure all of the angles.

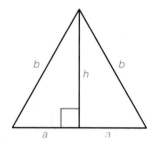

Discuss   Compare your results with those of your classmates. Are the results similar for all equilateral triangles regardless of size? What is the constant ratio $\frac{b}{a}$? Use your calculator to find $\sqrt{3}$, and compare this result with the ratio $\frac{h}{a}$.

The ratios that you found in the Explore suggest that you can find the length of the hypotenuse of a 30°-60° right triangle by multiplying a by 2, and that you can find the length of the third side by multiplying a by $\sqrt{3}$.

Theorem 10.6   The 30°-60° Right Triangle Theorem

For any 30°-60° right triangle, the hypotenuse is twice as long as the shorter leg, and the longer leg is $\sqrt{3}$ times as long as the shorter leg.

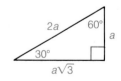

Since the angles of an equilateral triangle are each 60°, and the angle formed by the altitude is 30°, we can use the 30°-60° Right Triangle Theorem to find the height of an isosceles triangle.

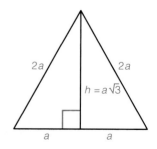

Examples Find the missing lengths to the nearest tenth.

1. We will make a table and fill in the given data.
The length of the hypotenuse is 8.

Hypotenuse AR	Shorter leg PR	Longer leg AP
$2a$	a	$a\sqrt{3}$
8	4	$4\sqrt{3} \approx 6.9$

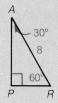

The shorter leg is half the hypotenuse.
So $PR = 4$. The longer leg is $\sqrt{3}$ times
the shorter leg. So $AP = 4\sqrt{3} \approx 6.9$.

2.

Hypotenuse GH	Shorter leg KG	Longer leg KH
$2a$	a	$a\sqrt{3}$
12	6	$6\sqrt{3} \approx 10.4$

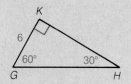

The hypotenuse is twice the shorter side.
So $GH = 12$. The longer leg is $\sqrt{3}$ times
the shorter leg. So $KH = 6\sqrt{3} \approx 10.4$.

3.

Hypotenuse MK	Shorter leg KA	Longer leg MA
$2a$	a	$a\sqrt{3}$
10	5	$5\sqrt{3}$

The longer leg is $\sqrt{3}$ times the shorter leg.
So $KA = 5$. The hypotenuse is twice the
shorter side. So $MK = 10$.

Try This... Find the missing lengths to the nearest tenth.

a.
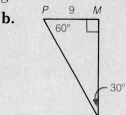

b.
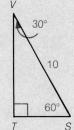

c.

Example 4. John wants to place an extension ladder against the base of a window that is 10 feet above the ground. The ladder is to be at a safe 60° angle. How far should John place it from the wall?

We have a 30°-60° right angle. The longer leg is 10 ft. We want to know the measure of the shorter leg.

By the 30°-60° Right Triangle Theorem,

long leg = (short leg) $\sqrt{3}$

10 = (short leg) $\sqrt{3}$

Therefore, we can find the short leg by dividing 10 by $\sqrt{3}$, or $\frac{10}{\sqrt{3}}$.

$10 \div 3\boxed{\sqrt{x}}\boxed{=} 5.77350$

The shorter leg is 5.8, to the nearest tenth. The ladder should be placed about 5.8 feet from the wall.

Try This... d. John wants to reach a window 12 ft from the ground. Assuming that the ladder can be extended to the proper length how far should it be placed from the wall, keeping the angle at 60°?

Discussion John's ladder can extend from 10 feet to 15 feet. How much was the ladder extended to reach the 10-ft-high window at a 60° angle? the 12-ft-high window? How high can the ladder reach?

Exercises

Practice

 Find the lengths of the other two sides of $\triangle PRT$ to the nearest tenth.

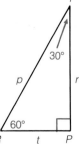

1. $p = 10$ **2.** $p = 18$

3. $p = 27$ **4.** $p = 35$

5. $t = 5$ **6.** $t = 11$

7. $t = 24$ **8.** $t = 5.6$

9. $r = 6\sqrt{3}$ **10.** $r = 10\sqrt{3}$

11. $r = 19\sqrt{3}$ **12.** $r = 15\sqrt{3}$

13. $r = 12$ **14.** $r = 20$

15. $r = 13$ **16.** $r = 32$

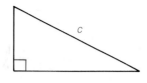

Extend and Apply

17. In the 30°-60°-90° triangle at the right, what is the length of the longer leg in terms of c?

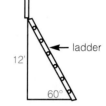

   (A) $\frac{c}{2}\sqrt{3}$ (B) $c\sqrt{3}$ (C) $2c$ (D) $\frac{1}{2}c$

18. The base of a ladder is placed 1.4 m from a wall. To what length must the ladder be extended in order to have an angle of 60° between the ladder and the ground?

19. An extension ladder forming a 60° angle with the ground is placed against an outside wall. The top of the ladder touches a window sill that is 12 ft high. To what length is the ladder extended? How far from the wall is the bottom of the ladder?

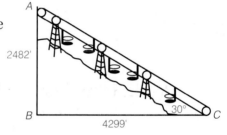

20. A ski lift is shown at the right. Find the distance from the bottom of the lift to the top of the lift.

21. In the ski lift in Exercise 20, find the length of the shortest cable that could be used. Assume that there is no length around the pulleys.

22. In the ski lift in Exercise 20, the actual length of the cable is 10,120 feet. About how much slack is in the cable?

23. The arch in the photograph shows an equilateral triangle, $\triangle ABC$. The sides of $\triangle ABC$ measure 3.6 m. How high is the arch?

24. A hockey player is 5 m from the center line of the rink. The angle to the center of the goal is 30°. How far must she hit the puck to reach the goal?

25. s is the length of the side of an equilateral triangle, and h is the length of an altitude. Find h in terms of s.

Use Mathematical Reasoning

26. One angle of a rhombus has a measure of 120°, and the length of each side is 18. Find the length of the longer diagonal.

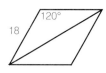

27. In this enlarged drawing of a bolt, the angle has a measure of 60°, and the distance from thread to thread is 12 mm. Find the depth of each thread.

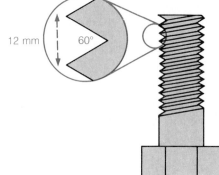

Mixed Review

28. Find the ratio of 30 inches to 6 feet.

29. Find the ratio of 6 feet to 78 inches.

30. Can 30 inches, 6 feet, and 78 inches be the lengths of sides of a right triangle?

Solve each proportion for x.

31. $\dfrac{x}{9} = \dfrac{1}{4}$ **32.** $\dfrac{440}{1760} = \dfrac{3}{x}$

33. Between which two whole numbers is $\sqrt{123}$?

Enrichment

The Möbius strip, introduced by the German mathematician and astronomer Augustus Ferdinand Möbius (1790–1868), can fool you.

Cut out a strip of paper about 11 inches long and 1 inch wide.

1. Take the strip and turn one end over. (Give the loop a half twist.) Tape the ends together.

Draw a pencil line down the middle, and shade one side of the line until you come back to your original point. Cut the strip in half along its center line. What happens?

2. Cut the strip in half again. What happens?

3. Take a second strip and connect it as you did the first one. Cut parallel to one edge and about one-third of the way from that edge. Continue cutting until you come back to your original position. What happens?

Making Measuring Instruments

Similar triangles can be used with some familiar objects to create homemade instruments for measurement. Work with a group or with a partner to build one or more of the following instruments. Try them out and see if you can find how similar triangles are used.

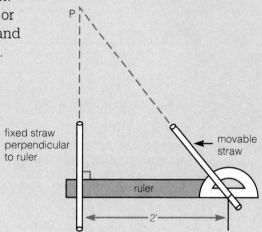

1. An instrument for measuring the distance to an object can be made from some straws, a pin, a protractor, and a yardstick as shown at the right. To find the distance to point P, first sight P through the fixed straw. Then sight P through the movable straw. Use the protractor to measure the angle formed by the movable straw. Finally, make a scale drawing and find the distance to P.

2. To measure the diameter of the moon, make a device using a paper clip or clothespin, a straw, and a meterstick or yardstick as shown. Close one eye and sight through the straw at the full moon. Slide the straw closer or farther until it just blocks out the moon. Determine the distance from your eye to the straw and also find the diameter of the straw. Use the fact that the distance from the earth to the moon is about 385,000 km to find the approximate diameter of the moon.

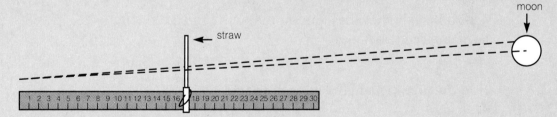

3. To build an instrument to measure heights, cut out a 10-in. square from heavy cardboard. Mark off the square in inches and attach a weighted string to a corner as shown below. To use the instrument to find PR, sight to point P and read off the distance AB on the scale. Multiply AB by the distance SR, divide by 10, and then add the height TS. The result will be PR.

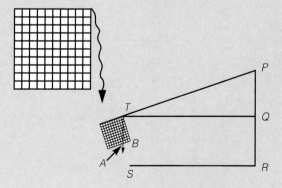

Chapter 10 Review

10-1 Name the similar triangles.

1.

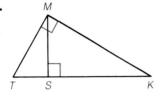

2.

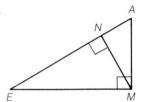

 Find the geometric mean of these numbers.

3. 4 and 16

4. 7 and 10

Find the missing lengths.

5. $RQ = 9$, $QF = 36$, $TQ = ?$

6. $XT = 6$, $XY = 9$, $XZ = ?$

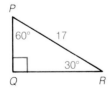

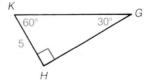

10-2 Use the given information to find the missing lengths.

7. $a = 20$, $b = 15$, $c = ?$

8. $b = 4$, $c = 5$, $a = ?$

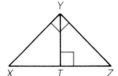

10-3 Determine whether these measurements can be the lengths of sides of a right triangle.

9. $p = 24$, $q = 36$, $r = 40$

10. $p = 5$, $q = 8$, $r = 10$

10-4 Find the missing lengths in each triangle.

11.

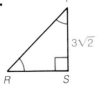

12.

13.

14.

15.

16.

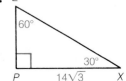 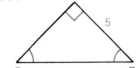

Chapter 10 Test

Name the similar triangles.

1.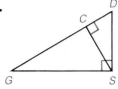

2.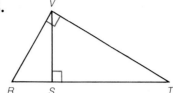

 Find the geometric mean of these numbers.

3. 9 and 36

4. 11 and 15

Find the missing lengths.

5. $MF = 16$, $FE = 25$, $SF = ?$

6. $RH = 5$, $RT = 8$, $RB = ?$

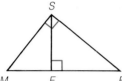

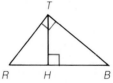

 Use the given information to find the missing lengths.

7. $b = 70$, $c = 74$, $a = ?$

8. $a = 8$, $b = 2$, $c = ?$

Determine whether these measurements can be the lengths of sides of a right triangle.

9. $p = 16$, $q = 12$, $r = 20$

10. $p = 10$, $q = 15$, $r = 29$

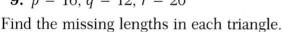 Find the missing lengths in each triangle.

11.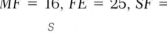

12.

13.

14.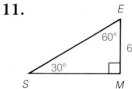

15.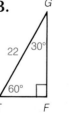

16.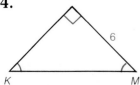

Before a quilt can be assembled, it must be carefully designed to ensure that its polygonal regions tessellate the plane.

11-1 Identifying Polygons

Objective: Classify polygons by their sides, and as convex or concave.

Examples of polygons can be found in nature, sports, and art. We have already studied two types of polygons: triangles and quadrilaterals. The Pentagon building near Washington, D.C., provides another example of a polygon.

Definition   A **polygon** is a figure formed by the line segments connecting three or more coplanar points. The segments are called *sides* and the endpoints are called *vertices*. Each side intersects exactly two other sides, one at each vertex. No three consecutive vertices can be collinear.

We usually name a polygon by naming its vertices in order. In polygon $QRSTV$, $\overline{QR}$, $\overline{RS}$, $\overline{ST}$, $\overline{TV}$, and $\overline{VQ}$ are sides. Points Q, R, S, T, and V are vertices.

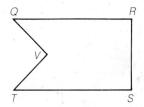

Examples Which figures are polygons? Tell why the others are not.

1.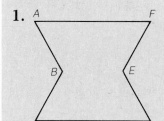

    Polygon

2.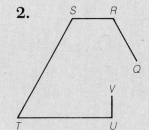

    Not a polygon. Sides $\overline{RQ}$ and $\overline{VU}$ do not intersect two other sides.

3.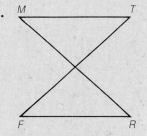

    Not a polygon. Side $\overline{MR}$ intersects three other sides ($\overline{MT}$, $\overline{FT}$, and $\overline{FR}$).

Try This... Which figures are polygons? Tell why the others are not.

a. 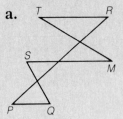 b. 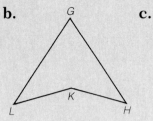 c.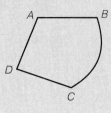

The table shows how we can classify polygons by the number of sides. Which polygons can you find in the photo below?

Number of Sides	Name of Polygon
3	Triangle
4	Quadrilateral
5	Pentagon
6	Hexagon
7	Heptagon
8	Octagon
9	Nonagon
10	Decagon
11	Undecagon
12	Dodecagon
n	n-gon

Examples Classify each polygon by the number of sides.

4. 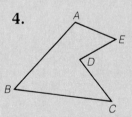 5. 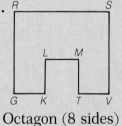 6.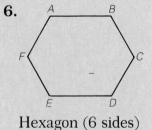

Pentagon (5 sides)   Octagon (8 sides)   Hexagon (6 sides)

Try This... Classify each polygon by the number of sides.

d. 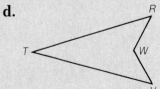   e.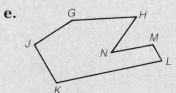

Polygons can also be classified as *convex* or *concave*. Consider the following examples.

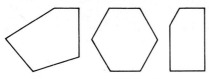

Convex polygons

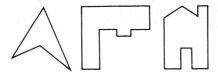

Concave polygons

Definition   A polygon is **convex** whenever no line containing a side of the polygon intersects the interior of the polygon. If a polygon is not convex, then it is **concave.**

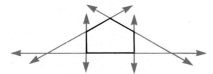

Convex: No lines contain points in the interior.

Concave: The line intersects the interior.

Examples   Classify each polygon as convex or concave.

7.

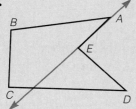

Concave; the line containing side $\overline{AE}$ intersects the interior.

8.

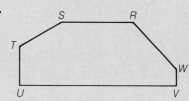

Convex; no line containing a side intersects the interior.

Try This...   Classify each polygon as convex or concave.

f.

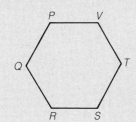

g.

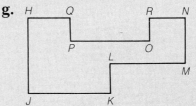

Exercises

Practice

Which figures in Exercises 1–6 are polygons? Tell why the other figures are not.

1.
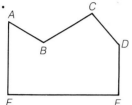

2.
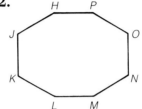

3.
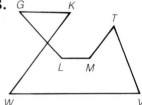

4.
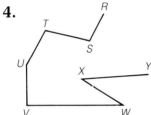

5.
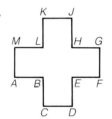

6.
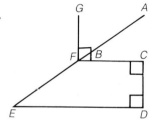

In Exercises 7–12, classify each polygon by the number of sides. Then classify each of the polygons as convex or concave.

7.
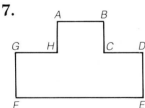

8.
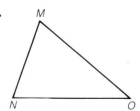

9.
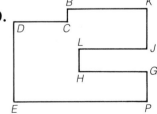

10.

11.
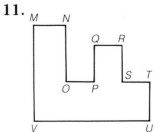

12.
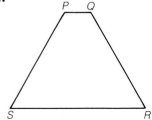

Extend and Apply

13. Which of the following is *not* a polygon?

(A)   (B)    (C)   (D)

Identify the polygon associated with each of the following.

14. Baseball diamond   **15.** Stop sign   **16.** Yield sign

Draw each of the following polygons.

17. A convex polygon with 5 sides   **18.** A concave polygon with 9 sides

19. A hexagon $ABCDEF$ with $AB = BC$ **20.** A concave polygon $PQRSTUVW$

Mixed Review

Can these numbers be the lengths of the sides of a triangle?

21. 7, 4, 11   **22.** 25, 13, 13   **23.** 9, 17, 7

Solve and check.

24. $6x + 9 = 10x - 19$ **25.** $10 - y = 4y - 30$ **26.** $6 + w = 30 - w$

Enrichment

Flexagons are polygons, folded from strips of paper, that change their faces when they are "flexed."

Make your own flexgon like this:

1. Prepare a strip of paper marked with equilateral triangles as shown.

2. Fold the strip backward along $\overline{AB}$.

3. Turn the figure upside down as shown.

4. Fold it backward again, along $\overline{CD}$.

5. Fold and paste or tape the last triangle on top of the first triangle.

6. Fold repeatedly along the dotted lines. Then pinch two adjacent triangles together and force them upward to turn the hexagon "inside out." If you mark the triangles with colors, you will see the changes that occur when you flex it.

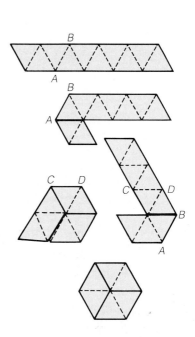

11-2 Diagonals of a Polygon

Objective: Calculate the number of diagonals of a polygon.

The design at the right was created by starting with a polygon and joining some of its vertices with string. A segment that joins two vertices of a polygon but is not a side of the polygon is called a *diagonal*.

Definition   A **diagonal** of a polygon is a segment that has vertices as endpoints and is not a side.

In pentagon *ABCDE*, $\overline{AC}$, $\overline{AD}$, $\overline{BD}$, $\overline{BE}$, and $\overline{CE}$ are diagonals.

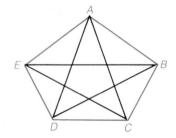

▬ *Explore* ▬

The following activity may be done with paper and pencil or by using the computer program at the end of this lesson.

1. Make a drawing of each of the following polygons: triangle, quadrilateral, pentagon, hexagon, heptagon, and octagon.
2. For each polygon, draw all the diagonals.
3. Prepare a chart that lists the polygons, the number of vertices of each polygon, and the number of diagonals of each polygon.
4. Predict the number of diagonals in a nonagon. Draw a nonagon and check your prediction.

We can find the number of diagonals of a polygon if we know
how many sides it has.

Theorem 11.1   For any polygon with n sides, the number of
diagonals, D, is $\frac{1}{2}n(n - 3)$.

The following Examples illustrate the use of Theorem 11.1.
A calculator may be helpful.

Examples **1.** How many diagonals does a 25-gon have?
The number of diagonals is $D = \frac{1}{2}n(n - 3)$.
Substituting 25 for n, we have

$$D = \frac{1}{2} \cdot 25(25 - 3)$$

$$= \frac{25}{2}(22)$$

$$= 25 \times 11$$

$$= 275$$

A 25-gon has 275 diagonals.

2. How many diagonals does a 30-gon have?
The number of diagonals is $D = \frac{1}{2}n(n - 3)$.
Substituting 30 for n, we have

$$D = \frac{1}{2} \cdot 30(30 - 3)$$

We can use a calculator to find the answer.
30 $\boxed{-}$ 3 $\boxed{=}$ $\boxed{\times}$ 30 $\boxed{\div}$ 2 $\boxed{=}$ $\boxed{\qquad 405}$

A 30-gon has 405 diagonals.

Try This... Find the number of diagonals in each of the
following polygns.

   a. Decagon
   b. 20-gon
   c. 35-gon

Discussion   Is it possible for two polygons with a different number
of sides to have the same number of diagonals?

Exercises

Practice

 Find the number of diagonals in each polygon.

1. 15-gon
2. 18-gon
3. 40-gon
4. 27-gon
5. 31-gon
6. 50-gon
7. 70-gon
8. 63-gon
9. 100-gon
10. 81-gon
11. 200-gon
12. 83-gon
13. 1000-gon
14. 405-gon

Extend and Apply

Find the number of diagonals of each polygon.

15.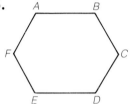

16.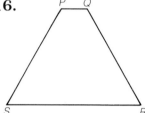

17.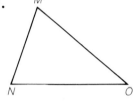

18. To make a piece of string art, Jason placed eye hooks at the vertices of an octagon. He used colored string to join each vertex to every other vertex. How many different string "segments" will be found in the string art? (Hint: Remember to count the sides.)

19. Twelve houses are arranged in a circle. The phone company wants to run a separate cable from each house to every other house. How many cables are needed?

20. Which of the following polygons has five diagonals?

    (A) Triangle   (B) Quadrilateral
    (C) Pentagon   (D) Hexagon

21. Find the number of sides of a polygon with no diagonals. How many vertices does the polygon have?

Use Mathematical Reasoning

22. How many diagonals of hexagon *ABCDEF* lie inside the polygon? How many diagonals lie outside the polygon?

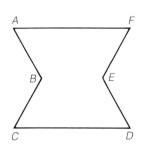

23. A polygon has exactly two diagonals. One diagonal lies inside the polygon and one diagonal lies outside the polygon. What type of polygon must this be?

Classify each polygon by the number of sides. Then classify each polygon as convex or concave.

24.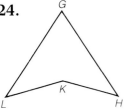

25.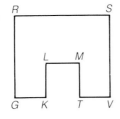

26.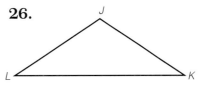

Solve and check.

27. $3z + 5 + 6z = 12z - 4$

28. $14 - x + 3x = 54 - 3x$

Enrichment

The following BASIC program draws regular polygons and their diagonals. It also tells how many sides and diagonals each polygon has.

```
10  HOME: HTAB 9: PRINT "DIAGONALS OF POLYGONS"
20  FOR D = 1 TO 3000: NEXT D
30  DIM X(25), Y(25)
40  FOR N = 4 TO 25
50  HGR: HCOLOR = 3
60  X(1) = 140: Y(1) = 150
70  FOR T = 1 TO N
80  X(T + 1) = 140 + 85 * SIN(6.3 * T/N)
90  Y(T + 1) = 80 + 70 * COS(6.3 * T/N)
100 NEXT T
110 VTAB 22: HTAB 5
120 PRINT N; " SIDES";
130 HTAB 25
140 PRINT N * (N - 3)/2; " DIAGONALS"
150 FOR T = 1 TO N + 1
160 FOR Q = T TO N + 1
170 HPLOT X(T), Y(T) TO X(Q), Y(Q)
180 NEXT Q: NEXT T
190 FOR D = 1 TO 3000: NEXT D
200 NEXT N
```

1. If a regular polygon has an odd number of sides, what do you notice about the center of the figure when its diagonals are drawn?

2. What happens if the regular polygon has an even number of sides? How does symmetry help explain this?

Simplify the Problem

Some problems seem difficult because of the large numbers involved. Sometimes we can solve such a problem by using the strategy *Simplify the Problem*. A problem can be simplified by considering a similar situation with smaller numbers.

Sample   There are 15 people in a room. Each person shakes hands with each other person one time. What is the total number of handshakes?

Simplify the problem by thinking of only four people in a room.

Draw a diagram of a rectangle, letting the vertices represent the people. The sides and the diagonals can represent the handshakes.

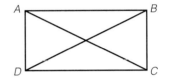

For example, diagonal $\overline{AC}$ represents persons A and C shaking hands.

There are four sides and two diagonals. So, there will be six handshakes.

Now we will apply the same idea to 15 people. Think of a 15-gon. There are 15 sides.

We can see from the simplified problem that we need to determine the number of diagonals of a 15-gon. The number of diagonals is given by Theorem 11.1.

$$D = \frac{1}{2}n(n - 3)$$
$$= \frac{1}{2} \cdot 15(15 - 3)$$
$$= \frac{1}{2} \cdot 15 \cdot 12$$
$$= 90$$

The total number of handshakes for 15 people is the number of sides (15) plus the number of diagonals (90), or 105 handshakes.

Problems

1. There are 25 people in a room. Each person shakes hands with each other person one time. What is the total number of handshakes?

2. How many telephone lines are needed to connect 10 buildings if each building is connected to each other building with a separate line?

11-3 Perimeter

Objective: Estimate and find the perimeter of polygons.

Roland Park is shaped like a hexagon. When Ann jogs once around the edge of the park, how far has she run? To answer this question, she needs to know the distance around the park, or its *perimeter*.

Definition   The **perimeter** of a polygon is the sum of the lengths of its sides.

The following Examples illustrate how to find and estimate perimeters.

Example   1. Find the perimeter of the polygon.

$p = AB + BC + CD + DA$
$= 8 + 4 + 7 + 3$
$= 22$ cm

The perimeter of polygon *ABCD* is 22 cm.

Try This... Find the perimeter of each polygon.

a.

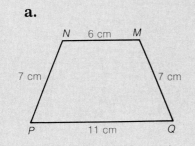

b.

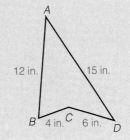

Example 2. First estimate the perimeter of the polygon in centimeters. Then measure its sides in centimeters to find the perimeter.

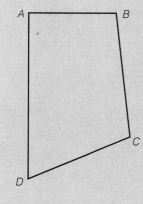

Estimate:

AB is about 2 cm, BC is about 3 cm, CD is about 3 cm, DA is about 4 cm

2 + 3 + 3 + 4 = 12 cm
So, the perimeter is about 12 cm.

By measuring, we find:
AB = 2.4 cm, BC = 3.3 cm, CD = 3 cm, DA = 4.4 cm

The perimeter is 2.4 + 3.3 + 3 + 4.4 = 13.1 cm.

Try This... First estimate the perimeter of each polygon in centimeters. Then measure its sides in centimeters to find the perimeter.

c.

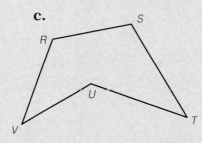

d.

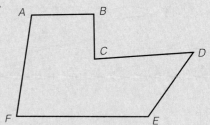

Discussion   Which of the following situations involve perimeter? the length of the coastline of an island; the amount of fertilizer needed for a lawn; the amount of carpet needed for a bedroom; the distance around a city block

Exercises

Practice

Find the perimeter of each polygon.

1.

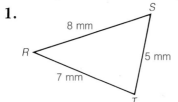

2.

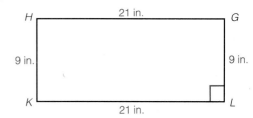

Find the perimeter of each polygon.

3.

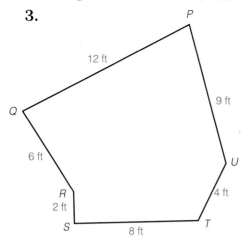

4.

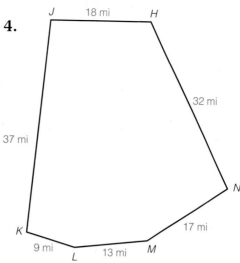

First estimate the perimeter of each polygon in centimeters. Then measure its sides in centimeters to find the perimeter.

5.

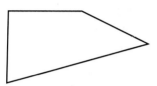

6.

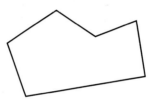

7.

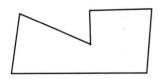

8.

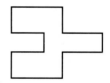

Extend and Apply

9. Find the cost of fencing the yard shown below. The fencing costs $1.35 per meter.

10. Find the cost of paneling for the room pictured. Omit 0.6 m for a doorway. The paneling costs $12.59 per meter.

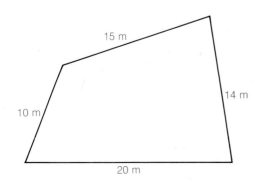

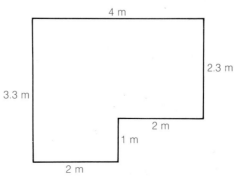

11. A room is to be decorated with paneling. The room is 12 ft by 16 ft with two doors that are each 3 ft wide. Find the total amount of paneling needed.

(A) 28 ft   (B) 25 ft   (C) 53 ft   (D) 50 ft

12. A wood frame is to be built for a king-size waterbed. The waterbed measures $6\frac{1}{2}$ ft by 7 ft. How many feet of wood are needed? Suppose the wood costs $1.12 per foot and is available only in lengths of 8 ft. How many lengths should be purchased and what will be the total cost? How much waste will there be?

13. An equilateral triangle has a perimeter of 48. What are the lengths of the sides of the triangle?

14. A square has a perimeter of 542. What are the lengths of the sides of the square?

Use Mathematical Reasoning

15. A rectangle has a perimeter of 66 in. The length of one side of the rectangle is 18 in. What are the lengths of the other sides?

16. An isosceles triangle has a perimeter of 40. One of the two congruent sides has a length of 15. What are the lengths of the other sides?

17. Can two congruent polygons have different perimeters? Why or why not?

18. A polygon has a perimeter of 7 cm. What is the perimeter of its image after a reflection or translation? Why?

19. A polygon has a perimeter of 7 cm. What is the perimeter of its image after a size transformation with a scale factor of 3? Why?

Mixed Review

 Find the number of diagonals in each polygon.

20. Decagon   **21.** 24-gon   **22.** 37-gon

Identify each pair of angles as corresponding angles, interior angles, alternate interior angles, or alternate exterior angles.

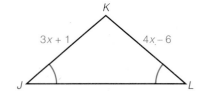

23. $\angle 1$ and $\angle 7$ **24.** $\angle 1$ and $\angle 8$

25. $\angle 5$ and $\angle 3$ **26.** $\angle 8$ and $\angle 4$

27. $\angle 2$ and $\angle 6$ **28.** $\angle 6$ and $\angle 4$

29. Find x.   **30.** Find KL.

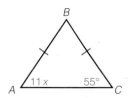

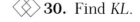

11-4 Angles of a Polygon

Objective: Solve problems involving the angles of a polygon.

Explore

Work in small groups. Computer software may be used.

1. Each member of the group should draw a different convex polygon (pentagon, hexagon, and so on).

2. In each polygon, draw all the diagonals from one vertex, as shown.

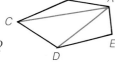

3. In each case, how many triangles are formed?

4. What is the sum of the measures of the angles of each triangle? of each polygon?

Discuss   Predict the sum of the angle measures of a polygon having n sides. Predict the sum of the exterior angles of a polygon having n sides.

If we know the number of sides of a convex polygon, we can determine the sum of its angle measures.

Theorem 11.2   The sum S of the angle measures of any convex polygon with n sides is given by the formula $S = (n - 2)180$.

Examples **1.** Find the sum of the angle measures of a 50-gon.

$$S = (n - 2)180$$
$$= (50 - 2)180 \quad \text{Substituting 50 for } n$$
$$= 48 \cdot 180 = 8640$$

The sum of the angle measures of a 50-gon is 8640°.

2. Find the sum of the angle measures of the polygon in a stop sign.

A stop sign has the shape of an octagon, so $n = 8$.

$$S = (n - 2)180$$
$$= (8 - 2)180$$
$$= 6 \cdot 180 = 1080$$

The sum of the angle measures in a stop sign is 1080°.

Try This... Find the sum of the angle measures of each polygon.

a. 70-gon **b.** 57-gon

c. Polygon *HJKLMNOP*

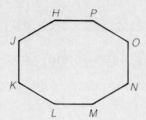

We have already discussed exterior angles of triangles. In the figure, $\angle CBF$ is an *exterior angle* of pentagon *ABCDE*. For clarity, we sometimes refer to the angles of a polygon as *interior angles*.

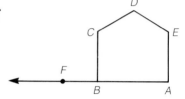

Definition An angle is an **exterior angle** of a convex polygon whenever it forms a linear pair with an angle of the polygon.

Explore

Work in small groups. Computer software may be used.

1. Each member of the group should draw a different convex polygon.
2. Draw an exterior angle at each vertex of the polygon, as shown.
3. Measure the exterior angles and find the sum of the measures.
4. Compare your results with those of classmates in your group. What do you notice?

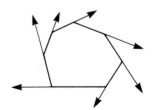

Theorem 11.3 summarizes the results of this Explore.

Theorem 11.3 For a convex polygon, the sum of the measures of the exterior angles, one at each vertex, is 360°.

For the heptagon shown,
$a + b + c + d + e + f + g = 360°$

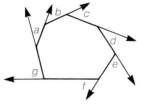

Example   **3.** Three of the four exterior angles of a quadrilateral have measures 45°, 73°, and 174°. Find the measure of the fourth exterior angle.

$$45 + 73 + 174 = 292$$

The sum of the three exterior angles is 292°.

By Theorem 11.3, we know that the sum of all four angles is 360°.

$$360 - 292 = x$$
$$68 = x$$

So, the measure of the fourth exterior angle is 68°.

Try This...   **d.** Four of the five exterior angles of a pentagon have measures 53°, 26°, 84°, and 103°. Find the measure of the fifth exterior angle.

Discussion   As the number of sides of a polygon increases, what happens to the sum of the measures of its angles? What happens to the measure of each angle? What happens to the sum of the measures of the exterior angles?

Exercises

Practice

 Find the sum of the angle measures of each polygon.

1. Decagon
2. Octagon
3. 40-gon
4. 38-gon
5. 100-gon
6. 200-gon
7. 267-gon
8. 320-gon
9. 800-gon
10. 903-gon
11. 1000-gon
12. 1021-gon

13. Six of the seven exterior angles of a heptagon have measures 18°, 12°, 42°, 75°, 38°, and 87°. Find the measure of the seventh exterior angle.

14. Nine of the ten exterior angles of a decagon have measures 12°, 22°, 16°, 20°, 10°, 18°, 23° 40°, and 51°. Find the measure of the tenth exterior angle.

15. Seven of the eight exterior angles of an octagon have measures 24°, 36.5°, 29.5°, 51.5°, 73°, 16°, and 38°. Find the measure of the eighth exterior angle.

16. Five of the six exterior angles of a hexagon have measures 83°, 24.5°, 75.75°, 39.6°, and 112.2°. Find the measure of the sixth exterior angle.

Extend and Apply

17. Which of the following is $m\angle A$?

(A) 70° (B) 30° (C) 90° (D) 65°

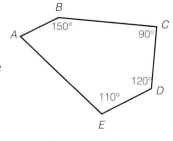

18. A hexagon is equiangular. Find the measure of each of its angles and each of its exterior angles.

19. One familiar pentagonal shape is home plate on a baseball diamond. The home plate has three right angles as shown. Find the measures of the other two angles.

20. The sum of the angles of a polygon is 1620°. How many sides does the polygon have?

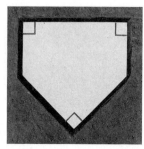

Use Mathematical Reasoning

Solve for x and find the measures of the exterior angles in each polygon.

21. The measures of the exterior angles of a heptagon are $2x$, $3x$, $4x$, $5x$, $7x$, $9x$, and $10x$.

22. The measures of the exterior angles of a nonagon are $6x$, $7x$, $8x$, $9x$, $10x$, $11x$, $12x$, $13x$, and $14x$.

23. The measures of the exterior angles of a hexagon are $4x$, $3x - 2$, $7x + 10$, $8x + 6$, $5x - 6$, and $2x + 4$.

24. The measures of the exterior angles of a pentagon are $3x$, $4x + 7$, $7x - 2$, $x + 12$, and $2x + 3$.

25. The sum of the angle measures of an equiangular polygon is 5040°. Find the measure of each angle.

26. Begin with the formula in Theorem 11.2 to develop a formula for finding the measure of one angle of an n-sided equiangular polygon.

Mixed Review

27. Find the perimeter of polygon $PQRST$.

Solve and check.

28. $3(x - 5) = x + 5$

29. $6(3x + 1) = 20x$

30. $16 = 8(6 - 2y)$

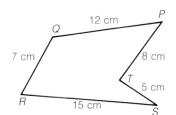

11-5 Similar Polygons

Objective: Identify similar polygons and solve problems related to similar polygons.

The two wrenches illustrate similar polygons. Similar polygons, like similar triangles, have the same shape but not necessarily the same size. The larger wrench is a magnification of the smaller wrench.

Recall the definition of similarity from Chapter 9. Two polygons are *similar* whenever their vertices can be matched so that the corresponding angles are congruent and the lengths of corresponding sides are proportional.

The polygons at the right are similar, since corresponding angles are congruent (all are 90°) and $\frac{8}{10} = \frac{4}{5} = \frac{8}{10} = \frac{4}{5}$.

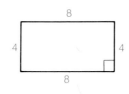
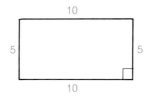

Similarity can be used to find the lengths of sides of polygons.

Example   **1.** Rectangle *ABCD* ~ rectangle *PQRS*. Find *PQ*.

Because the rectangles are similar, the lengths of corresponding sides are proportional.

Thus, $\frac{3}{6} = \frac{5}{PQ}$

$3(PQ) = 6 \cdot 5$

$3(PQ) = 30$

$PQ = 10$

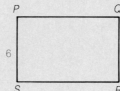

Try This... a. Trapezoid *RSTV* ~ trapezoid *DEFG*. Find *DE* and *GF*.

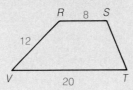

Exercises

Practice

In Exercises 1–8, the polygons shown are similar.
Find the indicated lengths.

1. Kite $RSTV \sim$ kite $ABCD$.
Find BC.

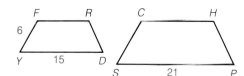

2. $\square GHJK \sim \square TVWX$. Find TV.

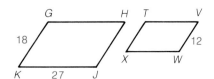

3. Trapezoid $FRDY \sim$ trapezoid $CHPS$. Find CS.

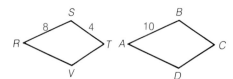

4. Rectangle $PQRS \sim$ rectangle $DEFG$. Find PQ.

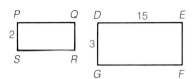

5. Pentagon $DEFGH \sim$ pentagon $RSTUV$. Find RS and UV.

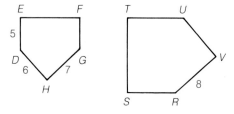

6. Hexagon $FGHJKL \sim$ hexagon $XRSTVW$. Find RX.

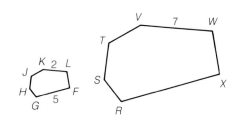

7. Octagon $PQRSTUVW \sim$ octagon $DEFGHJKL$. Find DL and WV.

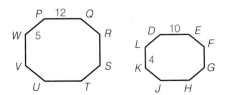

8. The two pentagons are similar. Find w, x, y, and z.

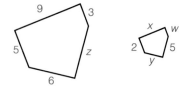

Extend and Apply

◇ **9.** Pentagon *ABCDE* is similar to pentagon *PQRST*. Which of the following is *TS*?

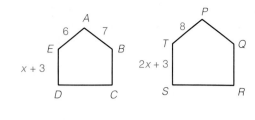

(A) $4\frac{1}{2}$ (B) $1\frac{1}{2}$

(C) 6   (D) 8

10. Draw two polygons that are not similar but whose sides have lengths that are proportional.

11. Can two polygons be similar if they do not have the same number of sides? Explain how you know.

12. A copy machine is to enlarge a diagram to 125% of its original size. A rectangle in the diagram is 9 cm by 12 cm. What are the dimensions of the enlarged rectangle?

13. A copy machine is to reduce a diagram to 75% of its original size. A right triangle in the diagram has a hypotenuse that is 20 cm long. What is the length of the hypotenuse of the reduced triangle?

14. Stan and Gina created a design 6 in. by 8 in. for a piece of cloth that is 1 ft wide. They plan to cross-stitch the same design on a piece of cloth that is 18 in. wide. What should be the measurements of the new design?

Mixed Review

Find the sum of the angle measures of each polygon.

15. 14-gon   **16.** Heptagon   **17.** 50-gon

18. Five of the six exterior angles of a hexagon have measures 33°, 50°, 68°, 80°, and 82°. Find the measure of the sixth exterior angle.

◇ Solve and check.

19. $5w + 6w = w + 40$ **20.** $9x + 12 - 4x = 6x + 2$

21. $y + 16 = 7y + 2y$

Visualization

Which piece at the right is not part of the design at the left?

(A)   (B)   (C)

11-6 Regular Polygons

Objective: Draw and find angle measures of regular polygons.

Explore

The following activity may be done using computer software.

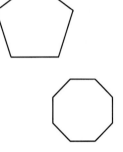

1. Draw or trace several polygons that are both equilateral and equiangular. You may use the polygons below and at the right.

2. For each polygon, calculate the sum of the angle measures.

3. For each polygon, use a protractor to measure one interior angle.

4. How is the angle measure in Step 3 related to the value you found in Step 2?

Discuss   How can you calculate the measure of an interior angle of an equilateral, equiangular polygon without measuring?

Many commonly occurring polygons are both equilateral and equiangular. These include octagonal stop signs, square tiles, and hexagonal bolts. Such polygons are called *regular*.

Definition   A convex polygon is **regular** whenever it is both equilateral and equiangular.

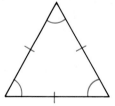 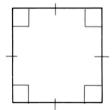 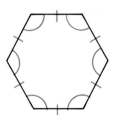

All angles of a regular polygon have the same measure. Thus, to find the measure of each angle, we can divide the sum of the angle measures by the number of angles. Recall from Theorem 11.2 that the sum of the angle measures of an n-sided polygon is $(n - 2)180$.

> **Theorem 11.4** The measure of an interior angle of a regular polygon with n sides is
> $$\frac{(n-2)180}{n}$$

Example 1. Find the measure of each interior angle of a regular decagon.

Substituting 10 for n, we have

$$\frac{(n-2)180}{n} = \frac{(10-2)180}{10}$$

$$= \frac{8 \times 180}{10}$$

$$= \frac{1440}{10} = 144$$

Each angle measures $144°$.

Try This... **a.** Find the measure of each interior angle of a regular octagon.

The following theorem about the exterior angles of a regular polygon follows directly from Theorem 11.4:

> **Theorem 11.5** The measure of each exterior angle of a regular polygon with n sides is $\frac{360}{n}$.

Example 2. Find the measure of each exterior angle of a regular heptagon.

$$\frac{360}{n} = \frac{360}{7} = 51\frac{3}{7}$$

Each exterior angle measures $51\frac{3}{7}°$.

Try This... **b.** Find the measure of each exterior angle of a regular nonagon.

Regular polygons can be drawn using computer graphics. Although some computer graphics programs draw regular polygons automatically, others do not. The instruction booklets often suggest a procedure similar to the one below.

Example 3. Draw a regular octagon.

Step 1 Draw a circle. We think of the circle as having 360 degrees.

Step 2 Because an octagon has eight sides, we divide 360 by 8.

$360 \div 8 = 45$

Draw eight angles, each with measure 45° and vertex at the center of the circle. This will divide the circle into eight congruent parts.

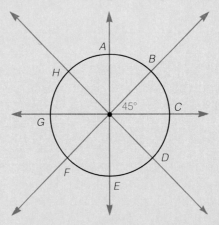

Step 3 Use a straightedge to connect the points where the rays intersect the circle.

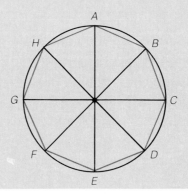

Try This... c. Draw a regular pentagon using the method of Example 3.

Exercises

Practice

Find the measure of each interior angle of these regular polygons. Then find the measure of each exterior angle.

1. Hexagon
2. Pentagon
3. Triangle
4. Quadrilateral
5. Heptagon
6. Nonagon
7. 20-gon
8. 50-gon
9. 100-gon
10. 1000-gon

Draw these regular polygons using the method of Example 3.

11. Hexagon
12. Decagon
13. Nonagon
14. Quadrilateral
15. Dodecagon
16. 15-gon

Extend and Apply

17. *ABCDE* is a regular pentagon. Which of the following is x?

    (A) 36   (B) 108   (C) 72   (D) 60

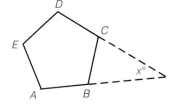

18. Mari is building a music box shaped like a regular hexagon with a side length of 12 cm. How many centimeters of metal trim will she need to fit around the box? Allow 3 cm for waste.

19. The Pentagon building near Washington, D.C., has sides that measure about 1050 ft each. About how far must a security guard walk to go around the outside of the building?

20. Draw an equilateral quadrilateral that is not regular. What is the polygon called?

21. Draw an equiangular quadrilateral that is not regular. What is the polygon called?

22. Draw an equiangular hexagon that is not regular.

23. Draw an equilateral hexagon that is not regular.

Use Mathematical Reasoning

Determine the number of lines of symmetry for each figure.

24. Equilateral triangle   **25.** Square   **26.** Regular pentagon

27. Regular hexagon   **28.** Regular 25-gon   **29.** Regular n-gon

30. Describe the lines of symmetry for a regular polygon with an odd number of sides.

31. Describe the lines of symmetry for a polygon with an even number of sides.

Mixed Review

Find the indicated lengths.

32. Quadrilateral *RSTV* ~ quadrilateral *DEFG*. Find *DE* and *GF*.

◈ **33.** Quadrilateral *ABCD* ~ quadrilateral *GHJK*. Find *DC* and *JK*.

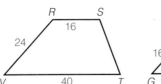

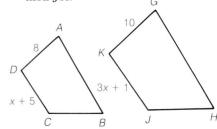

Enrichment

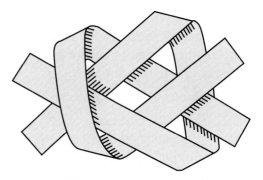

Some regular polygons can be constructed by tying knots in strips of paper. Use notebook paper or computer paper to cut out three strips that are about 11 in. long and $\frac{1}{2}$ in. wide.

1. Carefully tie a knot in one strip as shown.
2. Gently pull the ends tight and flatten the knot.
3. Cut off the extra paper to form a pentagon.

4. Tie a square knot with two strips as shown.
5. Gently pull the ends tight and flatten the knot.
6. Cut off the extra paper to form a hexagon.

Mean and Median

Statisticians often try to give information about a set of numbers in a concise way. Two of the most useful concepts for this purpose are the *mean* and the *median*.

The **mean**, or **average**, of a set of values is the sum of all the values divided by the number of values.

The **median** of a set of values is the middle value when all the values are arranged in order.

Sample Find the mean and the median of the angle measures of pentagon *ABCDE*.

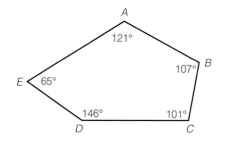

The sum of the values is
121 + 107 + 101 + 146 + 65 = 540
There are 5 values, so the mean is
$\frac{540}{5} = 108°$.

To find the median, we arrange the values in order: 65°, 101°, 107°, 121°, 146°. The middle value is 107°, so the median is 107°.

Notice that the mean of the angle measures of the pentagon, 108°, is the measure of any one angle of a *regular* pentagon. The Problems below explore this further.

In finding medians, there is always a middle value when there are an odd number of values. If the number of values is even, we take the average of the two middle values as the median. For example, the median of the values 1, 3, 4, and 8 is the average of 3 and 4. Thus, the median is $\frac{3 + 4}{2} = \frac{7}{2}$, or $3\frac{1}{2}$.

Problems

1. The angle measures of a quadrilateral are 48°, 157°, 71°, and 84°. Find the mean and the median of these values. How does the mean value relate to the measure of an angle of a regular quadrilateral?

2. What is the mean of the angle measures of any octagon? Is it possible to determine the median without more information?

11-7 Tessellations

Objective: Recognize and create tessellations.

Nature produces many patterns like the fish scales at the right. Mathematicians call these **tiling patterns**, or **tessellations.**

A tessellation is an arrangement of figures that fill the plane but do not overlap or leave gaps. When the same figure is used throughout, the tessellation is called a **pure tessellation.** A wall made of congruent rectangular bricks is an example of a pure tessellation. Here are some others.

Examples Name the figure used in each tessellation.

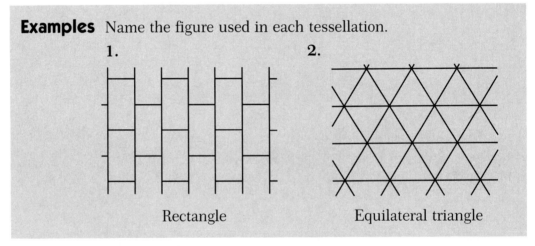

1.   2.

Rectangle   Equilateral triangle

Try This... Name the figure used in each tessellation.

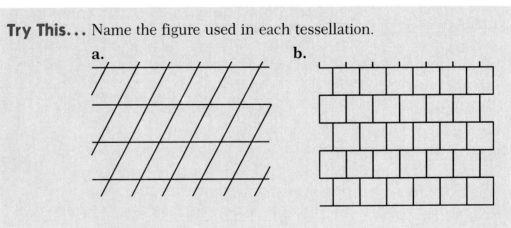

a.   b.

Many types of polygons can be used to create pure tessellations of the plane. However, the only *regular* polygons that can be used to create pure tessellations are equilateral triangles, squares, and regular hexagons.

The following Explore investigates tessellating the plane with nonregular polygons.

Explore

1. Stack two sheets of paper together and fold them twice to make eight layers.
2. Draw a large scalene triangle and cut it out so that you have eight congruent scalene triangles.
3. Try to create a tessellation with your triangles.
4. Repeat the activity with eight congruent quadrilaterals.
5. Can any triangle be used to create a pure tessellation? Can any quadrilateral be used to create a pure tessellation?

Theorem 11.6 Any triangle or quadrilateral can be used to create a pure tessellation.

Example **3.** Create a pure tessellation using a right triangle. Here is one way a right triangle may be used to tessellate the plane.

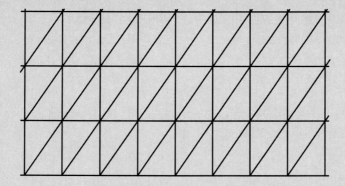

Try This... c. Create a pure tessellation using an isosceles trapezoid.

Exercises

Practice

Name the figure used in each tessellation.

1.

2.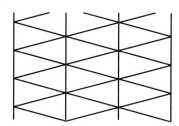

3.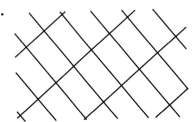

4.

5.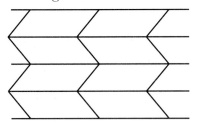

6.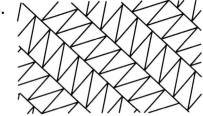

Create a pure tessellation using the following polygons.

7. Equilateral triangle

8. Parallelogram

9. Kite

10. Non-isosceles trapezoid

Extend and Apply

11. Which of the following polygons *cannot* be used to create a pure tessellation of the plane?

   (A) Square

   (B) Regular pentagon

   (C) Regular hexagon

   (D) Equilateral triangle

12. An interior designer wants to tile a kitchen floor. To control costs, only small, square tiles of one color will be used. Sketch three different tessellations that the designer might create.

Tessellations that use more than one type of polygon are called **semi-pure tessellations.** The tessellation at the right is an example. Use the following polygons to create a semi-pure tessellation. (Hint: You may wish to trace the polygons or make a template out of cardboard.)

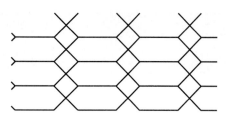

13.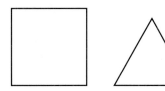

14.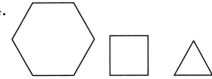

15.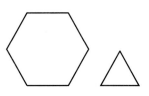

16.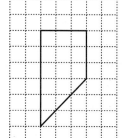

Use Mathematical Reasoning

Determine whether each statement is true or false.

17. Any polygon can be used to tessellate the plane.

18. Grid paper is an example of a pure tessellation.

19. The plane can be tessellated with regular octagons.

20. There is only one way to create a tessellation with squares.

21. A quadrilateral of any size or shape can be used to tessellate the plane.

Mixed Review

Find the measure of each interior angle of these regular polygons. Then find the measure of each exterior angle.

22. Octagon   **23.** Triangle   **24.** Decagon

Use grid paper to draw a figure similar to the given figure, with the indicated scale factor.

25. Scale factor of 2   **26.** Scale factor of $\frac{1}{3}$

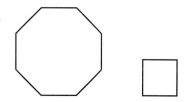

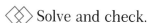 Solve and check.

27. $4(x + 9) = 10x + 6$ **28.** $3(12 - y) = 3y$ **29.** $1 + 5y = 4(10 - 2y)$

Tessellations with Translations

The tessellation at the right was created by
M. C. Escher. Escher may have made this tessellation
of birds and fish by starting with a rectangle, modifying
it to the shape of a fish, and then translating the fish.
Do you see how the artist might also have started by
drawing a bird in a rectangle and translating that
drawing?

Here is one way you can create a tessellation of flags.

1. Start with a rectangle and draw a curve on side $\overline{AB}$.

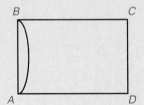

2. Trace the curve on side $\overline{AB}$ and translate it to side $\overline{CD}$.

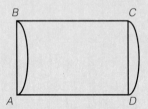

3. Now draw a curve on side $\overline{AD}$.

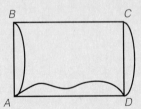

4. Trace the curve on side $\overline{AD}$ and translate it to side $\overline{BC}$.

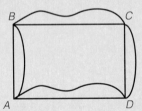

5. The figure can now be filled in and repeated (translated) horizontally and vertically to produce a tessellation.

Problems

1. Begin with a rectangle and follow the above steps to create your own tessellation.

2. Create a tessellation beginning with a parallelogram.

Chapter 11 Review

Which figures are polygons? Tell why the others are not. *11-1*

1.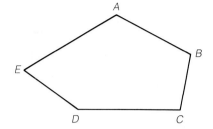

2.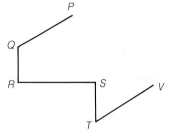

Classify each polygon by the number of sides. Then classify each polygon as convex or concave.

3.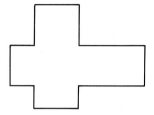

4.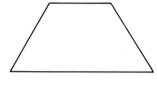

Find the number of diagonals in each polygon. *11-2*

5. Nonagon

6. 45-gon

Find the perimeter of each polygon. *11-3*

7.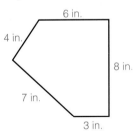

8.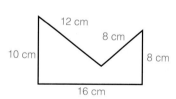

First estimate the perimeter of each polygon in centimeters. Then measure its sides in centimeters to find the perimeter.

9.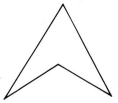

10.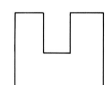

11-4 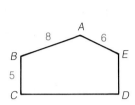 Find the sum of the angle measures of each polygon.

11. 18-gon   **12.** Pentagon

13. Three of the four exterior angles of a quadrilateral have measures 81°, 92°, and 69°. Find the measure of the fourth exterior angle.

11-5   **14.** Pentagon *ABCDE* ~ pentagon *GKLMN*. Find *GN* and *LK*.

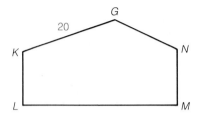

11-6 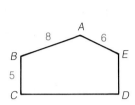 Find the measure of each interior angle of these regular polygons. Then find the measure of an exterior angle.

15. Octagon   **16.** 16-gon

11-7   Create a pure tessellation using the following polygons.

17. Isosceles triangle   **18.** Square

Chapter 11 Test

Which figures are polygons? Tell why the others are not.

1. 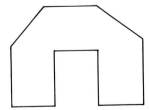   **2.**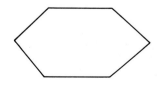

Classify each polygon by the number of sides. Then classify each polygon as convex or concave.

3. 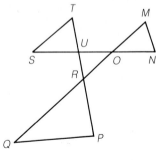   **4.**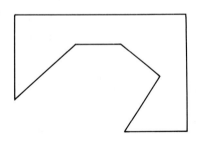

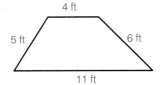
Find the number of diagonals in each polygon.

5. Decagon

6. 35-gon

Find the perimeter of each polygon.

7.

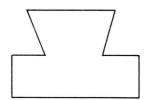

4 ft
5 ft
6 ft
11 ft

8.

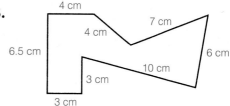

4 cm
7 cm
4 cm
6.5 cm
6 cm
10 cm
3 cm
3 cm

First estimate the perimeter of each polygon in centimeters. Then measure its sides in centimeters to find the perimeter.

9.

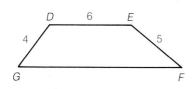

10.

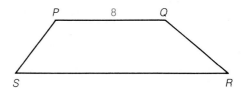

Find the sum of the angle measures of each polygon.

11. 14-gon

12. Hexagon

13. Four of the five exterior angles of a pentagon have measures 91°, 102°, 43°, and 16°. Find the measure of the fifth exterior angle.

14. Quadrilateral *DEFG* ~ quadrilateral *PQRS*. Find *PS* and *QR*.

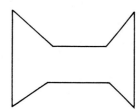

D   6   E
4   5
G   F

P   8   Q
S   R

Find the measure of each interior angle of these regular polygons. Then find the measure of an exterior angle.

15. Dodecagon

16. 30-gon

Create a pure tessellation using the following polygons.

17. Isosceles trapezoid

18. Regular hexagon

The solar energy collected by each rectangular panel is directly related to its area.
The total amount of energy collected depends on the total area of the panels.

12-1 The Meaning of Area

Objective: Estimate areas of simple polygonal regions.

Applications such as finding the amount of paint to cover a house or how much carpet to buy to cover a floor involve **area.** When we measure area, we measure the amount of surface covered by a **polygonal region.** A polygonal region is a polygon and its interior.

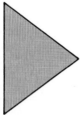
Triangular region

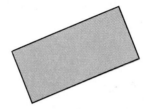
Rectangular region

Hexagonal region

Postulate 17   **The Area Postulate**   For every polygonal region, there corresponds one positive number called its **area.** The number is dependent upon the given unit.

For the area of a polygonal region *ABCD*, we write *A(ABCD)*. We usually use a square region as a *unit* of area. The area of a region is the number of square units contained within the region.

Examples   Estimate the area of each region. Use the small square as the unit.

1.
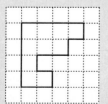

The area appears to be 10 square units.

2.
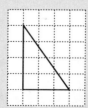

There are 3 complete units and 6 partial units. Estimate:
$3 + \frac{1}{2}(6) =$
6 square units.

3.
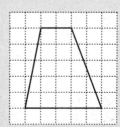

There are 12 complete units and 11 partial units. Estimate: $12 + \frac{1}{2}(11) =$ 17.5 square units.

Try This... Estimate the area of each region. Use the small square as the unit.

a.

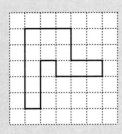

b.

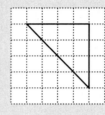

c.

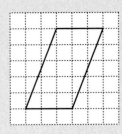

When we estimated the areas of the regions above we made use of the idea stated in the following postulate:

Postulate 18 The area of a polygonal region is the sum of the areas of the nonoverlapping regions that it contains.

 *Explore*

Use graph paper. Let the square on the paper be the unit of area.
1. Draw a figure with an area of 5 square units.
2. Draw a different figure with an area of 5 square units.
3. Draw two congruent figures. What is the area of each?
4. Can different noncongruent figures have the same area? What can you say about the areas of congruent figures?

Postulate 19 Congruent regions have the same area.

Example 4. *ABCE* is a rectangle and *A*(△*CEA*) = 12. Find *A*(*ABCD*).

Since *ABCE* is a rectangle, △*ABC* ≅ △*CEA*. By SAS, △*CEA* ≅ △*CED*. The three small triangles are congruent.

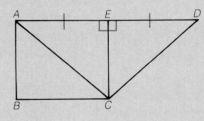

By Postulate 19,
A(△*ABC*) = *A*(△*CEA*) = *A*(△*CED*) = 12.

By Postulate 18,
A(*ABCD*) = 12 + 12 + 12 = 36.

Try This... d. $A(\triangle ADC) = 4.$
Find $A(\triangle ABC).$

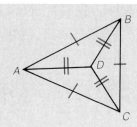

Area refers to a region. For convenience, we will refer to the area of a *triangle,* rather than always stating the complete term, *triangular region.* We will follow the same convention for other polygonal regions.

Discussion Suppose two polygons have the same perimeter. What can you say about their areas? Suppose two polygons have the same area. Must the two polygons be congruent?

Exercises

Practice

Estimate the area of each region. Use the square as the unit.

1.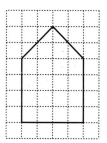

2.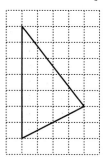

3.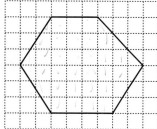

4.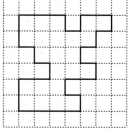

5.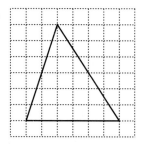

6.

7.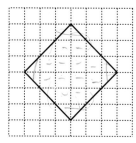

8.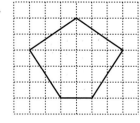

9.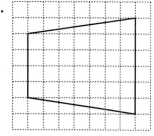

Find the area of each polygon.

10. *ABCD* is a square. $A(\triangle AED) = 7$. Find the area of *ABCD*.

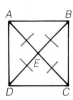

11. *ABCDE* is a regular pentagon. $A(\triangle DFC) = 3$. Find the area of *ABCDE*.

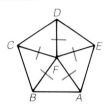

12. *ABCDEF* is a regular hexagon. $A(ABCO) = 7$. Find the area of *ABCDEF*.

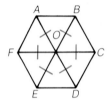

13. All the triangles in this figure are congruent. $A(BDFH) = 10$, and $A(\triangle ABH) = 7$. Find the area of *ABCDEFGH*.

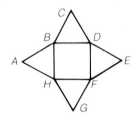

14. *ABCDEFGH* is a regular octagon. $A(\triangle ABO) = 12$. Find the area of *ABCDEFGH*.

15. All the triangles in this figure are congruent. $A(\triangle LAB) = 8$, and $A(BDHK) = 12$. Find the area of *ABCDEFGHIJKL*.

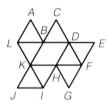

Extend and Apply

Determine whether each situation involves perimeter or area.

16. The amount of carpet for a room

17. The amount of fencing needed to surround a swimming pool

18. The length of the sidewalk around a house

19. The amount of paint needed to cover a house

20. The amount of wallpaper needed for a room

21. The length of a wallpaper border to go around the top of a room

22. The amount of roofing shingles needed to cover a roof

Use graph paper. Let the square on the paper be the unit of area.

23. Draw three different rectangles, each with an area of 24 square units. Find the perimeter of each.

24. Draw a rectangle with perimeter of 16 and area 12 square units.

25. Draw as many regions as you can that have an area of 5 square units.

Estimate the area of the shaded region.

26.

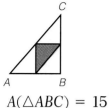

$A(\triangle ABC) = 15$

27.

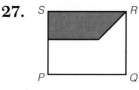

$A(PQRS) = 40$

28.

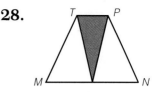

$A(MNPT) = 60$

29. Estimate the area of your desk using standard and metric measurements.

30. $A(ABCD) = 10$, and $A(EFGH) = 10$. Which of these could be the area of $AFGD$?

(A) 10   (B) 15   (C) 20   (D) 25

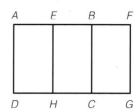

Use Mathematical Reasoning

31. $ABCD$ is a parallelogram with an area of 20 square units. $A(\triangle AED) = 5$. Find $A(\triangle AEB)$.

Mixed Review

Determine whether each statement is true or false.

32. A 17-gon has 17 vertices.

33. In an isosceles right triangle, the hypotenuse is twice the length of the shorter leg.

34. A triangle with side lengths 3, 4, and 5 is similar to a triangle with side lengths 5, 12, and 13, by SSS similarity.

35. Squares and rectangles are regular quadrilaterals.

◇ Simplify and solve.

36. $3x = \sqrt{36}$ **37.** $7^2 + y^2 = 25^2$

Visualization

The large square below represents one square unit.

Which of the following does *not* have an area of one-half of a square unit?

(A)   (B)   (C)   (D)

Algebra Connection

Exponents

The number 32 can be written as a product of 2s. That is,

$$32 = 2 \times 2 \times 2 \times 2 \times 2$$

There are 5 factors of 2, so 32 is called a **power** of 2. Another way we can write this product is 2^5. This is called **exponential notation**, where 5 is the **exponent** and 2 is the **base.** The exponent tells how many times the base is used as a factor.

$$b^n \text{ means } b \times b \times b \ldots \times b$$
$$\underleftarrow{\qquad n \text{ times} \qquad}$$

Samples Write using exponential notation.

1. $125 = 5 \times 5 \times 5 = 5^3$ (5 is used as a factor 3 times.)

2. $64 = 2 \times 2 \times 2 \times 2 \times 2 \times 2 = 2^6$

3. Write 10^5 using *standard* notation.
   $10^5 = 10 \times 10 \times 10 \times 10 \times 10 = 100,000$

   The $\boxed{x^2}$ key can be used to square numbers on a calculator.
   The $\boxed{y^x}$ key is used for exponents other than 2.

4. Evaluate 2^8.
   $2 \boxed{y^x} 8 \boxed{=} \boxed{\boxed{256}}$

Problems

Write using exponential notation.

1. 1,000,000,000
2. 10,000,000,000,000
3. 128
4. 81

Write using standard notation.

5. 10^2
6. 10^4
7. 3^5
8. 6^3

 Use a calculator to evaluate.

9. 5^{10}
10. 17^1
11. 412^2
12. 28^0

13. Can the $\boxed{y^x}$ key be used to square numbers?

14. Compare 2^4 and 4^2. Does this result hold true for any pair of numbers?

12-2 Area of Rectangles

Objective: Solve problems involving the area of a rectangle.

Explore

Use graph paper. Let the square on the paper be the unit of area.

1. Draw five rectangles with different areas on the graph paper.

2. Find the area of each rectangle by counting the squares.

3. Determine the lengths, in squares, of the sides of each rectangle.

Record your results in a chart.

Discuss Look for a pattern. How could you find the area of each of the rectangles without counting all of the squares?

In the discussion, you may have realized that, instead of counting square units, we can find the area of a square by multiplying the lengths of two adjacent sides. We often call these two sides the length and the width of the rectangle. We can also call these sides the **base** and the **height.** Notice that we can rotate the rectangle so that any side is the base and its adjacent side is the height.

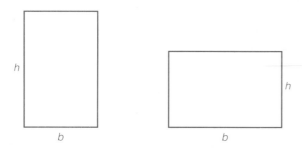

Postulate 20 The area of a rectangle is the product of the lengths of any two adjacent sides.

$A = bh$, where A is the area, b is the base length, and h is the corresponding height.

If the side of the square region that is being used as the unit of area is 1 centimeter in length, then the unit of area is called 1 square centimeter, which we abbreviate as cm^2. Similarly, we can have units of $in.^2$, ft^2, yd^2, m^2, and so on.

Because a square is a rectangle whose base length and height are equal, we have the following:

$$A = bh$$
$$A = bb \quad \text{Substituting } b \text{ for } h$$
$$A = b^2 \quad \text{Using exponential notation}$$

Theorem 12.1 The area of a square is the square of the length of a side s.

Examples Find the area of each rectangle.

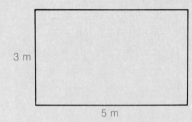

1. $A = bh$
   $A = (5 \times 3)$
   $= 15 \text{ m}^2$

2. $A = s^2$
   $A = (17.5)^2$
   $= 306.25 \text{ ft}^2$

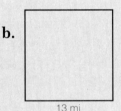

Try This... Find the area of each rectangle.

a.

b.

Applications of area often require that we first find the area of one or more rectangles.

Example 3. Tina plans to buy tiles to cover her patio floor. Each tile measures 1 foot on a side. How many tiles does she need?

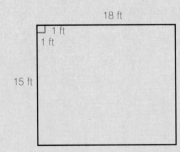

We need to find the area of a rectangle that is 18 ft × 15 ft.

$$A = bh$$
$$= (18)(15)$$
$$= 270 \text{ ft}^2$$

The area of the patio is 270 ft². Each tile covers 1 square foot, so Tina needs 270 tiles. We can see by the size of one tile that this is a reasonable answer.

Discussion Describe as many rectangles as you can that have an area of 18 cm².

Exercises

Practice

Find the area of each rectangle or square.

1. $b = 7$ km and $h = 14$ km

2. $b = 13$ km and $h = 8$ km

3. $b = 7$ cm and $h = 1.5$ cm

4. $b = 2.6$ cm and $h = 5$ cm

5. $b = 18$ yd and $h = 9$ yd

6. $b = 21$ ft and $h = 12$ ft

7. $b = 9\frac{1}{2}$ in. and $h = 10$ in.

8. $b = 3\frac{1}{4}$ ft and $h = 4\frac{1}{2}$ ft

9. $b = 12$ m (square)

10. $b = 1.5$ cm (square)

11. $b = 19.6$ m (square)

12. $b = 5\frac{1}{2}$ ft (square)

13. Flooring is to be installed in a rectangular-shaped room that is 22 ft long and 15 ft wide. How many square feet of flooring will cover the floor?

14. Fran plans to paint a rectangular-shaped wall that is 9 feet high and 16 feet long. How many square feet of wall must be painted?

15. A basketball court measures 26 m by 14 m. What is its area?

16. A baseball diamond is 90 ft by 90 ft. What is its area?

17. A softball diamond is 60 ft by 60 ft. What is its area?

18. A volleyball court is 30 ft by 60 ft. What is its area?

19. A doubles tennis court is 36 ft by 78 ft. What is its area?

Extend and Apply

Fill in the blanks to complete each statement.

20. _____ in.² = 1 ft²

21. _____ mm² = 1 cm²

22. _____ ft² = 1 yd²

23. _____ cm² = 1 m²

Find the missing length or width. Identify the unit.

24. $A = 78$ m², $b = 6$ meters

25. $A = 21.6$ cm², $h = 2.7$ cm

26. $A = 169$ km² (square)

27. $A = 5\sqrt{6}$ cm², $h = \sqrt{3}$ cm

28. $A = 441$ km² (square)

29. $A = 0.09$ m² (square)

30. Suppose a rectangle is translated. How does the area of its image compare with the area of the rectangle?

31. *ABCD* and *EFGH* are similar rectangles. Complete the sentences below.

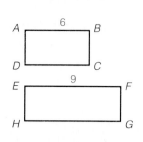

   a. The perimeter of *EFGH* is _____ times the perimeter of *ABCD*.

   b. The area of *EFGH* is _____ times the area of *ABCD*.

32. Draw as many rectangles as you can that have an area of 48 cm².

33. For the basketball backboard, what is the ratio of the area of the small rectangle to the area of the large rectangle?

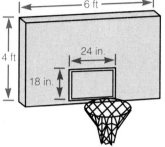

   (A) 1:18   (B) 1:8   (C) 1:6   (D) 1:4

34. Find the area of the shaded region.

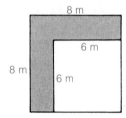

35. Find the area of the shaded region. Each white square is 1.5 cm wide.

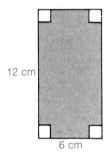

36. Beth plans to have her rec room retiled with Deluxe tiles. The first 10 cartons cost $32.95 each, and each additional carton over 10 cartons costs $29.95. A carton covers 45 ft². Find the cost of the tiles.

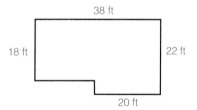

37. How many small squares does an 8½-by-11-in. piece of graph paper have if each square is $\frac{1}{8}$ in. wide?

38. A 4.5-by-6-m kitchen floor will be covered with square vinyl tiles that are 30 cm on a side. How many tiles will be needed?

Use Mathematical Reasoning

39. Tell two different ways that you could find the area. Then find the area.

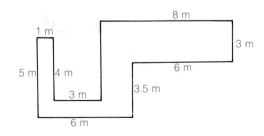

40. Chip said that all squares with the same perimeter have the same area. Do you agree or disagree? Explain your answer.

41. A chessboard has 8 × 8 small squares, each measuring 1.75 inches on a side. Will the board fit on a square table that has an area of 1.25 square feet? Why or why not?

42. A football field is 120 yards long and $53\frac{1}{3}$ yards wide. Suppose one bag of fertilizer covers 1000 square feet. How many bags of fertilizer are needed to cover the field?

Mixed Review

43. How many diagonals does a 3-sided figure have?

44. What is the perimeter of a regular octagon with side lengths of 4.5 cm?

◇ Solve and check.

45. $12(n - 9) = 72$ **46.** $12n - 9 = 72$ **47.** $12n - 9n = 72$

Enrichment

A perpendicular segment from the center of a regular polygon to any side is called the **apothem** (ap′ a thim). The area of a regular polygon is $\frac{1}{2}ap$, where a is the length of the apothem and p is the perimeter.

Find the area of this regular pentagon.
To find a, first find BA. Then use the Pythagorean Theorem.

$BA = \frac{1}{2}(6) = 3$

$a^2 = 5.1^2 - 3^2 = 26.01 - 9 = 17.01$

$a = \sqrt{17.01} \approx 4.12$

$p = 5 \times 6 = 30$ cm

$A = \frac{1}{2}ap \approx \frac{1}{2}(4.12)(30) \approx 61.8$ cm^2

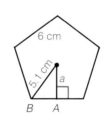

Problems

Find the area of each regular polygon.

1.

24 cm
20 cm

2.

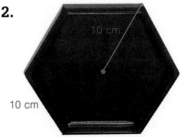

10 cm
10 cm

Make an Organized List

We can solve some problems by listing information in a systematic or organized way. This strategy for solving problems is called *Make an Organized List.*

Sample   A farmer has 220 yards of fence available to enclose a rectangular area. What is the largest rectangular region that the farmer can enclose?

We can first make an organized list of rectangles that have a perimeter of 220 yd. One such rectangle is shown. Then we can compare the areas of the rectangles.

80 yd

30 yd

Base	Height	Area
30	80	2400
35	75	2625
40	70	2800
45	65	2925
50	60	3000
55	55	3025
60	50	3000
65	45	2925

As we examine the list, we can see that the areas 3000 ← begin to decrease here.

Thus, the greatest area that can be enclosed is 3025 yd^2, a 55-by-55-yd square region.

Problems

1. A farmer has 380 ft of fence to enclose a rectangular area. What is the largest rectangular region that she can enclose?

2. What happens to the area of a rectangle if the height is doubled?

3. Suppose that the base and the height of a rectangle are each doubled. What happens to the area of the rectangle?

4. There are five entrants in a ping-pong tournament. Each entrant plays every other entrant once. How many games will be played?

5. Suppose you had an $8\frac{1}{2}$-by-11-in. piece of notebook paper. How could you use the piece of paper to create a 5-in. line segment?

6. Three boxes are labeled "Apples," "Oranges," and "Apples and Oranges." Each label is on the wrong box. You are allowed to pick exactly one piece of fruit from only one box. From which box would you select? How would you know what is in each box?

12-3 Area of Parallelograms

Objective: Solve problems involving the area of a parallelogram.

Explore

Use a piece of paper and scissors.

1. Draw and cut out a parallelogram such as the one shown.

2. Draw a segment perpendicular to and connecting the bases.

3. Cut along the segment and switch the positions of the pieces.

Discuss   What kind of figure is formed by the rearranged pieces?
How do you find the area of the new figure?
How does the area of the new figure compare to the area of the parallelogram?

The area of a parallelogram can be found by multiplying the base and the height, just as for a rectangle. Any side may be used as a base. The **height** is the length of any segment that is perpendicular to the lines containing the base and the side opposite the base. The perpendicular segment is called the **altitude** of the parallelogram.

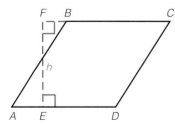

Base $\overline{AD}$ (or $\overline{BC}$)
Altitude $\overline{EF}$
Height h

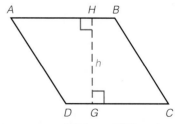

Base $\overline{AB}$ (or $\overline{DC}$)
Altitude $\overline{GH}$
Height h

Theorem 12.2   The area of a parallelogram is the product of a base and the corresponding height. Area = bh

---

Examples Find the area of each parallelogram.

1.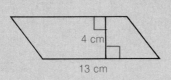

$$A = bh = 13 \times 4$$
$$= 52 \text{ cm}^2$$

2.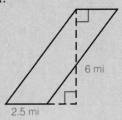

$$A = 2.5 \times 6 = 15 \text{ mi}^2$$

Try This... Find the area of each parallelogram.

a.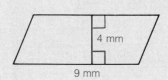

b.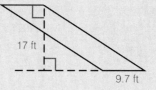

$\mathbf{K}$nowing the area, we can use it to find the base or height.

Example **3.** Find the height of the rhombus.

$$A = bh$$
$$52.7 = 8.5h$$
$$\frac{52.7}{8.5} = h$$
$$6.2 \text{ cm} = h$$

Area $= 52.7 \text{ cm}^2$

8.5 cm

Try This... **c.** Find the base of the parallelogram.

$A = 81 \text{ in.}^2$

$h = 6 \text{ in.}$

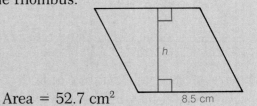

$\mathbf{I}$f an angle of a parallelogram is 30°, 45°, or 60°, we can use our knowledge of right triangles to find the height and the area of the parallelogram.

Example **4.** Find the area of $\square ABCD$.

Because $\triangle AEB$ is a 30°-60° right triangle,
$BE = \frac{1}{2}AB = \frac{1}{2} \times 6 = 3 \text{ m}.$
$A = 8 \times 3 = 24 \text{ m}^2$

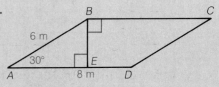

Try This... d. Find the area of ☐ *MNOP*.

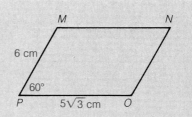

Discussion The base of a parallelogram is 12 in., and its corresponding height is 7 in. Another side of the parallelogram is 28 in. long. Can we find a corresponding height for this side?

Exercises

Practice

Find the area of each parallelogram from the given measures.

1. $b = 7$ m, $h = 13$ m **2.** $b = 8$ m, $h = 14$ m

3. $b = 1.2$ cm, $h = 5$ cm **4.** $b = 7.3$ cm, $h = 3$ cm

5. $b = 70$ mm, $h = 58$ mm **6.** $b = 120$ mm, $h = 70$ mm

7. $b = 2.5$ km, $h = 3.5$ km **8.** $b = 5.25$ m, $h = 7.75$ m

9. $b = 2\frac{1}{2}$ ft, $h = 6$ ft **10.** $b = 3\frac{1}{4}$ yd, $h = 4$ yd

11. $b = 3\frac{3}{4}$ in., $h = 4\frac{1}{2}$ in. **12.** $b = 5\frac{1}{8}$ in., $h = \frac{1}{4}$ in.

Find the missing base or height of each parallelogram.

13. $A = 2.25$ m², $b = 1.5$ m **14.** $A = 120$ mm², $h = 24$ mm

15. $A = 175$ cm², $h = 15$ cm **16.** $A = 14\frac{2}{3}$ ft², $b = 4$ ft

Find the area of each parallelogram.

17.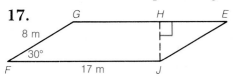

18. $AE = ED = 6$ cm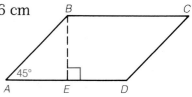

Extend and Apply

19. Draw two different parallelograms with an area of 32 square units.

20. Which parallelogram has an area of 30 square units?

(A) 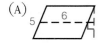 (B)  (C) 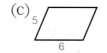 (D)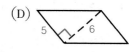

21. Suppose a parallelogram is reflected. How does the area of its image compare with the area of the parallelogram?

Estimate the area of the shaded region.

22.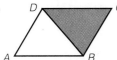

$A(ABCD) = 26$ in.^{2}

23.

$A(PQRS) = 50$ in.^{2}

24.

$A(HJKL) = 36$ ft^2

25. *ABCD* and *EFGH* are similar parallelograms. Fill in the blanks below.

    a. The perimeter of *EFGH* is _____ times the perimeter of *ABCD*.

    b. The area of *EFGH* is _____ times the area of *ABCD*.

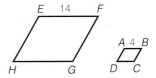

26. Five parallelograms like the one shown are to be used in a sculpture. What is the total area of the parallelograms? Could you order a piece of sheet metal with this area and be sure that you could cut out the five parallelograms? Explain.

27. Here is the top view of a rectangular box. The box is pushed on one side as shown.

For the open side shown, how does this change

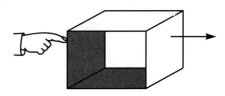

    a. the length of a side?   **b.** the perimeter?

    c. the height?   **d.** the area?

Use Mathematical Reasoning

28. A property had been a rectangular 100-by-200-ft parcel. The new zoning map changed the boundaries so that the property is now shaped like a parallelogram. The landowner complained that this represents a loss of area. Do you agree? If so, what is the loss in square feet?

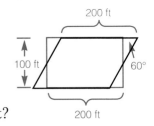

 Find the area of the shaded region.

29.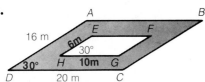

30.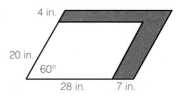

Mixed Review

31. Find the sum of the angle measures of a 12-sided figure.

32. What is the sum of the measures of the exterior angles of a pentagon?

◇ Solve and check.

    33. $3(h + 9) = 84$ **34.** $3h + 9 = 84$ **35.** $3h + 9h = 84$

12-4 Area of Triangles

Objective: Solve problems involving finding the area of a triangle.

Explore

1. Draw a large scalene triangle on one piece of paper.

2. Place a second piece of paper under the first, making sure that they do not slide.

3. Cut out the triangle. (You will have two congruent triangles, one from each piece of paper.)

4. Place the triangles next to each other, then rotate one triangle so that both touch along a congruent side. What kind of figure is formed?

Discuss How does the area of the triangle compare to the area of the figure that is formed?

Theorem 12.3 The area of a triangle is one-half the product of a base and the corresponding height.

$$\text{Area} = \frac{1}{2} bh$$

Examples Find the area of each triangle.

1.
$$A(\triangle XYZ) = \frac{1}{2} \times 9 \times 3$$
$$= 13.5 \text{ m}^2$$

2.
$$A(\triangle ABC) = \frac{1}{2} \times 6.2 \times 4.8$$
$$= 14.88 \text{ cm}^2$$

Try This... Find the area of each triangle.

a.
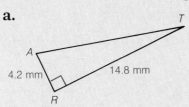

b.
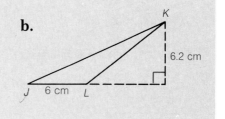

Example 3. A garage has a "hip" roof composed of four congruent isosceles triangles. Find the area of the roof.

Each section is shaped like a triangle.

$$A = \frac{1}{2}bh$$
$$= \frac{1}{2}(22)(12)$$
$$= 132 \text{ ft}^2$$

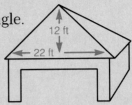

There are four congruent sections. $4 \times 132 = 528$

The area of the roof is 528 ft².

Try This... c. The gable end of a house is to be painted (the shaded region). Find the area of the gable end.

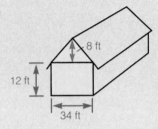

Discussion   Must all triangles with the same base and the same height have the same area? Explain.

Exercises

Practice

Find the area of each triangle.

1.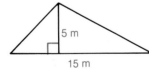
5 m
15 m

2.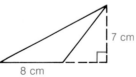
7 cm
8 cm

3.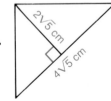
$2\sqrt{5}$ cm
$4\sqrt{5}$ cm

4.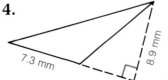
7.3 mm
8.9 mm

5.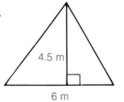
4.5 m
6 m

6.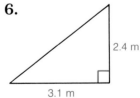
2.4 m
3.1 m

7.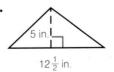
5 in.
$12\frac{1}{2}$ in.

8.
$4\frac{1}{2}$ ft   16 ft

9.
$5\frac{1}{2}$ ft $6\frac{1}{4}$ ft

Find the area of each triangle.

10.

11.

12. Jamaal needs to order bricks for the gable end of a house. Find the area of the gable end.

13. In Exercise 12, six bricks cover a square foot. About how many bricks should be ordered?

14. The shingles on a hip roof are to be replaced. Each of the four triangular sections has a base of 21 ft and a height of 11 feet. Find the area of the roof. A *square* of shingles covers 100 ft². How many squares of shingles must be ordered?

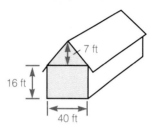

15. In Exercise 14, a square costs $42. The roofer charges $45 per square for labor to install the shingles. What is the cost of installation?

Extend and Apply

16. $A(\triangle ABC) = 18$ cm², and $\overline{AB}$ is the base of $\triangle ABC$. What is the height?

17. $A(\triangle DEF) = 144$ m². What is the length of the base of $\triangle DEF$?

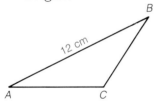

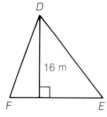

Find the missing base or height for a triangle with the given measures.

18. $A = 72$ in.², $h = 12$ in.

19. $A = 48.4$ cm², $h = 4$ cm

20. $A = 15$ cm², $b = 2.5$ cm

21. $A = 7.2$ m², $b = 0.6$ m

22. $b = h$, $A = 18$ m²

23. $b = h$, $A = 72$ ft²

Estimate the area of the shaded region.

24.

$A(\triangle ABC) = 24$ in.²

25.

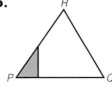

$A(\triangle PQR) = 56$ ft²

26.

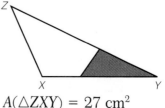

$A(\triangle ZXY) = 27$ cm²

27. What is the area of the large triangle?

(A) 300   (B) 150

(C) 250   (D) 187.5

28. Suppose a triangle is rotated. How does the area of its image compare with the area of the triangle?

29. $\triangle ABC$ and $\triangle DEF$ are similar triangles. Complete each sentence below.

    a. The perimeter of $\triangle DEF$ is _____ times the perimeter of $\triangle ABC$.

    b. The area of $\triangle DEF$ is _____ times the area of $\triangle ABC$.

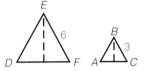

Use Mathematical Reasoning

30. A parallelogram and a triangle have the same area. Their bases have the same measure. How are their altitudes related?

31. Derive a formula for the area of an equilateral triangle whose side has measure n.

32. Construct a triangle whose area is equal to the area of a square. What are the dimensions of the triangle in relation to the square?

33. Draw a rhombus. The diagonals are 8 cm and 14 cm and are perpendicular bisectors of each other. Find the area of the rhombus.

Find the area of the shaded region.

34.

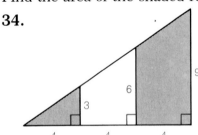

35. $\triangle ABC \sim \triangle DEF$. $GH = 8$, $AD = 2$, $DC = 3$, and $CF = 4$.

36. $ABCD$ is a rectangle. Explain why the sum of the areas of the two small triangles equals the area of the large triangle.

37. What is the total area of the material needed to make a triangular tent that has a rectangular base and sides as shown?

38. Write a convincing argument that the area of a rhombus is $\frac{1}{2}$ the product of the lengths of the diagonals.

    Hints: **a.** How can you find the area of $\triangle ABD$ and of $\triangle CBD$?

            b. What is $AE + EC$?

            c. What is the area of $ABCD$?

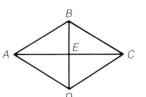

Mixed Review

39. A hexagon has 5 angles that are each 110°. What is the measure of the other angle?

⬥ Write using exponential notation.

40. 25   **41.** 49   **42.** 1,000,000

Enrichment

Heron of Alexandria, a mathematician who lived during the first century A.D., developed a formula for finding the area of a triangle from the lengths of the three sides, a, b, and c.

First, Heron found s, half of the sum of the lengths of the three sides.

$$s = \frac{1}{2}(a + b + c)$$

Then he developed the formula

$$A = \sqrt{s(s - a)(s - b)(s - c)}$$

to find the area.

🖩 To find the area of $\triangle ABC$, first find s.

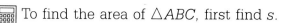

$s = 4\ \boxed{+}\ 5\ \boxed{+}\ 6\ \boxed{=}\ \boxed{\div}\ 2\ \boxed{=}\ \boxed{\quad\quad 7.5\quad}$

Store this number. 7.5 $\boxed{\text{STO}}$

Now evaluate the formula.

$\boxed{\text{RCL}}\ \boxed{\times}\ \boxed{(}\ \boxed{\text{RCL}}\ \boxed{-}\ 4\ \boxed{)}$

$\boxed{\times}\ \boxed{(}\ \boxed{\text{RCL}}\ \boxed{-}\ 5\ \boxed{)}$

$\boxed{\times}\ \boxed{(}\ \boxed{\text{RCL}}\ \boxed{-}\ 6\ \boxed{)}\ \boxed{=}\ \boxed{\quad 98.4375\quad}$

$\boxed{\sqrt{x}}\ \boxed{\quad 9.921567\quad}$

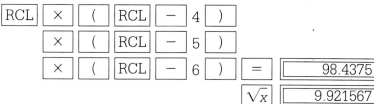

The area is about 9.9 m².

🖩 Use Heron's formula to find the areas of these triangles.

1.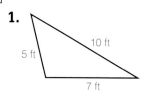

2.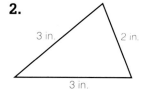

3.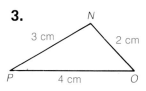

4. Draw a large triangle. Measure the lengths of the sides and find the area using Heron's formula. Now measure the height of the triangle and use the formula $A = \frac{1}{2}bh$ to find the area. How do the two calculations compare?

12-5 Area of Trapezoids

Objective: Solve problems involving areas of trapezoids.

Explore

Use two pieces of paper and a pair of scissors.

1. Draw a trapezoid on one piece of paper.

2. Place the second sheet under the first, making sure they do not slide.

3. Cut out two congruent trapezoids.

Discuss   How can you rearrange the trapezoids to get a figure whose area you can determine? How does the area of the trapezoid compare with the area of the figure that is formed?

Theorem 12.4   The area of a trapezoid is one-half the product of the sum of its bases and the height.

$$A = \frac{1}{2}h(b_1 + b_2)$$

Examples Find the area of each trapezoid.

1. Area of $ABCD = \dfrac{1}{2} \times 5 \times (12 + 16)$

$$= \frac{1}{2} \times 5 \times 28 = 70 \text{ cm}^2$$

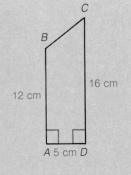

2. Area of $GHLK = \dfrac{1}{2}h(b_1 + b_2)$

$$= \frac{1}{2} \times 4 \times (9 + 13)$$

$$= \frac{1}{2} \times 4 \times 22 = 44 \text{ m}^2$$

Try This... Find the area of each trapezoid.

a.

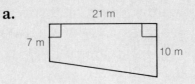

b.

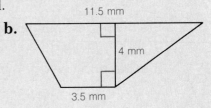

Example 3. In trapezoid $ABCD$, $m\angle CDA = 60°$, $BC = 8$ m, $CD = 6$ m, and $AD = 12$ m. Find the area of $ABCD$. $\triangle CED$ is a 30°-60° right triangle.

Thus, $CE = \dfrac{6}{2}\sqrt{3} = 3\sqrt{3}$.

Area of $ABCD = \dfrac{1}{2} \times (12 + 8)(3\sqrt{3})$

$\qquad = 30\sqrt{3}$

$\qquad \approx 52$ m^2

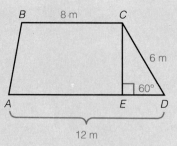

Try This... c. In trapezoid $MNQP$, $m\angle MPQ = 60°$, $MN = 21$ cm, $MP = 8$ cm, and $PQ = 18$ cm. Find the area of $MNQP$.

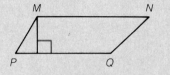

Discussion Daryl said that he thinks of finding the area of a trapezoid as multiplying the height by the average length of the two bases. Do you agree or disagree? Explain.

Exercises

Practice

Find the area of each trapezoid.

1.
$3\frac{1}{2}$ ft

6 ft $9\frac{1}{4}$ ft

2.
N

M 7.5 mm

2.5 mm

R 3.5 mm Q

3.
T A

18 cm 14 cm 23 cm

N

S

4.
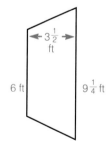
2.5 m

2.5 m

7.25 m

5.
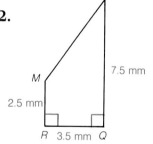
18 cm

24 cm

6 cm

6.
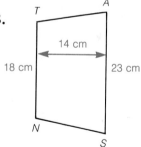
12 m

12 m

60°

16 m

7.
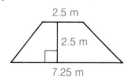
5 ft

4 ft

$10\frac{1}{2}$ ft

8.
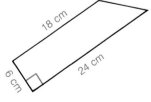
W 3 m X

2 m

Z 7 m Y

9.
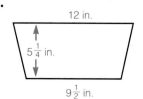
12 in.

$5\frac{1}{4}$ in.

$9\frac{1}{2}$ in.

10. In trapezoid $ABCD$, $AB = 12$ m, $AD = 10$ m, $BC = 14$ m, and $m \angle ABC = 60°$. Find the area of $ABCD$.

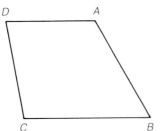

11. In isosceles trapezoid $RSTU$, $m \angle RUT = 45°$, $RS = 8$ ft, and $UT = 16$ ft. Find the area of $RSTU$.

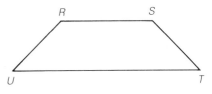

12. A facade of brick is to be placed over a fireplace. Find the number of square feet of surface to be covered.

13. A fluorescent light fixture has two ends that are in the shape of trapezoids. The bases are 19 in. and 15 in., with a height of 9 in. Find the number of square inches of metal needed for the two ends.

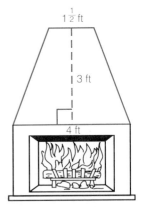

Extend and Apply

14. Use graph paper. Draw a rectangle, a parallelogram, a trapezoid, and a triangle each having an area of 12 square units.

15. Suppose a trapezoid is translated, reflected, and rotated. How does the area of its image compare with the area of the trapezoid?

16. $ABCD$ and $EFGH$ are similar trapezoids. Complete the sentences below.
   a. The perimeter of $ABCD$ is _____ times the perimeter of $EFGH$.
   b. The area of $ABCD$ is _____ times the area of $EFGH$.

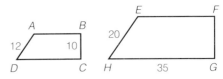

Find the missing base or altitude.

17.

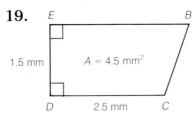

7 m

$A = 30$ m^2

8 m

18.

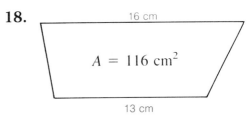

16 cm

$A = 116$ cm^2

13 cm

19.

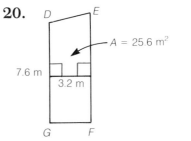

1.5 mm

$A = 4.5$ mm^2

2.5 mm

20.

$A = 25.6$ m^2

7.6 m

3.2 m

21. How much more area is contained in the international free-throw lane than in the American free-throw lane?

American free-throw lane

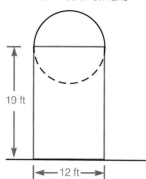

19 ft

12 ft

International free-throw lane

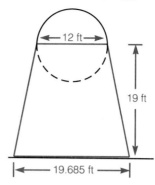

12 ft

19 ft

19.685 ft

Use Mathematical Reasoning

22. Find the area of this trapezoid using two different methods.

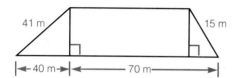

41 m · 15 m · 40 m · 70 m

23. In 1876, while serving as a member of the United States House of Representatives, future President James A. Garfield published an original proof of the Pythagorean Theorem by comparing the area of a trapezoid (like the one shown below) with the areas of the three triangles.

Find the area of the trapezoid.
What is the measure of $\overline{EB}$?
What is the measure of $\angle ACD$?
Find the sum of the areas of the three triangles and compare it with the area of the trapezoid. Explain how this shows that the Pythagorean Theorem is true.

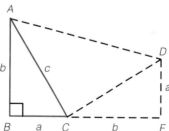

Mixed Review

24. The two triangles shown are _____ and _____ _____ triangles.

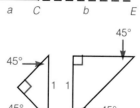

Solve these proportions.

25. $\dfrac{16}{44} = \dfrac{x}{5.5}$

26. $\dfrac{2.5}{8} = \dfrac{y}{60}$

27. $\dfrac{96}{72} = \dfrac{64}{p}$

Write using standard notation.

28. 4^3

29. 3^4

30. 23^1

Geometry In Technology

Computer-Assisted Design (CAD)

A designer is using a Computer-Assisted Design, or CAD, a computer program to create technical drawings. She is doing architectural drawings of a house. One of her sketches is a scale drawing of a floor plan.

1. Use a ruler to find the scale the designer used: 1 inch represents _____ feet.

Use the scale factor to determine the dimensions and the area of each room.

2. Family room

3. Dining room

4. Kitchen

5. Bedroom 1

6. Bedroom 2

7. Bedroom 3

8. Bathroom 1

9. Bathroom 2

10. What is the total square footage?

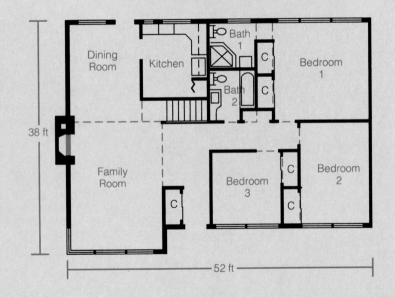

Chapter 12 Review

1. Estimate the area. Use the small square as the unit.

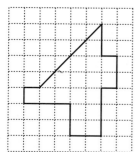

2. *ABCDEF* is a regular hexagon. *A*(△*AGF*) = 13. Find the area of the hexagon.

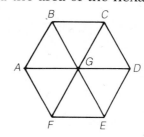

12-1

3. Find the height of a rectangle if *A* = 4.5 m² and *b* = 2.5 m.

4. Find the area of the shaded region.

12-2

9 m
3 m
6.5 m
2 m
11.5 m
2.5 m
14 m

5. Three sides of a shower will be tiled with square tiles that are 10 cm on a side. The width of the wall behind the tub is 1.5 m. The width of the wall at each end of the tub is 70 cm. The height of the walls to be tiled is 2 m. How many tiles are needed?

Find the area of each parallelogram.

12-3

6.

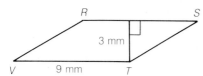

R   S
3 mm
V   9 mm   T

7.

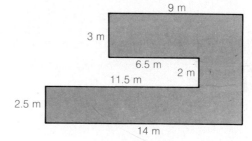

C   18 cm   *D*
18 cm
45°
F   *E*

8. Find the base of □ *GHJK*. Area = 48.75 mm².

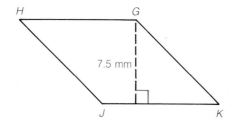

H   *G*
7.5 mm
J   *K*

9. *ABCD* and *WXYZ* are parallelograms. Find the area of the shaded region

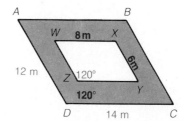

A   *B*
W   8 m   *X*
12 m   *Z*   120°   6m
120°   *Y*
D   14 m   *C*

12-4   Find the area of each triangle.

10.
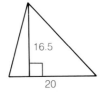
16.5
20

11.

3.5
2.3

12-5   Find the area of each trapezoid.

12.
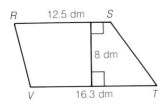
R   12.5 dm   S
8 dm
V   16.3 dm   T

13.
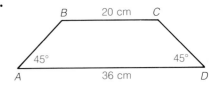
B   20 cm   C
45°   45°
A   36 cm   D

Chapter 12 Test

1. Estimate the area. Use the small square as the unit.

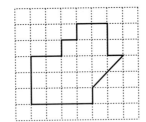

2. All small triangles in the figure are congruent. The area of each small square is 4 cm². Find the area of the figure.

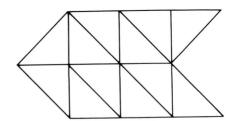

3. Find the base of a rectangle if $A = 3.6$ cm² and $h = 0.4$ cm.

4. Find the area of the shaded region.

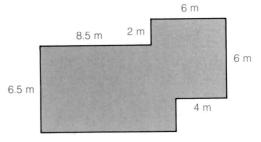
6 m
8.5 m   2 m
6 m
6.5 m
4 m

5. A kitchen floor measuring 4 m by 6 m will be tiled with square tiles that are 25 cm on a side. How many tiles are needed?

Find the area of each parallelogram.

6.

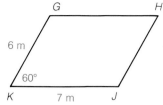

6 m
60°
K   7 m   J
G   H

7.

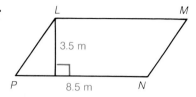

L   M
3.5 m
P   8.5 m   N

8. $A(\square WXYZ) = 5040$ mm². Find the base.

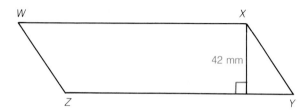

W   X
42 mm
Z   Y

9. *ABCD* and *EFGD* are parallelograms. $\overline{AE} \cong \overline{ED}$, and $DG = \frac{1}{3}DC$. Find the area of the shaded region.

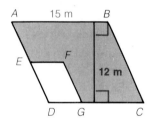

A   15 m   B
E   F
12 m
D   G   C

Find the area of each triangle.

10.

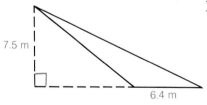

7.5 m
6.4 m

11.

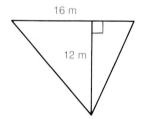

16 m
12 m

Find the area of each trapezoid.

12.

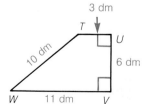

3 dm
T
U
10 dm
6 dm
W   11 dm   V

13.

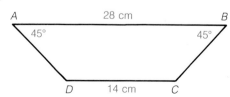

A   28 cm   B
45°   45°
D   14 cm   C

Cumulative Review Chapters 9–12

9-1 Express each ratio in lowest terms.

1. 45 to 9   **2.** 20 to 55   **3.** 8 months to 2 years

9-2 Tell whether each pair of ratios forms a proportion.

4. $\frac{5}{2}$ and $\frac{35}{14}$ **5.** $\frac{8}{5}$ and $\frac{25}{40}$ **6.** $\frac{9}{11}$ and $\frac{45}{44}$

9-3 Determine whether the figures are similar. If so, what is the scale factor that transforms the figure on the left to the figure on the right?

7.
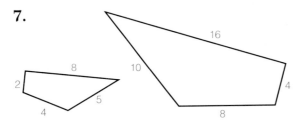

8.

9-4
9-5

9. $\triangle ABC \sim \triangle FED$. Find AC and BC.

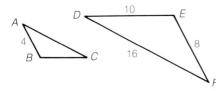

10. In $\triangle XYZ$, $\angle XRS \cong \angle XZY$, $XZ = 8$, $RZ = 5$, and $ZY = 14$. Find RS.

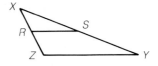

9-6 **11.** Which theorem (AA, SAS, or SSS) can be used to show that $\triangle DCE \sim \triangle BCA$?

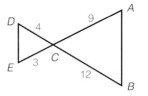

12. A support structure is designed as shown. $EB = 8$ ft, $AE = EC = 5$ ft, and $AB = BD = 7$ ft. Find CD.

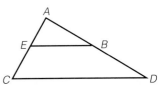

10-1 **13.** Find the geometric mean of 6 and 8.

10-2 **14.** Find c.

15. Find a.

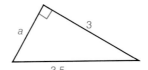

16. Can 6, 13, and 14 be the lengths of the sides of a right triangle?   10-3

Find the missing lengths in each triangle.   10-4

17.   **18.**   10-5

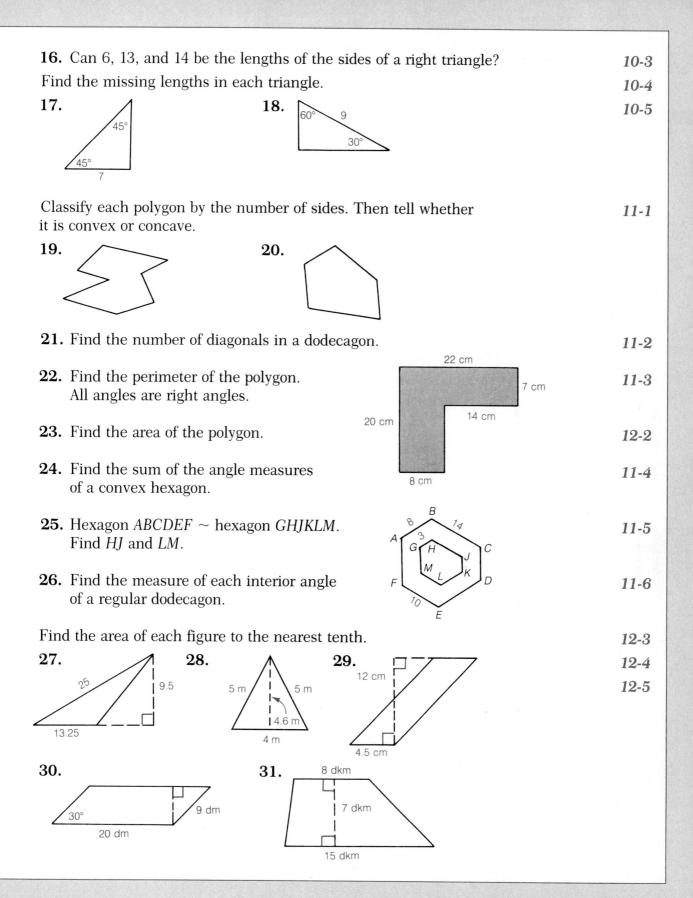

Classify each polygon by the number of sides. Then tell whether   11-1
it is convex or concave.

19.   **20.**

21. Find the number of diagonals in a dodecagon.   11-2

22. Find the perimeter of the polygon.   11-3
All angles are right angles.

22 cm

7 cm

14 cm

20 cm

23. Find the area of the polygon.   12-2

8 cm

24. Find the sum of the angle measures   11-4
of a convex hexagon.

25. Hexagon *ABCDEF* ~ hexagon *GHJKLM*.   11-5
Find *HJ* and *LM*.

26. Find the measure of each interior angle   11-6
of a regular dodecagon.

Find the area of each figure to the nearest tenth.   12-3

27.   **28.**   **29.**   12-4
12-5

25   9.5   5 m   5 m   12 cm

13.25   4.6 m   4.5 cm

4 m

30.   **31.**   8 dkm

30°   9 dm   7 dkm

20 dm   15 dkm

13 Circles

Bicycle wheels are just one example of circles in the world around us.
The circumference of a wheel determines the length of the spoke being replaced.

13-1 Circles and Chords

Objective: Solve problems involving radii, diameters, and chords of circles.

Circles are among the most familiar geometric figures. Everyday objects from bicycle wheels to watch faces are based on circles.

Recall that we discussed circles briefly in Chapter 2. We will now repeat the definition of *circle* and then give some new definitions.

Definition   A **circle** is the locus of all points in a plane a given distance (r) from a given point (O) in the plane. Point O is called the **center** of the circle.

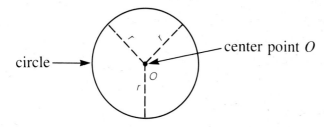

We will refer to this circle as $\odot O$ because O is the center. Here are some important segments associated with circles:

Term	Definition	Picture
Radius	A segment whose endpoints are the center of the circle and a point on the circle. The length of this segment is also called the radius (plural, *radii*).	
Chord	A segment with both endpoints on the circle.	
Diameter	A chord containing the center of the circle. The length of this segment is also called the diameter.	

Example **1.** The center of this circle is O.
Identify all radii, chords, and
diameters shown.

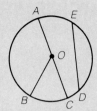

Radii: $\overline{OA}$, $\overline{OB}$, $\overline{OC}$
Chords: $\overline{DE}$ and $\overline{AC}$
Diameters: $\overline{AC}$

Try This... a. Identify all radii, diameters, and
chords shown in this photograph.

Explore

Computer software may be used for this activity.
1. Use your compass or computer to draw a circle.
2. Draw several diameters and radii.
3. Measure the diameters and radii. Compare their lengths.

The following two theorems summarize what you
may have discovered:

Theorem 13.1 All radii of a circle are congruent.

Theorem 13.2 A diameter of a circle is twice the length of a
radius of the circle.

$$d = 2r$$

Example **2.** The length of a radius is 9.5. Find the length of a
diameter of the same circle.

$d = 2r$
$\quad = 2 \times 9.5$
$\quad = 19$

Example   **3.** In $\odot O$, $DT = 16$. Find OK and OQ.

Because $\overline{DT}$ is a diameter, its length is twice that of a radius. Hence, $OK = 8$ and $OQ = 8$.

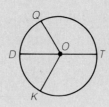

Try This...   **b.** The length of a diameter is 17. Find the length of a radius of the same circle.

c. The length of a radius is 4.6. Find the length of a diameter of the same circle.

Discussion   Given a circle, is there a maximum length for a chord? Is there a minimum length?

Exercises

Practice

Identify all radii, chords, and diameters in $\odot O$.

1.
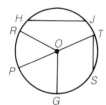

2.
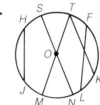

For each diameter, find the length of the radius.

3. 82.6   **4.** 75.4 cm   **5.** $4\frac{3}{4}$

6. 21 mm   **7.** 0.243   **8.** 0.056 m

For each radius, find the length of a diameter.

9. 8.7   **10.** $4\frac{1}{8}$ **11.** 0.08

12. 4.5 m   **13.** 1.9 cm   **14.** 1.1 cm

Extend and Apply

15. Which of the following segments is a radius of the circle?

(A) $\overline{UV}$ (B) $\overline{UY}$ (C) $\overline{XO}$ (D) $\overline{XY}$

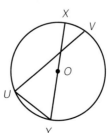

The figure at the right shows a radio signal that reaches 25 miles.

16. What is the maximum distance two people can live from each other and still receive the station's signal?

17. If Jeff's radio receives the station's signal, what can be said about his distance from the station?

18. What is the distance from the northernmost point of the station's signal to the southernmost point?

19. The Watsons and the Espinosas are each 25 miles from the radio station and also 25 miles from each other. Draw a picture showing how that can be.

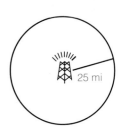

Determine whether each statement is true or false.

20. Every diameter of a circle is a line of symmetry.

21. Every chord of a circle is a line of symmetry.

22. Every line of symmetry of a circle must pass through the center of the circle.

Use Mathematical Reasoning

23. Draw two circles, $\odot A$ and $\odot B$, each with a radius of 5 cm and intersecting at two points C and D. What type of quadrilateral is $ACBD$?

24. Draw two circles, $\odot A$ and $\odot B$, each with a diameter of 6 cm, so that each circle passes through the center of the other. Label the points of intersection C and D. What is the length of $\overline{AB}$?

25. Explain why a diameter is the longest chord of a circle. (Hint: Consider the circle shown at the right and use the Triangle Inequality.)

Mixed Review

Find the area of each parallelogram from the given measures.

26. $b = 12$ m, $h = 6$ m

27. $b = 6\frac{1}{2}$ in., $h = 8$ in.

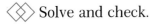 Solve and check.

28. $3x + 8x - 16 = 6x + 4 + x$

29. $6 + 5y - 7y + 9y = 10y - 24$

Problem Solving

What is the maximum number of pieces into which a circular pizza can be cut with 8 straight cuts? The pieces cannot be stacked or moved after cuts are made. Make an organized list and look for a pattern to help solve the problem.

13-2 More About Chords

Objective: Solve problems involving chords of circles and, given a circle, find the center of the circle.

Explore

Computer software may be used for this activity.

1. Draw a circle and any chord other than a diameter.

2. Draw a perpendicular from the center of the circle to the chord.

3. Draw another circle and a chord other than a diameter.

4. Find the midpoint of the chord and then draw a radius containing the midpoint. Measure the angle of intersection of the radius and the chord.

Discuss If a radius is perpendicular to a chord, what is true of the point of intersection? If a radius intersects a chord at its midpoint, what is true of the angle of intersection?

As you may have discovered in the Explore, chords of circles have some special properties. These are summarized in the next two theorems, which are converses of each other.

Theorem 13.3 If a radius of a circle is perpendicular to a chord, then it bisects the chord.

Theorem 13.4 If a radius of a circle bisects a chord that is not a diameter, then it is perpendicular to the chord.

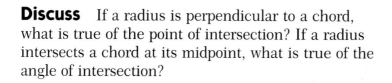

Example **1.** $\overline{OM} \perp \overline{AB}$, and $AM = 3$. Find AB.
Because $\overline{OM} \perp \overline{AB}$, M is the midpoint of $\overline{AB}$ by Theorem 13.3. Thus, $AB = 2AM = 2 \times 3 = 6$.

Example   2. Find the length of chord $\overline{AB}$.

By the Pythagorean Theorem,
$AM^2 + 4^2 = 5^2$. So, $AM = 3$.
Because M is a midpoint, $AB = 6$.

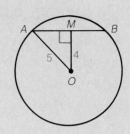

Try This... Find the length of chord $\overline{AB}$.

a.

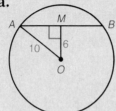

b.

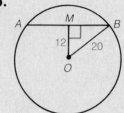

A chord's distance from the center of a circle depends on the length of the chord and the radius of the circle.

Example   3. Find the distance of chord $\overline{AB}$ from the center of the circle.

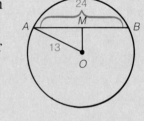

The distance from $\overline{AB}$ to the center is OM, the length of the perpendicular from $\overline{AB}$ to O.

This means $\triangle AMO$ is a right triangle.

So, we have
$$OM^2 + 12^2 = 13^2$$
$$OM^2 = 13^2 - 12^2 = 25$$
$$OM = 5$$

The distance from chord $\overline{AB}$ to the center O is 5.

Try This... c. In $\odot O$, $RT = 12$. Find the distance from chord $\overline{RT}$ to the center of the circle.

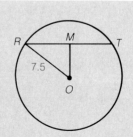

If we know the length of a chord and its distance from the
center of a circle, we can find the radius.

Example 4. The length of chord $\overline{CD}$ is 24, and
its distance from O is 9. Find the
radius of $\odot O$.

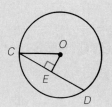

Because $\overline{OE} \perp \overline{CD}$, $\triangle OEC$ is a right
triangle. We also know $CE = 12$.

So, $OC^2 = 9^2 + 12^2$
            $= 225$
and $OC = 15$

The radius of the circle is 15.

Try This... d. The length of chord $\overline{XY}$ is 20, and
its distance from O is 7.5. Find the
radius of $\odot O$.

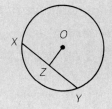

The following theorem is useful for finding the center of a circle:

Theorem 13.5   The perpendicular bisector of a chord of a circle
contains the center of the circle.

Example 5. Find the center of this circle.

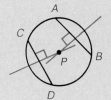

Draw any two chords $\overline{AB}$ and $\overline{CD}$. Draw the
perpendicular bisector of each chord.
Because the center of the circle lies on
both bisectors, it is the point at which they
meet, P.

Try This... e. Trace around a coin. Then use chords to find the center
of the circle.

Exercises

Practice

Find the length of each chord.

1.

2.

Find the distance of each chord from the center of the circle.

3.

$XY = 12$
$ZX = 10$

4.

$AB = 2.4$
$BD = 1.3$

Find the radius of each circle.

5.

$GJ = 32$
$FH = 12$

6.

$AC = 3.2$
$DB = 1.2$

7.

$TV = 10$
$PQ = 12$

8. Trace around a jar lid. Draw the perpendicular bisectors of two chords to locate the center of the circle. Test the accuracy of your drawing with a compass.

9. Trace around a coin. Draw the perpendicular bisectors of two chords to locate the center of the circle. Test the accuracy of your drawing with a compass.

Extend and Apply

10. Which of the following is x?

(A) 5   (B) 4   (C) 3   (D) 6

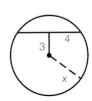

11. An archaeologist found a portion of an old clay plate. The archaeologist wants to determine the size of the complete plate. How can he determine the radius of the complete plate?

12. A circular garden has a radius of 25 ft. A path through the garden is 30 ft long. Find the distance from the path to the center of the garden.

Use Mathematical Reasoning

13. The diameter of a circle is 26 cm. A chord is perpendicular to a radius at a point 8 cm from its outer point. What is the length of the chord?

14. A circle has radius r. If a chord is the same length as the radius of the circle, what is the distance of the chord from the center of the circle?

Mixed Review

15. Identify all radii, chords, and diameters in the circle at the right.

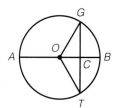

For each diameter, find the length of the radius.

16. 32   **17.** 5.9 in.   **18.** 6.08 m

For each radius, find the length of the diameter.

19. 7 yd   **20.** $7\frac{1}{4}$ cm   **21.** 19.8

Write using standard notation.

22. 10^5 **23.** 5^3 **24.** 2^4

Discover

1. Draw a large circle on a sheet of paper. Cut out the circle.

2. Make two folds in the circle to create two intersecting chords, $\overline{AD}$ and $\overline{BC}$, as shown. Label the point of intersection of the chords E.

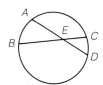

3. Carefully measure the segments $\overline{AE}$, $\overline{BE}$, $\overline{CE}$, and $\overline{DE}$.

4. Find $AE \cdot DE$ and $BE \cdot CE$. What do you notice?

Use your discovery to complete the following theorem.

Theorem 13.6 If two chords, $\overline{AD}$ and $\overline{BC}$, intersect at E in the interior of a circle, then _____ .

13-3 Tangents and Secants

Objective: Solve problems involving tangents and secants, and construct a tangent to a circle from a point not on the circle.

A wheelchair on a ramp is just one example of a *tangent* to a circle. Consider the following definitions:

> **Definitions** A line, coplanar with a circle, is **tangent** to the circle if it intersects the circle in exactly one point.
>
>   A **secant** is a line that intersects a circle in two points.

In the figure at the right, the tangent at *P* is line ℓ. Point *P* is called the **point of tangency.** Also, line *m* is a secant to the circle, since it intersects the circle at *Q* and *R*.

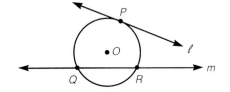

We sometimes refer to segments and rays as secants or tangents if they are contained in secant lines or tangent lines.

Example   1. Name the tangent lines and the secant lines to the circle.

The tangent lines are $\overleftrightarrow{AB}$, $\overleftrightarrow{BC}$, $\overleftrightarrow{CD}$, and $\overleftrightarrow{DA}$.

The secant lines are $\overleftrightarrow{EG}$, $\overleftrightarrow{EF}$, and $\overleftrightarrow{GF}$.

Try This... a. Name the tangent lines and secant lines to the circle.

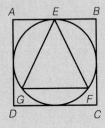

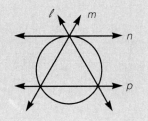

The following theorem is helpful in determining when a line is tangent to a circle:

Theorem 13.7   A line is tangent to $\odot O$ at a point P whenever the line is perpendicular to the radius $\overline{OP}$ at P.

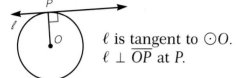 ℓ is tangent to $\odot O$.
$\ell \perp \overline{OP}$ at P.

ℓ is a secant line.
ℓ is *not* tangent to $\odot O$.
$\ell \not\perp \overline{OP}$ at P.

Example   **2.** $\overline{AB}$ is tangent to $\odot O$ at B, $\overline{OB} = 6$, and $\overline{OA} = 10$. Find AB.

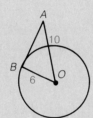   Since $\overline{AB}$ is tangent to $\odot O$ at B, $\overline{AB} \perp \overline{OB}$ by Theorem 13.7. Thus, $\angle ABO$ is a right angle.

By the Pythagorean Theorem,
$$6^2 + AB^2 = 10^2$$
$$AB^2 = 10^2 - 6^2$$
$$AB^2 = 64$$
and $\qquad AB = 8$

In Example 2, $\overline{AB}$ may be called a **tangent segment.**

Try This... b. $\overline{DE}$ is tangent to $\odot O$ at D,
$OD = 5$, and $DE = 12$.
Find OE.

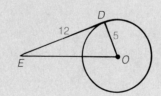

The following theorem is a consequence of the converse of Theorem 13.7 and the Hypotenuse-Leg Congruence Theorem:

Theorem 13.8   The tangent segments from a point to a circle are congruent.
$$\overline{AP} \cong \overline{BP}$$

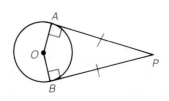

Example 3. $\overline{PA}$, $\overline{PB}$, and $\overline{PC}$ are tangent segments. If $PA = 12$, what is PC?

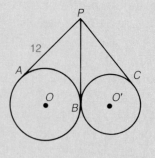

Since tangent segments from P are congruent, $\overline{PA} \cong \overline{PB}$ and $\overline{PB} \cong \overline{PC}$.

So, $\overline{PA} \cong \overline{PC}$. Therefore, $PA = PC$ and $PC = 12$.

In Example 3, $\odot O$ and $\odot O'$ intersect at one point, B. We say the circles are *tangent to each other*.

Example 4. $\overline{DE}$ and $\overline{DF}$ are tangent segments. $m\angle F = 70°$. Find $m\angle D$.

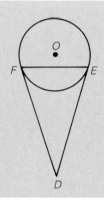

By Theorem 13.8, $\overline{DE} \cong \overline{DF}$. Hence, $\triangle DEF$ is isosceles and $\angle F \cong \angle E$.

Therefore, $m\angle F = m\angle E = 70°$, and $m\angle D = 180 - 70 - 70 = 40°$.

Try This... c. $\overline{RU}$ is tangent to the circles at R and U. $\overline{ST}$ is tangent to the circles at S and T. If $WS = 6$ and $WU = 4$, what is RU?

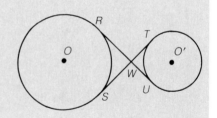

d. $\overline{PR}$ and $\overline{PS}$ are tangent segments and $m\angle P = 42°$. Find $m\angle R$.

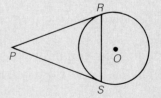

Discussion   Describe some common examples of tangents and secants to circles.

Exercises

Practice

Name the tangent lines and secant lines to each circle.

1.

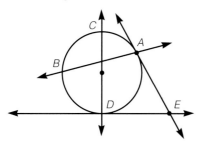

2.

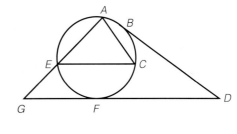

Use the figure at the right for Exercises 3–9.
$\overline{PQ}$ is tangent to $\odot O$ at Q.

3. $PQ = 12$ and $OP = 15$. Find OQ.

4. $OQ = 12$ and $OP = 20$. Find PQ.

5. $OQ = 0.9$ and $OP = 1.5$. Find PQ.

6. Diameter of $\odot O$ is 1, and $OP = 1.3$. Find PQ.

7. Diameter of $\odot O$ is 20, and $PQ = 24$. Find PS.

8. $OQ = QS = 1$. Find OP.

9. $PS = 80$, and the diameter of $\odot O$ is 100. Find PQ.

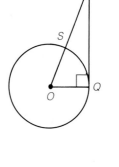

Use the figure at the right for Exercises 10–15.
$\overline{PA}$ and $\overline{PB}$ are tangent segments.

10. $m\angle BPA = 40°$. Find $m\angle PAB$ and $m\angle PBA$.

11. $m\angle PBA = 31°$. Find $m\angle PAB$ and $m\angle BPA$.

12. $m\angle APB = 35°$. Find $m\angle AOB$.

13. $m\angle AOB = 136°$. Find $m\angle BAP$.

Extend and Apply

14. $AB = x$, $PA = 2x + 3$, and $PB = x + 28$. Find AB.

15. $m\angle BAP = 2x + 3$, and $m\angle ABP = x + 38$. Find $m\angle ABP$.

16. $\overline{PC}$ and $\overline{PD}$ are tangent segments to $\odot O$. Which of the following angles *need not* be a right angle?

    (A) $\angle ODE$ (B) $\angle COD$ (C) $\angle OCP$ (D) $\angle PDO$

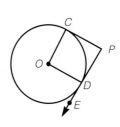

Use Mathematical Reasoning

The stand pictured at the right is used to hold a cylindrical drum having a diameter of 2 ft. The cross legs intersect at right angles.

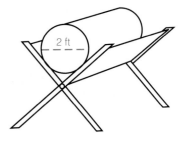

17. How far up from the intersection of the legs will the drum touch the legs?

18. How far is the bottom of the drum from the intersection of the legs?

19. How far is the top of the drum from the intersection of the legs?

20. The Sears Tower in Chicago is about 440 m tall. Suppose you are standing on the top of the tower. Given that the radius of the earth is 6400 km, find the distance of your line of sight to the horizon to the nearest kilometer.

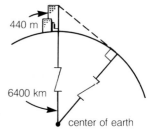

Mixed Review

Use the circle at the right for Exercises 21–23.

21. $AB = 24$ and $CD = 9$. Find DB.

22. $DB = 13$ and $CD = 12$. Find AB.

23. $DB = 25$ and $AB = 48$. Find CD.

Write using exponential notation.

24. 1000   **25.** 10,000   **26.** 1,000,000

Enrichment

Here is a way to use a compass and a straightedge to construct tangents to $\odot O$ from a point P not on the circle.

Step 1
Bisect $\overline{OP}$.

Step 2
Draw the circle with center A and radius OA.

Step 3
Draw segments $\overline{BP}$ and $\overline{CP}$ that are tangent to $\odot O$ at B and C.

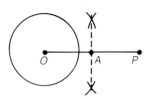
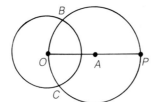
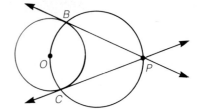

13-4 Angles and Arcs

Objective: Solve problems involving the degree measure of an arc.

The hands of a watch form an angle whose vertex is the center of the circle. Such angles are called *central angles*.

Definition A **central angle** is an angle whose vertex is the center of a circle.

∠*AOB* is a central angle.

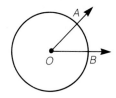

Central angles cut off part of the circle called an **arc**. Some special types of arcs and their notation are defined below.

Minor arc	Major arc	Semicircle
		$\overline{AB}$ is a diameter.
Minor arc $\overset{\frown}{AB}$	Major arc $\overset{\frown}{ACB}$	$\overset{\frown}{ACB}$ and $\overset{\frown}{ADB}$ are semicircles.

Example 1. $\overline{AC}$ and $\overline{BD}$ are diameters of ⊙*O*. Name the minor and major arcs and semicircles.

Minor arcs: $\overset{\frown}{AB}$, $\overset{\frown}{BC}$, $\overset{\frown}{CD}$, and $\overset{\frown}{DA}$

Major arcs: $\overset{\frown}{ADB}$ (or $\overset{\frown}{ACB}$), $\overset{\frown}{BAC}$ (or $\overset{\frown}{BDC}$), $\overset{\frown}{DCA}$ (or $\overset{\frown}{DBA}$), and $\overset{\frown}{DAC}$ (or $\overset{\frown}{DBC}$)

Semicircles: $\overset{\frown}{DAB}$, $\overset{\frown}{DCB}$, $\overset{\frown}{ABC}$, and $\overset{\frown}{ADC}$

Try This... **a.** $\overline{DE}$ is a diameter of $\odot O$. Name the minor and major arcs and semicircles.

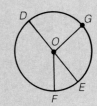

For any arc $\widehat{AB}$, the degree measure of the arc is written $m\widehat{AB}$. $m\widehat{AB}$ is the same as the *measure of its central angle* as defined below.

Minor arc	Semicircle	Major arc
$m\widehat{ACB} = m\angle AOB$ or $m\widehat{AB} = m\angle AOB$ $m\widehat{AB} < 180°$	$m\widehat{ACB} = 180°$	$m\widehat{ACB} = 360° - m\angle AOB$ $m\widehat{ACB} > 180°$

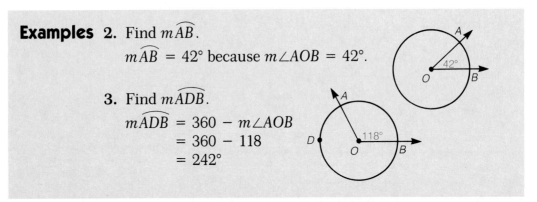

Congruent arcs are arcs with equal measure that lie in the same circle or in congruent circles.

Examples **2.** Find $m\widehat{AB}$.

$m\widehat{AB} = 42°$ because $m\angle AOB = 42°$.

3. Find $m\widehat{ADB}$.

$m\widehat{ADB} = 360 - m\angle AOB$
$= 360 - 118$
$= 242°$

Try This... **b.** Find $m\widehat{RS}$. **c.** Find $m\widehat{PRQ}$.

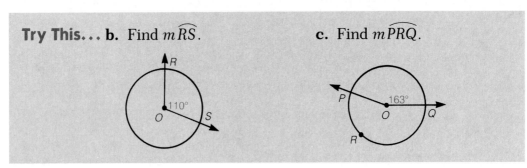

Arc addition is similar to angle addition. The following theorem is similar to the Angle Addition Postulate.

Theorem 13.9 **The Arc Addition Theorem**

If C is on $\overarc{AB}$, then $m\overarc{AC} + m\overarc{CB} = m\overarc{AB}$.

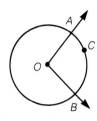

Example 4. $m\overarc{RS} = 24°$ and $m\overarc{ST} = 130°$.
Find $m\overarc{RST}$ and $m\overarc{RXT}$.

$$m\overarc{RST} = m\overarc{RS} + m\overarc{ST}$$
$$= 24 + 130$$
$$= 154°$$

$$m\overarc{RXT} = 360 - m\overarc{RST}$$
$$= 360 - 154$$
$$= 206°$$

Try This... d. $m\overarc{AB} = 37°$ and $m\overarc{BC} = 80°$.
Find $m\overarc{ABC}$ and $m\overarc{ADC}$.

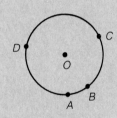

Example 5. $\overline{TP}$ and $\overline{GH}$ are diameters, and $m\overarc{GP} = 81°$.
Find $m\angle TOH$, $m\overarc{TH}$, and $m\overarc{TG}$.

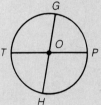

$m\overarc{GP} = 81°$. This means $m\angle GOP = 81°$. The Vertical Angle Theorem tells us that $m\angle TOH$ is also 81°. Thus, $m\overarc{TH} = 81°$. $\overline{GH}$ is a diameter, so $m\overarc{HTG} = 180°$. Thus,

$$m\overarc{TG} = 180 - m\overarc{TH}$$
$$= 180 - 81$$
$$= 99°$$

Try This... e. $\overline{AB}$ and $\overline{CD}$ are diameters, and $m\widehat{CB} = 24°$. Find $m\angle AOD$, $m\widehat{AD}$, and $m\widehat{DB}$.

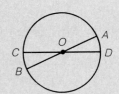

Discussion Roberta said, "If I draw a central angle of 50° in two different circles, the arcs they cut off will be congruent, since their central angles are congruent." Do you agree or disagree with this statement?

Exercises

Practice

In Exercises 1 and 2, name the minor arcs, major arcs, and semicircles.

1.

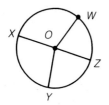

2.

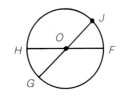

In Exercises 3–8, complete each statement by finding the measure of the indicated arc.

3. $m\widehat{AB} =$ _____

4. $m\widehat{CD} =$ _____

5. $m\widehat{PRQ} =$ _____

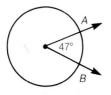

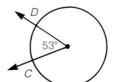

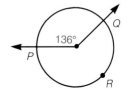

6. $m\widehat{KML} =$ _____

7. $\overline{ST}$ is a diameter. $m\widehat{TUS} =$ _____

8. $m\widehat{BCA} =$ _____

Chapter 13 Circles

9. $m\widehat{AB} = 37°$ and $m\widehat{BC} = 140°$. Find $m\widehat{ABC}$ and $m\widehat{ADC}$.

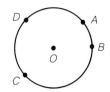

10. $m\widehat{EF} = 72°$ and $m\widehat{FG} = 150°$. Find $m\widehat{EFG}$ and $m\widehat{EHG}$.

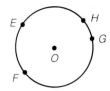

11. $\overline{RT}$ and $\overline{SU}$ are diameters, and $m\widehat{RU} = 125°$. Find $m\angle ROU$, $m\widehat{RS}$, and $m\widehat{TS}$.

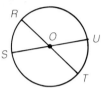

12. $\overline{AB}$ is a diameter, and $m\widehat{AD} = 40°$. Find $m\widehat{ACD}$ and $m\widehat{DB}$.

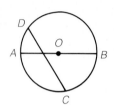

Extend and Apply

13. $\overline{TR}$ is tangent to $\odot O$, and $m\widehat{SR} = 55°$. Which of the following is $m\angle STR$?

(A) 55°   (B) 90°   (C) 45°   (D) 35°

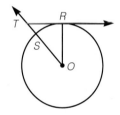

14. A carriage wheel has 12 spokes that are equally spaced. Find the degree measure of each of the arcs formed by adjacent spokes.

15. A sand dollar is a sea animal with a circular body. On its back is a pattern of five equally spaced lines. Find the degree measure of each of the central angles and each of the arcs in the pattern.

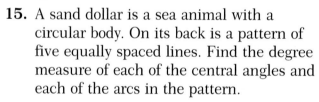

Pie charts are often used to show how a whole is divided into parts. For example, an amount that is 30% of the whole would be represented by a 108° central angle, since $0.3 \times 360° = 108°$. Use the data below and your protractor to draw pie charts.

16. A city's tax dollar is spent as follows: Schools, 48%; Public works, 14%; Police and fire, 18%; Administration, 8%; Debt retirement, 12%.

17. The North American Wholesale Lumber Association reports that our softwood lumber comes from the following sources: Western U.S., 42%; Eastern Canada, 13%; Western Canada, 21%; Southern U.S., 24%.

Use Mathematical Reasoning

In $\odot O$, $\overline{CE}$ is a diameter, $\overline{OF}$ bisects $\angle BOC$, and
$AE = BC = EO = 1$.

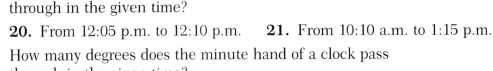

18. Find $m\overset{\frown}{AE}$, $m\overset{\frown}{ED}$, $m\overset{\frown}{DAB}$, and BF.

19. Find $m\overset{\frown}{AB}$, $m\angle BOF$, $m\overset{\frown}{ABC}$, and OF.

How many degrees does the hour hand of a clock pass
through in the given time?

20. From 12:05 p.m. to 12:10 p.m. **21.** From 10:10 a.m. to 1:15 p.m.

How many degrees does the minute hand of a clock pass
through in the given time?

22. From 12 noon to 12 midnight **23.** From 6 p.m. to 9 a.m.

24. From 4 p.m. to 3 p.m.

Mixed Review

$\overline{PA}$ and $\overline{PB}$ are tangent to $\odot O$.

25. $m\angle APB = 50°$. Find $m\angle AOB$.

26. $AP = 20$ and $AO = 15$. Find OP.

27. $m\angle PAB = 71°$. Find $m\angle APB$.

28. The radius of $\odot O$ is 5, and $AP = 12$. Find OP.

29. $m\angle QBP = 69°$. Find $m\angle QBO$.

Write using standard notation.

30. 3^3 **31.** 10^5 **32.** 4^4

Visualization

Which figure at the right is a turn, or rotation, of the figure at the left?

1. (A) (B) (C)

 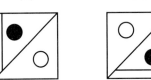

2. (A) (B) (C)

13-5 Rotations

Objective: Find the image of a figure under a rotation.

Artists use a device called a potter's wheel to make ceramic vases and bowls. The rapidly spinning wheel allows potters to create smooth, rounded surfaces. The potter's wheel is just one example of the transformation called a turn or rotation.

Definition Given a point *O* and an angle measure *x*, the **rotation image** of point *P* about *O* is a point *P'* such that *OP* = *OP'* and *m∠POP'* = *x*. *O* is called the **center** of rotation, and *x* is the **angle of rotation.**

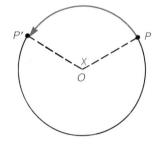

Recall that one degree is $\frac{1}{360}$ of a circle. Thus, a rotation that moves a figure around a point and back to its starting point is a 360° rotation.

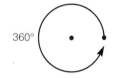

Examples Consider rotations of square *ABCD* about point *O*.

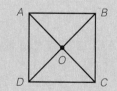

1. What is the 90° clockwise-rotation image of point *B*?
Point *C*

2. What is the 180° counterclockwise-rotation image of $\overline{OC}$?
$\overline{OA}$

3. Name two rotations that rotate point *C* to point *D*.
90° clockwise rotation;
270° counterclockwise rotation

Try This... Consider rotations of regular hexagon *QRSTUV* about point *P*.

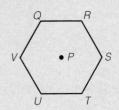

a. What is the 60° counterclockwise-rotation image of point *V*?

b. What is the 120° clockwise-rotation image of $\overline{RS}$?

c. Name two rotations that rotate point *Q* to point *T*.

Rotations preserve distances between points and collinearity of points. We can use these facts to find the rotation image of a figure. The following example shows how to begin with a figure and draw its image after rotation about a point.

Example 4. Draw the image of △*ABC* after a 100° clockwise rotation about *O*.

Step 1 Draw segments from *A*, *B*, and *C* to point *O*.

Step 2 Use a ruler and protractor to copy each segment so that it is rotated clockwise 100° from the original one.

Step 3 Connect the endpoints of the segments. The points corresponding to *A*, *B*, and *C* may be labeled *A'*, *B'*, and *C'*.

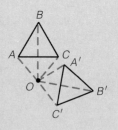

Try This... **d.** Trace or copy the figure. Draw the image of *PRST* after a 150° counterclockwise rotation about *N*.

Discussion A student claimed that rotating a figure clockwise 180° about a point always results in the same image as rotating the figure counterclockwise 180°. Do you agree or disagree?

Exercises

Practice

Consider rotations of regular hexagon *DEFGHI* about point *O* for Exercises 1–6.

1. What is the 60° clockwise-rotation image of point *G*?

2. What is the 120° counterclockwise-rotation image of point *I*?

3. What is the 240° clockwise-rotation image of $\overline{EF}$?

4. What is the 180° counterclockwise-rotation image of △*GOH*?

5. Name two rotations that rotate point *E* to point *F*.

6. Name two rotations that rotate $\overline{DE}$ to $\overline{FG}$.

Trace or copy each figure. Then draw the image of the figure after the specified rotation about point *O*.

7. 90° clockwise

8. 180° counterclockwise

9. 120° counterclockwise

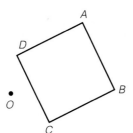

10. 200° clockwise

Extend and Apply

11. *ABCD* is a square. Which of the following rotations about point *O* will move point *A* to point *B*?

(A) 270° counterclockwise
(B) 90° counterclockwise
(C) 270° clockwise
(D) 180° clockwise

12. A hexagonal nut starts in the position shown. A mechanic rotates it clockwise 240°. Sketch its final position.

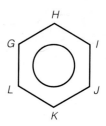

Use Mathematical Reasoning

The grid at the right shows the point $(3, 2)$ under a 90° counterclockwise rotation about the origin.

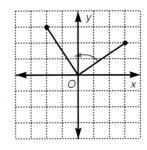

13. What is the image of $(3, 2)$?

14. Plot the point $(-1, 4)$ and carefully draw its image under a 90° counterclockwise rotation about the origin. What is the image?

15. In general, what is the image of (x, y) under a 90° counterclockwise rotation about the origin?

Determine whether each statement is true or false. If the statement is false, tell why.

16. The image of a figure after a rotation is always congruent to the original figure.

17. A 180° clockwise rotation results in the same image as a 180° counterclockwise rotation.

18. If an angle measures 30° before a rotation, it will measure 30° after the rotation.

19. No point remains fixed under a rotation.

20. Rotations preserve orientation.

21. If square $ABCD$ is rotated about its center, the image of point A cannot be point A.

Mixed Review

22. $\overline{AC}$ and $\overline{BD}$ are diameters, and $m\widehat{AB} = 88°$. Find $m\angle BOC$, $m\widehat{CD}$, and $m\widehat{BCA}$.

23. $\overline{AB}$ and $\overline{CD}$ are diameters, and $m\widehat{BD} = 161°$. Find $m\widehat{AD}$, $m\angle COB$, and $m\widehat{DAB}$.

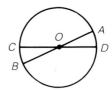

Find the missing base or altitude for a triangle with the given measures.

24. $A = 64$ m², $h = 8$ m

25. $A = 24$ cm², $b = 8$ cm

26. $A = 50$ in.², $h = 10$ in.

27. $A = 108$ m², $b = 1$ m

◇ Solve and check.

28. $6(y + 2y) = 54$

29. $99 = 3(6x - 3x)$

30. $5(z + z) = 4z + 30$

Geometry In Design

Tessellations with Rotations

Dutch artist M. C. Escher's tessellation at the right uses rotations to create a design. Escher may have made this tessellation by starting with an equilateral triangle, modifying it to the shape of a bird, and then rotating the bird.

Here is one way you can create a different tessellation of birds.

1. Start with an equilateral triangle, $\triangle ABC$, and draw a curve on side $\overline{AB}$.

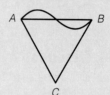

2. Rotate this curve 60° about point A as shown.

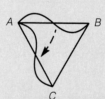

3. Locate point D, the midpoint of $\overline{BC}$. Draw a curve on $\overline{CD}$.

4. Rotate this curve 180° about point D as shown below.

5. The figure can now be filled in and rotated to produce a tessellation.

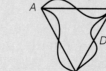

Problems

1. Begin with an equilateral triangle and follow the above steps to create your own tessellation.

2. Create a tessellation based on rotations beginning with a hexagon.

13-6 Inscribed Angles

Objective: Solve problems involving inscribed angles.

Explore

Computer software may be used for this activity.

1. Draw a circle and mark off an arc, $\overarc{AB}$, with measure 90°.

2. Choose points C, D, and E on the major arc and draw $\angle ACB$, $\angle ADB$, and $\angle AEB$ as shown.

3. Measure these angles and compare them with $m\overarc{AB}$.

Discuss What do you notice about the measures of $\angle ACB$, $\angle ADB$, and $\angle AEB$?

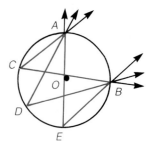

The angles you drew in the Explore have a special name, as well as some interesting properties.

Definition An **inscribed angle** is an angle whose vertex is on the circle and whose sides each intersect the circle in one other point.

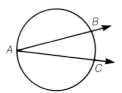

$\angle BAC$ **intercepts** $\overarc{BC}$.
$\angle BAC$ is **inscribed in** $\overarc{BAC}$.

The following theorem tells us how the measure of an inscribed angle is related to the measure of the intercepted arc:

Theorem 13.10 The measure of an inscribed angle is one-half the measure of its intercepted arc.

$$m\angle ABC = \frac{1}{2}m\overarc{AC}$$

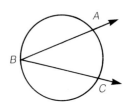

Example **1.** Identify all inscribed angles and their intercepted arcs.

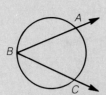 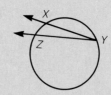 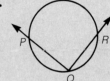 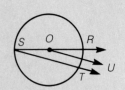

The inscribed angles are: $\angle ABC$, which intercepts $\widehat{AC}$,
$\angle XYZ$, which intercepts $\widehat{XZ}$,
and $\angle RST$, which intercepts $\widehat{RT}$.

Try This... Identify all inscribed angles and their intercepted arcs.

a.   **b.**   **c.**

Examples **2.** $m\widehat{AC} = 120°$. Find $m\angle ABC$.
$m\angle ABC = \dfrac{1}{2} \times 120 = 60°$

3. $m\angle DEF = 30°$. Find $m\widehat{DF}$.
$m\widehat{DF} = 60°$

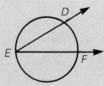

4. $m\widehat{SPQ} = 220°$. Find $m\angle P$ and and $m\angle R$.

$m\angle P = \dfrac{1}{2}(360 - 220) = 70°$

$m\angle R = \dfrac{1}{2} \times 220 = 110°$

5. $\overline{AB}$ is a diameter. Find $m\angle C$.
$m\angle C = \dfrac{1}{2} \times 180 = 90°$

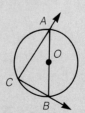

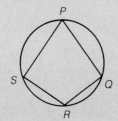

Try This... **d.** $\overline{DE}$ is a diameter. Find $m\angle A$, $m\angle B$, and $m\angle C$.

e. $m\widehat{AB} = m\widehat{BC} = 60°$. Find $m\angle ADB$, $m\angle ADC$, and $m\angle BEC$.

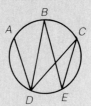

The following theorems are immediate consequences of Theorem 13.10:

Theorem 13.11 An angle inscribed in a semicircle is a right angle.

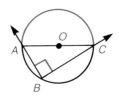

Theorem 13.12 Inscribed angles are congruent if they intercept the same arc or congruent arcs.

Example **6.** $\widehat{CD} \cong \widehat{AB}$, and $m\angle CAD = 47°$. Find $m\angle ADB$.
Since $\widehat{CD} \cong \widehat{AB}$, $\angle CAD \cong \angle ADB$ by Theorem 13.12.
Thus, $m\angle ADB = m\angle CAD = 47°$.

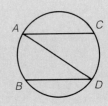

Try This... **f.** $\overline{SV}$ is a diameter. Find $m\angle SVT$.

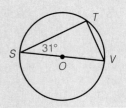

Discussion How can you use just an index card to draw a diameter of a circle?

Exercises

Practice

Identify each inscribed angle and its intercepted arc.

1.  **2.**   **3.**  **4.**

5. $m\widehat{AB} = 136°$.
Find $m\angle ACB$.

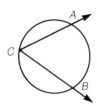

6. $m\angle DEF = 17°$.
Find $m\widehat{DF}$.

7. $\overline{XY}$ and $\overline{WZ}$ are
perpendicular
diameters.
Find $m\angle XWZ$.

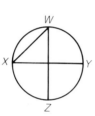

8. $\overline{PQ}$ is a diameter.
Find $m\angle PRQ$.

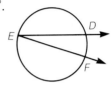

9. $m\angle MNP = 70°$, $m\angle QMN = 94°$, **10.** $m\angle C = 90°$, $m\widehat{AD} = 80°$, and
and $m\widehat{QP} = 85°$. Find $m\angle MQP$ $m\widehat{BC} = 125°$. Find $m\angle A$, $m\angle B$,
and $m\angle QPN$. and $m\angle D$.

Extend and Apply

Use $\odot O$ for Exercises 11–22.
$m\angle P = 24°$, $m\angle Q = 18°$, and $\overline{TR}$ is a diameter.
Find these measures.

11. $m\widehat{ST}$ **12.** $m\widehat{QR}$

13. $m\angle SUR$ **14.** $m\widehat{TQ}$

15. $m\angle PSQ$ **16.** $m\angle TRS$

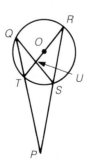

Suppose that $m\widehat{TS} = 40°$, $m\widehat{QR} = 78°$, and $\overline{TR}$
is a diameter. Find these measures in $\odot O$.

17. $m\angle QSR$ **18.** $m\angle P$
19. $m\angle TUQ$ **20.** $m\widehat{SR}$
21. $m\angle Q$ **22.** $m\angle PSQ$

23. Which of the following angles must be congruent to ∠WYZ?

(A) ∠WXZ (B) ∠WPZ (C) ∠XWY (D) ∠XZY

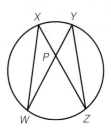

Use Mathematical Reasoning

24. △ABC is equilateral. Find the sum m∠D + m∠E + m∠F.

25. Explain how a carpenter's square can be used to check whether a semicircle has been cut in a board.

Mixed Review

Find the area of each rectangle.

26. $b = 6$ m, $h = 8$ m

27. $h = 4.1$ cm, $b = 6.2$ cm

◇ Write using exponential notation.

28. 10,000   **29.** 100,000   **30.** 100

Discover

Computer software may be used for this activity.

1. Use a compass to draw a circle with center O.
Draw two intersecting chords, $\overline{AB}$ and $\overline{CD}$. Label the point of intersection E.

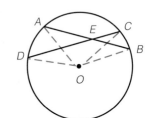

2. Draw angles ∠AOD and ∠COB and find their measures.

3. What are $m\overset{\frown}{AD}$ and $m\overset{\frown}{CB}$?

4. Measure ∠AED. How is m∠AED related to $m\overset{\frown}{AD}$ and $m\overset{\frown}{CB}$? (Hint: Consider the sum of $m\overset{\frown}{AD}$ and $m\overset{\frown}{CB}$.)

Use your discovery to complete the following theorem.

Theorem 13.13 The measure of the angle formed by two chords intersecting inside a circle is equal to _____ the sum of the measures of the intercepted arcs.

13-7 Inscribed Polygons

Objective: Construct regular polygons by inscribing them in a circle.

In the photo at the right, the square block of wood in the circular hole illustrates the idea of an *inscribed polygon*. In the figure, △ABC is inscribed in the circle. We can also say that the circle is circumscribed about △ABC.

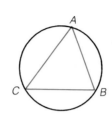

Definition   A polygon is **inscribed** in a circle if every vertex of the polygon lies on the circle. The circle is said to be **circumscribed** about the polygon.

Some regular polygons can be constructed by inscribing them in a circle.

Example   1. Inscribe a regular hexagon in a circle.

Step 1
Draw a circle with radius *r*.

Step 2
Choose a point *A* on the circle. Use a compass opening equal to *r* and mark off arcs around the circle.

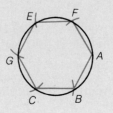

Step 3
Connect these points on the circle to obtain a regular hexagon.

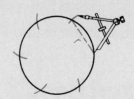

Example 2. Inscribe a regular pentagon in a circle.

Step 1

Draw a circle and a diameter $\overline{AB}$. Draw the perpendicular to $\overline{AB}$ at center O. Find the midpoint, D, of $\overline{OB}$.

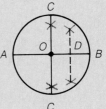

Step 2

Place the compass point at D. Use a compass opening CD and draw $\overset{\frown}{CE}$.

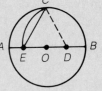

Step 3

Choose a point on the circle. Use compass opening CE and mark off arcs around the circle. Connect these points to obtain a regular pentagon.

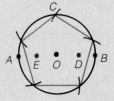

Discussion A student claimed that a circle could be circumscribed about every regular polygon, regardless of the number of sides. Do you agree? Is the same true for nonregular polygons?

Exercises

Practice

Draw a circle. Then use a compass and a straightedge to inscribe each polygon.

1. Square

2. Regular octagon

3. Regular hexagon

4. Regular dodecagon (12 sides)

5. Regular pentagon

6. Regular decagon

Extend and Apply

7. A square is inscribed in a circle. Which of the following is the measure of each of the intercepted minor arcs?

   (A) 60°   (B) 90°   (C) 45°   (D) 180°

8. Draw three circles and inscribe an acute, a right, and an obtuse triangle. Which triangle has a diameter for one side? Which triangle intersects every diameter of the circle? Which triangle is contained in the interior of a semicircle?

9. Judy cuts a circle out of paper. She wants to inscribe a regular octagon in the circle. How can she fold the circle to determine the eight vertices of the octagon?

Use Mathematical Reasoning

10. Draw several circles and inscribe some quadrilaterals. Try to discover a relationship among the four angles of an inscribed quadrilateral.

Mixed Review

11. $\overline{DF}$ is a diameter and $m\widehat{DG} = 80°$. Find $m\angle DEG$, $m\angle DFG$, and $m\widehat{FG}$.

12. $\overline{WY}$ is a diameter, and $m\angle WYX = 31°$. Find $m\widehat{XW}$, $m\widehat{XY}$, and $m\angle YZW$.

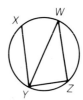

13. Identify all radii, chords, and diameters.

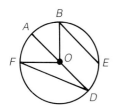

14. Diameter $AD = 12.6$ cm. Find OC.

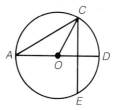

For each diameter, find the length of the radius.

15. 7.7 m   **16.** 19 ft   **17.** $3\frac{1}{2}$ in.

◇◇ Write using standard notation.

18. 2^5 **19.** 9^2 **20.** 3^4

13-8 Circumference of Circles

Objective: Solve problems involving the circumference of a circle.

Carlos wanted to find the distance around the edge of a quarter. He did this by rolling the quarter along a table and measuring the distance shown.

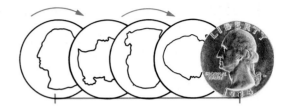

We know that the distance around a polygon is its *perimeter* (p). The distance around a circle is its **circumference.** We can use perimeters of inscribed regular polygons to approximate the circumference of a circle.

$p = 5.90$ cm $p = 6$ cm $p = 6.20$ cm $p = 6.24$ cm

As the number of sides of the inscribed polygons increases, the perimeters approach a number called their *limit*. We can find perimeters as close as we want to this limit.

Definition   The **circumference** of a circle is the limit of the perimeters of all inscribed regular polygons.

■ *Explore*

1. Collect several circular objects such as coins, jar lids, soup cans, and so on.
2. Measure the circumference (c) of each object by first wrapping a piece of string around it and then measuring the string. Measure the diameter (d) of each object.
3. For each object calculate the ratio $\dfrac{c}{d}$. What do you notice?

The Explore leads to the following theorem:

> **Theorem 13.14**   For all circles, the ratio $\dfrac{\text{circumference}}{\text{diameter}}$ is the same number.

We use the Greek letter π (pi) to represent this number.

$$\pi = \frac{c}{d} = \frac{\text{circumference}}{\text{diameter}}, \text{ or } c = \pi d$$

When written as a decimal, π does not terminate or repeat in a cycle. Such a number is called *irrational*. A good rational-number approximation of π is 3.1415927.

With calculations involving π, we can either express the exact answer in terms of π, or we can use 3.14 for π to get an approximate answer.

Example   **1.** Find the circumference of the circle.

$$\begin{aligned} c &= \pi d \\ &= \pi \times 4 \\ &= 4\pi \text{ cm} \end{aligned}$$

We can also use 3.14 for π and approximate the answer as $4 \times 3.14 = 12.56$ cm.

Try This...   **a.** Find the circumference of a circle whose diameter is 7.5 m.

Example   **2.** Find the circumference of a circle whose radius is 10 m. Use 3.14 for π.

$$\begin{aligned} c &= \pi d \\ &= \pi \cdot 2r \ (d = 2r) \\ &\approx 3.14 \times 2 \times 10 \\ &\approx 62.8 \text{ m} \end{aligned}$$

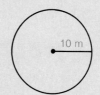

Try This...   **b.** Find the circumference of a circle whose radius is 15 cm. Use 3.14 for π.

The following Examples show how a calculator can be helpful in problems involving π.

Example 3. A flagpole has a circumference of 30 cm. Find its diameter.

$$\pi d = c$$
$$d = \frac{c}{\pi} = \frac{30}{\pi}$$

It is helpful to use a calculator for this computation. Use the π key on your calculator.

| 30 | ÷ | π | = | 9.5492966 |

Rounding to the nearest tenth, the diameter is 9.5 cm.

Try This... c. The circumference of a tree is 76 cm. Find its diameter to the nearest tenth of a centimeter.

Example 4. The sides of a square are 10 cm. Find the circumference of the circumscribed circle to the nearest centimeter.

Because the diagonal of the square contains the center of the circle, it is a diameter.

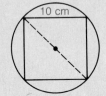
10 cm

By the Pythagorean Theorem,
$d^2 = 10^2 + 10^2 = 200$

Thus, $d = \sqrt{200}$ and $c = \pi d = \pi \times \sqrt{200} = \pi\sqrt{200}$ cm.

We can use a calculator as follows:

| π | × | 200 | $\sqrt{x}$ | = | 44.428829 |

The circumference, to the nearest centimeter, is 44 cm.

To find an approximate value in calculations involving π, use the π key on your calculator. If your calculator does not have a π key, use 3.14. Note that this will give a less precise estimate.

Try This... d. A rectangle 40 cm by 30 cm is inscribed in a circle. Find the circumference of the circle to the nearest centimeter.

Discussion Before the invention of decimals, $\frac{22}{7}$ was used for π. Use your calculator to find the decimal for $\frac{22}{7}$. Compare this to the value given by the π key. Is the rational number $\frac{22}{7}$ greater than or less than π?

Exercises

Practice

Find the circumference of each circle. Express your answer in terms of π. Then use your calculator to give an approximate answer to the nearest tenth.

1.

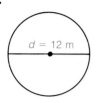

$d = 12$ m

2.

$r = 9.5$ m

3.

$d = 0.5$ km

4.

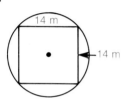

14 m

14 m

5.

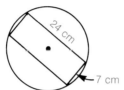

24 cm

7 cm

6.

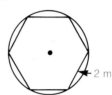

2 m

7. $d = 75$ cm   **8.** $r = 0.75$ m   **9.** $r = 0.86$ m

Round to the nearest tenth.

10. The girth (circumference) of a tree is 63 cm. Find its diameter.

11. The circumference of a pipe is 42 cm. Find its diameter.

Extend and Apply

12. A circular flower bed has a diameter of 12 ft. A gardener wants to place a small fence around the edge of the bed. Which of the following is the best estimate of the amount of fencing needed?

(A) 24 ft   (B) 18 ft   (C) 74 ft   (D) 37 ft

13. Suppose that the earth travels in a nearly circular orbit around the sun. The average distance from the sun is about 150,000,000 km. Find the distance the earth travels in one orbit around the sun.

14. The radius of a circle is doubled. How does the circumference change?

15. The diameter of a circle is halved. How does the circumference change?

Round to the nearest whole number in Exercises 16–18.

16. The diameter of a wheel is 45 cm. How far does a wheel travel in 100 revolutions?

17. The radius of a wheel is 30 cm. How far does a wheel travel in 100 revolutions?

18. Find the length of a circular orbit of a satellite if the satellite is 800 km above the surface of the earth. (The radius of the earth is approximately 6400 km.)

19. Here is a formula for calculating the distance that a bicycle travels in one revolution of the pedals:

$$D = \pi \cdot d \cdot \frac{a}{b}, \text{ where } \begin{array}{l} d = \text{diameter of the wheel (in centimeters)} \\ a = \text{number of teeth on the pedal sprocket} \\ b = \text{number of teeth on the rear sprocket} \end{array}$$

If $d = 70$ cm, $a = 52$, and $b = 18$, find D to the nearest centimeter.

Use Mathematical Reasoning

20. The circumference of a circle is 40π cm. Find the perimeter of an inscribed regular hexagon.

21. Imagine a steel band wrapped around the equator of the earth. A 10-meter piece is added to the band. How does this affect its radius? Suppose the space between the band and the equator is equidistant around the sphere. Could you slip a piece of paper under the band, crawl under the band, or walk under the band?

Mixed Review

Find the area of each trapezoid.

22.

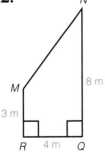

23.

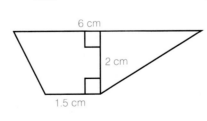

24.

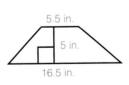

Solve and check.

25. $3z + 9z = 48 + 6z$

26. $x - 5x + 9x = 55$

27. $16 + 4w = w + 1 + 8w$

Probability Connection

Probability and Pi

The number π occurs in many unexpected situations. You can do the following experiment to see how π is related to the probability of a tossed toothpick landing on a line. The experiment was first carried out in the 18th century by French mathematician Georges Buffon.

1. On a large sheet of stiff paper or cardboard, draw parallel segments that are as far apart as the toothpick is long.

2. Place the sheet of paper or cardboard on the floor and, while standing, toss the toothpick onto the paper or cardboard. Note whether or not the toothpick crosses a segment.

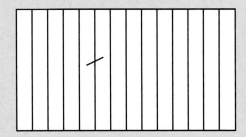

3. Repeat the process several times and record the number of tosses (t) and the number of times the toothpick crosses a segment (c). Be sure to count only the tosses in which the toothpick lands entirely on the paper or cardboard, as shown in the figure above. You may wish to compile your results in a chart.

4. Double the grand total of tosses and divide by the grand total of crosses. The quotient should be an approximation of π. The more tosses you make, the better the approximation should be.

$$\pi \approx \frac{2t}{c}$$

Problems

1. Combine your results from Step 3 with those of classmates, if necessary, so that there is a total of at least 100 tosses. Calculate $\frac{2t}{c}$ and compare your results with the actual value of π. Is your approximation greater than or less than the actual value of π?

2. Could the fraction $\frac{2t}{c}$ ever be exactly equal to π? Why? (Hint: Recall that π is irrational.)

13-9 Area of Circles

Objective: Solve problems involving the area of a circle and the area of a sector of a circle.

Explore

1. Draw on a sheet of paper a circle with a radius of 4 in. Cut it out and fold it into 16 equal sections as shown.

2. Cut out the sections and assemble them as shown below. The resulting figure should be close to a parallelogram.

3. What is the height of the "parallelogram"?

4. What is the base length of the "parallelogram"? (Hint: Consider the circumference of the circle.)

Discuss What is the area of the "parallelogram"? Develop a formula for the area of a circle.

The Explore describes one way to approximate the area of a circle. Another way to approximate the area of a circle is to consider the areas of inscribed polygons. As the number of sides of the inscribed polygons increases, the areas approach a number called their limit.

> **Definition** The **area** of a circle is the limit of the areas of all inscribed regular polygons.

$A_3 = 1.3$

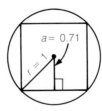

$A_4 = 2$

$A_6 = 2.6$

$A_{12} = 3.0$

As you may have discovered in the Explore, the area of a circle is equal to the area of a parallelogram with height r and base length $\frac{1}{2}\pi d$, or πr. The area of the parallelogram is therefore $r \cdot \pi r = \pi r^2$. This leads to the following theorem:

> **Theorem 13.15** The area of a circle with radius r is πr^2.

The following Examples illustrate the use of Theorem 13.15.

 Examples Find the area of each of the following in terms of π. Then find an approximation to the nearest tenth.

1. A circle with a radius of 6 cm
   $$A = \pi r^2 = \pi \cdot 6^2 = 36\pi \text{ cm}^2 \approx 113.1 \text{ cm}^2$$

2. A circle with a diameter of 4.8 m
   Because $r = \dfrac{d}{2}$, we have $r = \dfrac{4.8}{2} = 2.4$ m.
   Thus, $A = \pi r^2 = \pi \cdot (2.4)^2 = 5.76\pi \text{ m}^2 \approx 18.1 \text{ m}^2$.

3. A circular flower bed with a radius of 7.2 ft
   $$A = \pi r^2 = \pi \cdot (7.2)^2 = 51.84\pi \text{ ft}^2 \approx 162.9 \text{ ft}^2$$

 Try This... Find the area of each circle in terms of π. Then find an approximation to the nearest hundredth.

a. $r = 8$ cm   b. $d = 0.14$ in.   c. $r = 2.8$ km

Sometimes we need to find the area of a wedge-shaped part of a circular region. Such regions are called *sectors*.

> **Definition** A **sector** of a circle is a region bounded by two radii and an arc.
>
>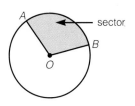

As the figures below show, the area of a sector is proportional to its arc (or central angle) measure.

$m\widehat{AB} = 180°$ $m\widehat{AB} = 90°$ $m\widehat{AB} = 60°$

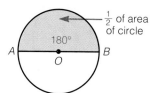   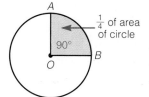   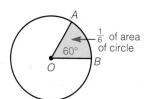

> **Theorem 13.16** The area of a sector with radius r and arc (or central angle) measure n is $\dfrac{n}{360} \cdot \pi r^2$.

Example 4. Find the area of the shaded sector to the nearest tenth.

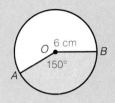

$$m\widehat{AC} = 150°, \quad r = 6 \text{ cm}$$

$$\text{Area} = \frac{150}{360} \cdot \pi(6)^2$$

$$\approx 47.1 \text{ cm}^2$$

Try This... d. Find the area of the shaded sector to the nearest tenth.

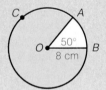

The next Example shows how the area of some regions can be found by adding or subtracting the areas of other regions.

Example 5. Find the area of the shaded region to the nearest tenth.

The area of the shaded region is the area of the circle minus the area of the rectangle.

$$A_\odot - A_\square$$

Diagonal $\overline{DB}$ is a diameter. So, by the Pythagorean Theorem,

$$DB^2 = 4^2 + 3^2 = 25$$

Thus, $DB = 5$ cm and the radius is $\dfrac{5}{2}$ cm.

Now, $A_{\text{shaded region}} = \dfrac{25}{4}\pi - (4 \times 3)$

$$\approx 19.6 - 12, \text{ or } 7.6 \text{ cm}^2$$

Try This... e. Find the area of the shaded region to the nearest tenth. The centers are O and P, and the radius of the smaller circle is 4 cm.

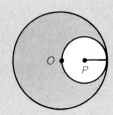

Exercises

Practice

In Exercises 1–15, first find the answer using π; then find an approximation to the nearest tenth.

Given the radius, find the area of the circle.

1. 14 m   **2.** 5.7 cm   **3.** 10 mm

4. 0.3 in.   **5.** 15 ft   **6.** 8.1 yd

7. 9 km   **8.** 3 cm   **9.** 12.9 m

Given the diameter, find the area of the circle.

10. 24 cm   **11.** 18 in.   **12.** 1.2 m

13. 3 mm   **14.** 74 m   **15.** 4.6 cm

Find the area of each shaded sector to the nearest tenth.

16.   **17.**

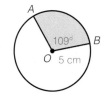

18.   **19.**

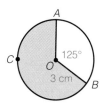

Find the area of the shaded regions to the nearest tenth.

20.   **21.** All radii are 1 m.

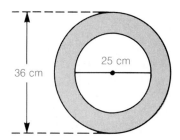
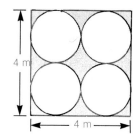

Extend and Apply

22. A local radio station can reach a radius of 30 miles. How many square miles of listening audience does the station have?

23. A circle has a circumference of 12π cm. Which of the following is the area of the circle?

(A) 18π cm^2 (B) 144π cm^2 (C) 36π cm^2 (D) 6π cm^2

Find the area of each shaded region to the nearest tenth.

24. All diameters are $\frac{2}{3}$ m.

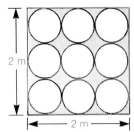

2 m

2 m

25. $ABCD$ is a square with $AB = 5$ m.

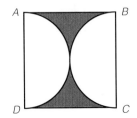

26. A glass company charges $0.20 per square inch of glass for circular windows, and $0.10 per inch for the circular cutting. What will they charge for these three windows if the diameter of each window is 9 inches?

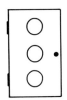

27. A circular table top made of oak weighs 2.5 grams per square centimeter of surface. The top weighs no more than 20 kilograms. What is the maximum diameter of the table top?

Use Mathematical Reasoning

28. $\triangle ABC$ has sides measuring 6, 8, and 10. Show that the area of the semicircle on the hypotenuse is equal to the sum of the areas of the other two semicircles.

29. The Andersons tied their goat to the outside corner of their barn with a 15-ft rope. Assuming the sides of the barn are longer than 15 ft, how many square feet of grazing area does the goat have?

Mixed Review

Round to the nearest whole number.

30. The circumference of a tree is 23 ft. Find its diameter.

31. The circumference of a planet is 21,000 mi. Find its radius.

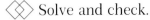 Solve and check.

32. $7(x + 9) = 98$ **33.** $48 = 6(16 - 2y)$ **34.** $138 = 3(1 + 5y)$

Chapter 13 Review

For each diameter, find the length of the radius.

13-1

1. 62.2 mm   **2.** 9 ft

For each radius, find the length of the diameter.

3. 7.9 cm   **4.** 0.07 m

Use the figure at the right for Problems 5–7.

13-2

5. $AD = 39$ and $AB = 15$. Find CD.

6. $CD = 8$ and $AB = 3$. What is the radius of the circle?

7. $CD = 24$ and $AD = 13$. Find AB.

Use the figure at the right for Problems 8–10. $\overline{AB}$ and $\overline{AC}$ are tangent segments.

13-3

8. $OA = 10$ and $AC = 8$. Find OC.

9. Diameter of $\odot O$ is 10, and $AC = 12$. Find OA.

10. $m\angle BAC = 80°$. Find $m\angle BOC$.

Use $\odot K$ for Problems 11–13. $m\angle AKD = 72°$.

13-4

11. Name the minor arcs, major arcs, and semicircles.

12. Find $m\,\widehat{BC}$.

13. Find $m\,\widehat{DCB}$.

Consider rotations about point O in the regular polygon at the right for Problems 14–16.

13-5

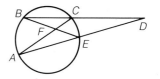

14. What is the 90° counterclockwise-rotation image of point Q?

15. What is the 225° clockwise-rotation image of $\overline{TL}$?

16. Name two rotations that rotate point P to point S.

Use the figure at the right for Problems 17–19.

13-6

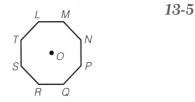

17. $m\,\widehat{AB} = 82°$. Find $m\angle BEA$.

18. $m\angle CBE = 29°$. Find $m\angle CAE$ and $m\,\widehat{CE}$.

19. $m\,\widehat{BEA} = 290°$. Find $m\angle BCA$.

20. Draw a circle. Then use a compass and a straightedge to inscribe a regular hexagon.

13-7

13-8 Find the circumference of each circle. Express your answer in terms of π. Then given an approximate answer to the nearest tenth.

21.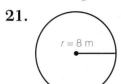

r = 8 m

22.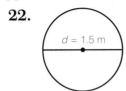

d = 1.5 m

23.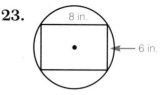

8 in.

6 in.

13-9 Find the area of each circle. First use π and then find an approximation to the nearest tenth.

24. r = 6 in. **25.** d = 4 m **26.** d = 6.4 cm

27. Find the area of the shaded sector to the nearest tenth.

O
10 m
60
A
B

28. Find the area of the shaded region to the nearest tenth. Diameter AB = 10 m, and diameter AC = 12 m.

A
B
C

Chapter 13 Test

For each diameter, find the length of the radius.

1. 11 ft **2.** 80.8 m

For each radius, find the length of the diameter.

3. 8.8 cm **4.** $1\frac{1}{4}$ in.

Use the figure at the right for Problems 5–7.

5. SP = 10 and TP = 26. Find RT.

6. PQ = 30 and TR = 48. Find PS.

7. PS = 24 and RT = 14. What is the radius of the circle?

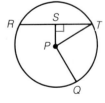

R
S
T
P
Q

Use the figure at the right for Problems 8–10. $\overline{PA}$ and $\overline{PB}$ are tangent segments.

8. PA = 12 and AO = 5. Find PO.

9. Diameter of ⊙O is 12, and OP = 10. Find PA.

10. m∠APB = 44°. Find m∠PBA.

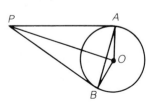

P
A
O
B

Use ⊙O for Problems 11–13. $m\overset{\frown}{CD} = 52°$.

11. Name the minor arcs, major arcs, and semicircles.

12. Find $m\angle COE$.

13. Find $m\overset{\frown}{CFD}$.

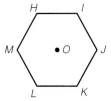

Consider rotations about point O in the regular polygon at the right for Problems 14–16.

14. What is the 120° counterclockwise-rotation image of point L?

15. What is the 240° clockwise-rotation image of $\overline{MH}$?

16. Name two rotations that rotate point I to point M.

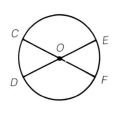

Use the figure at the right for Problems 17–19.

17. $m\overset{\frown}{SU} = 46°$. Find $m\angle SWU$

18. $m\angle RUW = 45°$. Find $m\angle RSW$ and $m\overset{\frown}{RW}$.

19. $m\overset{\frown}{SRU} = 310°$. Find $m\angle SRU$.

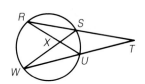

20. Draw a circle. Then use a compass and a straightedge to inscribe a regular pentagon.

 Find the circumference of each circle. Express your answer in terms of π. Then given an approximate answer to the nearest tenth.

21.

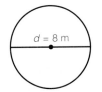

$d = 8$ m

22.

$r = 2.4$ m

23.

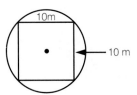

10m

10 m

 Find the area of each circle. First use π and then find an approximation to the nearest tenth.

24. $r = 8$ ft

25. $d = 10$ m

26. $d = 12.2$ cm

 27. Find the area of the shaded sector to the nearest tenth.

 28. Find the area of the shaded region to the nearest tenth. Diameter $SV = 24$ m, and diameter $ST =$ diameter $TV = 12$ m.

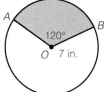

120°

O 7 in.

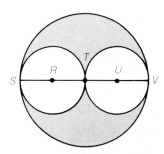

14 Space Figures

Computers are capable of creating realistic images of space figures. The use of
perspective makes this two-dimensional drawing appear three dimensional.

14-1 Polyhedrons

Objective: Recognize and classify polyhedrons.

Most of the figures we have worked with so far have been flat and two-dimensional. However, many of the most common geometric figures around us are three-dimensional solids (or space figures). Cardboard boxes, soup cans, ball bearings, and the pyramids of Egypt are all examples of space figures.

The space figures below were all drawn by a computer using a program called *Mathematica.* What do these solids have in common?

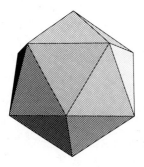

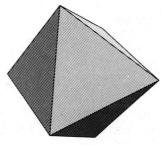

Space figures like those shown above, which are made up of only polygonal regions, are called **polyhedrons.** The polygonal regions are the **faces** of the polyhedron. We call the sides and vertices of the faces **edges** and **vertices** of the polyhedron, respectively.

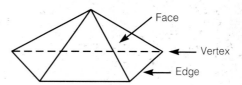

Face

Vertex

Edge

Example   **1.** Name the faces, edges, and vertices of this polyhedron.

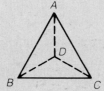

Faces: $\triangle ABC$, $\triangle ADC$, $\triangle ADB$, and $\triangle BDC$
Edges: $\overline{AB}$, $\overline{AC}$, $\overline{AD}$, $\overline{BD}$, $\overline{BC}$, and $\overline{DC}$
Vertices: A, B, C, and D

Try This... a. Name the faces, edges, and vertices of this polyhedron.

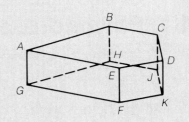

In the same way that we classified polygons by their number of sides, we can classify polyhedrons by their number of faces. The table below summarizes some of the more common polyhedrons.

Number of Faces	Name	Number of Faces	Name
4	Tetrahedron	9	Nonahedron
5	Pentahedron	10	Decahedron
6	Hexahedron	11	Undecahedron
7	Heptahedron	12	Dodecahedron
8	Octahedron	20	Icosahedron

Example 2. Identify the polyhedron at the right.

Since the polyhedron has eight faces, it is an octahedron.

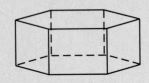

Try This... b. Identify the polyhedron at the right.

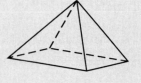

Certain polyhedrons have more common names. For example, a hexahedron with square faces is a *cube*.

Discussion Give examples of space figures that are *not* polyhedrons.

Exercises

Name the faces, edges, and vertices of each polyhedron.

1.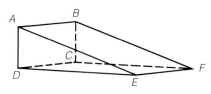

2.

3.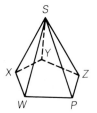

4.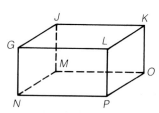

Identify each polyhedron.

5.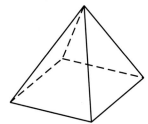

6.

7.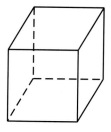

8.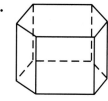

Extend and Apply

9. Which of the following is *not* an example of a polyhedron?

    (A) Cube   (B) Square   (C) Pyramid   (D) Pentahedron

Determine whether each statement is true or false.

10. A pentahedron may have six faces.

11. A cube is a hexahedron.

12. Every space figure is a polyhedron.

13. A polyhedron with hexagonal faces is a hexahedron.

14. The pyramids of Egypt are pentahedrons.

15. A polyhedron with 20 faces is called an icosahedron.

For each polyhedron, determine which one of the patterns could be cut out and folded to produce a model of it.

16. Tetrahedron

17. Pentahedron

18. Hexahedron

19. Octahedron

a.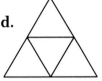

b.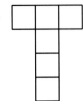

c.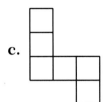

d.

e.

Use Mathematical Reasoning

20. Prepare a table using drawings of four different polyhedrons from this lesson. For each polyhedron, list the number of faces (F), the number of edges (E), and the number of vertices (V). Then calculate $F - E + V$ for each polyhedron. What do you notice? Do you think this is true for every polyhedron?

Mixed Review

 Given the diameter, find the area of the circle to the nearest tenth.

21. 14 m   **22.** 6 cm   **23.** 9 in.

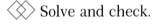 Solve and check.

24. $15x = 9(x + 2)$ **25.** $7(8 - y) = 7y$ **26.** $3(4 + 2z) = 7z + 2$

Enrichment

Here is a way to use an envelope to make a model of a polyhedron.

1. On a sealed envelope, construct an equilateral triangle, △ABC, as shown.

2. Cut along $\overline{DE}$, through C, parallel to $\overline{AB}$.

3. Fold the envelope back and forth along $\overline{AC}$ and $\overline{BC}$.

4. Let C' be the point on the reverse side corresponding to point C.

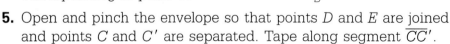

5. Open and pinch the envelope so that points D and E are joined and points C and C' are separated. Tape along segment $\overline{CC'}$.

6. What type of polyhedron have you created?

14-2 Prisms and Their Surface Area

Objective: Classify prisms, identify parts of prisms, and find lateral and surface areas of prisms.

The space figure at the right is a special type of polyhedron called a **prism.** It is a prism because two of its faces (*ABCD* and *EFGH*) are congruent and lie in parallel planes. The congruent and parallel faces are called **bases.**

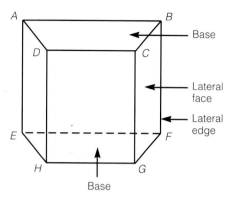

The vertices of the bases are joined to form parallelogram-shaped regions called **lateral faces.** The intersection of two lateral faces is a **lateral edge.**

If the lateral faces of a prism are rectangles, the prism is a **right prism.** Otherwise, the prism is an **oblique prism.** The length of a lateral edge of a right prism is its **height.**

Type of Prism	Right	Oblique
Triangular		
Quadrangular		
Pentagonal		

Example

1. Classify the prism, and name the bases, the lateral faces, and the lateral edges.

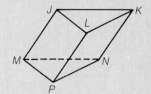

Oblique triangular prism
Bases: △*JKL* and △*MNP*
Lateral faces: ▱*JLPM*, ▱*JKNM*, and ▱*LKNP*
Lateral edges: $\overline{LP}$, $\overline{JM}$, and $\overline{KN}$

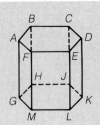

The box below is called a **rectangular prism**, since its bases are rectangles. In this case, any pair of opposite faces could be considered bases.

Suppose a box manufacturer wants to construct the box shown at the right out of cardboard and to cover the lateral faces with decorative paper. How much cardboard does the manufacturer need? How much decorative paper does the manufacturer need?

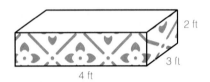

The amount of cardboard needed is the surface area of the box. The **surface area** of a polyhedron is the sum of the areas of all its faces.

The amount of decorative paper needed is the lateral area of the box. The **lateral area** of a prism is the sum of the areas of its lateral faces.

Example **2.** Find the lateral area and surface area of the box above.

Lateral area:
Area of front and back $2 \times 4 \times 2 = 16$
Area of left and right sides $2 \times 3 \times 2 = 12$

Sum of areas of lateral faces $= 16 + 12 = 28$ ft^2

Surface area:
Area of front and back $2 \times 4 \times 2 = 16$
Area of left and right sides $2 \times 3 \times 2 = 12$
Area of top and bottom $2 \times 4 \times 3 = 24$

Sum of areas of faces $= 16 + 12 + 24 = 52$ ft^2

Try This... b. Find the lateral area and surface area of the right triangular prism.

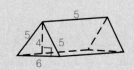

▰ *Explore* ▰▰▰▰▰▰▰▰▰▰▰

1. Fold over the ends of a rectangular sheet of paper as shown.
2. Tape the edges together to construct a model of the lateral faces of a triangular prism.
3. Measure the edges of the prism and find the lateral area of the prism.
4. Find the perimeter of a base of the prism. Multiply this by the height of the prism. What do you notice?

The Explore leads to the following theorem:

Theorem 14.1   The lateral area of a right prism is the product of the perimeter p of a base and the height h of the prism.

$$\text{Lateral Area} = ph$$

$$\text{Surface Area} = \text{Lateral Area} + 2 \times \text{Base Area} = ph + 2B$$

The next Example shows how to use Theorem 14.1.

Example   **3.** Find the lateral area of the right pentagonal prism.

$$p = 5 + 3 + 3 + 6 + 4$$
$$= 21 \text{ cm}$$
$$h = 3 \text{ cm}$$

$$\text{Lateral area} = ph = 21 \times 3$$
$$= 63 \text{ cm}^2$$

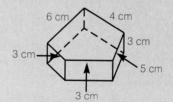

Try This...   **c.** Find the lateral area of the right prism if the bases are equilateral triangles.

Discussion   What is the minimum number of faces that a prism can have? Is there a maximum number of faces?

Exercises

Classify each prism, and name the bases, the lateral faces, and the lateral edges.

1.

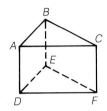

2. *URST* is a base.

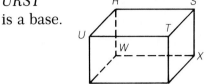

3. *MNPQ* is a base.

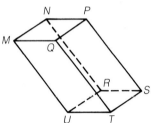

4.

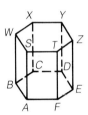

Find the lateral area and surface area of each right prism.

5.

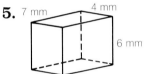

6.

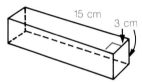

7.

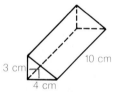

8.

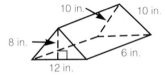

Use the formula Lateral Area = *ph* to find the lateral area of each right prism.

9.

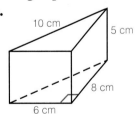

10.

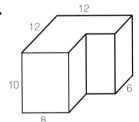

11.

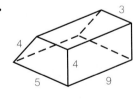

Extend and Apply

12. Which of the following is the lateral area of the right prism shown at the right?

  (A) 468 cm^2 (B) 78 cm^2

  (C) 540 cm^2 (D) 234 cm^2

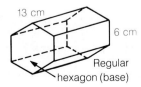

Regular hexagon (base)

13. a. Estimate the surface area of the box of detergent.

b. Sketch all the faces to find the surface area of the box.

c. Use the formula Lateral Area = ph to find the lateral area of the box. Add the areas of the bases to find the surface area.

d. How does your answer in part **a.** compare to your answers in parts **b.** and **c.**?

33.5 cm

8 cm   25 cm

Suppose each small cube in the drawings has edges of length 1 in. There are no missing cubes in the back of the solids. Find the surface area of each solid.

14.
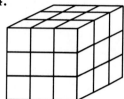

15.
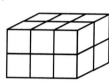

16.
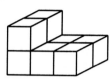

Use Mathematical Reasoning

17. First estimate and then find the surface area of the solid. (Hint: Be sure to count the areas of the "inside" faces.)

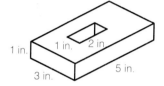

1 in.   1 in.   2 in.

3 in.   5 in.

Mixed Review

Identify each polyhedron by the number of faces.

18. 4 faces   **19.** 8 faces   **20.** 20 faces

Write using standard notation.

21. 3^3 **22.** 9^4 **23.** 10^6

Visualization

In each of the Problems below, if the cube on the left is unfolded, which figure on the right could it become?

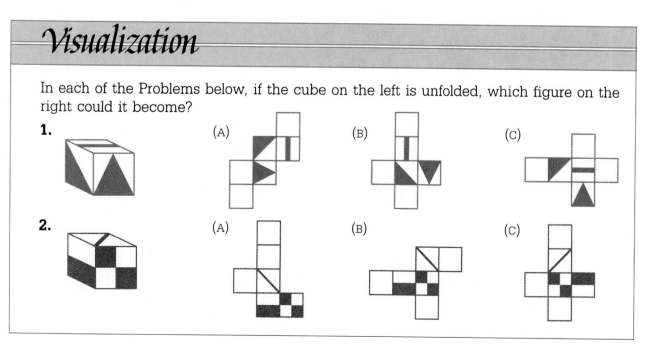

1.   (A)   (B)   (C)

2.   (A)   (B)   (C)

14-3 Pyramids and Their Surface Area

Objective: Classify pyramids, identify parts of pyramids, and find lateral and surface areas of pyramids.

Pyramids have been used in architecture since the days of the ancient Egyptians. The building at the right, designed by I. M. Pei, is part of the Louvre Museum in Paris.

A pyramid is a special type of polyhedron. It consists of a base and three or more triangular regions called lateral faces. The lateral faces meet at a common point, called the **vertex** of the pyramid. The **slant height**, k, is the length of the perpendicular from the vertex to an edge of the base.

The segment from the vertex of a pyramid perpendicular to the plane of the base is called the **altitude** of the pyramid. The length of the altitude is the **height** of the pyramid.

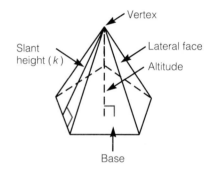

Most pyramids seen in everyday life are regular pyramids. A **regular pyramid** is a pyramid whose base is a regular polygon and whose lateral faces are congruent. We will assume all of the pyramids in this chapter are regular pyramids unless otherwise specified.

As shown below, pyramids are often classified by their bases.

Triangular pyramid
(Tetrahedron)

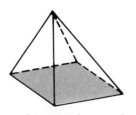

Regular quadrangular
pyramid

Regular hexagonal
pyramid

Example 1. Classify the pyramid, and name the base, the lateral faces, and the lateral edges.

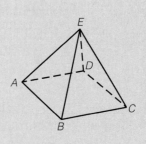

Regular quadrangular pyramid
Base: $ABCD$
Lateral faces: $\triangle AEB$, $\triangle BEC$, $\triangle CED$, and $\triangle DEA$
Lateral edges: $\overline{EA}$, $\overline{EB}$, $\overline{EC}$, and $\overline{ED}$

Try This... **a.** Classify the pyramid, and name the base, the lateral faces, and the lateral edges.

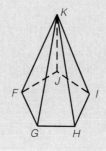

As shown in the following Example, we can find the lateral area and the surface area of a pyramid in a manner similar to that used for prisms.

Example 2. Find the lateral area and surface area of the regular quadrangular pyramid.

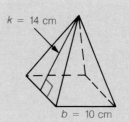

Because the base is a square region, its area is 10×10, or 100 cm^2.

The lateral faces are congruent, and each has area $\frac{1}{2}bk$, or $\frac{1}{2} \times 10 \times 14$.

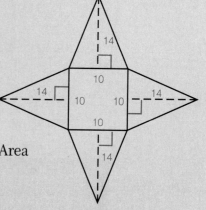

Thus, Lateral Area $= 4 \times \frac{1}{2}bk$

$= 4 \times \frac{1}{2} \times 10 \times 14$

$= 280 \text{ cm}^2$

And, Surface Area $=$ Lateral Area $+$ Base Area
$= 280 \text{ cm}^2 + 100 \text{ cm}^2$
$= 380 \text{ cm}^2$

Try This... b. Find the lateral area and surface area of the regular pyramid.

10 cm

4 cm

Note that 4×10 is the *perimeter* of the base in Example 2. Thus, the lateral area is also $\frac{1}{2} \times 14 \times 40$, or

$$\frac{1}{2} \times \text{slant height} \times \text{perimeter of base} = \frac{1}{2}kp.$$

Theorem 14.2 For any regular pyramid, where k is the slant height and p is the perimeter of the base,

Surface Area = Lateral Area + Base Area

$$= \frac{1}{2}kp + B$$

Example 3. Find the lateral area of this regular pyramid.

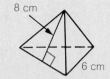

8 cm

6 cm

Because this is a regular pyramid, the base is bounded by an equilateral triangle with sides 6 cm. Thus, its perimeter is 18.

$$\text{Lateral Area} = \frac{1}{2}kp$$

$$= \frac{1}{2} \times 8 \times 18$$

$$= 72 \text{ cm}^2$$

Try This... c. Find the lateral area of this regular pyramid.

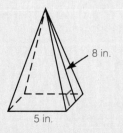

8 in.

5 in.

Discussion What type of triangles bound the lateral faces of a regular pyramid?

Exercises

Practice

Classify each pyramid, and name the base, the lateral faces, and the lateral edges.

1.

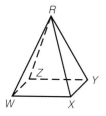

2.

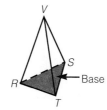

3.

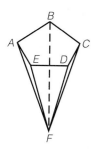

4.

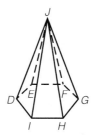

Sketch all faces to find the lateral area and the surface area of each regular pyramid. Round to the nearest whole number.

5.

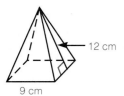

12 cm

9 cm

6.

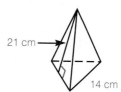

21 cm

14 cm

7.

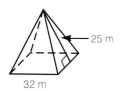

25 m

32 m

8.

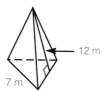

12 m

7 m

Use the formula in Theorem 14.2 to find the lateral area of each regular pyramid.

9.

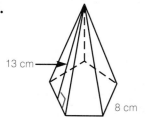

13 cm

8 cm

10.

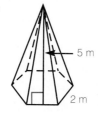

5 m

2 m

Use the formula in Theorem 14.2 to find the lateral area of each regular pyramid.

11.
3.8 m
7.2 m

12.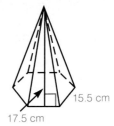
15.5 cm
17.5 cm

Extend and Apply

13. A manufacturer makes grain loaders like the one shown at the right. Estimate the lateral area of the grain loader. Then use the formula from Theorem 14.2 to find the lateral area of the grain loader. How reasonable was your estimate?

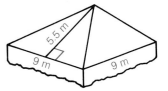
10 ft
10 ft
$11\frac{1}{2}$ ft

14. The cost of roofing is $6 per square meter. Which of the following is the cost of roofing the garage shown?

(A) $594   (B) $1188   (C) $99   (D) $297

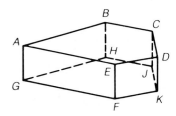
5.5 m
9 m
9 m

Use Mathematical Reasoning

15. The Great Pyramid is approximately 137 m tall. The square base measures 225 m on each edge. Find the lateral area of the pyramid.

Complete each statement with *always, sometimes,* or *never.*

16. The lateral faces of a pyramid are _____ triangular regions.

17. The number of lateral edges is _____ the same as the number of vertices of the base of a regular pyramid.

18. The lateral faces of a pyramid are _____ congruent.

19. The base of a regular pyramid is _____ congruent to each lateral face.

20. The lateral faces of a regular pyramid are _____ scalene triangles.

Mixed Review

21. Classify the prism at the right. Name the bases, the lateral faces, and the lateral edges.

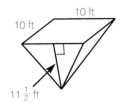
B
C
A
H
D
E
J
G
F
K

◇ Add or subtract.

22. $-15 + (-21)$ **23.** $-15 - (-21)$ **24.** $-21 - (-15)$

More on Perspective Drawing

The prism at the right is positioned so that an edge is facing the viewer. To make a perspective drawing of such a figure, we need two vanishing points on the horizon.

Sample Make a perspective drawing of a rectangular prism with an edge facing the viewer.

Draw a vertical segment $\overline{AB}$ for the front edge of the box. Then draw a horizon line perpendicular to $\overline{AB}$ and select two vanishing points.

Draw four vanishing lines as shown to create the two visible sides of the prism.

Draw segments $\overline{CD}$ and $\overline{EF}$ parallel to $\overline{AB}$ to determine the length and width of the prism.

Now draw the remaining vanishing lines from points C, D, E, and F to determine the other edges. Note that only the four vertical edges of the prism are parallel to each other in the drawing.

Hidden edges are usually drawn with dashed lines.

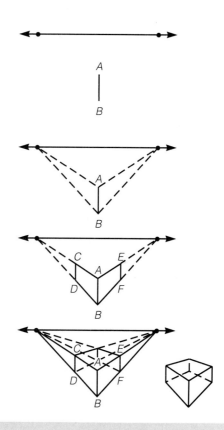

Problems

1. Make your own perspective drawing of a rectangular prism with an edge facing the viewer.

2. Make a perspective drawing of a rectangular prism with an edge facing the viewer as it would look from below. (Hint: Begin with the horizon line below $\overline{AB}$.)

14-4 Volume of Prisms

Objective: Find the volume of prisms.

While setting up a new aquarium, Paula calculated the amount of water she would need to fill the tank. This is a situation in which the volume of a prism is needed.

The **volume** of a polyhedron is the number of cubic units contained in it. We agree that a cube whose edges are all of length 1 has a volume of 1. We use such a cube as a unit of volume. The cube on the right has a volume of 1 cubic centimeter, abbreviated 1 cm³.

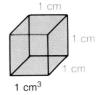

Postulate 21   For every polyhedron, there corresponds a positive number called its *volume*.

In the following Example, we find the volume of polyhedrons simply by counting cubic units.

Example   **1.** How many cubic centimeters are contained in this right rectangular prism?

There are 5 × 2, or 10 cm³, in each layer. Because there are 3 layers, the prism contains 10 × 3, or 30 cm³.

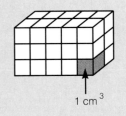

Try This... a. How many cubic inches are contained in this right rectangular prism?

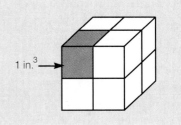

Note that the base area of the prism in Example 1 is 5 × 2, or 10 cm², and that 10 cm² × 3 cm = 30 cm³. This suggests the following assumption:

Postulate 22   The volume of a right rectangular prism is the product of its base area and its height.

$$\text{Volume} = \text{Base Area} \times \text{Height}$$
$$V = Bh$$

Example   **2.** Find the volume of this right rectangular prism.

Volume = Base area × height
$$V = Bh$$
$$V = \ell \cdot w \cdot h$$
$$V = 12 \cdot 6 \cdot 8.5$$
$$V = 612 \text{ in.}^3$$

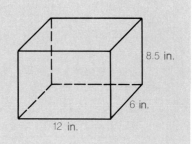

8.5 in.
6 in.
12 in.

Try This... b. Find the volume of this right rectangular prism.

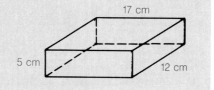

17 cm
5 cm
12 cm

To find the volumes of other types of prisms, we need the following idea. Suppose there are two stacks of paper, each containing 500 identical sheets. Note that the height, h, of each stack is the same. Since each stack consists of the same number of sheets, the volume of each stack should be the same.

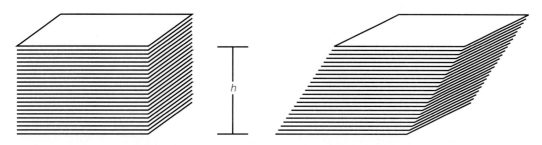

h

This idea led the Italian mathematician Bonaventura Cavalieri (1598–1647) to the following postulate. (You may wish to review the definition of *cross section* on page 37.)

Postulate 23 Cavalieri's Principle

Suppose R and S are two space figures, and $\mathcal{M}$ is a plane. If every plane parallel to $\mathcal{M}$ and intersecting R or S also intersects the other space figure, and the resulting cross sections have the same area, then R and S have the same volume.

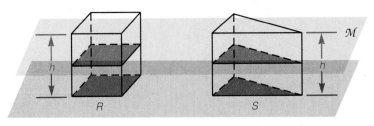

Cavalieri's Principle can be used to justify our next theorem.

Theorem 14.3 The volume of any prism is the product of its base area, B, and its height, h.

$$V = Bh$$

Example **3.** Find the volume of this right prism.

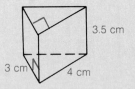

The base is a triangle. Therefore,
$B = \frac{1}{2} \cdot 3 \cdot 4 = 6.$
$V = Bh$
$V = 6 \cdot 3.5$
$V = 21 \text{ cm}^3$

Try This... Find the volume of each right prism.

c.

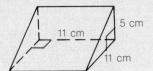

d.

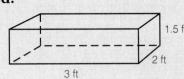

Discussion Suppose each dimension of a right rectangular prism is doubled. How does this affect the volume? Consider specific examples and draw a conclusion.

Exercises

Practice

Find the volume of each right prism.

1.
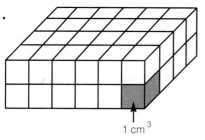
1 cm³

2.
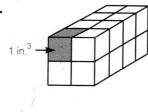
1 in.³

3.
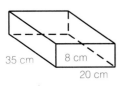
35 cm   8 cm
20 cm

4.
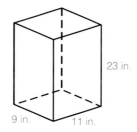
23 in.
9 in.   11 in.

5.
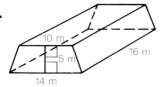
10 m
5 m
16 m
14 m

6.
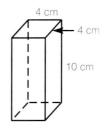
4 cm
4 cm
10 cm

Extend and Apply

7. Which of the following is the volume of the right prism at the right?

(A) 84 m³   (B) 21 m³   (C) 42 m³   (D) 49 m³

3 m   4 m
7 m

Estimate the volume of each right prism. Then use a calculator to find the volume to the nearest tenth and evaluate your estimate.

8.
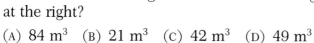
0.8 m
3.2 m   10.6 m

9.
$1\frac{1}{4}$ ft
1 ft
$1\frac{1}{4}$ ft
$3\frac{1}{2}$ ft   1 ft

10.
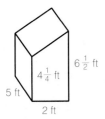
$6\frac{1}{2}$ ft
$4\frac{1}{4}$ ft
5 ft
2 ft

11.
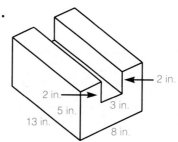
2 in.
2 in.
5 in.   3 in.
13 in.   8 in.

Match each item with the best estimate of its volume.

12. Swimming pool   (A) 120 cm³

13. Soap box   (B) 7200 cm³

14. Test tube   (C) 380 m³

15. Bar of soap   (D) 5000 mm³

In estimating the volume of an object, we try to select a unit that is appropriate. For example, to estimate the volume of a room, we would use a cubic meter (1 m³), and to estimate the volume of a cereal box, we would use a cubic centimeter (1 cm³). Complete each statement with the most appropriate unit (m³, cm³, or mm³).

16. The volume of a 10-gallon fish tank is about 40,000 _____ .

17. The volume of a gymnasium is about 30,000 _____ .

18. The volume of a refrigerator is about 1.1 _____ .

19. The volume of a can of condensed milk is about 354 _____ .

20. The volume of an allergy capsule is about 784 _____ .

Use Mathematical Reasoning

21. During a snowstorm, 5 cm of snow fell. At the end of the storm, how many liters of snow covered a lawn that is 32 m by 18 m? (1 m³ = 1000 liters.)

22. A rock is submerged in a rectangular tank that is 25 cm by 40 cm. The rock raises the water level by 2 cm. What is the volume of the rock?

Mixed Review

Can these numbers be the lengths of the sides of a triangle?

23. 5, 12, 7   **24.** 15, 17, 16   **25.** 1, 1, 2

Write using standard notation.

26. 2^5 **27.** 4^3 **28.** 3^4

Enrichment

Dot paper is useful for drawing figures from different perspectives. The figures at the right show two different views of the same solid. For each of the figures below, draw the solid from a different perspective.

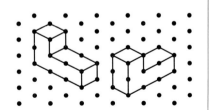

1.

2.

14-5 Volume of Pyramids

Objective: Find the volume of pyramids.

Explore

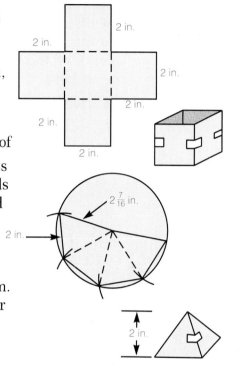

1. On a piece of stiff paper, draw a figure with the dimensions shown. Cut it out, fold on the dotted lines, and tape the sides to make an open, 2-inch prism.

2. Use a compass to draw a circle with radius $2\frac{7}{16}$ inches on a piece of stiff paper. Then open your compass to 2 inches and mark off four chords on the circle. Cut out as shown and tape the seam to form the lateral faces of a pyramid.

3. Check to be sure that your pyramid fits snugly inside the prism. The bases of your pyramid and your prism are both 2-by-2-in. square regions. The height of each is also 2 in.

4. Now fill the pyramid with small beans, unpopped popcorn, or a similar material, and empty it into the prism. Continue until the prism is filled. Compare the volume of the pyramid and the volume of the prism.

Discuss How does the volume of the pyramid compare to the volume of the prism? In general, if a pyramid and a prism have the same height and have bases with the same area, how do you think their volumes compare?

You may have discovered the following theorem in the Explore:

Theorem 14.4 The volume of any pyramid is one-third the product of its base area and its height.

$$V = \frac{1}{3}Bh$$

The following Examples use Theorem 14.4.

Examples **1.** Find the volume of this pyramid.

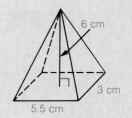

$$V = \frac{1}{3}Bh$$

$$= \frac{1}{3} \times 3 \times 5.5 \times 6$$

$$= 33 \text{ cm}^3$$

2. The distance from the vertex P of the building to the floor is 22 ft. Find the total volume of the building.

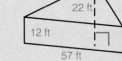

The height of the pyramid is $22 - 12$, or 10 ft.

Volume of pyramid

$$= \frac{1}{3} \times 57 \times 35 \times 10$$

$$= 6650 \text{ ft}^3$$

Volume of prism
$$= 57 \times 35 \times 12$$
$$= 23{,}940 \text{ ft}^3$$

Total Volume
$$= 23{,}940 + 6650$$
$$= 30{,}590 \text{ ft}^3$$

Try This... **a.** Find the volume of this pyramid.

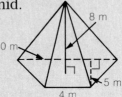

b. The distance from the vertex P of the building to the floor is 15 meters. Find the total volume of the building.

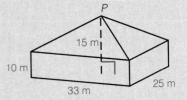

Discussion Two students drew pyramids and stated that their pyramids had the same volume. When the students compared the pyramids, they found that both had the same height. What can you say about the bases of the two pyramids?

Exercises

Practice

Find the volume of each pyramid.

1.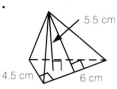
5.5 cm
4.5 cm
6 cm

2.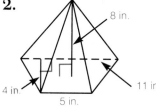
8 in.
4 in.
5 in.
11 in.

3.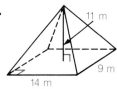
11 m
9 m
14 m

4.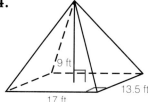
9 ft
13.5 ft
17 ft

5.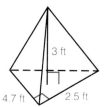
3 ft
4.7 ft
2.5 ft

6.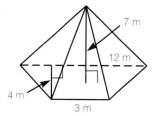
7 m
12 m
4 m
3 m

Find the total volume of each solid.

7.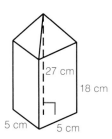
27 cm
18 cm
5 cm
5 cm

8.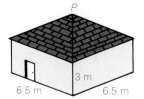
18 ft
8 ft
16 ft
21 ft

9.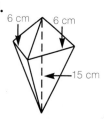
6 cm   6 cm
15 cm

Extend and Apply

10. The area of the base of a pyramid is 56 in². The height of the pyramid is 12 in. Which of the following is the volume of the pyramid?

(A) 448 in.³   (B) 672 in.³   (C) 336 in.³   (D) 224 in.³

11. The distance from vertex *P* of the building to the floor is 5 m. Estimate the total volume of the building. Then calculate the volume to check the reasonableness of your estimate.

P
3 m
6.5 m   6.5 m

12. The Great Pyramid in Egypt is approximately 137 m tall. The square base measures 225 m on each edge. Find the volume of the pyramid.

13. The area of the base of a pyramid is 237 cm², and the height of the pyramid is 1 m. Find its volume in cubic centimeters.

14. The height of a pyramid is 1.5 ft. The base is a right triangle whose legs are 9 in. and 12 in. long. Find the volume of the pyramid in cubic inches.

15. The two pyramids at the right have bases with the same area. Compare their volumes.

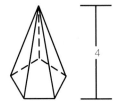

16. If the area of the base of a pyramid is doubled, how does that affect the volume?

Use Mathematical Reasoning

17. A regular pyramid has a base area of 289 ft² and a volume of 867 ft³. What is the height of the pyramid?

18. Two regular pyramids have square bases and equal heights. If the length of a side of one base is 1 m, and the length of a side of the other is 3 m, how will the volumes compare?

19. A cube is broken up into six identical pyramids as shown. Each face of the cube is a base of a pyramid. An edge of the cube is 10 cm. What is the volume of a pyramid to the nearest tenth?

Mixed Review

Find the volume of each right prism.

20.

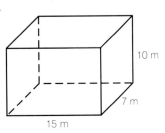

21.

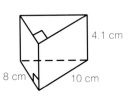

Find the lateral area of each right prism.

22. Cube

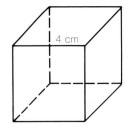

23.

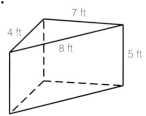

◇ Write using exponential notation.

24. 1000   25. 10,000,000   26. 10,000

Scientific Notation

Scientists often have to work with very large or very small numbers. For example, astronomers estimate that the mass of the moon is 73,700,000,000,000,000,000,000 kg. Scientists also estimate that the weight of an oxygen molecule is 0.000000000000000000000053 g. **Scientific notation** is often used to represent numbers like these.

When we use scientific notation, we express the number as a product of a power of 10 and a number between 1 and 10 (including 1).

$$\underline{\text{Standard Notation}} \qquad \underline{\text{Scientific Notation}}$$

$$93,000,000 \quad = \quad 9.3 \quad \times \quad 10^7$$
$$\qquad\qquad\qquad\qquad \uparrow \qquad\qquad \uparrow$$
$$\qquad\qquad 1 \le 9.3 < 10 \quad \text{Power of 10}$$

The pattern below will help you see how standard notation and scientific notation are related. Notice how the power of 10 is related to the position of the decimal point.

$$9.3$$

$93 = 9.3 \times 10^1$		$0.93 = 9.3 \times 10^{-1}$
$930 = 9.3 \times 10^2$		$0.093 = 9.3 \times 10^{-2}$
$9300 = 9.3 \times 10^3$		$0.0093 = 9.3 \times 10^{-3}$
$93,000 = 9.3 \times 10^4$		$0.00093 = 9.3 \times 10^{-4}$

To find the power of ten, begin at the right of the first nonzero digit and count the number of digits to the decimal point. Then determine whether the power is positive or negative. How can you determine this based on the direction in which you counted?

Problems

Write each of the following using scientific notation.

1. 84,000   **2.** 6,000,000   **3.** 94,000,000,000

4. 0.006   **5.** 0.00000073   **6.** 0.0000025

Write each of the following using standard notation.

7. 9.8×10^5 **8.** 1.7×10^2 **9.** 3.9×10^6

10. 7.6×10^{-3} **11.** 4×10^{-4} **12.** 2.8×10^{-6}

13. Write the mass of the moon, given above, in scientific notation.

14. Write the weight of an oxygen molecule, given above, in scientific notation.

14-6 Space Figures with Curved Surfaces

Objective: Identify cylinders, cones, and spheres.

In previous lessons we have studied three-dimensional figures whose surfaces were parts of a plane. We shall now look at some space figures with curved surfaces.

Soup cans, farm silos, and pipes of all sizes are everyday examples of geometric solids known as **cylinders.**

Similar to prisms, cylinders have two congruent bases that lie in parallel planes. Unlike prisms, these bases are circular regions. The segment joining the centers of the circular bases is called the **axis** of the cylinder. The region joining the two circles makes up the **lateral surface** of the cylinder.

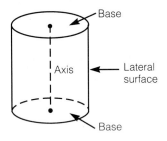

If the axis is perpendicular to the bases, the cylinder is a **right cylinder.** We will assume cylinders are right cylinders unless otherwise stated. The **height** of a right cylinder is the length of its axis.

Another familiar geometric solid is a **cone**. Ice cream cones and funnels are familiar examples of cones. Cones are like pyramids except that the base of a cone is a circular region.

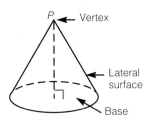

The circular region is called the **base** of the cone. The point *P* is called the **vertex** of the cone. The region joining the vertex and the base of a cone is the **lateral surface.**

The **height** of a cone is the length of the perpendicular from the vertex to the plane of the base. If this perpendicular intersects the plane at the center of the base, the cone is a **right cone.** We will assume cones are right cones unless otherwise stated.

Example 1. How are cones and pyramids alike? How are they different?

Cones and pyramids both have a vertex and a base. The base of a pyramid is a polygonal region. The base of a cone is a circle.

Try This... **a.** How are cylinders and prisms alike? How are they different?

Perhaps the most common geometric solid of all is the **sphere.** The earth we live on is essentially spherical. Baseballs, oranges, ball bearings, and marbles are just a few common examples of spheres in the world around us.

The definition of a sphere is like the definition of a circle. A **sphere** is the locus of all points *in space* a given distance from a given point.

The given distance is called the **radius,** and the given point is called the **center** of the sphere.

A plane that contains the center of the sphere divides the sphere into two **hemispheres.** The intersection of such a plane with the sphere is called a **great circle.**

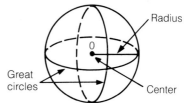

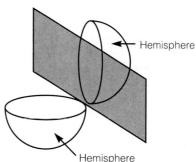

Examples What concept is suggested by each of the following?

2. The label on a soup can
   The lateral surface of a cylinder

3. The earth's equator
   A great circle of a sphere

Try This... What concept is suggested by each of the following?

b. The point of a party hat

c. The length of a drinking straw

Exercises

Practice

1. How are spheres and circles alike?

2. How are spheres different from circles?

What concept is suggested by each of the following?

3. Half of an orange

4. The ends of a can of vegetables

5. The bottom tip of an ice cream cone

6. The length of a wooden dowel

7. The equator on a globe

8. The label on a can of tuna

Extend and Apply

9. Suppose the lateral surface of a cone is peeled open and laid flat. Which of the following regions will result?

    (A) Rectangle   (B) Circle
    (C) Triangle   (D) Sector of a circle

Sketch each of the following.

10. A right cone   11. A right cylinder

12. A sphere and a great circle   13. A hemisphere

14. A silo is made up of a cylinder and a hemisphere as shown in the figure. The radius of the hemisphere is 30 feet. The height of the cylinder is 50 feet. Find the height of the silo.

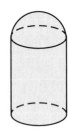

Determine whether each statement is true or false.

15. The bases of a cylinder are always congruent.

16. The intersection of a plane with a sphere is always a great circle.

17. The base of a cone is a circle.

18. The segment joining the centers of the bases of a cylinder is the radius of the cylinder.

19. The intersection of a plane with a cylinder is always a circle.

20. All points on a sphere are equidistant from the center of the sphere.

Mixed Review

 Find the volume of each pyramid.

21.

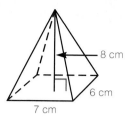

8 cm

6 cm

7 cm

22.

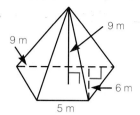

9 m

9 m

6 m

5 m

 Given the radius, find the area of the circle. First find the answer using π; then find an approximation to the nearest tenth.

23. 7 m   **24.** 1.2 cm   **25.** 8.4 in.

Write each of the following using scientific notation.

26. 6500   **27.** 3,700,000   **28.** 0.000011

Visualization

Some geometric solids can be obtained by rotating a two-dimensional figure 360° about an axis of revolution. The two-dimensional figure sweeps out a three-dimensional figure.

For example, imagine rotating a semicircular region 360° about a vertical axis as shown at the right. The result of this rotation is a sphere.

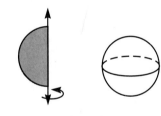

Describe the solid you would get as a result of rotating the two-dimensional figure 360° about the vertical axis.

1.

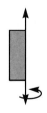

2.

3.

4.

5.

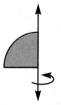

6.

14-7 Surface Area of Cylinders and Cones

Objective: Find the surface area of cylinders and cones.

Explore

1. Remove the label from a can of soup or vegetables.

2. Measure the length and width of the label. Find the area of the label. How is this related to the lateral area of the can?

3. Measure the radius of the can. How is this related to the length of the label?

Discuss How can you use only the height and radius of the can to find its lateral area?

In the Explore, you discovered that the lateral area of a cylinder is the product of the circumference of the cylinder and the height of the cylinder. If we add the areas of the bases, we can find the total surface area of the cylinder. This is summarized in the following theorem:

> **Theorem 14.5** The lateral area of a right circular cylinder with radius r and height h is $2\pi rh$.
>
> Surface Area $= 2 \times$ Base Area $+$ Lateral Area
> $= 2\pi r^2 + 2\pi rh$

It is usually best to find the surface area of a cylinder by finding the areas of its parts rather than by substituting values into the formula.

Example　1. Find the surface area of this cylinder to the nearest tenth.

Lateral Area $= 2\pi rh$
$\qquad = 2\pi \cdot 4 \cdot 12 = 96\pi$

Base Area $\quad = \pi r^2$
$\qquad = \pi \cdot 4^2 = 16\pi$

Surface Area $= 96\pi + 2 \cdot 16\pi$
$\qquad = 128\pi$ cm^2

Using the π key on the calculator gives 128π cm$^2 \approx 402.1$ cm^2.

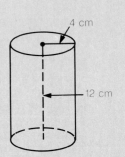

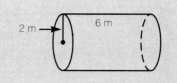

2 m →   6 m

The lateral area and surface area of a cone can be found in a similar manner. We first need the concept of *slant height*. The **slant height**, *s*, of a cone is the distance along the surface from the vertex to the base.

To find the lateral area of a cone, consider cutting along the length of the cone and spreading out the lateral surface. As shown in the figure, this results in a sector of a circle.

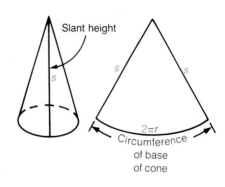

Slant height

s s s

$2\pi r$
Circumference
of base
of cone

It can be shown that the area of a sector is $\frac{1}{2}$ the product of the radius and the arc length.

So, Lateral Area = Area of the sector

$$= \frac{1}{2} \times \text{radius} \times \text{arc length}$$

$$= \frac{1}{2} \times s \times 2\pi r$$

$$= \pi s r$$

Theorem 14.6 For any right cone with base radius r and slant height s, the lateral area is $\pi r s$.

$$\text{Surface Area} = \text{Lateral Area} + \text{Base Area}$$
$$= \pi r s + \pi r^2$$

Example 2. Find the surface area of this cone to the nearest tenth.

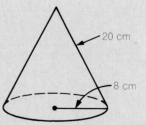

20 cm

8 cm

Lateral Area $= \pi r s$
$$= \pi \cdot 8 \cdot 20 = 160\pi$$

Base Area $= \pi r^2$
$$= \pi \cdot 8^2 = 64\pi$$

Surface Area = Lateral Area + Base Area
$$= 160\pi + 64\pi$$
$$= 224\pi \text{ cm}^2$$
$$\approx 703.7 \text{ cm}^2$$

Try This... b. Find the area of a right cone with a radius of 4 in. and a slant height of 15 in. Round your answer to the nearest tenth.

Discussion Suppose you have a right cylinder and a right cone, both with congruent bases of radius r. How is the height of the cylinder related to the slant height of the cone if the lateral areas are equal?

Exercises

Practice

Find the surface area of each cylinder. Express your answer in terms of π. Then use your calculator to obtain an approximate answer to the nearest tenth.

1.

3 cm

7 cm

2.

4.25 m

12 m

3.

42 cm

18 cm

Find the surface area of each cone. Express your answer in terms of π. Then use your calculator to obtain an approximate answer to the nearest tenth.

4.

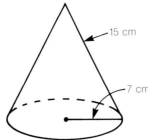

15 cm

7 cm

5.

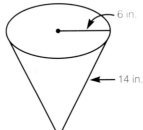

6 in.

14 in.

6.

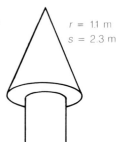

$r = 1.1$ m
$s = 2.3$ m

7. Find the surface area of a cone with radius 3 m and slant height 5.5 m. Round your answer to the nearest tenth.

8. Find the surface area of a cylinder with radius 8 in. and height 6.4 in. Round your answer to the nearest tenth.

Extend and Apply

9. A cylinder has height 2 cm and radius 3 cm. Which of the following is the best estimate of the lateral area of the cylinder?

(A) 18 cm^2 (B) 36 cm^2 (C) 12 cm^2 (D) 6 cm^2

Round your answers to the nearest tenth in Exercises 10 and 11.

10. A cement roller has a length of 3.5 m and a radius of 0.7 m. What is the area of road surface covered in one revolution of the roller?

11. A cylindrical paint can has a diameter of 16 cm and a height of 21 cm. What is the area of the paper label needed to cover the side of the can?

12. A cylindrical storage tank is 23.8 m tall and has a radius of 6.2 m. Find the surface area of the tank to the nearest tenth.

13. A manufacturer plans to produce 500 tin funnels. Each funnel has a 7-cm radius and is 15 cm deep. Tin costs $5 per square meter. What is the cost per funnel for the tin?

Use Mathematical Reasoning

14. Find the surface area of the cone at the right to the nearest tenth.

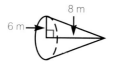

Mixed Review

What concept is suggested by each of the following?

15. The point of a party hat

16. The dome of the U.S. Capitol building

◇ Write each of the following using standard notation.

17. 5.4×10^{-3} **18.** 9.2×10^{6} **19.** 7×10^{3}

Enrichment

The following BASIC program calculates the lateral area of a right prism with a regular polygonal base and height 1. It assumes the radius of the polygonal base is 1. The program uses larger and larger values for the number of sides of the base. As this value increases, the lateral area approaches the lateral area of a cylinder of height 1 and radius 1.

```
10  REM LATERAL AREA OF PRISM
20  REM N = NUMBER OF SIDES OF BASE,
    A = LATERAL AREA
30  PI = 3.1416
40  N = 3:REM START WITH TRIANGULAR BASE
50  A = 2*N*COS(PI/2*(N-2)/N)
60  PRINT "NUMBER OF SIDES OF BASE ";N;",
    LATERAL AREA ";A
70  FOR I = 1 TO 1000:NEXT I
80  N = N + 1:REM ADD 1 TO NUMBER OF SIDES
90  GO TO 50
```

14-8 Volume of Cylinders and Cones

Objective: Find the volume of cylinders and cones.

We have seen that cylinders and prisms are alike in that both have bases that are congruent and lie in parallel planes. Prisms have bases that are polygonal regions. Cylinders have bases that are circular regions.

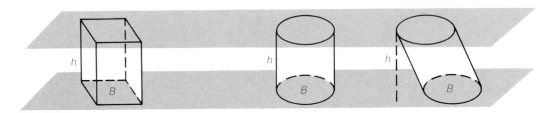

Suppose the cylinders and the prism shown have the same base area and the same height. It can be shown that every cross section of a cylinder parallel to its base has the same area as the base. By Cavalieri's Principle, each cylinder has the same volume as the prism. Because the volume of the prism is Bh, we have the following theorem:

Theorem 14.7   The volume of a cylinder is the product of its base area, B, and its height, h.

$$V = Bh$$

Notice that this is the same as the formula for the volume of a prism. However, for a cylinder, $B = \pi r^2$ where r is the radius of a base.

Example   1. Find the volume of this cylinder to the nearest tenth.

$$\begin{aligned}
V &= Bh \\
&= \pi r^2 h \\
&= \pi \cdot 4^2 \cdot 12 \\
&= 192\pi \text{ cm}^3
\end{aligned}$$

Using a calculator, we find 192π cm$^3 \approx 603.2$ cm^3.

Try This... Find the volume of each cylinder to the nearest tenth.

a.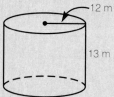
12 m
13 m

b.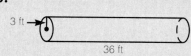
3 ft
36 ft

We have also seen that cones and pyramids are alike in that both have a base and a vertex point not in the plane of the base. A pyramid has a base that is a polygonal region. A cone has a base that is a circular region.

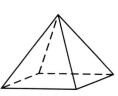

It can be shown using Cavalieri's Principle that if a pyramid and a cone have the same height and base area, then they have the same volume.

Because the volume of a pyramid is $\frac{1}{3}Bh$, we have the following theorem:

Theorem 14.8   The volume of a cone is one-third the product of its base area, B, and its height, h.

$$V = \frac{1}{3}Bh$$

Notice that this is the same as the formula for the volume of a pyramid. For a cone, however, $B = \pi r^2$ where r is the radius of the base.

Example   2. Find the volume of this cone to the nearest tenth.

$$V = \frac{1}{3}Bh$$

$$= \frac{1}{3}\pi r^2 h$$

$$= \frac{1}{3}\pi \cdot 3^2 \cdot 7$$

$$= 21\pi \text{ cm}^3 \approx 66.0 \text{ cm}^3$$

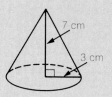
7 cm
3 cm

Try This... c. Find the volume of this cone to the nearest tenth.

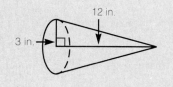
12 in.
3 in.

Exercises

 Practice

Find the volume of each cylinder. Express your answer in
terms of π. Then use your calculator to obtain an approximate
answer to the nearest tenth.

1.

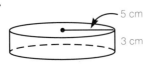

5 cm

3 cm

2.

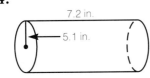

13 m

4 m

3.

3 m

32 m

4.

7.2 in.

5.1 in.

5.

d = 6 ft

2.4 ft

6.

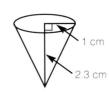

15 cm

d = 6 cm

Find the volume of each cone. Express your answer in terms
of π. Then use your calculator to obtain an approximate
answer to the nearest tenth.

7.

31 m

12 m

8.

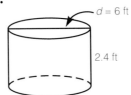

13 m

7 m

9.

1 cm

2.3 cm

10.

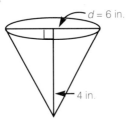

d = 6 in.

4 in.

11.

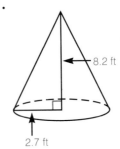

8.2 ft

2.7 ft

12.

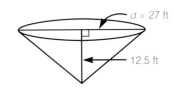

d = 27 ft

12.5 ft

Extend and Apply

13. The volume of the cylinder at the right is 36π m³. Which of the following is the height of the cylinder?

(A) 4 m   (B) 12 m   (C) 6 m   (D) 2 m

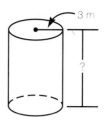

14. Find the volume of the toy top shown at the right. Round your answer to the nearest tenth.

15. The container at the right is a right cylinder with a right cone attached at the bottom. Find the volume of the container to the nearest tenth.

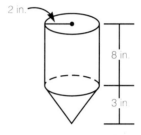

16. Estimate which holds more, the cylindrical glass or the cone-shaped drinking cup. Then calculate the volumes to check your estimate.

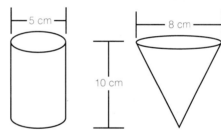

Use Mathematical Reasoning

17. If cylinder A holds 1 liter, how many liters does cylinder B hold?

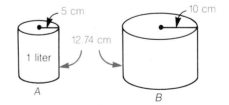

18. The graduated cylinder shown at the right is a device for measuring liquids. Suppose a cylinder has an inside radius of 4 cm. If 10 cm³ of liquid is poured into it, how much will the water level rise? Round your answer to the nearest tenth.

19. Find the quantity of water needed to fill a circular swimming pool that is 7 m in diameter and 1.5 m deep. Round your answer to the nearest whole number. (Hint: 1 m³ = 1000 liters.)

20. A piston in an automobile engine moves up and down in a cylinder as shown in the figure. When it moves up, it displaces a certain volume of the gas mixture. Find the volume displaced in this cylinder. Round your answer to the nearest whole number.

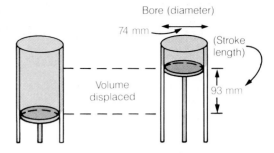

Bore (diameter)

74 mm

(Stroke length)

Volume displaced

93 mm

In Exercises 21 and 22, refer to Problem 20. Round your answers to the nearest tenth.

21. An engine with four cylinders has a combined gas displacement of 1600 cm³. The bore of each cylinder is 74 mm. What is the stroke length?

22. An eight-cylinder engine has a combined displacement of 5 liters. The stroke length of each cylinder is 103 mm. What is the bore?

23. A can is packed securely in a box as shown. Find the ratio of the volume of the can to the volume of the box. Ignore the thickness of the can and the box. (Hint: Begin by letting h be the height of the can and letting r be the radius of the can.)

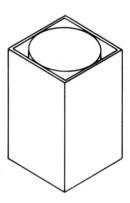

Mixed Review

Find the surface area of each cylinder or cone. Express your answer in terms of π. Then use your calculator to obtain an approximate answer to the nearest tenth.

24.

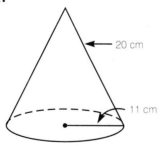

20 cm

11 cm

25.

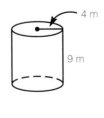

4 m

9 m

26. Find the surface area of a cone with radius 8 ft and slant height 6 ft. Round your answer to the nearest tenth.

Write each of the following using scientific notation.

27. 4300   **28.** 50,000,000   **29.** 0.009

14-9 Surface Area and Volume of Spheres

Objective: Find the surface area and the volume of spheres.

The utility department wants to paint the outside of its spherical water-storage tank. In order to estimate the amount of paint required, the department will need to find the surface area of a sphere.

The formula for the surface area of a sphere can be derived with a series of approximating polyhedrons as shown at the right. This is similar to how we used a series of inscribed polygons to find the circumference of a circle. Theorem 14.9 summarizes the results of such a procedure.

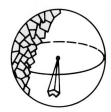

Theorem 14.9 The surface area of a sphere with radius r is $4\pi r^2$.

The following Example illustrates how to find the surface area of a sphere:

Example 1. Find the surface area of this sphere to the nearest tenth.

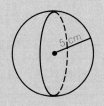

Surface Area $= 4\pi r^2$
$= 4 \cdot \pi \cdot 5^2$
$= 100\pi$ cm$^2 \approx 314.2$ cm^2

Try This... a. Find the surface area of a sphere with radius 8 in. Round your answer to the nearest tenth.

The formula for the volume of a sphere can also be determined by a series of approximations. The result of such a process is summarized in Theorem 14.10.

Theorem 14.10 The volume of a sphere with radius r is $\frac{4}{3}\pi r^3$.

Example **2.** Find the volume of a sphere with radius 6 m. Round your answer to the nearest tenth.

$$V = \frac{4}{3}\pi r^3$$

$$= \frac{4}{3} \cdot \pi \cdot 6^3$$

$$= 288\pi \text{ m}^3 \approx 904.8 \text{ m}^3$$

Try This... **b.** Find the volume of a sphere with radius 3 cm. Round your answer to the nearest tenth.

Example **3.** The volume of a ball is 36π cm³. Find the dimensions of a box that is just large enough to hold the ball.

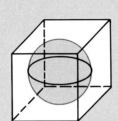

$$V = \frac{4}{3}\pi r^3 = 36\pi$$

$$r^3 = 27$$

$$r = 3 \text{ cm}$$

Thus, the diameter of the sphere is 6 cm, and each edge of the box is 6 cm.

Try This... **c.** The volume of a ball is 288π cm³. Find the dimensions of a box that is just large enough to hold the ball.

Example **4.** A cardboard package consists of a cylinder and hemisphere as shown at the right. Find the volume of the package to the nearest tenth.

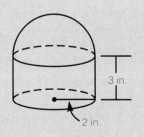

3 in.

2 in.

Volume of package = Volume of cylinder + Volume of hemisphere

$$= \pi r^2 h + \frac{1}{2} \cdot \frac{4}{3}\pi r^3$$

$$= \pi \cdot 2^2 \cdot 3 + \frac{1}{2} \cdot \frac{4}{3} \cdot \pi \cdot 2^3$$

$$= 12\pi + \frac{16}{3}\pi = 17\frac{1}{3}\pi \text{ in.}^3 \approx 54.5 \text{ in.}^3$$

Try This... d. Find the volume of the space figure at the right. Round your answer to the nearest tenth.

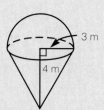

3 m

4 m

Exercises

Practice

In Exercises 1–12, find the volume and surface area of a sphere with the given radius.

In Exercises 1–9, round your answers to the nearest tenth.

1. 4 cm
2. 9 m
3. 1.5 in.

4. 1.3 cm
5. 7 m
6. 8.7 cm

7. 2.1 cm
8. 3.2 cm
9. 3.5 cm

In Exercises 10–12, leave your answers in terms of π.

10. 17 cm
11. 1600 km
12. 6400 km

13. A cardboard box in the shape of a cube has edges 18 in. long. The box is just large enough to hold a beach ball. Find the volume of the beach ball to the nearest tenth.

14. The surface area of a ball is 16π cm². Find the dimensions of a box just large enough to hold the ball.

Find the volume of each space figure. Round your answer to the nearest tenth.

15.

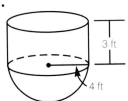

3 ft

4 ft

16.

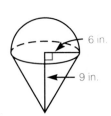

6 in.

9 in.

17.

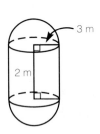

3 m

2 m

Extend and Apply

18. A sphere has a radius of 3 m. Which of the following is the best estimate of the volume of the sphere?

(A) 36 m³   (B) 110 m³   (c) 325 m³   (D) 650 m³

19. Find the radius of a sphere whose surface area is 36π cm². Find the volume of the sphere. Leave your answer in terms of π.

20. Find the volume of a sphere whose surface area is 196π cm². Round your answer to the nearest tenth.

21. Find the surface area of a sphere whose volume is 36π cm³. Leave your answer in terms of π.

The aluminum buoy at the right has an outer radius of 2 ft and an inner radius of 1 ft $11\frac{1}{2}$ in.

22. What is the thickness of the buoy?

23. How many cubic feet of aluminum, to the nearest cubic foot, were required to make the buoy?

24. What is the volume of air space inside the buoy? Round your answer to the nearest tenth.

25. What is the outer surface area of the buoy to the nearest tenth?

Use Mathematical Reasoning

26. Find the surface areas of spheres having radii of 1, 2, and 4 in. How does a change in the radius affect the surface area of a sphere?

27. How is the surface area of a sphere related to the area of a great circle of the sphere? (Hint: Begin by considering a sphere of radius r.)

28. A can contains three tennis balls, each with radius r. The radius of the base of the container is also r. The height of the container is $6r$. Find the total volume of the tennis balls and the volume of the container. Compare the volumes.

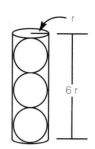

Mixed Review

 Find the volume of each cylinder or cone. Express your answer in terms of π. Then use your calculator to obtain an approximate answer to the nearest tenth.

29.

4 m
41 m

30.

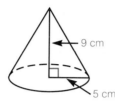
9 cm
5 cm

31.

3 m
1.1 m

32. Find the volume of the right rectangular prism shown at the right.

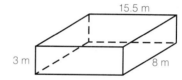
15.5 m
3 m
8 m

 In Exercises 33–35, find the circumference of each circle given the diameter. Express your answer in terms of π. Then use your calculator to obtain an approximate answer to the nearest tenth.

33. 12 in.   **34.** 22 m   **35.** 9.4 m

Solve and check.

36. $5x + 4 + 5x = 1 + 11x$ **37.** $16y - 6 - 9y = y + 18$

38. $6w = w + 32 + w$

Problem Solving

What should be the dimensions of the bottom layer of a pyramid that continues to build like the one shown, if there are 100 blocks to work with and as many blocks as possible are to be used?

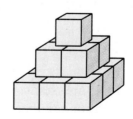

14-10 Surface Areas and Volumes of Similar Solids

Objective: Compare volumes and surface areas of similar solids.

In Chapter 11 we learned that polygons are similar whenever they can be matched so that corresponding angles are congruent and corresponding sides are proportional. The photo shows an example of similar space figures.

Space figures, like polygons, are *similar* if they have the same shape but not necessarily the same size. In particular, if space figures are similar, then all corresponding edge lengths are proportional. The following Example illustrates how we can use this fact:

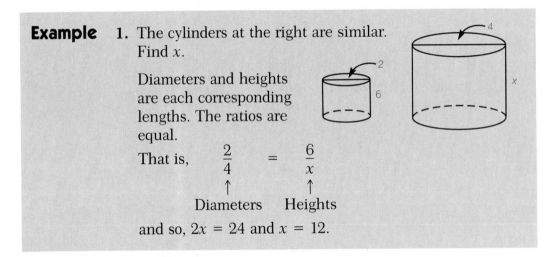

Example **1.** The cylinders at the right are similar. Find x.

Diameters and heights are each corresponding lengths. The ratios are equal.

That is, $\dfrac{2}{4}$ $=$ $\dfrac{6}{x}$

↑ Diameters   ↑ Heights

and so, $2x = 24$ and $x = 12$.

Try This... a. The prisms shown are similar. Find x and y.

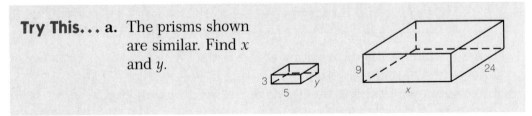

As in two dimensions, similar solids can be thought of as *size transformations* of each other. In Example 1, a scale factor of 2 transforms the cylinder on the left to the cylinder on the right.

The ratios of surface areas and the ratios of volumes of similar solids are also related to the ratio of the corresponding lengths.

▪ *Explore*

The three cubes at the right are similar.
1. Find the surface area of each cube.
2. Find the volume of each cube.
3. Prepare a table showing the ratios of corresponding edge lengths, surface areas, and volumes for Cube A and Cube B, Cube A and Cube C, and Cube B and Cube C. Be sure to express ratios in lowest terms.
4. What do you notice about the ratios?

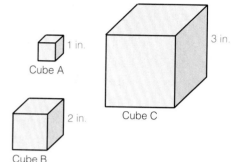

The Explore leads to the following theorem:

Theorem 14.11   If the ratio of two corresponding lengths of a pair of similar solids is $a:b$, then

(a) the ratio of *any* two corresponding lengths is $a:b$,
(b) the ratio of their surface areas is $a^2:b^2$, and
(c) the ratio of their volumes is $a^3:b^3$.

Example   **2.** Here are two similar right triangular prisms. Find the ratios of corresponding lengths, surface areas, and volumes. Express each ratio in lowest terms.

The ratio of corresponding lengths is $4:1$.

The ratio of surface areas is $4^2:1^2$, or $16:1$.

The ratio of volumes is $4^3:1^3$, or $64:1$.

Try This... **b.** Here are two similar right triangular prisms. Find the ratios of corresponding lengths, surface areas, and volumes. Express each ratio in lowest terms.

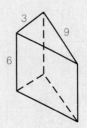

Theorem 14.11 gives us a way of finding surface areas and volumes using proportions.

Example 3. If the surface area of the larger of two similar cylinders is 54π, find the surface area of the smaller cylinder.

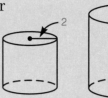

Since the ratio of the radii is $2{:}3$, the ratio of the surface areas is $2^2{:}3^2$, or $4{:}9$.

Thus, $\dfrac{4}{9} = \dfrac{x}{54\pi}$.

$9x = 4 \cdot 54\pi$ and $x = 24\pi$

Therefore, the surface area of the smaller cylinder is 24π.

Try This... c. If the volume of the larger of two similar cones is 686π, find the volume of the smaller cone. Leave your answer in terms of π.

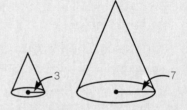

Discussion   When helium is added to a spherical balloon, the balloon's radius doubles. How does this affect the balloon's volume? Why is this so?

Exercises

Practice

The space figures in each Exercise are similar. Find x. Then find the ratios of corresponding lengths, surface areas, and volumes. Express each ratio in lowest terms.

1.

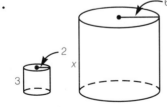

2.

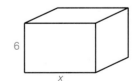

The space figures in each Exercise are similar. Find x. Then
find the ratios of corresponding lengths, surface areas, and
volumes. Express each ratio in lowest terms.

3.

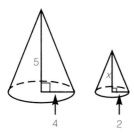

4.

   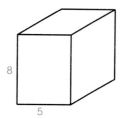

In Exercises 5–9, suppose the cylinders at the
right are similar. Leave answers in terms of π.

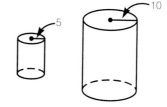

5. The height of the smaller cylinder is 8.
   What is the height of the larger cylinder?

6. The surface area of the larger cylinder is
   288π. What is the surface area of the
   smaller cylinder?

7. The volume of the smaller cylinder is 125π. What is the volume
   of the larger cylinder?

8. The volume of the larger cylinder is 136π. What is the volume
   of the smaller cylinder?

9. What is the scale factor that transforms the cylinder on the left
   to the cylinder on the right?

Extend and Apply

10. Two similar pyramids have heights in the ratio $2:5$. Which of the
   following is the ratio of their surface areas?

   (A) $2:5$ (B) $5:2$ (C) $4:25$ (D) $8:125$

11. A scientist has two cylindrical beakers that are similar in shape. The
   larger beaker is twice as tall as the smaller beaker. How much greater is
   the volume of the larger beaker?

12. A manufacturer ships instant rice in two boxes that are similar in shape.
   The edges of the larger box are twice the length of the corresponding
   edges of the smaller box. The price of the smaller box is $0.69, and the
   price of the larger box is $4.99. Which is the better buy? (Hint:
   Compare the volumes of the boxes.)

13. A grocer stocks one brand of juice in two cans that are similar in shape.
   The height of one can is 14 cm, and the height of the other is 21 cm.
   The larger can sells for $1.99, and the smaller can sells for $0.65.
   Which is the better buy?

14. One sphere is transformed to another by a scale factor of 4. What is the
   ratio of the spheres' surface areas? What is the ratio of the spheres'
   volumes?

Determine whether each statement is true or false.

15. If each edge length of a space figure is doubled, then the surface area will quadruple.

16. If each edge length of a space figure is doubled, then the volume becomes eight times greater.

17. If the ratio of the radii of two spheres is $5:7$, then the ratio of the volumes of the spheres is $25:49$.

18. If the ratio of the volumes of two similar cylinders is $216:27$, then the ratio of the diameters is $36:9$.

19. If the edge of a cube is 5 times longer than another cube, then its volume will be 25 times greater than the volume of the smaller cube.

Use Mathematical Reasoning

20. Which would take more paint, the surface of six balls each with a 2-in. radius, or one ball with a 5-in. radius?

21. Which would take more paint, the surface of 36 cubes one inch on each edge, or the surface of one cube six inches on each edge?

Mixed Review

 Find the volume and surface area of a sphere with the given radius. Round your answers to the nearest tenth.

22. 3 m **23.** 10 ft **24.** 7.5 cm

Identify each pair of angles as interior angles, alternate interior angles, alternate exterior angles, or corresponding angles.

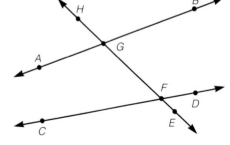

25. $\angle DFG$ and $\angle BGF$

26. $\angle HGB$ and $\angle EFC$

27. $\angle AGF$ and $\angle CFE$

28. $\angle AGF$ and $\angle GFD$

List three inequalities for each triangle.

29.
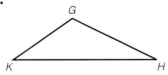

30.
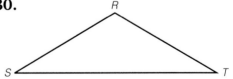

Can these numbers be the lengths of the sides of a triangle?

31. 4, 8, 4 **32.** 15, 30, 12 **33.** 7, 9, 15

Write each of the following using standard notation.

34. 3.4×10^6 **35.** 1×10^{-5} **36.** 8.4×10^{-1}

Chapter 14 Review

1. Name the faces, edges, and vertices of the polyhedron at the right.

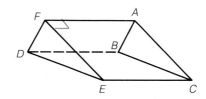

14-1

2. Find the lateral area and the surface area of the prism at the right.

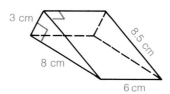

14-2

3. Sketch all faces to find the lateral area and the surface area of the regular pyramid at the right.

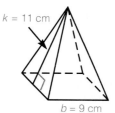

14-3

Find the volume of each right prism.

14-4

4.

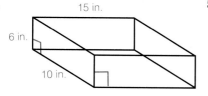

5.

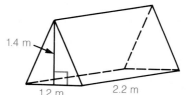

Find the volume of each pyramid.

14-5

6.

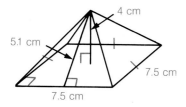

7.

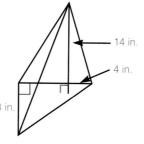

What concept is suggested by each of the following?

14-6

8. The label on a can of soup

9. The part of the earth north of the equator

14-7 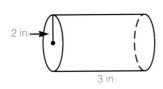 Find the surface area of each cylinder or cone. Express your answer in terms of π. Then use your calculator to obtain an approximate answer to the nearest tenth.

10.

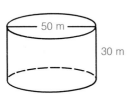

2 in.

3 in.

11.

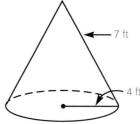

7 ft

4 ft

14-8 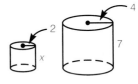 Find the volume of each cylinder or cone. Express your answer in terms of π. Then use your calculator to obtain an approximate answer to the nearest tenth.

12.

50 m

30 m

13.

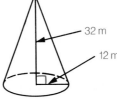

32 m

12 m

14-9   **14.** Find the volume and surface area of a sphere with radius 8 m. Round your answers to the nearest tenth.

14-10 **15.** The space figures at the right are similar. Find x. Then find the ratio of corresponding lengths, surface areas, and volumes. Express each ratio in lowest terms.

2

4

x

7

Chapter 14 Test

1. Name the faces, edges, and vertices of the polyhedron at the right.

B

A

C

E

D

F

2. Find the lateral area and the surface area of the prism at the right.

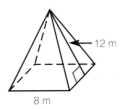

5 m

12 m

8 m

3. Sketch all faces to find the lateral area and the surface area of the regular pyramid at the right.

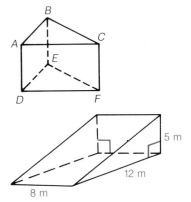

12 m

8 m

Find the volume of each right prism.

4. Cube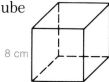

8 cm

5.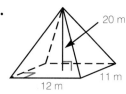

3 cm

8 cm

15 cm

Find the volume of each pyramid.

6.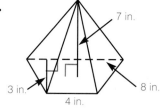

7 in.

3 in.

4 in.

8 in.

7.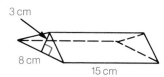

20 m

12 m

11 m

What concept is suggested by each of the following?

8. The ends of a can of juice

9. The length of the side of a funnel

Find the surface area of each cylinder or cone. Express your answer in terms of π. Then use your calculator to obtain an approximate answer to the nearest tenth.

10.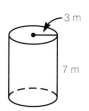

3 m

7 m

11.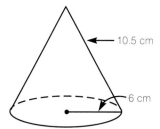

10.5 cm

6 cm

Find the volume of each cylinder or cone. Express your answer in terms of π. Then use your calculator to obtain an approximate answer to the nearest tenth.

12.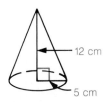

12 cm

5 cm

13.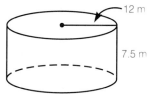

12 m

7.5 m

14. Find the volume and surface area of a sphere with radius 6 in. Round your answers to the nearest tenth.

15. The space figures at the right are similar. Find x. Then find the ratio of corresponding lengths, surface areas, and volumes. Express each ratio in lowest terms.

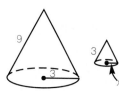

9

3

3

x

15 Coordinate Geometry

Air-traffic controllers use a coordinate system to locate planes and plan their routes. Each dot on the screen represents the position of an airplane in flight.

15-1 The Distance Formula

Objective: Solve problems using the Distance Formula.

Explore

There is a method for finding distances on lines that are neither horizontal nor vertical.

1. Use graph paper to graph $A(4,5)$ and $B(1,1)$.

2. Draw the horizontal line that contains B.

3. Draw the vertical line that contains A.

4. Label the intersection of the lines C.

5. Find AC and BC.

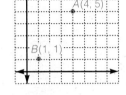

Discuss   What kind of triangle is $\triangle ABC$? How can you find AB?

The distance between any two points is the square root of the sum of the squares of the horizontal and vertical distances between them. The following formula states this mathematically:

Theorem 15.1   The Distance Formula

The distance (d) between any two points $A(x_1, y_1)$ and $B(x_2, y_2)$ is given by the formula

$$d = \sqrt{\underbrace{(x_1 - x_2)^2}_{\substack{\text{horizontal} \\ \text{distance}}} + \underbrace{(y_1 - y_2)^2}_{\substack{\text{vertical} \\ \text{distance}}}}$$

Example   1. Find the distance between $P(-5,3)$ and $Q(7,8)$.

We will choose $(-5,3)$ as (x_1, y_1) and $(7,8)$ as (x_2, y_2).

$$d = \sqrt{(x_1 - x_2)^2 + (y_1 - y_2)^2}$$

$$= \sqrt{(-5 - 7)^2 + (3 - 8)^2} \quad \text{Substituting for } x_1, x_2, y_1, \text{ and } y_2$$

$$= \sqrt{(-12)^2 + (-5)^2} \quad \text{Subtracting}$$

$$= \sqrt{144 + 25} \quad \text{Squaring}$$

$$= \sqrt{169} = 13$$

Thus, the distance between the points is 13 units.

When using the Distance Formula, *either* point can be (x_1, y_1).

Example   2. Find the distance between $P(-2, 5)$ and $Q(3, 7)$.

We will choose $(3, 7)$ as (x_1, y_1) and $(-2, 5)$ as (x_2, y_2).

$$d = \sqrt{(x_1 - x_2)^2 + (y_1 - y_2)^2}$$
$$= \sqrt{(3 - (-2))^2 + (7 - 5)^2} \quad \text{Substituting for } x_1, x_2, y_1, \text{ and } y_2$$

We can use a calculator to evaluate.

| 3 | − | 2 | +/− | = | x^2 | | 25 | STO |

| 7 | − | 5 | = | x^2 | | 4 |

| + | RCL | = | | 29 |

| $\sqrt{x}$ | | 5.385164807 |

The distance is about 5.4 units.

Try This... Find the distance between each pair of points.

a. $(3, 2)$ and $(-1, -1)$ **b.** $(5, 9)$ and $(1, -6)$

c. $(5, 7)$ and $(-4, 4)$ **d.** $(3, 8)$ and $(4, -3)$

The Distance Formula may be used to solve problems.

Example   3. A ship at coordinates $A(1, 4)$ can dock at $B(7, 5)$ or $C(4, 9)$. Which point is closer? What is the distance to the closer point?

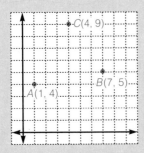

First, we find AB and AC using the distance formula.

$$AB = \sqrt{(1 - 7)^2 + (4 - 5)^2} = \sqrt{(-6)^2 + (1)^2}$$
$$= \sqrt{36 + 1} = \sqrt{37} \approx 6.1$$
$$AC = \sqrt{(1 - 4)^2 + (4 - 9)^2} = \sqrt{(-3)^2 + (-5)^2}$$
$$= \sqrt{9 + 25} = \sqrt{34} \approx 5.8$$

The distance from A to B is 6.1. The distance from A to C is 5.8. C is nearer.

Try This... e. A boat at coordinates $A(-6, 6)$ can travel to $B(-2, -7)$ or $C(7, 3)$. Which point is closer? What is the distance to the closer point?

Exercises

Practice

Find the distance between each pair of points.

1. $(6, 1)$ and $(9, 5)$ **2.** $(1, 10)$ and $(7, 2)$

3. $(2, 2)$ and $(7, 3)$ **4.** $(5, 6)$ and $(3, 2)$

5. $(4, 4)$ and $(4, -4)$ **6.** $(-2, 1)$ and $(2, 4)$

7. $(-5, -2)$ and $(1, -5)$ **8.** $(3, -4)$ and $(0, -7)$

9. $(-5, 9)$ and $(1, -6)$ **10.** $(7, 0)$ and $(-6, 4)$

11. $(35, 27)$ and $(91, -18)$ **12.** $(3, -21)$ and $(57, 44)$

Solve.

13. A boat at coordinates $A(4, -1)$ can travel to $B(-5, 8)$ or $C(16, 2)$. Which point is closer? What is the distance to the closer point?

14. A traveler at coordinates $A(-3, 3)$ can go to $B(-14, 12)$ or $C(-17, 7)$. Which point is closer? What is the distance to the closer point?

Extend and Apply

15. Find the perimeter of a triangle with vertices $P(5, 6)$, $Q(-1, -2)$, and $R(6, -3)$.

16. Find the perimeter of a triangle with vertices $A(-2, 7)$, $B(-3, 9)$, and $C(2, 6)$.

17. Find the lengths of the diagonals of a quadrilateral with vertices $X(3, 7)$, $Y(-1, -2)$, $Z(-5, 1)$, and $W(-2, 9)$.

18. Find the lengths of the diagonals of a quadrilateral with vertices $R(2, 9)$, $S(-2, -3)$, $T(-6, 2)$, and $V(-3, 10)$.

19. A ship at coordinates $A(5, 4)$ needs to sail to both $B(6, 13)$ and $C(11, 11)$. What is the shortest distance for the trip? What route should the ship travel?

20. An airplane at coordinates $A(70, 10)$ needs to fly to both $B(10, 50)$ and $C(100, 75)$. What is the shortest distance for the trip? What route should the airplane travel?

21. How far is it from point A to point B?

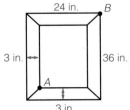

24 in. B

3 in.   36 in.

A

3 in.

22. Which is not a correct use of the Distance Formula for finding the distance between $(-3, 4)$ and $(2, -10)$?

(A) $\sqrt{(2 - (-3))^2 + (-10 - 4)^2}$ (B) $\sqrt{(-3 - 2)^2 + (4 - (-10))^2}$

(C) $\sqrt{(-3 - (-10))^2 + (4 - 2)^2}$ (D) None of these

Use Mathematical Reasoning

23. A ship at coordinates $A(5, 12)$ needs to go to $B(3, 16)$, $C(8, 7)$, and $D(6, 8)$. What is the shortest distance? What route should the ship travel?

Mixed Review

24. Find the volume of a room that is a right rectangular prism with length 18 ft, width 16 ft, and height 8 ft.

25. Find the volume of a pyramid that has a base area of 6000 m^2 and a height of 150 m.

26. Find the volume of a cylindrical silo that has a radius of 14 ft and a height of 25 ft.

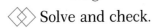 Solve and check.

27. $3x + 18 = x + 18$

Enrichment

Just as inequalities can describe a set of points on a line, they can also describe points in the coordinate plane.

For which point is $x > 3$ and $y < 2$?

Points I and H satisfy $x > 3$. Points A, D, E, and H satisfy $y < 2$. The only point that satisfies both conditions is H.

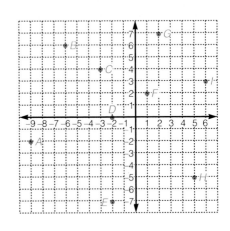

1. For which point is $x < 3$ and $y > 6$?
2. For which point is $y > 2$ and $x > 2$?
3. For which point is $y < 4$ and $x < -2$?
4. For which point is $y < -3$ and $x < 5$?
5. For which point is $y > -2$ and $y < 2$?
6. For which point is $x > -9$ and $x < -3$?
7. For which point is $x > -2$ and $x < 5$ and $y < 6$?
8. Use inequality symbols to describe point C only.
9. For which point is $y > x + 10$?

Multiplying and Dividing with Negative Numbers

Find the answer to each multiplication. Use a calculator if necessary.
Look for a pattern.

$5 \times 4 = 20$	$5 \times -1 = \underline{\hphantom{00}}$	$-5 \times 4 = \underline{\hphantom{00}}$	$-5 \times -1 = \underline{\hphantom{00}}$
$5 \times 3 = \underline{\hphantom{00}}$	$5 \times -2 = \underline{\hphantom{00}}$	$-5 \times 3 = \underline{\hphantom{00}}$	$-5 \times -2 = \underline{\hphantom{00}}$
$5 \times 2 = \underline{\hphantom{00}}$	$5 \times -3 = \underline{\hphantom{00}}$	$-5 \times 2 = \underline{\hphantom{00}}$	$-5 \times -3 = \underline{\hphantom{00}}$
$5 \times 1 = \underline{\hphantom{00}}$	$5 \times -4 = \underline{\hphantom{00}}$	$-5 \times 1 = \underline{\hphantom{00}}$	$-5 \times -4 = \underline{\hphantom{00}}$
$5 \times 0 = \underline{\hphantom{00}}$		$-5 \times 0 = \underline{\hphantom{00}}$	

Notice that

- A positive number $\times$ a positive number = a positive number.
- A positive number $\times$ a negative number = a negative number.
- A negative number $\times$ a positive number = a negative number.
- A negative number $\times$ a negative number = a positive number.

Samples   Multiply.

1. $8 \times -3 = -24$ **2.** $-16 \times \dfrac{1}{2} = -8$ **3.** $-10 \times -3 = 30$

Notice that $-16 \times \dfrac{1}{2}$ is the same as $-16 \div 2$. Therefore, the same rules
hold for division.

- A positive number $\div$ a positive number = a positive number.
- A positive number $\div$ a negative number = a negative number.
- A negative number $\div$ a positive number = a negative number.
- A negative number $\div$ a negative number = a positive number.

Samples   Divide.

4. $-50 \div 2 = -25$ **5.** $-72 \div -9 = 8$ **6.** $\dfrac{24}{-8} = -3$

Problems

Multiply or divide.

1. -9×-2 **2.** -15×3 **3.** -0.4×-2 **4.** $-\dfrac{1}{3} \times 12$

5. $12 \div -3$ **6.** $-81 \div -9$ **7.** $18 \div -6$ **8.** $-0.64 \div 0.4$

9. $\dfrac{-56}{7}$ **10.** $\dfrac{-72}{-24}$ **11.** $\dfrac{80}{-25}$ **12.** $\dfrac{-0.01}{100}$

15-2 Slope

Objective: Find the slope of a line that contains a given pair of points.

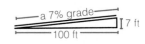

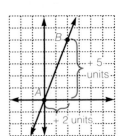

The measure of the steepness of a road is called its *grade*. The grade of a road is the ratio of the vertical *rise* to its horizontal *run*. A road with a 7% grade rises 7 feet for each 100 feet of horizontal distance.

Similarly, a line also has steepness. In the graph shown, as we move from A to B, the line rises 5 units for a horizontal run of 2 units. The $\dfrac{\text{rise}}{\text{run}} = \dfrac{5}{2}$. The ratio of rise to run is called the *slope* of the line. The letter m is commonly used to denote the slope.

Definition   **Slope of a line**

$$\text{Slope } m = \frac{\text{rise}}{\text{run}} = \frac{\text{change in } y\text{-coordinates}}{\text{change in } x\text{-coordinates}}$$

Upward-sloping lines have positive slopes; downward-sloping lines have negative slopes.

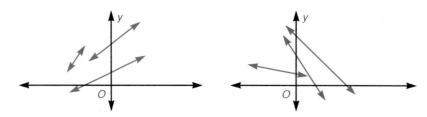

Example   1. Use graph paper or a computer. Plot $A(2, 1)$ and $B(5, -3)$. Draw $\overleftrightarrow{AB}$ and find its slope.

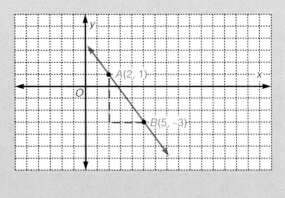

Moving from A to B, the change in the y-coordinates is $-3 - 1 = \mathbf{-4}$ units. The change in the x-coordinates is $5 - 2 = \mathbf{3}$ units. So, $m = \dfrac{-4}{3}$, or $-\dfrac{4}{3}$ units.

Notice that we can also find the slope moving from B to A.

$$\begin{array}{l}\text{change in } y\text{-coordinates} \rightarrow \\ \text{change in } x\text{-coordinates} \rightarrow\end{array} \frac{\mathbf{1 - (-3)}}{\mathbf{2 - 5}} = \frac{4}{-3} = -\frac{4}{3} \text{ units}$$

Note that we can find the slope without drawing the line.
For any two points (x_1, y_1) and (x_2, y_2),

$$m = \frac{y_2 - y_1}{x_2 - x_1}$$

Example   **2.** Find the slope of the line containing the points $(5, 2)$ and $(1, 0)$.

$$m = \frac{y_2 - y_1}{x_2 - x_1}$$

$$m = \frac{2 - 0}{5 - 1} \qquad \text{Substituting } (1, 0) \text{ for } (x_1, y_1) \text{ and } (5, 2) \text{ for } (x_2, y_2)$$

$$m = \frac{2}{4} = \frac{1}{2}$$

What are the slopes of horizontal and vertical lines?

Example   **3.** Find the slope of the line that contains points $L(-3, 2)$ and $G(4, 2)$.

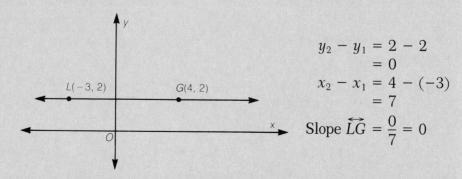

$$y_2 - y_1 = 2 - 2$$
$$= 0$$
$$x_2 - x_1 = 4 - (-3)$$
$$= 7$$

$$\text{Slope } \overleftrightarrow{LG} = \frac{0}{7} = 0$$

Any two points on a horizontal line have the same y-coordinate. Because the change in y is 0, the slope of a horizontal line is 0.

Example 4. Find the slope of the line that contains points $N(-3, 3)$ and $F(-3, -2)$.

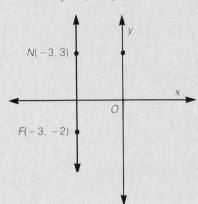

$y_2 - y_1 = 3 - (-2)$
$ = 5$
$x_2 - x_1 = -3 - (-3)$
$ = 0$
Slope $\overleftrightarrow{NF} = \dfrac{5}{0}$

Since we do not define division by zero, slope for this line is not defined.

Theorem 15.2   A horizontal line has a slope of 0. Slope for a vertical line *is not defined.*

Try This... Find the slope of the line containing the given pair of points.

g. $(5, 7)$ and $(-3, 7)$ **h.** $(2.1, -9)$ and $(2.2, 9)$

i. $(3, 3)$ and $(3, -1)$

Discussion   Slope can be thought of as the change in y for a one-unit increase in x. A slope of $\frac{5}{2}$ means that if x increases by 1, then y increases by $\frac{5}{2}$. Use this reasoning to explain why slope is not defined for a vertical line.

Exercises

Practice

Find the slope of each line.

1.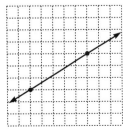
2.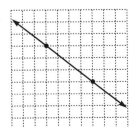
3.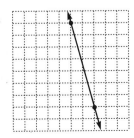

Graph the lines containing these points and find their slopes.

4. $(4, 1)$ and $(-2, -3)$ **5.** $(-2, 4)$ and $(4, 0)$ **6.** $(-4, 2)$ and $(2, -3)$

7. $(3, 2)$ and $(-2, 3)$ **8.** $(0, 4)$ and $(-4, -2)$ **9.** $(1, 6)$ and $(-2, -3)$

Find the slope of the line that contains the given pair of points.

10. $(6, 0)$ and $(7, 9)$ **11.** $(5, 0)$ and $(6, 2)$

12. $(0, 9)$ and $(-3, 6)$ **13.** $(-2, 1)$ and $(-4, 5)$

14. $(0, 0)$ and $(-6, -9)$ **15.** $(0, 0)$ and $(-8, -12)$

16. $(4, -2)$ and $(8, -3)$ **17.** $(5, -3)$ and $(9, -4)$

18. $(-3, -6)$ and $(-7, -4)$ **19.** $(-2, -5)$ and $(-8, -6)$

20. $(8, 9)$ and $(-4, 9)$ **21.** $(2, 4)$ and $(2, -6)$

22. $(-2, 6)$ and $(-2, 21)$ **23.** $(0, 11)$ and $(5, 11)$

24. $(4, -10)$ and $(18, -10)$ **25.** $(-5, -4)$ and $(16, -4)$

26. $(3, 6)$ and $(3, -11)$ **27.** $(-11, 9)$ and $(16, 9)$

28. $\left(\frac{1}{2}, \frac{1}{4}\right)$ and $\left(\frac{1}{4}, \frac{3}{4}\right)$ **29.** $\left(\frac{1}{4}, \frac{1}{8}\right)$ and $\left(\frac{3}{4}, \frac{1}{2}\right)$

30. $\left(\frac{1}{8}, \frac{1}{2}\right)$ and $\left(\frac{3}{8}, \frac{1}{4}\right)$ **31.** $\left(\frac{1}{3}, -\frac{1}{5}\right)$ and $\left(\frac{2}{3}, \frac{1}{10}\right)$

Extend and Apply

The slope of a line and the coordinates of two points on the line are given. Find the missing coordinate.

32. $m = 2$, $A(4, 3)$ and $B(x, 7)$

33. $m = -3$, $C(2, y)$ and $D(6, 4)$

34. Suppose a plane climbs 11 feet for every 44 feet it moves horizontally. At what grade is the plane climbing?

(A) 25%   (B) $\frac{1}{4}$%   (C) 4%   (D) 40%

35. The vertices of a triangle are $A(2, 2)$, $B(2, -2)$, and $C(-1, 4)$. Draw the triangle and find the slope of each side.

Use Mathematical Reasoning

36. Use the definition of slope to determine if $(1, 1)$, $(10, 8)$, and $(-1, 3)$ are collinear. Write a convincing argument to explain your answer.

37. Three nonvertical lines, each containing the origin, have slopes 5, -2, and n. What is the y-coordinate of the point on each line if the x-coordinate is 1?

Mixed Review

38. A field hockey ball has a diameter of 3 in. What is its volume?

39. A bowling ball has a circumference of 27 in. What is its volume?

Solve and check.

40. $5(x + 3) = 3(x + 5)$

Numerical Control

A laser can be controlled by a computer. It may be directed
to go to certain coordinates called its *absolute coordinates.*
Or it may be directed to move *relative* to its current position.
To use *relative coordinates,* we describe each point in relation
to the previous point.

Sample   Use relative coordinates
from $(0, 0)$ to direct a laser
to cut along a rectangle
with vertices $(1, 1)$, $(6, 1)$,
$(6, 5)$, and $(1, 5)$, then
return to $(0, 0)$ without
cutting.

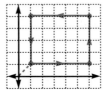

We can use the notation $[r, u]$ to describe the number of
units to move right and up from the previous point.

Laser Up	We want the laser to move without cutting.
$[1, 1]$	To go from $(0, 0)$ to $(1, 1)$ is 1 unit right and 1 unit up.
Laser Down	We want the laser to begin cutting.
$[5, 0]$	To go from $(1, 1)$ to $(6, 1)$ is 5 units right.
$[0, 4]$	To go from $(6, 1)$ to $(6, 5)$ is 4 units up.
$[-5, 0]$	To go from $(6, 5)$ to $(1, 5)$ is 5 units left.
$[0, -4]$	To go from $(1, 5)$ to $(1, 1)$ is 4 units down.
Laser Up	We want the laser to stop cutting.
$[-1, -1]$	To go from $(1, 1)$ to $(0, 0)$ is 1 unit left and 1 unit down.

Problems

Classify the figure cut by the laser.

1. Laser Down, $[4, 0]$, $[-2, 3]$, $[-2, -3]$

2. Laser Down, $[3, 0]$, $[1, 3]$, $[-2, 0]$, $[-2, -3]$

3. Laser Down, $[2, 7]$, $[2, -7]$, $[-5, 4]$, $[6, 0]$, $[-5, -4]$

4. a. In Problems 1–3, what is the sum of the *r*-coordinates?
   b. In Problems 1–3, what is the sum of the *u*-coordinates?
   c. Explain the results from parts **a.** and **b.**

5. Give instructions to move a laser from $(2, 8)$ to $(3, 5)$ without
cutting, to then cut out a square of side length 6, and to
move to $(0, 0)$ without cutting.

15-3 The Midpoint Theorem

Objective: Solve problems involving midpoints.

Explore

Use graph paper or computer software. Graph the segments with these endpoints.

$A(1, 4)$ and $B(3, 8)$ $P(0, 5)$ and $Q(4, 3)$

$H(10, 3)$ and $F(6, 7)$ $K(8, 2)$ and $M(-6, 10)$

Find or estimate the coordinates of the midpoint of each segment.

Discuss   What relationship exists between the coordinates of the endpoints and the coordinates of the midpoint of each segment?

Theorem 15.3   **The Midpoint Theorem**

The coordinates of the midpoint M of a segment with endpoints $A(x_1, y_1)$ and $B(x_2, y_2)$ are

$$\left(\frac{x_1 + x_2}{2}, \frac{y_1 + y_2}{2}\right)$$

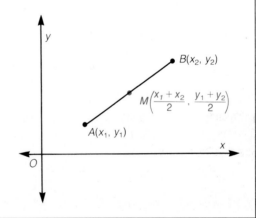

Example   1. Find the coordinates of the midpoint of $\overline{PQ}$ if the coordinates of the endpoints are $P(5, 7)$ and $Q(9, -3)$.

We can substitute the coordinates of P and Q for (x_1, y_1) and (x_2, y_2) in the midpoint formula. Then the coordinates of the midpoint are

$$\left(\frac{x_1 + x_2}{2}, \frac{y_1 + y_2}{2}\right) = \left(\frac{5 + 9}{2}, \frac{7 + (-3)}{2}\right)$$

$$= \left(\frac{14}{2}, \frac{4}{2}\right) = (7, 2)$$

Thus, the coordinates of the midpoint of $\overline{PQ}$ are $(7, 2)$.

Try This... Given P and Q, find the coordinates of the midpoint of $\overline{PQ}$.
  a. $P(5, 10)$ and $Q(7, -2)$ **b.** $P(6, -12)$ and $Q(-8, 2)$

Example   2. Find the length of the median from A to $\overline{BC}$.

Suppose M, the midpoint of $\overline{BC}$, has coordinates (x, y).

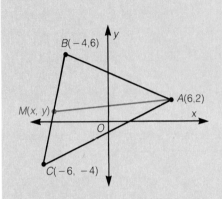

Then $x = \dfrac{-4 + (-6)}{2} = \dfrac{-10}{2} = -5$

and $y = \dfrac{6 + (-4)}{2} = \dfrac{2}{2} = 1$

The coordinates of M are $(-5, 1)$.

Using the Distance Formula, we can find AM.

$$AM = \sqrt{(-5 - 6)^2 + (1 - 2)^2}$$
$$= \sqrt{(-11)^2 + (-1)^2}$$
$$= \sqrt{121 + 1}$$
$$= \sqrt{122} \approx 11.0$$

Try This... c. Consider $\triangle ABC$ in Example 2. Find the length of the median from B to $\overline{AC}$.

Discussion   Loic rewrote the Midpoint Theorem as, "The coordinates of the midpoint of a segment are the average of the x-coordinates of the endpoints and the average of the y-coordinates of the endpoints." Is he correct?

Exercises

Practice

Given P and Q, find the coordinates of the midpoint of $\overline{PQ}$.

1. $P(2, 5)$, $Q(6, 3)$ **2.** $P(8, 2)$, $Q(6, 10)$ **3.** $P(5, 4)$, $Q(-3, 6)$

4. $P(7, 5)$, $Q(3, 3)$ **5.** $P(0, 2)$, $Q(6, 3)$ **6.** $P(0, 5)$, $Q(8, 2)$

7. $P(9, 10)$, $Q(-6, -4)$ **8.** $P(17, 12)$, $Q(-8, -5)$

9. $P(-8, -5)$, $Q(-7, -6)$ **10.** $P(0, 9)$, $Q(-11, 0)$

11. $P(0, -8)$, $Q(-6, 0)$ **12.** $P(a, 6)$, $Q(5a, 9)$

13. $P(s, t)$, $Q(6s, 9t)$ **14.** $P(6x, 3)$, $Q(-2x, -4)$

15. Suppose $\triangle PQR$ has vertices $P(8, 4)$, $Q(-4, 10)$, and $R(-8, -6)$.
  a. Find the length of the median from P to $\overline{QR}$.
  b. Find the length of the median from Q to $\overline{PR}$.

16. The coordinates of the vertices of a rectangle are $(0, 0)$, $(6, 0)$, $(6, 4)$, and $(0, 4)$. Find the coordinates of the point of intersection of the diagonals.

17. A right triangle has vertices of $(0, 0)$, $(8, 0)$, and $(0, 6)$. Find the coordinates of the midpoint of the hypotenuse. Find the distance from the midpoint of the hypotenuse to the vertex of the right angle.

Extend and Apply

18. Consider $\overline{PQ}$. The coordinates of P are $(7, 2)$, and the coordinates of the midpoint M are $(-6, 4)$. Find the coordinates of Q.

19. Consider $\overline{AB}$. The coordinates of A are $(5, 8)$, and the coordinates of the midpoint M are $(-2, 3)$. Find the coordinates of B.

20. The coordinates of the endpoints and the midpoint of a segment are known. Two of the points are $(3, 9)$ and $(12, 16)$. Which could *not* be the other point?
  (A) $(21, 23)$ (B) $(-6, 2)$ (C) $(7\frac{1}{2}, 12\frac{1}{2})$ (D) $(15, 25)$

Use Mathematical Reasoning

21. Find the point on the x-axis that is equidistant from $A(1, 3)$ and $B(8, 4)$.

22. Draw a triangle with vertices $(1, 4)$, $(5, 10)$, and $(9, 2)$. Form a new triangle by connecting the midpoints of the sides. Repeat to form a third triangle. Relate the slopes of the sides of the third triangle to those of the original triangle.

Mixed Review

23. How many vertices does a heptagon have?

24. What is -2000 divided by -50?

Enrichment

The following BASIC program uses midpoints to draw a figure. Lines 10, 20, 60, and 80 are specific to *Applesoft* BASIC. Line 10 displays graphics; line 50 randomly chooses 0, 1, or 2; and line 70 plots points on the graphics screen. Modify these lines for your computer if necessary.

Enter and run the program, and describe or draw the resulting figure.

```
10 HGR: HCOLOR = 3
20 X(0) = 0: Y(0) = 128
30 X(1) = 128: Y(1) = 0
40 X(2) = 256: Y(2) = 128
50 J = INT(3*RND(1))
60 X = (X + X(J))/2: Y = (Y + Y(J))/2
70 HPLOT X,Y: GO TO 50
```

15-4 Graphing Equations

Objective: Determine solutions and graph equations.

▪ Explore

1. Find y in the equation $y = 3x + 1$, if $x = 0$.

2. Find y in the equation $y = 3x + 1$, if $x = 1$.

3. Find y in the equation $y = 3x + 1$, if $x = 2$.

Discuss   Does an equation like $y = 3x + 1$ have a single numerical solution for y?

An equation like $7 = 3x + 1$, which has one variable, has one solution. An equation like $y = 3x + 1$ has an unlimited number of solutions, each dependent upon the value of x. When $x = 0$, $y = 1$; when $x = 1$, $y = 4$; and so on. We can write any solution as an ordered pair (x, y), such as $(0, 1)$, $(1, 4)$, and so on.

Examples 1. Determine whether $(5, 2)$ is a solution of $4y = 3x - 7$.

Substitute the ordered pair in the equation.

$$4y = 3x - 7$$
$$4(\mathbf{2}) = 3(\mathbf{5}) - 7 \qquad \text{Substituting}$$
$$8 = 15 - 7$$
$$8 = 8 \checkmark$$

We obtain a true equation, so $(5, 2)$ is a solution of $4y = 3x - 7$.

2. Determine whether $(5, 2)$ is a solution of $4y = 3(x - 7)$.

$$4y = 3(x - 7)$$
$$4(\mathbf{2}) = 3(\mathbf{5} - 7) \qquad \text{Substituting}$$
$$8 = 3(-2)$$
$$8 \neq -6 \qquad \text{A false equation}$$

$(5, 2)$ is *not* a solution of $4y = 3(x - 7)$.

Try This... Is the ordered pair a solution of $7x = 3y + 5$?

    a. $(3, 5)$ **b.** $(2, 3)$ **c.** $(-1, -4)$

Is the ordered pair a solution of $-3(y - 4) = 2x$?

    d. $(3, 6)$ **e.** $(2, 3)$ **f.** $(6, 0)$

By finding solutions, we can plot points and find the line given by an equation. The graph of $y = -5x + 8$ is the graph of all of its solutions.

Example 3. Graph $y = -5x + 8$.

First find three points.

When $x = 0$, $y = -5(0) + 8$
$\quad\quad\quad\quad\quad = 0 + 8$
$\quad\quad\quad\quad\quad = 8$ This gives us the point $(\mathbf{0, 8})$.

When $x = 1$, $y = -5(1) + 8$
$\quad\quad\quad\quad\quad = -5 + 8$
$\quad\quad\quad\quad\quad = 3$ This gives us the point $(\mathbf{1, 3})$.

When $x = 2$, $y = -5(2) + 8$
$\quad\quad\quad\quad\quad = -10 + 8$
$\quad\quad\quad\quad\quad = -2$ This gives us the point $(\mathbf{2, -2})$.

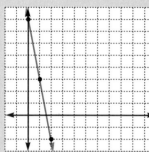

The three points are collinear when graphed. The line drawn through these points is the graph of all the equation's solutions.

Try This... **g.** Graph $y = -x - 3$.

Discussion   How can you find the slope of the line given by the equation $y = 4x - 7$?

Exercises

Practice

Determine whether the ordered pair is a solution of the equation.

1. $(9, -4)$, $4x + 9y = 0$
2. $(2, 5)$, $y = 3x - 13$
3. $(-3, -8)$, $3y - x = -27$
4. $(2, 0)$, $5y = 3(x - 2)$
5. $(-2, 1)$, $5(y - x) = 15$
6. $(-2, 2)$, $3(y - 4) = 2(5 - x)$

Find three solutions of each equation.

7. $y - x = 5$
8. $-4x = 2y + 3$
9. $7y = 4x - 4$

Make a table of solutions and graph each equation.

10. $y = 4x - 5$ **11.** $y = 2x + 2$ **12.** $y = x + 1$ **13.** $y = 3x - 2$

14. $y = -x - 2$ **15.** $y = -2x + 1$ **16.** $y = \frac{1}{2}x + 1$ **17.** $y = \frac{1}{3}x$

Extend and Apply

18. The graph shown is the graph of

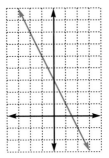

(A) $y = 2x + 3$ (B) $y = -2x + 3$

(C) $y = 2x - 3$ (D) $y = -2x - 3$

Use Mathematical Reasoning

19. Suppose you want to graph the equation of a line. Your table of solutions gives you $(1, 5)$, $(5, 9)$, and $(-5, -9)$. Can you graph the equation of the line? Why or why not?

Mixed Review

For Exercises 20–23, find each distance.

20. from $(0, 30)$ to $(30, 0)$.   **21.** from $(0, 30)$ to $(-30, 0)$.

22. from $(0, -30)$ to $(30, 0)$.   **23.** from $(0, -30)$ to $(-30, 0)$.

24. What is the slope of the line through $(0, 30)$ and $(30, 0)$?

25. What is the slope of the line through $(0, 30)$ and $(-30, 0)$?

26. What is the slope of the line through $(0, -30)$ and $(30, 0)$?

27. What is the slope of the line through $(0, -30)$ and $(-30, 0)$?

 Multiply or divide.

28. -7×-9 **29.** $90 \div -18$ **30.** $-90 \div 18$

Visualization

The coordinates of A are $(-10, 10)$.
The coordinates of B are $(-10, -10)$.
The center has coordinates $(0, 0)$.
Which is the best estimate for the coordinates of C?

(A) $(-11, 0)$
(B) $(-10, 0)$
(C) $(-9, 0)$
(D) $(0, -9)$

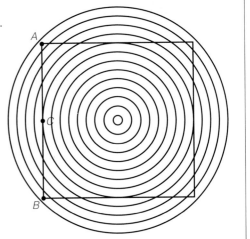

15-5 Lines and Equations

Objective: Use the slope-intercept equation and recognize parallel and perpendicular lines from their slopes.

Explore

1. Use graph paper or computer software to graph the lines $y = 2x - 5$ and $y = -3x + 4$.

2. What is the slope of each line?

3. What is the value of y where each line crosses the y-axis?

Discuss Can you find the answers to **2.** and **3.** without graphing?

The value of y where a line crosses the y-axis is called the **y-intercept.** An equation in the form $y = mx + b$ is called the **slope-intercept** equation of a line, where m is the slope and b is the y-intercept.

Example **1.** Find the slope and y-intercept of the line with equation $y = 4x - \dfrac{1}{5}$.

We can read the numbers directly from the slope-intercept equation.

$$y = 4x - \frac{1}{5}$$

$$\text{slope} = 4 \qquad y\text{-intercept} = -\frac{1}{5}$$

Try This... a. Find the slope and y-intercept of the line with equation $y = -\dfrac{3}{4}x + 3\dfrac{1}{4}$.

Example **2.** Write an equation for a line that has a slope of -2 and a y-intercept of -3.

$y = mx + b$
$y = -2x + (-3)$ Substituting
$y = -2x - 3$

Explore

Use graph paper or computer software.
1. Graph these two lines using the same set of axes:
   $y = -2x + 2$ and $y = -2x - 3$.
2. Determine the slope of each line.
3. Graph these two lines on another set of axes:
   $y = -2x + 2$ and $y = \frac{1}{2}x - 3$.
4. Determine the slope of each line.
5. What is the product of the slopes in **4.**?
6. How are the lines in **1.** related? What do you think is true of lines with the same slopes?
7. How are the lines in **3.** related? What do you think is true of two lines whose slopes have a product of -1?

The Explore suggests the following theorems:

Theorem 15.4 Two nonvertical lines are parallel if and only if they have the same slope.

Theorem 15.5 Two nonvertical lines are perpendicular if and only if the product of their slopes is -1.

All vertical lines are parallel, all horizontal lines are parallel, and vertical and horizontal lines are perpendicular.

Example **3.** Determine whether line ℓ, which has an equation $y = 2x - 5$, and line m, containing $(-2, 1)$ and $(0, 5)$, are parallel, perpendicular, or neither.

From the equation, the slope of line ℓ is 2.

Line m has slope $\dfrac{5 - 1}{0 - (-2)} = \dfrac{4}{2} = 2$.

Since their slopes are the same, the lines are parallel.

Try This... **c.** Determine whether line ℓ, containing $(-3, 1)$ and $(3, -2)$, and line m, containing $(0, -3)$ and $(2, 1)$, are parallel, perpendicular, or neither.

Exercises

Practice

Find the slope and y-intercept of the line described by each equation.

1. $y = 8x + 3$
2. $y = 9x + 2$
3. $y = -5x + 4$
4. $y = -4x + 7$
5. $y = -6x - 9$
6. $y = -8x - 12$
7. $y = \frac{2}{3}x + 8$
8. $y = \frac{3}{5}x + 10$
9. $y = -0.9x$

Write an equation of each line.

10. Slope 3, y-intercept 2
11. Slope -3, y-intercept -2
12. y-intercept $\frac{2}{3}$, slope -2
13. y-intercept -3, slope $\frac{5}{3}$
14. Slope 5, y-intercept 0
15. Slope 0, y-intercept 4

Determine whether the lines are parallel, perpendicular, or neither.

16. $y = 4x - 3$ and $y = \frac{1}{4}x + 3$
17. $y = -3x + 5$ and $y = -3x$
18. $y = -x + 2$ and $y = x + 5$
19. $y = 8x + 5$ and $y = -8x - \frac{1}{5}$
20. $y = -2x + 7$ and line ℓ, containing $(-1, 1)$ and $(3, -7)$
21. $y = -2x + 7$ and line ℓ, containing $(-1, 1)$ and $(3, 3)$
22. line ℓ through $(0, 0)$ and $(3, 2)$, and line m through $(5, -3)$ and $(3, -6)$
23. line ℓ through $(2, -2)$ and $(3, 3)$, and line m through $(-3, 3)$ and $(2, 2)$

Extend and Apply

24. Determine whether line ℓ, containing $(3, -5)$ and $(-3, -5)$, and line m, containing $(-4, 7)$ and $(-4, -1)$, are parallel, perpendicular, or neither.

25. If the slope of line ℓ is less than 1 and greater than $\frac{1}{2}$, which could be the slope of the line that is perpendicular to ℓ?
    (A) $\frac{3}{4}$
    (B) $-\frac{3}{4}$
    (C) $-\frac{5}{3}$
    (D) -2

Use Mathematical Reasoning

26. In plane $\mathcal{R}$, line ℓ is perpendicular to line m, and line m is perpendicular to line n. How are lines ℓ and n related?

Mixed Review

Find the coordinates of the midpoint of the segment with these endpoints.

27. $(50, 100)$ and $(-100, 50)$
28. $(99, -99)$ and $(-99, 99)$

Multiply or divide.

29. $-76 \div -19$
30. -4×-19
31. $\frac{-1}{-1}$

Chapter 15 Review

1. Find the distance between $P(-2, 3)$ and $Q(8, 0)$.

2. A ship at coordinates $A(-4, 4)$ needs to go to either $B(13, 4)$ or $C(-16, 16)$. Which point is closer? What is the distance to the closer point?

15-2   Find the slope of the line containing each pair of points.

3. $R(4, 2)$ and $S(3, 4)$

4. $J(8, -1)$ and $L(-3, -1)$

5. $U(5, 2)$ and $K(-1, 4)$

6. $J(6, -2)$ and $L(6, -1)$

7. Find the slope of the line shown.

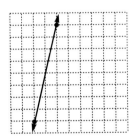

15-3   The coordinates of the endpoints of a segment are given. Find the coordinates of the midpoint.

8. $(2, 3)$ and $(2, -3)$

9. $(-4, 6)$ and $(-7, 7)$

15-4   Determine whether the ordered pair is a solution of $-2x + 6 = -3y$.

10. $(-6, 6)$ **11.** $(6, 2)$ **12.** $(3, 0)$

13. Make a table of solutions and graph $y = -\dfrac{3}{2}x + 1$.

15-5   **14.** Find the slope and y-intercept of the line with equation $y = 5x - 2$.

15. Write an equation for a line that has a slope of -3 and a y-intercept of 3.

16. Determine whether line ℓ, which has an equation $y = \dfrac{2}{3}x + 5$, and line m, containing $(-2, 1)$ and $(1, 7)$, are parallel, perpendicular, or neither.

Chapter 15 Test

1. Find the distance between $K(5, -3)$ and $M(9, 2)$.

2. A ship at coordinates $A(1, -2)$ can go to either $B(20, 5)$ or $C(21, 2)$. Which point is closer? What is the distance to the closer point?

Find the slope of the line containing each pair of points.

3. $T(1, 1)$ and $S(-3, 6)$

4. $V(7, 7)$ and $Z(-5, -1)$

5. $A(0, 9)$ and $D(-8, -7)$

6. $Q(10, -12)$ and $L(10, 3)$

7. Find the slope of the line shown.

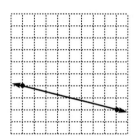

The coordinates of the endpoints of a segment are given. Find the coordinates of the midpoint.

8. $(4, 3)$ and $(2, -7)$

9. $(45, -26)$ and $(-22, 27)$

Determine whether the ordered pair is a solution of $-3x + 12 = 3y$.

10. $(4, 0)$ 11. $(6, 2)$ 12. $(-9, 13)$

13. Make a table of solutions and graph $y = -\dfrac{2}{3}x - 1$.

14. Find the slope and y-intercept of the line with equation $y = -5x - 4$.

15. Write an equation for a line that has a y-intercept of -1 and a slope of -4.

16. Determine whether line ℓ, which has an equation $y = -\dfrac{2}{3}x + 1$, and line m, containing $(-2, 1)$ and $(1, 7)$, are parallel, perpendicular, or neither.

16 *Trigonometric Ratios*

Surveyors use a transit to measure distances along the ground. Trigonometric ratios are often used to calculate distances that are difficult to measure directly.

16-1 Trigonometric Ratios

Objective: Find trigonometric ratios.

We used the Pythagorean Theorem in Chapter 10 to find the length of one side of a right triangle when the lengths of the other two sides were known. In this chapter, we will see how to use **trigonometric ratios** to find missing lengths such as the height of the tree at the right.

For any right triangle six ratios of pairs of sides are possible.

$$\frac{a}{c}, \frac{b}{c}, \frac{a}{b}, \frac{b}{a}, \frac{c}{a}, \frac{c}{b}$$

These six ratios are called *trigonometric ratios*. Any right triangle similar to $\triangle ABC$ would have six trigonometric ratios equal to these.

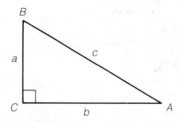

Example   1. Find the trigonometric ratios for $\triangle PQR$.

The six trigonometric ratios are

$$\frac{15}{39}, \frac{36}{39}, \frac{15}{36}, \frac{36}{15}, \frac{39}{15}, \frac{39}{36}$$

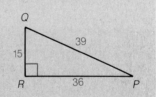

Try This... a. Find the six trigonometric ratios for $\triangle PQR$.

b. Find the six trigonometric ratios for $\triangle XYZ$.

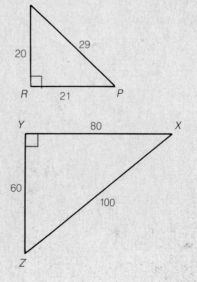

The six trigonometric ratios are usually referenced to a specific angle. We give each side a name depending upon the *reference angle*.

Suppose the reference angle is ∠X in △XYZ at the right.

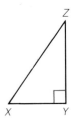

$\overline{YZ}$ is called the **side opposite** ∠X.
$\overline{XY}$ is called the **side adjacent** to ∠X.
$\overline{XZ}$ is the **hypotenuse.**

Example 2. Let the reference angle be ∠Z in △XYZ above. Name the side opposite ∠Z, the side adjacent to ∠Z, and the hypotenuse.

$\overline{XY}$ is the side opposite ∠Z.
$\overline{ZY}$ is the side adjacent to ∠Z.
$\overline{XZ}$ is the hypotenuse.

Try This... Use △STU to name the hypotenuse, the side opposite, and the side adjacent to each reference angle.

c. ∠S   **d.** ∠T

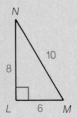

We can form each of the six trigonometric ratios by referencing an angle and its respective sides. For example,

$$\frac{a}{b} = \frac{\text{the length of the side opposite } \angle A}{\text{the length of the side adjacent to } \angle A}.$$

We will abbreviate this as $\dfrac{\text{opposite } \angle A}{\text{adjacent } \angle A}$.

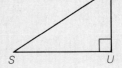

Example 3. Find the following trigonometric ratio. Then use a calculator to find an approximation to 4 decimal places.

$$\frac{\text{opposite } \angle M}{\text{adjacent } \angle M}$$

$$\frac{\text{opposite } \angle M}{\text{adjacent } \angle M} = \frac{8}{6} \approx 1.3333$$

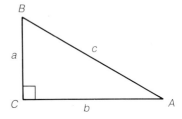

4. Find the following trigonometric ratio. Then use a calculator to find an approximation to 4 decimal places.

$$\frac{\text{adjacent } \angle N}{\text{hypotenuse}}$$

$$\frac{\text{adjacent } \angle N}{\text{hypotenuse}} = \frac{8}{10} = 0.8000$$

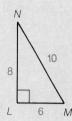

■ **Try This...** Find each trigonometric ratio. Then use a calculator to find an approximation to 4 decimal places.

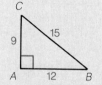

e. $\dfrac{\text{side opposite } \angle B}{\text{hypotenuse}}$ f. $\dfrac{\text{side adjacent } \angle C}{\text{side opposite } \angle C}$

Discussion   What are the six trigonometric ratios in terms of reference angles B and C for $\triangle ABC$ above? Are any of these ratios the same?

Exercises

Practice

1. Find the six trigonometric ratios for $\triangle XYZ$.

2. Find the six trigonometric ratios for $\triangle RST$.

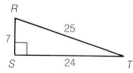

Use $\triangle VTU$ to name the hypotenuse, the side opposite, and the side adjacent to each reference angle.

3. $\angle V$

4. $\angle U$

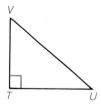

Use $\triangle MKL$ to name the hypotenuse, the side opposite, and the side adjacent to each reference angle.

5. $\angle M$

6. $\angle L$

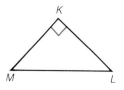

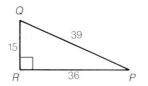 Use △QPR to find each trigonometric ratio. Then use a calculator to find an approximation to 4 decimal places.

7. $\dfrac{\text{opposite } \angle Q}{\text{hypotenuse}}$

8. $\dfrac{\text{opposite } \angle P}{\text{adjacent } \angle P}$

9. $\dfrac{\text{opposite } \angle P}{\text{hypotenuse}}$

10. $\dfrac{\text{opposite } \angle Q}{\text{adjacent } \angle Q}$

Extend and Apply

11. If $\dfrac{\text{opposite } \angle A}{\text{adjacent } \angle A} = \dfrac{5}{4}$, then which of the following ratios is also $\dfrac{5}{4}$?

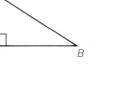

(A) $\dfrac{\text{adjacent } \angle A}{\text{opposite } \angle A}$

(B) $\dfrac{\text{opposite } \angle B}{\text{hypotenuse}}$

(C) $\dfrac{\text{adjacent } \angle B}{\text{opposite } \angle B}$

(D) None of these

12. A support for a freeway overpass contains a right triangle, △LMN. The blueprints show that $\dfrac{\text{opposite } \angle L}{\text{adjacent } \angle L} = \dfrac{7}{24}$ and that $MN = 21$ ft. Find LN and LM.

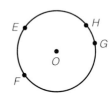

Use Mathematical Reasoning

13. In △RST, $\angle T$ is a right angle and $\dfrac{\text{opposite } \angle R}{\text{adjacent } \angle R} = 1$. What else can you conclude about △RST? Find $m\angle R$ and $m\angle S$.

Mixed Review

14. In ⊙O, $m\widehat{EF} = 80°$ and $m\widehat{EG} = 132°$. Find $m\widehat{FHG}$ and $m\widehat{FG}$.

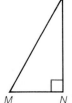

Add or subtract.

15. $13 - (-15)$ **16.** $14 + (-15)$ **17.** $12 - 18$ **18.** $-6 + (-4)$

Visualization

A circular plate is broken into two pieces. One piece is shown on the left. Which of the pieces on the right completes the plate?

(A)   (B)   (C)

16-2 The Sine, Cosine, and Tangent Ratios

Objective: Given a right triangle, find the sine, cosine, and tangent ratios of its angles.

Explore

1. Measure the sides of triangles ABC and XYZ to the nearest tenth of a centimeter.

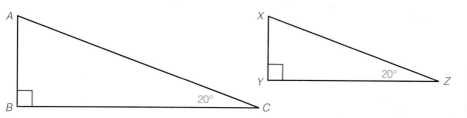

2. Using your calculator, compute the following ratios to 2 decimal places.

$$\frac{\text{opposite } \angle C}{\text{hypotenuse}} \qquad \frac{\text{adjacent } \angle C}{\text{hypotenuse}} \qquad \frac{\text{opposite } \angle C}{\text{adjacent } \angle C}$$

$$\frac{\text{opposite } \angle Z}{\text{hypotenuse}} \qquad \frac{\text{adjacent } \angle Z}{\text{hypotenuse}} \qquad \frac{\text{opposite } \angle Z}{\text{adjacent } \angle Z}$$

Discuss Are triangles ABC and XYZ similar? Why? Do you notice anything special about the above ratios? Suppose right triangle $\triangle RST$ has a 20° angle at $\angle S$.

What do you think the ratio $\dfrac{\text{opposite } \angle S}{\text{hypotenuse}}$ will be?

Since all right triangles with a 20° angle are similar, their sides are proportional and their trigonometric ratios will be equal to those ratios you found in the Explore. These trigonometric ratios are given special names.

Definitions For any acute angle, $\angle A$, of a right triangle,

sine of $\angle A$ = sin A = $\dfrac{\text{opposite } \angle A}{\text{hypotenuse}}$

cosine of $\angle A$ = cos A = $\dfrac{\text{adjacent } \angle A}{\text{hypotenuse}}$

tangent of $\angle A$ = tan A = $\dfrac{\text{opposite } \angle A}{\text{adjacent } \angle A}$

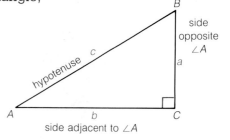

Example   **1.** In △ABC, find sin B.

$$\sin B = \frac{\text{length of side opposite } \angle B}{\text{length of the hypotenuse}}$$

$$\sin B = \frac{15}{17}$$

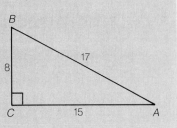

Try This... a. In △PQR, find sin R and sin P.

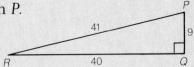

Examples   Use your calculator to find each of the following to four decimal places.

2. sin R

$$\sin R = \frac{\text{opposite } \angle R}{\text{hypotenuse}} = \frac{20}{52}$$

$$\approx 0.3846$$

3. tan Q

$$\tan Q = \frac{\text{opposite } \angle Q}{\text{adjacent } \angle Q} = \frac{48}{20}$$

$$= 2.4000$$

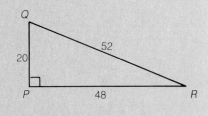

Try This... In △PQR, use your calculator to find each of the following to four decimal places.

b. cos R   **c.** tan R

Example   **4.** Find the sine of the 30° angle in each triangle and compare the ratios.

   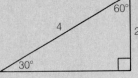   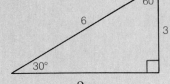

$\sin 30° = \dfrac{1}{2} = 0.5$ $\sin 30° = \dfrac{2}{4} = 0.5$ $\sin 30° = \dfrac{3}{6} = 0.5$

Because the triangles are similar, corresponding ratios are the same. Thus, the sine of 30° is the same number in any such triangle.

Try This... d. Find the cosine of the 60° angle in each of the triangles in Example 4. Compare the ratios.

As illustrated in Example 4, the sine, cosine, and tangent of an angle depend only on the size of the angle, not the triangle it is contained in.

Discussion   Julia drew an isosceles right triangle. She said this triangle could be used to show that the tangent of 45° is 1. Explain her reasoning.

Exercises

Practice

In each triangle, find the sine ratio for each acute angle. Then use your calculator to find an approximation to four decimal places.

1.

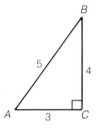

2.

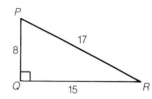

In each triangle, find the cosine ratio for each acute angle. Then use your calculator to find an approximation to four decimal places.

3.

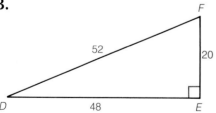

4.

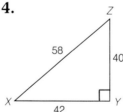

In each triangle, find the tangent ratio for each acute angle. Then use your calculator to find an approximation to four decimal places.

5.

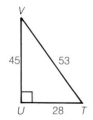

6.

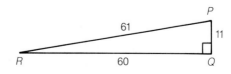

7. Find the cosine of the 30° angle in each of the triangles below. Use a calculator to find an approximation for the ratios to 4 decimal places. Compare the ratios.

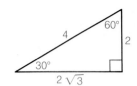

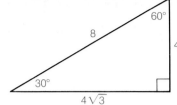

Extend and Apply

8. In right triangle ABC, $m\angle B = 30°$ and $\sin B = 0.5$. What is $\sin X$ in right triangle XYZ if $m\angle X = 30°$?

(A) 5   (B) 0.5   (C) 1   (D) Not enough information

9. A section of a bridge contains right triangle PQR. The lengths shown are in meters. Use your calculator to find the cosine of $\angle R$ to four decimal places.

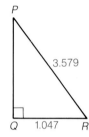

Use Mathematical Reasoning

10. What type of right triangle will have the same tangent ratio for each acute angle?

11. What is the measure of an angle whose sine and cosine ratios are the same?

Mixed Review

12. Find the six trigonometric ratios for $\triangle DEF$.

13. Find the six trigonometric ratios for $\triangle KLM$.

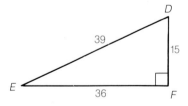

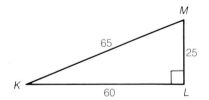

Find the volume of a sphere with the given radius. Round your answer to the nearest tenth.

14. 1 m   **15.** 5.5 cm   **16.** 3.7 in.

◇ Solve and check.

17. $4x + 5 + 3x = 9x - 9$

18. $8w - 5w = w + 22$

19. $7y - 8 = 2y + 3y$

The Speed of Raindrops

Have you ever been riding in a car in a rainstorm and noticed that the rear window is completely clear?

When a car is not moving, all the raindrops in the volume above the window will hit the window.

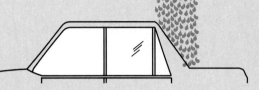

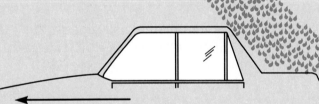

When a car is moving forward quickly, however, the window is moving out of the path of each raindrop faster than the raindrop falls.

How fast does a car need to move for the raindrops to just miss the window? (The figures at the right show a single raindrop that just misses the window.) At what speed does a raindrop fall?

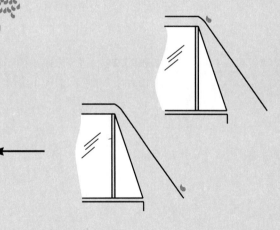

Work with a group or partner. On a rainy day that is not windy, use a car with a slanted rear window, as shown above. Find the angle the window makes with the horizontal (measure the height and length of the window and use trigonometry, or use a protractor).

Without exceeding the speed limit, increase the speed of the car until a passenger notices that the rain stops hitting the rear window. Note the speed of the car.

Use trigonometric ratios to solve for the speed of the raindrop.

16-3 Using the Sine, Cosine, and Tangent Ratios

Objective: Find the sine, cosine, and tangent ratios using a table or calculator, and use these ratios to solve problems.

Explore

1. Draw three different right triangles, each having a 30° angle.

2. Measure the sides of each triangle.

3. Compute sin 30° for each triangle.

4. Use your calculator to find sin 30° by entering 30 $\boxed{\text{sin}}$.

Discuss   Are each of your three triangles similar? Why?
What is true about sin 30° for each of your triangles?
How does this value compare to that given by your calculator?

Since the trigonometric ratios are the same for all similar triangles, the ratios for any angle measure need to be calculated only once. The ratios have been computed and arranged in the table provided on page 583.

Example   1. Find sin 58° using a calculator. Then look up sin 58° in the table. Compare the two answers.

Using a calculator, we have 58 $\boxed{\text{sin}}$ $\boxed{\text{0.8480481}}$
So, sin 58° ≈ 0.8480481.

To use a table, first find 58° in the *Degrees* column, as shown in the partial table below. Then find the entry in the *Sin* column. This is 0.8480.

Degrees	Sin	Cos	Tan
57°	0.8387	0.5446	1.5399
58°	0.8480	0.5299	1.6003
59°	0.8572	0.5150	1.6643

Using the table, sin 58° ≈ 0.8480.

The answers are the same. However, the calculator gives the value with more precision. If we round the calculator's value to 4 decimal places, we have the same answer as with the table: sin 58° ≈ 0.8480.

Try This... Find the indicated ratios using a calculator and then using the table. Compare your answers.

    a. tan 33°

    b. cos 81°

    c. sin 8°

We can now find the missing sides of a right triangle if we know the length of one side and the measure of one acute angle. Examples 2 and 3 show how this may be done using a calculator or table.

Example **2.** In right triangle ABC, $m\angle B = 61°$ and $c = 10$ cm. Find b to the nearest tenth.

Since we know the measure of $\angle B$, it will be the reference angle. Side b is opposite $\angle B$, and we know the length of side c, the hypotenuse. Therefore, we will use the sine ratio for $\angle B$.

$$\sin B = \frac{\text{opposite } \angle B}{\text{hypotenuse}}$$

$$\sin 61° = \frac{b}{10}$$

We can multiply both sides by 10 to solve for b.

$$10 \times \sin 61° = b$$

Using a calculator, we have

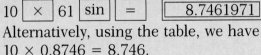

Alternatively, using the table, we have
$10 \times 0.8746 = 8.746$.

In both cases we can say that $b \approx 8.7$ cm, to the nearest tenth.

Try This... d. In right triangle ABC, $m\angle A = 44°$ and $c = 20$ cm. Find a to the nearest tenth.

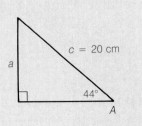

Example 3. In right triangle DEF, $m\angle D = 25°$ and $f = 18$ km. Find e to the nearest tenth.

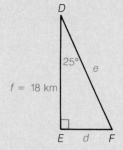

In this case, $\angle D$ is the reference angle. Side e is the hypotenuse, and we know the length of side f, the side adjacent to $\angle D$. Thus, we will use the cosine ratio for $\angle D$.

$$\cos D = \frac{\text{adjacent } \angle D}{\text{hypotenuse}}, \text{ so } \cos 25° = \frac{18}{e}.$$

Thus, $e \times \cos 25° = 18$, or $e = \dfrac{18}{\cos 25°}$.

 Using a calculator, we have

$$18 \boxed{÷} \; 25 \boxed{\cos} \boxed{=} \quad \boxed{19.860803}$$

Using the table, we have $18 ÷ 0.9063 = 19.860973$.

In both cases we can say $e \approx 19.9$ km.

Try This... e. In right triangle DEF, $m\angle D = 36°$ and $f = 30$ m. Find e to the nearest tenth.

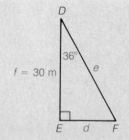

Discussion Use your calculator or the table to find the tangent of different angles. What do you think is the range of values for the tangent ratio? What is the range for the sine and cosine ratios?

Exercises

 Practice

Find the indicated ratios using a calculator and then using the table. Compare your answers.

1. $\sin 40°$	**2.** $\cos 55°$	**3.** $\tan 14°$
4. $\sin 79°$	**5.** $\tan 81°$	**6.** $\tan 27°$
7. $\cos 30°$	**8.** $\sin 9°$	**9.** $\cos 48°$
10. $\tan 56°$	**11.** $\sin 1°$	**12.** $\cos 73°$
13. $\cos 3°$	**14.** $\tan 45°$	**15.** $\sin 29°$

In Exercises 16–19, round your answers to the nearest tenth.

16. $m\angle B = 38°$ and $c = 37$ cm.
Find b.

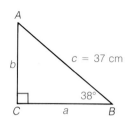

17. $m\angle B = 57°$ and $c = 24$ cm.
Find b.

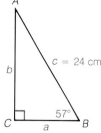

18. $m\angle D = 39°$ and $f = 42$ cm.
Find e.

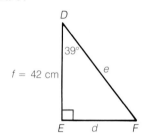

19. $m\angle D = 18°$ and $f = 16$ cm.
Find e.

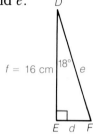

Extend and Apply

20. In right triangle STU, $ST = 10$ in. and
$m\angle T = 30°$. Which of the following is the best
estimate of b?

(A) 9 in.   (B) 5 in.   (C) 12 in.   (D) 3 in.

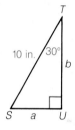

21. A 25-ft ladder rests against the top of a building.
The ladder makes an angle of 65° with the ground.
Find the height of the building to the nearest tenth.

Mixed Review

Find the area of each triangle, given the base and the height.

22. $b = 14$ cm, $h = 6$ cm

23. $b = 6.5$ m, $h = 8$ m

24. $b = 9$ m, $h = 4.6$ m

 Write each of the following using scientific notation.

25. 400

26. 53,000,000

27. 0.00092

Problem Solving

The figure at the right was made with
4 triangles, each having sides 1 in. long. If a
longer figure is made in the same way with
70 triangles, what will its perimeter be?

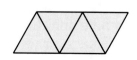

16-4 Solving Triangle Problems

Objective: Use the trigonometric ratios to find angle measures.

The figure at the right illustrates the concepts of **angle of elevation** and **angle of depression.** These angles are useful in many situations, and the trigonometric ratios can often be used to calculate their measures.

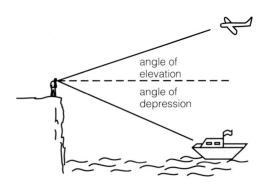

We will begin by considering the following question. Suppose we know that the cosine of ∠A is 0.5594. How can we find the measure of ∠A? A calculator or the table may be used as shown in Example 1.

Example   **1.** Suppose that cos A = 0.5594. Find $m\angle A$ to the nearest degree.

 Using a calculator, we have

0.5594 |2nd| |cos| |⎡ 55.985686 ⎤|

Thus, to the nearest degree, $m\angle A \approx 56°$.

Using the table, find the entry closest to 0.5594 in the *Cos* column. The closest entry is 0.5592. Then find the entry in the *Degrees* column opposite 0.5592. This is 56°.

In both cases, $m\angle A \approx 56°$.

Try This... Use a calculator or the table to find $m\angle A$ to the nearest degree.

   a. sin A = 0.6294

   b. cos A = 0.2755

   c. tan A = 1.4283

   d. tan A = 0.2127

We can now find the missing sides or angles in a right triangle whenever we know certain measurements. This is called *solving for a right triangle.* Example 2 illustrates the process.

Example 2. Find $m\angle K$ to the nearest degree.

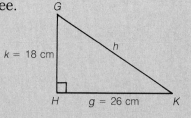

We know k, the side opposite $\angle K$, and g, the side adjacent to $\angle K$. Therefore, we will use the tangent ratio for $\angle K$.

$$\tan K = \frac{\text{opposite } \angle K}{\text{adjacent } \angle K} = \frac{18}{26} \approx 0.6923077$$

 Using a calculator, we have

0.6923077 | 2nd | | tan | | 34.695154 |

So, to the nearest degree, $m\angle K \approx 35°$.

Notice that in Example 2, we can use the fact that the measures of the angles of a triangle add up to 180° to find $m\angle G$.
$m\angle G = 180 - 90 - 35 = 55°$.

Try This... e. Find $m\angle G$ to the nearest degree. Then find $m\angle K$.

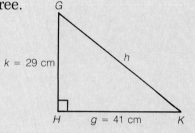

We will now look at some applications involving the angle of depression and angle of elevation.

Example 3. An air traffic controller determines that the angle of elevation to an airplane is 12°. The distance to the plane is 16 km. How high is the airplane to the nearest tenth?

$$\sin 12° = \frac{h}{16}$$
$$16 \times \sin 12° = h$$

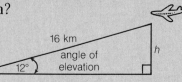

 Using a calculator, we have

16 | × | 12 | sin | | = | | 3.3265871 |

Thus, rounding to the nearest tenth, the altitude of the plane is about 3.3 km.

A lookout tower is 43 m tall. The angle of depression from the top of the tower to a forest fire is 5°. How far away from the base of the tower is the fire? (Hint: Find d in the figure.)

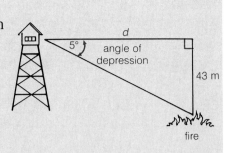

Discussion   Suppose the lengths of all three sides of a right triangle are known. A student claimed that any of the three trigonometric ratios could be used to find the measure of one of the acute angles. Do you agree or disagree?

Exercises

Practice

Use a calculator or the table to find $m\angle A$ to the nearest degree.

1. $\cos A = 0.0874$ **2.** $\sin A = 0.7982$ **3.** $\tan A = 0.2127$

4. $\tan A = 14.3005$ **5.** $\cos A = 0.3910$ **6.** $\sin A = 0.5300$

7. $\cos A = 0.7073$ **8.** $\tan A = 0.3444$ **9.** $\sin A = 0.9997$

10. $\sin A = 0.9614$ **11.** $\tan A = 1.0001$ **12.** $\cos A = 0.8989$

In Exercises 13–16, round your answers to the nearest degree.

13. Find $m\angle S$. Then find $m\angle T$.   **14.** Find $m\angle D$. Then find $m\angle F$.

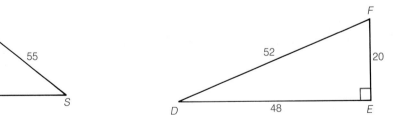

15. Find $m\angle X$. Then find $m\angle Z$.   **16.** Find $m\angle R$. Then find $m\angle M$.

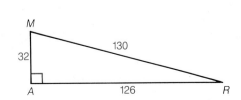

In Exercises 17–20, draw a picture before solving the problem. Round your answers to the nearest tenth.

17. The angle of elevation of an airplane is 9°. The distance to the plane is 21 km. How high is the plane?

18. A kite is flown with 210 m of string. The angle of elevation of the kite is 61°. How high is the kite?

19. The top of a lighthouse is 110 m above the level of the water. The angle of depression from the top of the lighthouse to a fishing boat is 18°. How far from the base of the lighthouse is the fishing boat?

20. An observation tower is 98 m tall. The angle of depression from the top of the tower to a historical marker is 23°. How far from the base of the tower is the marker?

Extend and Apply

21. Suppose tan X = 1.1. Which of the following is the best approximation for $m\angle X$?

   (A) 43°   (B) 27°   (C) 48°   (D) 86°

22. Find $m\angle B$ in the triangle at the right. Then find $m\angle A$. Round your answers to the nearest degree.

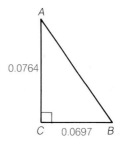

23. A flagpole casts a shadow 4.6 m long. The angle of elevation of the sun is 49°. How tall is the flagpole? Round your answer to the nearest tenth.

Use Mathematical Reasoning

24. In the figure at the right, $AC = CE = 4$, $EG = 6$, and $m\angle A = 27°$. Find BC, DE, and FG. Round your answers to the nearest tenth.

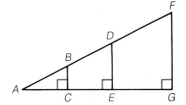

Mixed Review

Find the indicated ratios using a calculator and then using the table. Compare your answers.

25. sin 70°   **26.** cos 34°   **27.** tan 54°

28. cos 3°   **29.** tan 68°   **30.** sin 40°

◇ Solve and check.

31. $6(y + 5) = 10y + 14$ **32.** $5x - 32 = 2(x - 1)$

33. $7(9 - w) = 2w$

Compound Probability

If two or more results are unrelated, they are called **independent events.** The probability of two independent events occurring at the same time, called their **compound probability**, is just the product of their probabilities.

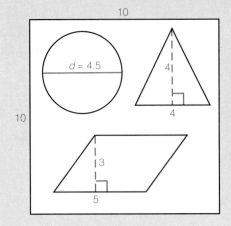

The dartboard at right contains geometric shapes. Suppose a dart has an equal probability of hitting anywhere on the dartboard. The probability of hitting a region is

$$\frac{\text{area of the region}}{\text{area of the dartboard}}$$

Sample Suppose two darts are thrown. What is the probability of hitting the parallelogram-shaped region with both throws?

The two throws are independent events. The area of the parallelogram is 15 and the area of the dartboard is 100, so the probability of hitting the parallelogram-shaped region with both throws is

$$\frac{15}{100} \times \frac{15}{100} = \frac{225}{10,000} = 0.0225$$

Problems

For Problems 1–3, suppose two darts are thrown.

1. What is the probability of hitting the triangular region and then the parallelogram-shaped region?

2. What is the probability of hitting the circular region both times?

3. What is the probability that both darts miss the three regions altogether?

4. What is the probability that a person was born on a Monday morning (between midnight and noon)?

5. Suppose the probability that Hoffman School is closed is $\frac{1}{18}$, and the probability that it snows in Hoffman is $\frac{1}{25}$. Can you calculate the probability that it snows in Hoffman and Hoffman School is closed?

6. An event has a 99% chance of occurring on any particular day. What is the probability that it will occur every day during a 30-day month?

Chapter 16 Review

Use $\triangle PQR$ for Problems 1–3.

1. Find the six trigonometric ratios for $\triangle PQR$.

2. Name the hypotenuse, the side opposite, and the side adjacent to $\angle P$.

16-1

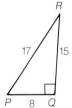

3. Find the ratio $\dfrac{\text{opposite } \angle R}{\text{hypotenuse}}$. Then use a calculator to find an approximation to 4 decimal places.

Use $\triangle RST$ to find the ratios in Problems 4–6. Then give an approximation to four decimal places.

16-2

4. $\sin R$

5. $\cos T$

6. $\tan R$

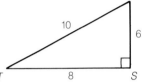

Find the indicated ratios using a calculator and then using the table. Compare your answers.

16-3

7. $\sin 38°$ 8. $\cos 80°$

9. $\tan 55°$ 10. $\sin 30°$

11. $m\angle C = 36°$ and $BC = 15$ cm. Find b to the nearest tenth.

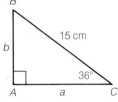

Use a calculator or a table to find $m\angle A$ to the nearest degree.

16-4

12. $\cos A = 0.9658$ 13. $\tan A = 0.7810$

14. Find $m\angle Z$ to the nearest degree. Then find $m\angle X$.

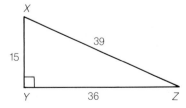

15. The angle of elevation of an airplane is $8°$. The distance to the plane is 18 km. How high is the plane? Round your answer to the nearest tenth.

Chapter 16 Test

Use $\triangle LMN$ for Problems 1–3.

1. Find the six trigonometric ratios for $\triangle LMN$.

2. Name the hypotenuse, the side opposite, and the side adjacent to $\angle L$.

3. Find the ratio $\dfrac{\text{adjacent } \angle M}{\text{hypotenuse}}$. Then give an approximation to 4 decimal places.

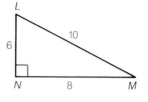

Use $\triangle PQR$ to find the ratios in Problems 4–6. Then use a calculator to find an approximation to four decimal places.

4. sin Q

5. cos Q

6. tan R

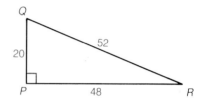

Find the indicated ratios using a calculator and then using the table. Compare your answers.

7. sin 87°

8. cos 21°

9. tan 18°

10. cos 57°

11. $m\angle Y = 15°$ and $WY = 10$ m. Find a to the nearest tenth.

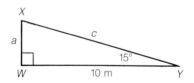

Use a calculator or the table to find $m\angle A$ to the nearest degree.

12. sin $A = 0.6562$

13. tan $A = 7.1101$

14. Find $m\angle L$ to the nearest degree. Then find $m\angle J$.

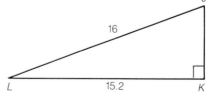

15. The angle of elevation of an airplane is 10°. The distance to the plane is 22 km. How high is the plane? Round your answer to the nearest tenth.

Cumulative Review Chapters 13–16

1. If $AO = 15$ and $BO = 9$, find CD.

13-2

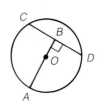

2. $\overline{RS}$ and $\overline{WS}$ are tangent segments. $OS = 13$ and $OR = 5$. Find RS, OW, and WS.

13-3

3. $\overline{AC}$ is a diameter. Identify and find the measure of each minor arc.

13-4

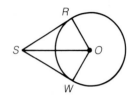

4. Name two rotations that rotate $\overline{AO}$ to $\overline{BO}$.

13-5

5. $m\angle A = 93°$, $m\overset{\frown}{BC} = 126°$, and $m\overset{\frown}{AD} = 78°$. Find $m\angle B$, $m\angle C$, and $m\angle D$.

13-6

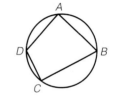

6. The circumference of a pipe is 66 cm. Find the diameter of the pipe to the nearest tenth.

13-8

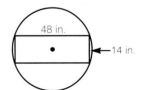

7. Find the area of the circle that is not covered by the inscribed rectangle. Use 3.14 for π.

13-9

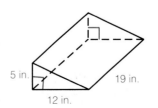

8. Find the lateral area and surface area of the right prism.

14-2

14-3 Find the lateral area and the surface area of each regular pyramid.

9.
12 cm
8 cm

10.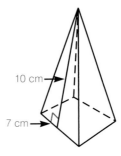
10 cm
7 cm

14-4 Find the volume of each right prism.

11.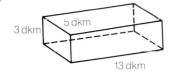
5 dkm
3 dkm
13 dkm

12.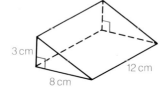
3 cm
8 cm
12 cm

14-5 **13.** Find the volume of this pyramid.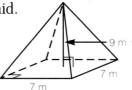
9 m
7 m
7 m

14-7 **14.** A cylindrical container has a radius of 7.5 in. and a height of 19 in. Find the area of the paper label, to the nearest square inch, needed to cover the lateral surface of the container.

14-8 **15.** Find the volume, to the nearest cubic centimeter, of this cone.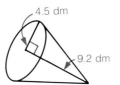
4.5 dm
9.2 dm

14-9 **16.** The diameter of a ball is 7 cm. Find its surface area and volume to the nearest tenth.

14-10 **17.** The prisms shown are similar. Find x and y.
y
7
12
3
6 x

18. In the prisms above, find the ratios of corresponding lengths, surface areas, and volumes. Express each ratio in lowest terms.

Find the distance between the points P and Q to the nearest tenth.

15-1

19. $P(3, -2)$ and $Q(-4, 3)$

20. $P(-3, 14)$ and $Q(-4, 3)$

Find the slope of each line containing the points T and V.

15-2

21. $T(-4, 3)$ and $V(3, -2)$

22. $T(-4, 2)$ and $V(6, 4)$

23. $T(5, 3)$ and $V(5, -3)$

24. $T(-4, -1)$ and $V(14, -1)$

25. Find the coordinates of the midpoint of $\overline{PQ}$ if the coordinates of the endpoints are $P(2, 7)$ and $Q(-1, 5)$.

15-3

Determine whether each ordered pair is a solution of the given equation.

15-4

26. $(2, 4)$, $3x + 4y = 20$

27. $(-1, -3)$, $4y - 2 = 14x$

28. $(0, -5)$, $6x = 2y + 10$

29. $(-8, 1)$, $2y - x + 3 = -20$

30. Find three solutions of the equation $x = \frac{1}{2}y - 5$.

31. Write an equation of a line that has y-intercept -2 and slope 6.

15-5

32. Find the slope and y-intercept of the line with equation $y = \frac{1}{2}x - 4$.

Determine whether each pair of lines is parallel, perpendicular, or neither.

33. $y = 3x + 1$ and $y = 3x - 1$

34. $y = \frac{1}{2}x + 1$ and $y = 2x - 1$

35. Determine whether the line containing $W(3, 6)$ and $R(-6, 3)$ and the line containing $K(3, 3)$ and $L(9, 9)$ are parallel, perpendicular, or neither.

36. In $\triangle RST$, find $\sin S$, $\cos S$, and $\tan S$ to four decimal places.

16-2

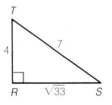

37. From 268 ft away, it is a 62° angle to the top of the United Nations building. How tall is the building?

16-3

38. Since going into a climb, an airplane has traveled 22,000 ft, but only 17,000 ft horizontally. At what angle has the plane been climbing?

16-4

Squares and Square Roots

Number n	Square n^2	Square root $\sqrt{n}$	Number n	Square n^2	Square root $\sqrt{n}$
1	1	1.000	51	2,601	7.141
2	4	1.414	52	2,704	7.211
3	9	1.732	53	2,809	7.280
4	16	2.000	54	2,916	7.348
5	25	2.236	55	3,025	7.416
6	36	2.449	56	3,136	7.483
7	49	2.646	57	3,249	7.550
8	64	2.828	58	3,364	7.616
9	81	3.000	59	3,481	7.681
10	100	3.162	60	3,600	7.746
11	121	3.317	61	3,721	7.810
12	144	3.464	62	3,844	7.874
13	169	3.606	63	3,969	7.937
14	196	3.742	64	4,096	8.000
15	225	3.873	65	4,225	8.062
16	256	4.000	66	4,356	8.124
17	289	4.123	67	4,489	8.185
18	324	4.243	68	4,624	8.246
19	361	4.359	69	4,761	8.307
20	400	4.472	70	4,900	8.367
21	441	4.583	71	5,041	8.426
22	484	4.690	72	5,184	8.485
23	529	4.796	73	5,329	8.544
24	576	4.899	74	5,476	8.602
25	625	5.000	75	5,625	8.660
26	676	5.099	76	5,776	8.718
27	729	5.196	77	5,929	8.775
28	784	5.292	78	6,084	8.832
29	841	5.385	79	6,241	8.888
30	900	5.477	80	6,400	8.944
31	961	5.568	81	6,561	9.000
32	1,024	5.657	82	6,724	9.055
33	1,089	5.745	83	6,889	9.110
34	1,156	5.831	84	7,056	9.165
35	1,225	5.916	85	7,225	9.220
36	1,296	6.000	86	7,396	9.274
37	1,369	6.083	87	7,569	9.327
38	1,444	6.164	88	7,744	9.381
39	1,521	6.245	89	7,921	9.434
40	1,600	6.325	90	8,100	9.487
41	1,681	6.403	91	8,281	9.539
42	1,764	6.481	92	8,464	9.592
43	1,849	6.557	93	8,649	9.644
44	1,936	6.633	94	8,836	9.695
45	2,025	6.708	95	9,025	9.747
46	2,116	6.782	96	9,216	9.798
47	2,209	6.856	97	9,409	9.849
48	2,304	6.928	98	9,604	9.899
49	2,401	7.000	99	9,801	9.950
50	2,500	7.071	100	10,000	10.000

Trigonometric Ratios

Degrees	Sin	Cos	Tan	Degrees	Sin	Cos	Tan
0°	0.0000	1.0000	0.0000				
1°	0.0175	0.9998	0.0175	46°	0.7193	0.6947	1.0355
2°	0.0349	0.9994	0.0349	47°	0.7314	0.6820	1.0724
3°	0.0523	0.9986	0.0524	48°	0.7431	0.6691	1.1106
4°	0.0698	0.9976	0.0699	49°	0.7547	0.6561	1.1504
5°	0.0872	0.9962	0.0875	50°	0.7660	0.6428	1.1918
6°	0.1045	0.9945	0.1051	51°	0.7771	0.6293	1.2349
7°	0.1219	0.9925	0.1228	52°	0.7880	0.6157	1.2799
8°	0.1392	0.9903	0.1405	53°	0.7986	0.6018	1.3270
9°	0.1564	0.9877	0.1584	54°	0.8090	0.5878	1.3764
10°	0.1736	0.9848	0.1763	55°	0.8192	0.5736	1.4281
11°	0.1908	0.9816	0.1944	56°	0.8290	0.5592	1.4826
12°	0.2079	0.9781	0.2126	57°	0.8387	0.5446	1.5399
13°	0.2250	0.9744	0.2309	58°	0.8480	0.5299	1.6003
14°	0.2419	0.9703	0.2493	59°	0.8572	0.5150	1.6643
15°	0.2588	0.9659	0.2679	60°	0.8660	0.5000	1.7321
16°	0.2756	0.9613	0.2867	61°	0.8746	0.4848	1.8040
17°	0.2924	0.9563	0.3057	62°	0.8829	0.4695	1.8807
18°	0.3090	0.9511	0.3249	63°	0.8910	0.4540	1.9626
19°	0.3256	0.9455	0.3443	64°	0.8988	0.4384	2.0503
20°	0.3420	0.9397	0.3640	65°	0.9063	0.4226	2.1445
21°	0.3584	0.9336	0.3839	66°	0.9135	0.4067	2.2460
22°	0.3746	0.9272	0.4040	67°	0.9205	0.3907	2.3559
23°	0.3907	0.9205	0.4245	68°	0.9272	0.3746	2.4751
24°	0.4067	0.9135	0.4452	69°	0.9336	0.3584	2.6051
25°	0.4226	0.9063	0.4663	70°	0.9397	0.3420	2.7475
26°	0.4384	0.8988	0.4877	71°	0.9455	0.3256	2.9042
27°	0.4540	0.8910	0.5095	72°	0.9511	0.3090	3.0777
28°	0.4695	0.8829	0.5317	73°	0.9563	0.2924	3.2709
29°	0.4848	0.8746	0.5543	74°	0.9613	0.2756	3.4874
30°	0.5000	0.8660	0.5774	75°	0.9659	0.2588	3.7321
31°	0.5150	0.8572	0.6009	76°	0.9703	0.2419	4.0108
32°	0.5299	0.8480	0.6249	77°	0.9744	0.2250	4.3315
33°	0.5446	0.8387	0.6494	78°	0.9781	0.2079	4.7046
34°	0.5592	0.8290	0.6745	79°	0.9816	0.1908	5.1446
35°	0.5736	0.8192	0.7002	80°	0.9848	0.1736	5.6713
36°	0.5878	0.8090	0.7265	81°	0.9877	0.1564	6.3138
37°	0.6018	0.7986	0.7536	82°	0.9903	0.1392	7.1154
38°	0.6157	0.7880	0.7813	83°	0.9925	0.1219	8.1443
39°	0.6293	0.7771	0.8098	84°	0.9945	0.1045	9.5144
40°	0.6428	0.7660	0.8391	85°	0.9962	0.0872	11.4301
41°	0.6561	0.7547	0.8693	86°	0.9976	0.0698	14.3007
42°	0.6691	0.7431	0.9004	87°	0.9986	0.0523	19.0811
43°	0.6820	0.7314	0.9325	88°	0.9994	0.0349	28.6363
44°	0.6947	0.7193	0.9657	89°	0.9998	0.0175	57.2900
45°	0.7071	0.7071	1.0000	90°	1.0000	0.0000	

Customary Units of Measure

Length

$$12 \text{ inches (in.)} = 1 \text{ foot (ft)}$$

$$\left.\begin{array}{r} 36 \text{ inches} \\ 3 \text{ feet} \end{array}\right\} = 1 \text{ yard (yd)}$$

$$\left.\begin{array}{r} 5280 \text{ feet} \\ 1760 \text{ yards} \end{array}\right\} = 1 \text{ mile (mi)}$$

Area

$$144 \text{ square inches (in.}^2) = 1 \text{ square foot (ft}^2)$$

$$9 \text{ square feet} = 1 \text{ square yard (yd}^2)$$

$$\left.\begin{array}{r} 43{,}560 \text{ square feet} \\ 4840 \text{ square yards} \end{array}\right\} = 1 \text{ acre (A)}$$

Volume

$$1728 \text{ cubic inches (in.}^3) = 1 \text{ cubic foot (ft}^3)$$

$$27 \text{ cubic feet} = 1 \text{ cubic yard (yd}^3)$$

Liquid Capacity

$$8 \text{ fluid ounces (fl oz)} = 1 \text{ cup (c)}$$

$$2 \text{ cups} = 1 \text{ pint (pt)}$$

$$2 \text{ pints} = 1 \text{ quart (qt)}$$

$$4 \text{ quarts} = 1 \text{ gallon (gal)}$$

Weight

$$16 \text{ ounces (oz)} = 1 \text{ pound (lb)}$$

$$2000 \text{ pounds} = 1 \text{ ton (t)}$$

Temperature:
Degrees Fahrenheit (°F)

$$32°F = \text{freezing point of water}$$

$$98.6°F = \text{normal body temperature}$$

$$212°F = \text{boiling point of water}$$

Time

$$60 \text{ seconds (s)} = 1 \text{ minute (min)}$$

$$60 \text{ minutes} = 1 \text{ hour (h)}$$

$$24 \text{ hours} = 1 \text{ day (d)}$$

$$7 \text{ days} = 1 \text{ week}$$

$$\left.\begin{array}{r} 365 \text{ days} \\ 52 \text{ weeks} \\ 12 \text{ months} \end{array}\right\} = 1 \text{ year}$$

$$10 \text{ years} = 1 \text{ decade}$$

$$100 \text{ years} = 1 \text{ century}$$

Metric Units of Measure

Length

$$10 \text{ millimeters (mm)} = 1 \text{ centimeter (cm)}$$
$$10 \text{ centimeters} = 1 \text{ decimeter (dm)}$$

$$\left.\begin{array}{r} 10 \text{ decimeters} \\ 100 \text{ centimeters} \\ 1000 \text{ millimeters} \end{array}\right\} = 1 \text{ meter (m)}$$

$$1000 \text{ meters} = 1 \text{ kilometer (km)}$$

Area

$$100 \text{ square millimeters (mm}^2) = 1 \text{ square centimeter (cm}^2)$$
$$10{,}000 \text{ square centimeters} = 1 \text{ square meter (m}^2)$$
$$10{,}000 \text{ square meters} = 1 \text{ hectare (ha)}$$

Volume

$$1000 \text{ cubic millimeters (mm}^3) = 1 \text{ cubic centimeter (cm}^3)$$
$$1{,}000{,}000 \text{ cubic centimeters} = 1 \text{ cubic meter (m}^3)$$

Liquid Capacity

$$1000 \text{ milliliters (mL)} = 1 \text{ liter (L)}$$
$$1000 \text{ cubic centimeters} = 1 \text{ liter}$$
$$1000 \text{ liters} = 1 \text{ kiloliter (kL)}$$

Mass

$$1000 \text{ milligrams (mg)} = 1 \text{ gram (g)}$$
$$1000 \text{ grams} = 1 \text{ kilogram (kg)}$$
$$1000 \text{ kilograms} = 1 \text{ metric ton (t)}$$

Temperature:
Degrees Celsius (°C)

$$0°C = \text{freezing point of water}$$
$$37°C = \text{normal body temperature}$$
$$100°C = \text{boiling point of water}$$

Metric Prefixes:

$$\text{milli} = \text{thousandth}$$
$$\text{centi} = \text{hundredth}$$
$$\text{deci} = \text{tenth}$$
$$\text{kilo} = \text{thousand}$$

Comparisons of Metric and Customary Measures

2.5 centimeters $\approx$ 1 inch   28 grams $\approx$ 1 ounce   0.95 liter $\approx$ 1 quart

90 centimeters $\approx$ 1 yard   0.5 kilogram $\approx$ 1 pound   3.8 liters $\approx$ 1 gallon

1.6 kilometers $\approx$ 1 mile

Postulates

Chapter 2   Organizing Geometry

Postulate 1   Given any two points, there is exactly one line containing the two points.   [p. 44]

Postulate 2   Given any three noncollinear points, there is exactly one plane containing them.   [p. 44]

Postulate 3   Any line contains at least two points. Any plane contains at least three noncollinear points. Space contains at least four noncoplanar points.   [p. 45]

Postulate 4   If two points lie in a plane, then the line containing them is in the plane.   [p. 46]

Postulate 5   If two planes intersect, then their intersection is a line.   [p. 46]

Chapter 3   Distance and Angle Measure

Postulate 6   **The Distance Postulate**   [p. 64]
To every pair of distinct points A and B, there corresponds a positive number, AB, called the distance between the points.

Postulate 7   **The Angle Measure Postulate**   [p. 84]
For each angle there is exactly one number n, called its measure, where $0° < n \leq 180°$.

Postulate 8   **The Angle Addition Postulate**   [p. 89]
For any nonstraight angle $\angle ABC$, if D is a point in its interior, then $m\angle ABD + m\angle DBC = m\angle ABC$.

For any straight angle $\angle ABC$, if D is a point not on $\angle ABC$, then $m\angle ABD + m\angle DBC = m\angle ABC$.

Chapter 4   Angle Relationships

Postulate 9   The reflection of a set of collinear points over a line is also a set of collinear points. (Reflections preserve collinearity.)   [p. 138]

Chapter 5   Triangles and Congruence

Postulate 10   **The SSS (Side-Side-Side) Postulate**   [p. 167]
If three sides of one triangle are congruent to three sides of another triangle, then the triangles are congruent.

Postulate 11   **The SAS (Side-Angle-Side) Postulate**   [p. 168]
Two triangles are congruent if two sides and the included angle of one triangle are congruent to two sides and the included angle of the other triangle.

Postulate 12   **The ASA (Angle-Side-Angle) Postulate**   [p. 169]
If two angles and the included side of a triangle are congruent to two angles and the included side of another triangle, then the triangles are congruent.

Postulate 13   **The AAS (Angle-Angle-Side) Postulate**   [p. 182]
If two angles and a non-included side of one triangle are congruent to two angles and the corresponding non-included side of another triangle, then the triangles are congruent.

Postulate 14   **The HL (Hypotenuse-Leg) Postulate**   [p. 183]
If the hypotenuse and a leg of one right triangle are congruent to the hypotenuse and a leg of another right triangle, then the two right triangles are congruent.

Chapter 7   Parallel Lines

Postulate 15   **The Parallel Postulate**   [p. 244]
Given a line ℓ and a point P not on ℓ, there is exactly one line that contains P and is parallel to ℓ.

Chapter 9   Similarity and Scale Change

Postulate 16   **The AAA Similarity Postulate**   [p. 326]
For any two triangles, if the corresponding angles are congruent, then the triangles are similar.

Chapter 12   Area of Polygons

Postulate 17   **The Area Postulate**   [p. 405]
For every polygonal region, there corresponds one positive number called its *area*. The number is dependent upon the given unit.

Postulate 18   The area of a polygonal region is the sum of the areas of the nonoverlapping regions that it contains.   [p. 406]

Postulate 19   Congruent regions have the same area.   [p. 406]

Postulate 20   The area of a rectangle is the product of the lengths of any two adjacent sides. $A = bh$, where A is the area, b is the base length, and h is the corresponding height.   [p. 411]

Chapter 14   Space Figures

Postulate 21   For every polyhedron, there corresponds a positive number called its *volume*.   [p. 500]

Postulate 22   The volume of a right rectangular prism is the product of its base area and its height. Volume = Base Area × height. $V = Bh$.   [p. 501]

Postulate 23   **Cavalieri's Principle**   [p. 502]
Suppose R and S are two space figures, and $\mathcal{M}$ is a plane. If every plane parallel to $\mathcal{M}$ and intersecting R or S also intersects the other space figure, and the resulting cross sections have the same area, then R and S have the same volume.

Theorems

Chapter 2   Organizing Geometry

Theorem 2.1   If two lines intersect, then they intersect in exactly one point.   [p. 49]

Theorem 2.2   If a line and a plane intersect and the line is not in the plane, then they intersect in a point.   [p. 49]

Theorem 2.3   If ℓ is a line and P is a point not on the line, then ℓ and P are contained in exactly one plane.   [p. 50]

Theorem 2.4   If two lines, ℓ and m, intersect, then the two lines are contained in exactly one plane.   [p. 54]

Chapter 3   Distance and Angle Measure

Theorem 3.1   Every segment has exactly one midpoint.   [p. 68]

Chapter 4   Angle Relationships

Theorem 4.1   If two angles are congruent, then their supplements are congruent.   [p. 111]

Theorem 4.2   If two angles are congruent, then their complements are congruent.   [p. 111]

Theorem 4.3   If two angles form a linear pair, then they are supplementary.   [p. 118]

Theorem 4.4   **The Vertical Angle Theorem**   [p. 123]
Vertical angles are congruent.

Theorem 4.5   If two lines form one right angle, then they form four right angles.   [p. 130]

Theorem 4.6   If A' and B' are the reflections of A and B over a line, then $AB = A'B'$.   [p. 138]

Theorem 4.7   If $\angle A'B'C'$ is the reflection of $\angle ABC$ over a line, then $m\angle ABC = m\angle A'B'C'$.   [p. 138]

Theorem 4.8   **The Perpendicular Bisector Theorem**   [p. 141]
The perpendicular bisector of a segment is the locus of points that are equidistant from the endpoints of the segment.

Chapter 6 Triangle Relationships

Theorem 6.1 **The Isosceles Triangle Theorem** [p. 193]
Two sides of a triangle are congruent whenever the angles opposite them are congruent.

Two angles of a triangle are congruent whenever the sides opposite them are congruent.

Theorem 6.2 **The Exterior Angle Theorem** [p. 201]
The measure of an exterior angle of a triangle is greater than the measure of either of its remote interior angles.

Theorem 6.3 **The Opposite Parts Theorem** [p. 204]
In any $\triangle ABC$, if $CA > CB$, then $m\angle B > m\angle A$.

Theorem 6.4 In any $\triangle ABC$, if $m\angle C > m\angle B$, then $AB > AC$. [p. 205]

Theorem 6.5 **The Triangle Inequality** [p. 208]
The sum of the lengths of any two sides of a triangle is greater than the length of the third side.

$AB + BC > AC$
$AC + CB > AB$
$AB + AC > BC$

Theorem 6.6 **The Hinge Theorem** [p. 213]
In $\triangle ABC$ and $\triangle DEF$, if $AB = DE$ and $AC = DF$, and $m\angle A < m\angle D$, then $BC < EF$.

Theorem 6.7 In $\triangle ABC$ and $\triangle DEF$, if $AB = DE$, $AC = DF$, and $BC < EF$, then $m\angle A < m\angle D$. [p. 214]

Theorem 6.8 The angle bisectors of a triangle are concurrent in a point that is equidistant from the sides of a triangle. [p. 217]

Theorem 6.9 The perpendicular bisectors of the sides of a triangle are concurrent in a point that is equidistant from the vertices of the triangle. [p. 218]

Theorem 6.10 **The Angle Bisector Theorem** [p. 221]
The bisector of an angle is the locus of points that are equidistant from the sides of the angle.

Chapter 7   Parallel Lines

Theorem 7.1   If two lines and a transversal form congruent alternate interior angles, then the lines are parallel.   [p. 239]

Theorem 7.2   If two lines and a transversal form congruent corresponding angles, then the lines are parallel.   [p. 240]

Theorem 7.3   If two lines and a transversal form congruent alternate exterior angles, then the lines are parallel.   [p. 240]

Theorem 7.4   If two lines and a transversal form supplementary interior angles on the same side of the transversal, then the lines are parallel.   [p. 240]

Theorem 7.5   If a transversal intersects two parallel lines, then the alternate interior angles are congruent.   [p. 245]

Theorem 7.6   If a transversal intersects two parallel lines, then the corresponding angles are congruent.   [p. 245]

Theorem 7.7   If a transversal intersects two parallel lines, then the alternate exterior angles are congruent.   [p. 245]

Theorem 7.8   If a transversal intersects two parallel lines, then the interior angles on the same side of the transversal are supplementary.   [p. 245]

Theorem 7.9   **The Angle Sum Theorem**   [p. 249]
The sum of the measures of the angles of a triangle is 180°.

Theorem 7.10   If two angles of one triangle are congruent to two angles of another triangle, then the third angles are congruent. [p. 250]

Theorem 7.11   The acute angles of a right triangle are complementary. [p. 251]

Theorem 7.12   The measure of an exterior angle of a triangle is the sum of the measures of the two remote interior angles.   [p. 251]

Theorem 7.13   Translations preserve collinearity of points, distance, and angle measure.   [p. 257]

Chapter 8   Quadrilaterals

Theorem 8.1   The sum of the measures of the angles of a quadrilateral is 360°.   [p. 265]

Theorem 8.2   A diagonal of a parallelogram determines two congruent triangles.   [p. 272]

Theorem 8.3   The opposite angles of a parallelogram are congruent. [p. 272]

Theorem 8.4   The opposite sides of a parallelogram are congruent. [p. 272]

Theorem 8.5   Consecutive angles of a parallelogram are supplementary. [p. 274]

Theorem 8.6   The diagonals of a parallelogram bisect each other.   [p. 275]

Theorem 8.7   If the opposite angles of a quadrilateral are congruent, then the quadrilateral is a parallelogram.   [p. 281]

Theorem 8.8   If the opposite sides of a quadrilateral are congruent, then the quadrilateral is a parallelogram.   [p. 281]

Theorem 8.9   If the diagonals of a quadrilateral bisect each other, then the quadrilateral is a parallelogram.   [p. 281]

Theorem 8.10   If two sides of a quadrilateral are both parallel and congruent, then the quadrilateral is a parallelogram. [p. 281]

Theorem 8.11   A parallelogram is a rhombus if its diagonals are perpendicular.   [p. 285]

Theorem 8.12   A parallelogram is a rectangle if its diagonals are congruent.   [p. 286]

Theorem 8.13   A parallelogram is a square if its diagonals are perpendicular and congruent.   [p. 286]

Theorem 8.14   A median of a trapezoid is parallel to the bases and is half as long as the sum of the lengths of the two bases. [p. 292]

Theorem 8.15   In an isosceles trapezoid, the angles of each pair of base angles are congruent.   [p. 293]

Theorem 8.16   The diagonals of an isosceles trapezoid are congruent. [p. 293]

Theorem 8.17   A midsegment connecting any two sides of a triangle is parallel to the third side and half as long as the third side. [p. 297]

Chapter 9   Similarity and Scale Change

Theorem 9.1   If $\dfrac{a}{b} = \dfrac{c}{d}$, then $ad = bc$.

If $ad = bc$ with $b \neq 0$ and $d \neq 0$, then $\dfrac{a}{b} = \dfrac{c}{d}$.   [p. 310]

Theorem 9.2   **The AA Similarity Theorem**   [p. 326]
For any two triangles, if two pairs of corresponding angles are congruent, then the triangles are similar.

Theorem 9.3	**The SAS Similarity Theorem** [p. 331] For any two triangles, if one pair of corresponding angles is congruent and the sides that include these angles are proportional, then the triangles are similar.

If $\dfrac{AB}{PQ} = \dfrac{AC}{PR}$, and $\angle A \cong \angle P$,

then $\triangle ABC \sim \triangle PQR$.

Theorem 9.4	**The SSS Similarity Theorem** [p. 332] For any two triangles, if all three pairs of corresponding sides are proportional, then the triangles are similar.

If $\dfrac{AB}{PQ} = \dfrac{AC}{PR} = \dfrac{BC}{QR}$,

then $\triangle ABC \sim \triangle PQR$.

Theorem 9.5	If a line is parallel to one side of a triangle and intersects the other sides at any point except a vertex, then a triangle similar to the given triangle is formed. [p. 335]

Chapter 10   Using Similar Triangles

Theorem 10.1	In a right triangle, the altitude to the hypotenuse forms two triangles, with both triangles similar to the right triangle and each one similar to the other. [p. 341]

Theorem 10.2	In a right triangle, the length of the altitude to the hypotenuse is the geometric mean of the lengths of the segments on the hypotenuse. [p. 343]

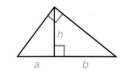

$$\frac{a}{h} = \frac{h}{b}$$

Theorem 10.3	**The Pythagorean Theorem** [p. 346] In a right triangle, the sum of the squares of the lengths of the two legs is equal to the square of the length of the hypotenuse.

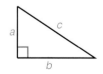

$$a^2 + b^2 = c^2$$

Theorem 10.4	**The Converse of the Pythagorean Theorem** [p. 352] If the lengths of the sides of a triangle are a, b, and c, and $a^2 + b^2 = c^2$, then the triangle is a right triangle with right angle opposite the longest side, whose length is c.

Theorem 10.5   **The Isosceles Right Triangle Theorem**   [p. 356]
For any isosceles right triangle, the hypotenuse is $\sqrt{2}$
times as long as either leg.

Theorem 10.6   **The 30°-60° Right Triangle Theorem**   [p. 360]
For any 30°-60° right triangle, the hypotenuse is twice as
long as the shorter leg, and the longer leg is $\sqrt{3}$ times as
long as the shorter leg.

Chapter 11   Polygons

Theorem 11.1   For any polygon with n sides, the number of diagonals is
$\frac{1}{2}n(n - 3)$.   [p. 375]

Theorem 11.2   The sum S of the angle measures of any convex polygon
with n sides is given by the formula $S = (n - 2)180$.
[p. 383]

Theorem 11.3   For a convex polygon, the sum of the measures of the
exterior angles, one at each vertex, is 360°.   [p. 384]

Theorem 11.4   The measure of an interior angle of a regular polygon with
n sides is $\frac{(n - 2)180}{n}$.   [p. 391]

Theorem 11.5   The measure of each exterior angle of a regular polygon
with n sides is $\frac{360}{n}$.   [p. 391]

Theorem 11.6   Any triangle or quadrilateral can be used to create a pure
tessellation.   [p. 397]

Chapter 12   Area of Polygons

Theorem 12.1   The area of a square is the square of the length of a side.
Area = s^2.   [p. 412]

Theorem 12.2   The area of a parallelogram is the product of a base and the
corresponding height. Area = bh.   [p. 417]

Theorem 12.3   The area of a triangle is one-half the product of a base and
the corresponding height. Area = $\frac{1}{2}bh$.   [p. 421]

Theorem 12.4   The area of a trapezoid is one-half the product of the sum
of its bases and the height. Area = $\frac{1}{2}h(b_1 + b_2)$.   [p. 426]

Chapter 13   Circles

Theorem 13.1   All radii of a circle are congruent.   [p. 438]

Theorem 13.2   A diameter of a circle is twice the length of a radius of the
circle. $d = 2r$.   [p. 438]

Theorem 13.3	If a radius of a circle is perpendicular to a chord, then it bisects the chord. [p. 441]
Theorem 13.4	If a radius of a circle bisects a chord that is not a diameter, then it is perpendicular to the chord. [p. 441]
Theorem 13.5	The perpendicular bisector of a chord of a circle contains the center of the circle. [p. 443]
Theorem 13.6	If two chords, $\overline{AD}$ and $\overline{BC}$, intersect at E in the interior of a circle, then $AE \cdot DE = BE \cdot CE$. [p. 445]
Theorem 13.7	A line is tangent to $\odot O$ at a point P whenever the line is perpendicular to the radius $\overline{OP}$ at P. [p. 447]
Theorem 13.8	The tangent segments from a point to a circle are congruent. [p. 447]
Theorem 13.9	**The Arc Addition Theorem** [p. 453] If C is on $\overset{\frown}{AB}$, then $m\overset{\frown}{AC} + m\overset{\frown}{CB} = m\overset{\frown}{AB}$.
Theorem 13.10	The measure of an inscribed angle is one-half the measure of its intercepted arc. [p. 462]
Theorem 13.11	An inscribed angle is a right angle if its intercepted arc is a semicircle. [p. 464]
Theorem 13.12	Inscribed angles are congruent if they intercept the same arc or congruent arcs. [p. 464]
Theorem 13.13	The measure of the angle formed by two chords intersecting inside a circle is equal to one-half the sum of the measures of the intercepted arcs. [p. 466]
Theorem 13.14	For all circles, the ratio $\dfrac{\text{circumference}}{\text{diameter}}$ is the same number. [p. 471]
Theorem 13.15	The area of a circle with radius r is πr^2. [p. 477]
Theorem 13.16	The area of a sector with radius r and arc (or central angle) measure n is $\dfrac{n}{360} \cdot \pi r^2$. [p. 478]

Chapter 14   Space Figures

Theorem 14.1	The lateral area of a right prism is the product of the perimeter p of a base and the height h of the prism. Lateral Area $= ph$. [p. 491]
Theorem 14.2	For any regular pyramid, where k is the slant height and p is the perimeter of the base, Surface Area $=$ Lateral Area $+$ Base Area $= \dfrac{1}{2}kp + B$. [p. 496]
Theorem 14.3	The volume of any prism is the product of its base area, B, and its height, h. $V = Bh$. [p. 502]

Theorem 14.4   The volume of any pyramid is one-third the product of its base area and its height. $V = \frac{1}{3}Bh$. [p. 505]

Theorem 14.5   The lateral area of a right circular cylinder with radius r and height h is $2\pi rh$. The surface area is
$2 \times$ Base Area + Lateral Area $= 2\pi r^2 + 2\pi rh$. [p. 514]

Theorem 14.6   For any right cone with base radius r and slant height s, the lateral area is πrs and the surface area is
Lateral Area + Base Area $= \pi rs + \pi r^2$. [p. 515]

Theorem 14.7   The volume of a cylinder is the product of its base area, B, and its height, h. $V = Bh$. [p. 518]

Theorem 14.8   The volume of a cone is one-third the product of its base area, B, and its height, h. $V = \frac{1}{3}Bh$. [p. 519]

Theorem 14.9   The surface area of a sphere with radius r is $4\pi r^2$. [p. 523]

Theorem 14.10   The volume of a sphere with radius r is $\frac{4}{3}\pi r^3$. [p. 523]

Theorem 14.11   If the ratio of two corresponding lengths of a pair of similar solids is $a{:}b$, then

(a) the ratio of any two corresponding lengths is $a{:}b$,
(b) the ratio of their surface areas is $a^2{:}b^2$, and
(c) the ratio of their volumes is $a^3{:}b^3$. [p. 529]

Chapter 15   Coordinate Geometry

Theorem 15.1   **The Distance Formula** [p. 537]
The distance (d) between any two points $A(x_1, y_1)$ and $B(x_2, y_2)$ is given by the formula
$d = \sqrt{(x_1 - x_2)^2 + (y_1 - y_2)^2}$.

Theorem 15.2   A horizontal line has a slope of 0. Slope for a vertical line *is not defined.* [p. 544]

Theorem 15.3   **The Midpoint Theorem** [p. 547]
The coordinates of the midpoint M of a segment with endpoints $A(x_1, y_1)$ and $B(x_2, y_2)$ are $\left(\dfrac{x_1 + x_2}{2}, \dfrac{y_1 + y_2}{2}\right)$.

Theorem 15.4   Two nonvertical lines are parallel if and only if they have the same slope. [p. 554]

Theorem 15.5   Two nonvertical lines are perpendicular if and only if the product of their slopes is -1. [p. 554]

Glossary

absolute value of a number   Its distance from 0 on the number line. (p. 15)

acute angle   An angle whose measure is less than 90°. (p. 12)

acute triangle   A triangle with all angles acute. (p. 150)

adjacent angles   Angles having a common side, a common vertex, and interiors that do not intersect. ∠1 and ∠2 are adjacent angles, as are ∠1 and ∠4, ∠4 and ∠3, ∠3 and ∠2. (p. 116)

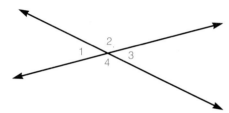

alternate exterior angles   ∠1 and ∠7 or ∠2 and ∠8 in the figure below. (p. 234)

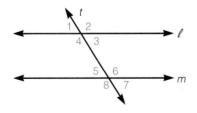

alternate interior angles   ∠4 and ∠6 or ∠3 and ∠5 in the figure above. (p. 234)

altitude of a figure   The height of the figure or the segment representing the height. (p. 341).

angle   A figure consisting of two rays with a common endpoint. (p. 11, 78)

angle bisector   A ray, located in the interior of an angle, whose endpoint is the vertex and that separates the angle into two angles with equal measure. (p. 91)

angle of depression   The angle formed by the line of sight and the horizontal when looking down as shown in the figure below. (p. 572)

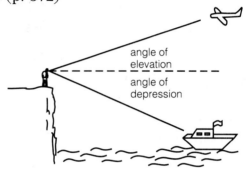

angle of elevation   The angle formed by the line of sight and the horizontal when looking up as shown in the figure above. (p. 572)

apothem   A segment from the center of a regular polygon perpendicular to any side. (p. 415)

arc   Central angles cut off part of a circle called an arc. (p. 451)

area   The number of square units contained by a plane figure. (p. 405)

associative property   The sum or product of three or more numbers is the same regardless of grouping.
$(a + b) + c = a + (b + c)$
$(a \times b) \times c = a \times (b \times c)$
(p. 188)

axis in the *x-y* plane   The horizontal number line (*x*-axis) or the vertical number line (*y*-axis). (p. 72)

axis of a cylinder   The segment joining the centers of the circular bases of a cylinder. (p. 510)

base of a cone   The circular region of the cone. (p. 510)

base of a parallelogram   Any side of the parallelogram; the corresponding altitude must form a 90° angle with the base. (p. 417)

base of a rectangle   Any side of the rectangle; the corresponding altitude must form a 90° angle with the base. (p. 411)

base of a triangle   Any side of a triangle; the corresponding altitude must form a 90° angle with the base. (p. 421)

bases of a cylinder   The circular regions of the cylinder. (p. 510)

bases of a prism   The congruent and parallel faces of the prism. (p. 489)

bases of a trapezoid   The parallel sides of the trapezoid. (p. 291)

bisector of a segment   The midpoint of the segment. (p. 17)

bisector of an angle   *See* angle bisector.

central angle   An angle whose vertex is the center of a circle. (p. 451)

centroid of a triangle   The point of concurrency of the medians of the triangle; the point at which the figure balances. (p. 222)

chord of a circle   A segment with both endpoints on the circle. (p. 437)

circle   The locus of all points in a plane a given distance from a given point in the plane. The given distance is called the *radius* of the circle. (p. 437)

circumcenter of a triangle   The point of concurrency of the perpendicular bisectors of the sides of the triangle. (p. 219)

circumference of a circle   The distance around the circle. (p. 470)

circumscribed circle   A circle is circumscribed about a polygon if every vertex of the polygon lies on the circle. (p. 467)

collinear points   Points that are on the same line. (p. 38)

commutative property   The sum or product of any two numbers is the same regardless of the order they are added or multiplied. (p. 188) $a + b = b + a, a \times b = b \times a$

complementary angles   Two angles whose measures total 90°. (p. 109)

concave polygon   A polygon for which there is a line containing a side that intersects the interior. The following are examples of concave polygons. (p. 371)

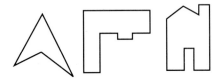

conclusion   In a conditional sentence, the part following "then" that tells the result of the statement. (p. 54)

concurrent lines Lines intersecting in a single point. (p. 217)

conditional sentence An "if-then" sentence consisting of a hypothesis (if) and a conclusion (then). (p. 54)

cone A space figure with a vertex and a circular base. (p. 510)

congruent figures Figures that have the same size and shape. (p. 158)

congruent segments Segments that have the same length. (p. 103)

converse of a conditional sentence A conditional sentence in which the hypothesis and the conclusion have been interchanged. (p. 56)

convex polygon A polygon for which no line containing a side intersects the interior of the polygon. The following are examples of convex polygons. (p. 371)

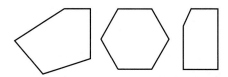

coordinate of a point on a line A number that corresponds to the point on the number line. (p. 63)

coordinate plane *See x-y* plane.

coordinates of a point on a plane An ordered pair of numbers that corresponds to the point. (p. 72)

coplanar Points or lines that are in the same plane. (p. 39)

corresponding angles $\angle 2$ and $\angle 6$, $\angle 3$ and $\angle 7$, $\angle 1$ and $\angle 5$, $\angle 4$ and $\angle 8$ in the figure. (p. 234)

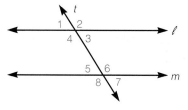

cosine ratio For any acute angle, $\angle A$, of a right triangle,
$$\cos A = \frac{\text{length of side adjacent to } \angle A}{\text{length of the hypotenuse}}$$
(p. 563)

counterexample An example that shows that a statement is not always true. (p. 26)

cross products The products ad and bc in the proportion $\frac{a}{b} = \frac{c}{d}$. (p. 310)

cross section of a solid The resulting figure when a solid is cut by a plane. (p. 37)

cube A hexahedron with square faces. (p. 486)

cylinder A space figure having congruent circular bases in a pair of parallel planes. (p. 510)

decagon A polygon with ten sides. (p. 370)

decahedron A polyhedron with ten faces. (p. 486)

deductive reasoning To reach a conclusion on the basis of assumptions and rules of logic. (p. 33)

degree A unit of angle measure. (p. 84)

diagonal of a polygon A segment that has vertices as endpoints and is not a side. (p. 374)

diameter of a circle A chord of a circle that contains the center of the circle. (p. 437)

distributive property   Connects multiplication and addition of numbers. $a(b + c) = ab + ac$ (p. 238)

dodecagon   A polygon with twelve sides. (p. 370)

dodecahedron   A polyhedron with twelve faces. (p. 486)

edges of a polyhedron   The sides of each face of the polyhedron. (p. 485)

equiangular triangle   A triangle with all three angles congruent. (p. 150)

equilateral triangle   A triangle with all three sides congruent. (p. 150)

exterior angle of a convex polygon An angle that forms a linear pair with an interior angle of the polygon. $\angle ACD$ is an exterior angle. (p. 384)

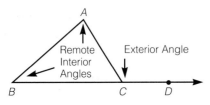

faces of a polyhedron   The polygonal regions that comprise the polyhedron. (p. 485)

geometric mean   For any positive numbers a and b, the geometric mean is the positive number x whenever $\dfrac{a}{x} = \dfrac{x}{b}$. (p. 342)

great circle   A circle that is the intersection of a sphere and a plane that contains the center of the sphere. (p. 511)

hemisphere   Half of a sphere. (p. 511)

heptagon   A polygon with seven sides. (p. 370)

heptahedron   A polyhedron with seven faces. (p. 486)

hexagon   A polygon with six sides. (p. 370)

hexagonal region   A hexagon together with its interior. (p. 405)

hexahedron   A polyhedron with six faces. (p. 486)

hypotenuse   The side opposite the right angle in a right triangle. (p. 183)

hypothesis   In a conditional sentence, the part following "if" that states the conditions necessary for the result to be true. (p. 54)

icosahedron   A polyhedron with twenty faces. (p. 486)

if-then sentence   *See* conditional sentence.

image   The figure resulting from a transformation. (p. 136, 255, 457)

incenter of a triangle   The point of concurrency of the angle bisectors. (p. 218)

included angle   An angle formed by two given sides of a triangle. (p. 149)

included side   A side common to two given angles of a triangle. (p. 149)

inductive reasoning   Reaching a conclusion on the basis of a series of examples. (p. 25)

inscribed angle   An angle whose vertex is on a circle and whose sides each intersect the circle in one other point. (p. 462)

inscribed polygon A polygon whose vertices all lie on a circle. (p. 467)

intercepted arc The arc that is in the interior of an inscribed angle. (p. 462)

interior angles ∠3, ∠4, ∠5, and ∠6 in the figure. (p. 234)

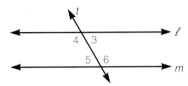

interior of an angle The shaded region in the figure. (p. 80)

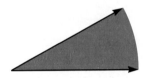

intersection The intersection of two figures is the set of points that is common to both figures. (p. 39)

isosceles trapezoid A trapezoid with two congruent non-parallel sides. (p. 293)

isosceles triangle A triangle with at least two congruent sides; the congruent sides are called the *legs* and the third side is called the *base*. (p. 150)

kite A quadrilateral with exactly two pairs of congruent consecutive sides. (p. 266)

lateral area of a prism The sum of the areas of the prism's lateral faces. (p. 490)

lateral edge of a prism The intersection of two lateral faces. (p. 489)

lateral faces of a prism The faces that are not bases. (p. 489)

lateral surface of a cylinder The region joining the circular bases of the cylinder. (p. 510)

legs of a right triangle The perpendicular sides of a right triangle. (p. 183)

legs of a trapezoid The two nonparallel sides of a trapezoid. (p. 291)

legs of an isosceles triangle The two congruent sides of the triangle. (p. 193)

line of symmetry A line of reflection is called a line of symmetry whenever the reflection image of every point in the figure about this line is also a point of the figure. (p. 155)

linear pair Adjacent angles whose noncommon sides are opposite rays. (p. 117)

locus The set of all the points that satisfy a given condition. (p. 40)

major arc An arc whose measure is greater than 180°, but less than 360°. (p. 451)

mean of a set of values The sum of all the values divided by the number of values; the average. (p. 395)

median of a set of values The middle value when all the values are arranged in order. (p. 395)

median of a trapezoid The segment whose endpoints are the midpoints of the legs. (p. 291)

median of a triangle A segment from a vertex to the midpoint of the opposite side. (p. 25)

midpoint A point halfway between the endpoints of a line segment. (p. 17, 68)

midsegment of a triangle Any segment connecting the midpoints of two sides of a triangle. (p. 297)

minor arc An arc whose measure is less than 180°. (p. 451)

mosaic *See* tessellation.

***n*-gon** A polygon with n sides. (p. 370)

nonagon A polygon with nine sides. (p. 370)

nonahedron A polyhedron with nine faces. (p. 486)

noncollinear points Points that are not all contained in the same line. (p. 38)

noncoplanar Points or lines that are not all contained in the same plane. (p. 39)

oblique prism A prism that is not a right prism. (p. 489)

obtuse angle An angle whose measure is greater than 90° but less than 180°. (p. 12).

obtuse triangle A triangle with one obtuse angle. (p. 150)

octagon A polygon with eight sides. (p. 370)

octahedron A polyhedron with eight faces. (p. 486)

opposite rays Given a line $\overleftrightarrow{CB}$ with A between C and B, the rays $\overrightarrow{AB}$, and $\overrightarrow{AC}$ are called opposite rays. (p. 67)

orientation of a figure The direction in which a set of points is ordered. (p. 135)

origin A point with coordinate zero; also the point at which the x-axis and y-axis intersect. (p. 63, 72)

pantograph An instrument used to reduce or enlarge a drawing. (p. 320)

parallel lines Lines in the same plane that do not intersect. (p. 227)

parallel rays Two rays contained in two parallel lines. (p. 227)

parallel segments Two segments contained in two parallel lines. (p. 227)

parallelogram A quadrilateral with two pairs of parallel sides. (p. 266)

pentagon A polygon with five sides. (p. 370)

pentagonal prism A prism with a pentagonal base. (p. 489)

pentahedron A polyhedron with five faces. (p. 486)

perimeter of a polygon The sum of the lengths of the sides; the distance around the polygon. (p. 379)

perpendicular bisector of a segment A line, segment, or plane that is perpendicular to a segment at its midpoint. (p. 131)

perpendicular lines Two lines that intersect to form a right angle. (p. 129)

perspective drawing A two-dimensional drawing of a three-dimensional object that appears to be three-dimensional. (p. 232)

Pi (π) A Greek letter that represents the ratio of the circumference of any circle to its diameter. $\pi = 3.1415926535897\ldots$ continuing endlessly. An acceptable approximation for π is 3.14. (p. 471)

polygon A figure formed by the line segments connecting three or more coplanar points. (p. 369)

polygonal region Any polygon and its interior. (p. 405)

polyhedron A three-dimensional space figure formed by polygonal regions. (p. 485)

postulate A statement that is assumed to be true. (p. 44)

prism A polyhedron that has two congruent faces that lie in parallel planes. The faces are called *bases*. (p. 489)

probability The likelihood that an event will occur. If an event can occur m ways out of a possible n equally likely ways, the probability of the event (E) is
$$P(E) = \frac{m}{n}.$$
(p. 351)

proportion A statement that two ratios are equal. (p. 310)

pure tessellation A tessellation using only one figure. (p. 396)

pyramid A space figure whose base is a polygonal region and whose lateral faces are triangles. (p. 494)

Pythagorean triple Three positive integers that can be the lengths of the sides of a right triangle. (p. 359)

quadrangular prism A prism with quadrangular bases. (p. 489)

quadrilateral A polygon with four sides. (p. 370)

radius of a circle *See* circle.

radius of a sphere *See* sphere.

ratio The comparison of two numbers by division. The ratio of a and b is $\frac{a}{b}$ or $a:b$. (p. 305)

ray The subset of a line $\overleftrightarrow{AB}$ that contains A and all the points on the same side of A as B. This is called ray AB or $\overrightarrow{AB}$. (p. 67)

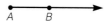

rectangle A parallelogram with four right angles. (p. 266)

rectangular prism A prism with a rectangular base. (p. 490)

rectangular region A rectangle together with its interior. (p. 405)

reflection A transformation that reflects a figure across a given line; a flip. (p. 22, 136)

reflection image The figure resulting from a reflection. (p. 136)

regular polygon A convex polygon that is both equilateral and equiangular. (p. 390)

regular pyramid A pyramid whose base is a regular polygon and whose lateral edges are congruent. (p. 494)

remote interior angles The angles in a triangle that are not adjacent to a given exterior angle of the triangle. $\angle A$ and $\angle B$ are remote interior angles. (p. 200)

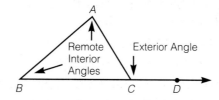

rhombus A parallelogram with four congruent sides. (p. 266)

right angle An angle whose measure is 90°. (p. 11)

right cone A cone whose segment joining the vertex to the center of the base is perpendicular to the base. (p. 510)

right cylinder A cylinder whose axis is perpendicular to the bases. (p. 510)

right prism A prism whose lateral faces are rectangles. (p. 489)

right triangle A triangle with one right angle. (p. 150)

rotation A transformation that rotates a figure about a given point; a turn. (p. 22, 457)

rotation image The figure resulting from a rotation. (p. 457)

scale of a map A statement that equates a given distance on a map with a given distance on the ground, such as 1 inch = 12 miles. (p. 312)

scalene triangle A triangle with no congruent sides. (p. 150)

secant A line that intersects a circle in two points. (p. 446)

sector of a circle A region bounded by two radii and an arc of a circle. (p. 477)

segment The set of points of a line containing two points and all the points between them. (p. 67)

semi-pure tessellation A tessellation created with more than one type of polygon. (p. 399)

semicircle An arc whose measure is 180°. (p. 451)

similar figures Figures with the same shape, but not necessarily the same size. (p. 315)

similar polygons Polygons with the same shape, but not necessarily the same size. (p. 387)

sine ratio For any acute angle, $\angle A$, of a right triangle,
$$\sin A = \frac{\text{length of side opposite } \angle A}{\text{length of the hypotenuse}}$$
(p. 563)

skew lines Two lines that do not lie in the same plane. (p. 228)

skew segments Two segments that lie on two skew lines. (p. 228)

slant height of a cone The distance along the surface of the cone from the vertex to the base. (p. 515)

slant height of a pyramid The length of the perpendicular from the vertex to an edge of the base. (p. 494)

slope of a line The ratio of the rise to run. Slope $m =$
$$\frac{\text{rise}}{\text{run}} = \frac{\text{change in } y\text{-coordinates}}{\text{change in } x\text{-coordinates}}$$
(p. 542)

sphere The locus of all points in space that are a given distance from a given point. The given distance is called the *radius* of the sphere. (p. 511)

square A rectangle with four congruent sides; a rhombus with four right angles. (p. 266)

straight angle An angle measuring 180° formed by opposite rays. (p. 11, 78)

supplementary angles Two angles whose measures total 180°. (p. 110)

symmetry A figure is symmetric with respect to a line m if the reflection of every point of the figure across line m is also a point of the same figure. (p. 155)

tangent   A line, coplanar with a circle, that intersects the circle in exactly one point. (p. 446)

tangent circles   Coplanar circles that are tangent to the same line at the same point. (p. 448)

tangent ratio   For any acute angle, $\angle A$, of a right triangle,

$$\tan A = \frac{\text{length of side opposite } \angle A}{\text{length of side adjacent to } \angle A}$$

(p. 563)

tessellation   An arrangement of figures that fill the plane and do not overlap or leave a gap. (p. 396)

tetrahedron   A polyhedron with four triangular faces. (p. 486)

theorem   A statement that can be proved. (p. 49)

tiling patterns   *See* tessellation.

transformation   Some type of uniform change to a figure such as a translation, rotation, or reflection. (p. 22)

translation   A transformation that moves all the points of a figure the same distance and direction; a slide. (p. 22, 255)

translation image   *See* image.

transversal   A line that intersects two or more coplanar lines in different points. (p. 234)

trapezoid   A quadrilateral with exactly one pair of parallel sides. (p. 266)

triangle   A figure consisting of three noncollinear points joined by segments. (p. 18)

triangular prism   A prism with triangular bases. (p. 489)

triangular pyramid   A pyramid whose base is a triangle. *See also* tetrahedron. (p. 494)

triangular region   A triangle together with its interior. (p. 405)

undecagon   A polygon with eleven sides. (p. 370)

undecahedron   A polygon with eleven faces. (p. 486)

vanishing lines   In a perspective drawing, lines drawn from a figure to the vanishing point. (p. 232)

vanishing point   In a perspective drawing, a point on the horizon at which lines appear to meet. (p. 232)

vertex of a cone   *See* cone.

vertex of a polygon   A point at which two sides intersect. (p. 369)

vertex of a polyhedron   A point at which three or more edges intersect. (p. 485)

vertex of a pyramid   The intersection of the lateral faces of a pyramid. (p. 494)

vertex of an angle   The common endpoint of the two rays forming the angle. (p. 11)

vertical angles   Two non-straight angles whose sides form two pairs of opposite rays. $\angle 1$ and $\angle 3$ are vertical angles; $\angle 2$ and $\angle 4$ are vertical angles. (p. 122)

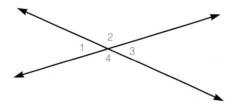

volume   The number of cubic units contained in a space figure. (p. 500)

x-axis The horizontal number line in the coordinate system. (p. 72)

x-coordinate The first number of the coordinates, which tells the distance right (positive) or left (negative) from the vertical axis. (p. 72)

x-y plane The plane determined by the x-axis and the y-axis. (p. 72)

y-axis The vertical number line in the coordinate system. (p. 72)

y-coordinate The second number of the coordinates, which tells the distance up (positive) or down (negative) from the horizontal axis. (p. 72)

Selected Answers

Chapter 1

Lesson 1-1

Try This **a.** They are both flat; a triangle has three sides, and a square has four sides. **b.** They are both triangular; a triangle is flat, and a pyramid is a space figure containing triangles. **c.**

Exercises **1.** octagon **3.** cylinder **5.** cone **7.** They are both square in shape; a square is flat, and a cube is a space figure containing squares. **9a.** 12 **b.** **21.** a cone **23.** a circle **27.** $\frac{3}{4}$ **29.** 1 **31.** 3.27 **33.** 11.46 **35.** 0.05 **37.** 0.8 **39.** 70% **41.** 10%

Lesson 1-2

Try This **a.** plane **b.** point **c.** line **d.** line **e.** $\overline{MN}$ or $\overline{NM}$ **f.** $\overleftrightarrow{WM}$ or $\overleftrightarrow{MW}$ **g.** $\overrightarrow{EF}$ or $\overrightarrow{FE}$ **h.** $\overrightarrow{KJ}$

Exercises **1.** plane **3.** line **5.** point **7.** plane **9.** line **11.** $\overline{TS}$ or $\overline{ST}$ **13.** $\overline{MN}$ or $\overline{NM}$ **15.** $\overrightarrow{DE}$ or $\overrightarrow{ED}$ **33.** 4 lines **35.** $\frac{1}{2}$ **37.** $1\frac{1}{4}$ **39.** $4\frac{7}{8}$ **41.** $\frac{1}{4}$ **43.** $\frac{9}{10}$ **45.** 0.67 **47.** 16 **49.** 16 **51.** 15 **53.** 100

Lesson 1-3

Try This **a.** $3\frac{10}{16}$ in., 9.2 cm **b.** $1\frac{13}{16}$ in., 4.6 cm **f.** neither **g.** right **h.** straight **i.** neither **j.** obtuse **k.** acute **l.** acute **m.** right

Exercises **1.** 3 in., 7.6 cm **3.** $2\frac{1}{8}$ in., 5.4 cm **5.** $2\frac{11}{16}$ in., 6.8 cm **7.** $2\frac{1}{2}$ in., 6.4 cm **9.** $3\frac{1}{16}$ in., 7.8 cm **21.** straight **23.** right **25.** acute **27.** $\frac{1}{4}$ in. **31a.** right **c.** straight **d.** acute **33.** 13 mi **35a.** 29 mi **b.** Bass to Clinton to Avon to Fline; 32 mi **c.** 84 min; 58 min **37.** point **39.** point **41.** 23 ft **43.** 60 mm **45.** 6000 m **47.** 2.5 mi

Algebra Connection, p. 15

1. -450 **3.** -4 **5.** $-2\frac{5}{8}$ **7.** 20 **9.** 2.4 **11.** $-1\frac{1}{2}$ **13.** 7 **15.** 12 **17.** 56 **19.** 100 **21.** $3\frac{3}{4}$

Lesson 1-4

Try This **a.** The copy should be as long as the original. **b.** Each smaller segment should have the same length and be about 3.5 cm long. **c.** The copy should fit precisely over $\triangle ABC$. **d.** The sides of the copy should fit precisely over $\overline{MN}$, $\overline{NT}$, and $\overline{TM}$.

Exercises **17.** Length should be $3\frac{1}{16}$ in. **19.** Length should be $4\frac{3}{16}$ in. **21.** Length should be $7\frac{1}{4}$ in. **25.** ray **27.** bisect **29.** 20

Lesson 1-5

Try This   **a.** The figure has been rotated.   **b.** 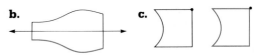   **c.**

Exercises   **1.** reflected (or rotated)   **3.** rotated   **5.**
13. The figure returns to its original position.
17. 8.2   **19.** 2.0   **21.** -14

Lesson 1-6

Try This   **a.** The medians in each triangle meet in one point.   **b.** *Super Mario Bros.*
c. 0

Exercises   **1.** The angles are the same size   **3.** The largest angle is opposite the longest side, and so on.   **5.** *Friday the 13th*   **7.** a yellow rose   **9.** a cheese pizza   **11.** 4 divided by 2   **13.** sugarless candy   **15.** They form a complete, 360° circle.   **19.** $\frac{1}{6}$ in.   **21.** reflection   **23.** reflection

Problem-Solving Strategies, p. 29

1. 55 cubes   **3.** 31

Chapter 1 Review

1. They are both flat; one has 6 sides, the other has 8.   **2a.** 11
b.   **c.**   **d.**   **3.** plane   **4.** line   **5.** $\overline{KS}$ or $\overline{SK}$ **6.** $\overrightarrow{AZ}$ or $\overrightarrow{ZA}$
12. right   **13.** obtuse   **14.** acute
17. translated   **20.** a triangle with two sides the same length

Chapter 2

Lesson 2-1

Try This   **a.** $180° + 180° + 180° = 540°$ **b.** All squares are quadrilaterals.   **c.** The number 346 is an even number.   **d.** John arrived at 8 a.m.

Exercises   **1.** All elephants have hair.   **3.** The grass will grow.   **5.** Pythagoras was a good mathematician.   **7.** Jody lives $6\frac{1}{2}$ miles from school.   **9.** Carrie did not order a yearbook.   **11.** 900°   **13.** Marty was not driving his car at 8:30 and did not run the red light.   **15.** The measures of the other two angles add up to 90°.   **19.** a table with three legs   **21.** a four-door sedan
23. $+27,325$ **25.** -5

Visualization, p. 37

1. C   **3.** $\overline{BC}, \overline{AC}, \overline{CE}$ **5.** B

Lesson 2-2

Try This   **a.** Collinear: T, P, Q. Noncollinear: R, S, Q; etc.   **b.** coplanar: B, H, F, G; etc. noncoplanar: B, H, F, A; etc.   **c.** Does not satisfy (1).   **d.** Does not satisfy (1).

Exercises   **1.** dot on computer screen, nail hole in wall, etc.   **3.** table top, sheet of glass, etc.   **5.** three birds on a wire   **7.** bottom tips of tripod   **9.** three birds on ground and the fourth in flight   **11.** true   **13.** false   **15.** true   **17.** V, R, S; T, P, V; etc.   **19.** W, V, T, P; W, Q, R, T; etc.   **21.** noncoplanar   **23.** coplanar   **25.** coplanar   **27.** Figure is not named.   **31.** point I **33.** point E **35.** Members are six-sided figures.   **37.** The angle measures of a square add up to 360°.   **39.** 201   **41.** -99

Algebra Connection, p. 43

1. -12 **3.** -26 **5.** -24 **7.** -58 **9.** 3   **11.** 5   **13.** 11   **15.** 19   **17.** -6 **19.** -27 **21.** -17 **23.** -33 **25.** 12   **27.** 17

Lesson 2-3

Try This   **a.** Postulate 3   **b.** Postulate 1   **c.** Postulate 2   **d.** 6   **e.** Postulate 4

Exercises   **1.** Postulate 2   **3.** Postulate 3   **5.** Postulate 1   **7.** Postulate 1   **9.** Postulate 2   **11.** Postulate 5   **13.** Postulate 5   **15.** Postulate 2   **17.** 4 lines   **19.** line   **21.** $\overrightarrow{DE}$ **25.** always   **27.** sometimes   **29.** C, F, G; etc.   **31.** point F **33.** 0   **35.** 17.1

Lesson 2-4

Try This   **a.** Theorem 2.2   **b.** Theorem 2.1   **c.** Theorem 2.2

Exercises   **1.** Theorem 2.2   **3.** Theorem 2.3   **5.** Theorem 2.3   **7.** Theorem 2.1   **9.** false   **11.** true   **13.** Theorem 2.1   **15.** sometimes   **17.** never   **19.** sometimes   **21.** true   **23.** false   **25.** point A **27.** acute   **29.** obtuse   **31.** -14 **33.** -13

Algebra Connection p. 53

1. -5 **3.** -9 **5.** -11 **7.** -21 **9.** 9   **11.** 12   **13.** 35   **15.** 30   **17.** 2   **19.** -4 **21.** -8 **23.** -13 **25.** -30 **27.** -38

Lesson 2-5

Try This   **a.** You go out in the rain. You will get wet.   **b.** A number is odd. It is not even.   **c.** Jamal works hard. He will succeed.   **d.** She earns enough money. She will go to the concert.   **e.** You go out in the rain. You will get wet.   **f.** If a figure has four sides, then it is a square.   **g.** If Hal can vote, then he is 18 or older.

Exercises   **1.** A number is odd. It is not divisible by 2. If a number is not divisible by 2, then it is odd.   **3.** Water freezes. The temperature is low. If the temperature is low, then water freezes.   **5.** Rita saves enough money. She will buy the dress. If she buys the dress, then Rita saves enough money.   **7.** You change the oil. Your car runs better. If your car runs better, then you change the oil.

9. The birds are flying south. Winter is coming. If winter is coming, then the birds are flying south. **11.** You practice. It is easy. If it is easy, then you practice. **13.** $3y + 2 = 17$. $y = 5$. If $y = 5$, then $3y + 2 = 17$. **15.** If you spare the rod, then you will spoil the child. **17.** If a figure is a square, then it is a rectangle. **19.** If you eat an apple a day, then you will keep the doctor away. **21.** true; false **23.** true; true **25.** false; true **27.** Theorem 2.1 **29.** Fran will have a good morning. **31.** -7 **33.** -4

Geometry in Communication, p. 58

1. You're out running a couple of miles a day. You're on your way to building a better body. If you're on your way to building a better body, then you're out running a couple of miles a day. **3.** Freshness is what counts. Shop at the Farm Mart. If you shop at the Farm Mart, then freshness is what counts.

Chapter 2 Review

1. All trout have gills. **2.** Jim did not go to the game. **3.** You can afford that stereo. **4.** $1260°$ **5.** R, S, P **6.** R, S, T; etc. **7.** T **8.** R **9.** D, L, E **10.** D, E, F, G; etc. **11.** D, E, F, K; etc. **12.** Not all circles fit definition **13.** Postulate 2 **14.** Postulate 1 **15.** Postulate 4 **16.** Postulate 5 **17.** one **18.** Theorem 2.2 **19.** Theorem 2.2 **20.** Theorem 2.1 **21.** Theorem 2.3 **22.** The rain falls. Weeds grow. If weeds grow, then the rain falls. **23.** Two lines are parallel. They do not intersect. If they do not intersect, then two lines are parallel. **24.** There is lightning. I am staying inside. If I am staying inside, then there is lightning. **25.** His sister helps him. Frank mows the lawn. If Frank mows the lawn, then his sister helps him.

Chapter 3

Lesson 3-1

Try This **a.** $-2, 0$ **b.** G **c.** 9 **d.** 5 **e.** 5 **f.** 4

Exercises **1.** 5 **3.** 1 **5.** -8 **7.** -2 **9.** -10 **11.** L **13.** U **15.** B **17.** Coded messages are fun **19.** 4 **21.** 1 **25.** $0°$ **27.** $-15°$ **29.** false **31.** true **33.** $A = 1, C = 7, E = 13, F = 16, G = 19, H = 22$ **35.** $A = 1, B = 7, D = 19, E = 25, F = 31, G = 37$ **37.** $B = 10\frac{1}{2}, C = 15, D = 19\frac{1}{2}, E = 24, F = 28\frac{1}{2}, H = 37\frac{1}{2}$ **39.** 0 or 14 **41.** If you are late, then it is 11:00. **43.** If the ice melts, then it is too hot. **45.** -3 **47.** 2

Lesson 3-2

Try This **a.** point D **b.** $\overrightarrow{EC}$ **c.** $\overrightarrow{CA}$ and $\overrightarrow{CE}$; $\overrightarrow{CB}$ and $\overrightarrow{CD}$ **d.** $\overline{BD}$ **e.** -2 **f.** 1.31

Exercises **1.** point T **3.** $\overrightarrow{TS}$ **5.** $\overrightarrow{SR}$ and $\overrightarrow{ST}$; $\overrightarrow{SP}$ and $\overrightarrow{SV}$; $\overrightarrow{TQ}$ and $\overrightarrow{TW}$ **7.** point S **9.** $AB = 4$; -2 **11.** $AB = 6$; -6.5 **13.** $AB = 17.8$; -1.18 **15.** 5 **17.** 5.7 **19.** 9 ft **21.** $\overline{MN}$ **23.** $10\frac{3}{4}$ **25.** never **27.** never **29.** sometimes **31.** sometimes **33.** always **35.** Theorem 2.2 **37.** Theorem 2.2 **39.** J, K, L; N, M, L **41.** point K **43.** 5.3 **45.** 123

Lesson 3-3

Try This **a–f.**

g. $T(-3, 2)$; $Q(-6, -1)$; $R(0, 6)$; $P(6, 3)$; $D(2, -6)$
h. $(-2, 3)$ **i.** $(4, -5)$ **j.** $(-6, 2)$

Exercises **13.** $A(3, 1)$, $B(-4, 2)$, $C(-6, 0)$, $D(-2, -4)$, $E(6, -6)$ **15.** $(3, -1)$ **17.** $(3, 2)$
19. $(1, -4)$ **21.** $(-3, 1)$ **23.** $(-3, -2)$ **25.** $(-1, 4)$ **27.** $(3, -4)$
29. $(1, 5)$ **33.** It does not move; $(0, -5)$ **39.** Points are on a line.
41. $\overrightarrow{PE}$ **43.** $\overrightarrow{PA}$ and $\overrightarrow{PC}$, $\overrightarrow{EA}$ and $\overrightarrow{EC}$ **45.** point P **47.** $RS = 19$; $-4\frac{1}{2}$
49. -14 **51.** -1 **53.** -3

Geometry in Communication, p. 77

1. 2500   **3.** 4   **5.** 150 students work 5 hours on weekends
7. 225   **9.** 75

Lesson 3-4

Try This **a.** $\angle DEF$ or $\angle FED$ or $\angle E$ **b.** $\angle FGH$ or $\angle HGF$; $\angle FGJ$ or $\angle JGF$; $\angle HGJ$ or
$\angle JGH$ **c.** vertex B, sides $\overrightarrow{BA}$ and $\overrightarrow{BC}$ **d.** $\angle 1$: vertex T, sides $\overrightarrow{TR}$ and $\overrightarrow{TS}$;
$\angle 2$: vertex T, sides $\overrightarrow{TQ}$ and $\overrightarrow{TS}$; $\angle RTQ$: vertex T, sides $\overrightarrow{TR}$ and $\overrightarrow{TQ}$ **e.** B
f. B, E, C

Exercises **1.** $\angle B$ or $\angle ABC$ or $\angle CBA$, vertex B, sides $\overrightarrow{BA}$ and $\overrightarrow{BC}$ **3.** $\angle QRS$ or $\angle SRQ$,
vertex R, sides $\overrightarrow{RQ}$ and $\overrightarrow{RS}$; $\angle QRT$ or $\angle TRQ$, vertex R, sides $\overrightarrow{RQ}$ and $\overrightarrow{RT}$;
$\angle SRT$ or $\angle TRS$, vertex R, sides $\overrightarrow{RT}$ and $\overrightarrow{RS}$ **5.** $\angle ABD$ or $\angle DBA$, vertex B,
sides $\overrightarrow{BA}$ and $\overrightarrow{BD}$; $\angle DBC$ or $\angle CBD$, vertex B, sides $\overrightarrow{BD}$ and $\overrightarrow{BC}$; $\angle ABC$ or
$\angle CBA$, vertex B, sides $\overrightarrow{BA}$ and $\overrightarrow{BC}$ **11.** interior: D; exterior: F; on $\angle ABC$: A,
B, C, E **13.** interior: W; exterior: T; on $\angle XYZ$: X, Y, Z **15.** $\angle A$, $\angle C$
19. 18   **23.** 7   **25.** 0

Algebra Connection, p. 82

1. 54	**3.** 33	**5.** 212	**7.** 234	**9.** 27	**11.** 426
13. 154	**15.** 168	**17.** 68	**19.** 35	**21.** 48	**23.** 113
25. 38	**27.** 67	**29.** 350	**31.** 6	**33.** 28	**35.** 9
37. 26	**39.** 58	**41.** 272	**43.** 133	**45.** 429	**47.** 238
49. 10	**51.** 310	**53.** 7	**55.** 14	**57.** 104	**59.** 70

Lesson 3-5

Try This **a.** $m\angle PQR = 60°$; $m\angle PQS = 100°$; $m\angle PQT = 160°$
b. $m\angle ABC = 20°$; $m\angle ABD = 150°$ **c.** $71°$ **d.** $127°$

Exercises **1.** $120°$ **3.** $40°$ **5.** $75°$ **7.** $140°$ **9.** $19°$ **15.** $80°$ **17.** $(3, -2)$
19. $(5, 1)$ **21.** 42

Algebra Connection, p. 88

1. 7 **3.** 16 **5.** 2 **7.** 4 **9.** 2 **11.** 9 **13.** 11 **15.** 1 **17.** 5

Lesson 3-6

Try This **a.** $m\angle BFD$ **b.** $m\angle FCE$ **c.** $m\angle BFD$ **d.** $83°$ **e.** 50 **f.** 13 **g.** $74°$
h. Both are $73°$. **i.** 8

Exercises **1.** $m\angle AFE$ **3.** $m\angle AFE$ **5.** $m\angle ACD$ **7.** $m\angle GPK$ **9.** $m\angle MPL$ **11.** $66°$
13. $45°$ **15.** $67°$ **17.** $83°$ **19.** 25 **21.** $m\angle DBC = m\angle ABD = 87°$
23. $75°, 75°$ **25.** 9 **27.** $90°$ **29.** $40°$ **31.** both $22°$ **33.** $28°$ **35.** $\overrightarrow{MT}$
and $\overrightarrow{ML}$; M **37.** S, P, U **39.** L, M, N, T **41.** 2

Lesson 3-7

Try This **a–b.** Answers will vary. Fold original angle to check. **c–d.** Answers will
vary. Match angles to check.

Exercises **13.** Tie the chalk to the end of the string and use it like a compass.
19.
21.
23.
29. $m\angle CMA$ (or $160°$) **31.** $m\angle AMF$ (or $20°$) **33.** $90°$ **35.** $10°$ **37.** 4
39. 12 **41.** 3

Chapter 3 Review

1. $-3, 5, 0$ **2.** D, L **3.** $7, 5, 3$ **4.** point Y **5.** $\overrightarrow{YX}$ and $\overrightarrow{YZ}$ **6.** $-4\frac{1}{2}$

7. $A(-5, 1)$; $B(4, 0)$; $C(-3, -5)$ **8.** $(-5, -1)$ **9.** $(-3, -3)$ **10.** $\angle CBA$
11. vertex: B; sides: $\overrightarrow{BA}$ and $\overrightarrow{BD}$ **12.** point C **13.** $40°$ **14.** $140°$ **15.** $115°$
16. $47°$ **17.** $154°$ **18.** $48°$ **19.** 7 **20.** Check by folding. **21.** Check by
matching.

Chapter 4

Lesson 4-1

Try This **a.** $\overline{FG} \cong \overline{HK}$ **b.** $\angle A \cong \angle E$

Exercises **1.** $\overline{HK} \cong \overline{JV}$ **3.** $\angle A \cong \angle C$, $\angle B \cong \angle D$ **5.** $48°$ **7.** $\overline{TV}$ **9.** $\angle S$
11. "$m\angle ABC = m\angle STR$" means that $\angle ABC$ and $\angle STR$ have the same
measure. "$\angle ABC \cong \angle STR$" means that the two angles are the same shape.
13. $\angle AGB$, $15°$ **15.** $\angle BGD$, $105°$ **17.** $\angle CGE$ and $\angle DGF$, $60°$ **19.** $\angle BAD$
and $\angle BCD$; $\angle BAE$ and $\angle BCE$; $\angle DAE$ and $\angle DCE$; $\angle ABE$ and $\angle CBE$; $\angle ADB$
and $\angle CDB$; $\angle AEB$, $\angle AED$, $\angle CEB$, and $\angle CED$ **21.** true **23.** Postulate 6—
the Distance Postulate **25.** 4 **27.** 5

Algebra Connection, p. 108

1. 1 **3.** 3 **5.** 4 **7.** 7 **9.** 4 **11.** 2 **13.** 19.4 **15.** 1

Lesson 4-2

Try This **a.** ∠1 and ∠2, ∠1 and ∠3, ∠2 and ∠4, ∠3 and ∠4 **b.** 45° **c.** 72°
d. 5° **e.** ∠1 and ∠2, ∠1 and ∠3, ∠2 and ∠4, ∠3 and ∠4 **f.** 142°
g. 23° **h.** 90° **i.** 35°, 125° **j.** 19°, 109°

Exercises **1.** ∠2 and ∠3, ∠1 and ∠4 **3.** ∠1 and ∠2, ∠1 and ∠4, ∠2 and ∠3, ∠3 and
∠4 **5.** 68° **7.** 34° **9.** 1° **11.** ∠1 and ∠2, ∠3 and ∠4, ∠2 and ∠3, ∠1
and ∠4 **13.** none **15.** 61° **17.** 37° **19.** 106° **21.** 12°, 102° **23.** 27°
25. 45° and 45° **27.** $m\angle AEB = m\angle CED$ **31.** ∠AEF and ∠BED, ∠AEB
and ∠DEF **37.** 77, 76.5 **39.** (3, −9) **41.** 3 **43.** 2

Problem-Solving Strategies, p. 115

1. 30°, 60° **3.** 80°, 100° **5.** no solution

Lesson 4-3

Try This **a.** no **b.** yes **c.** no **d.** no **e.** 157°

Exercises **1.** no **3.** no **5.** yes **7.** yes **9.** no **11.** no **13.** 48° **15.** always
17. yes **19.** yes **23.** If two angle measures add to 180° and they have the
same measure, they must measure 90° each, so they are both right angles.
25. ∠APB and ∠FPE, ∠DPE and ∠APF **27.** 149° **29.** $\vec{PB}$ and $\vec{PD}$ **31.** 25
33. −25 **35.** 11 **37.** 8.5

Algebra Connection, p. 121

1. 6 **3.** 4 **5.** 5 **7.** 6 **9.** $\frac{3}{4}$ **11.** 12 **13.** 9 **15.** 4 **17.** 1 **19.** 2
21. 2 **23.** $\frac{1}{6}$

Lesson 4-4

Try This **a.** $m\angle 2 = 82°$, $m\angle 3 = 98°$, $m\angle 4 = 82°$ **b.** $m\angle 1 = 10°$, $m\angle 3 = 129°$,
$m\angle 5 = 41°$, $m\angle 6 = 129°$ **c.** $m\angle PQR = 32° = m\angle TQS$

Exercises **1.** $m\angle 2 = 142°$, $m\angle 3 = 38°$ **3.** $m\angle 2 = 153°$, $m\angle 3 = 27°$, $m\angle 4 = 153°$
5. $m\angle 1 = 107°$, $m\angle 2 = 26°$, $m\angle 3 = 107°$, $m\angle 4 = 47°$ **7.** $m\angle 1 = 51°$,
$m\angle 3 = 49°$, $m\angle 4 = 52°$, $m\angle 5 = 51°$, $m\angle 6 = 28°$ **9.** $m\angle PQR =$
$m\angle TQS = 135°$ **11.** $m\angle TQS = m\angle GQR = 22°$ **13.** $m\angle ABC =$
$m\angle DBF = 85°$ **15.** $\angle 1 \cong \angle 4$ **19.** always true **21.** never true
23. sometimes true **25.** Since the angles are vertical, they are congruent. If
the measure of one angle is x, $x + x = 90$, so $x = 45$. Each measures 45°.
27. No, intersecting lines form vertical angles, so pairs of angles must have
the same measure. **29.** If two angles have the same measure, then they are
congruent. **31.** If two angles form a linear pair, then they are adjacent and
supplementary. **33.** ∠1 and ∠2, ∠2 and ∠3, ∠3 and ∠4, ∠1 and ∠4, ∠5
and ∠6 **35.** 14 **37.** 7

Lesson 4-5

Try This **a.** $\ell \perp q$ and $m \perp p$ **b-c.** Check by measuring the angles formed. They should measure 90°.

Exercises **1.** $q \perp t$ **7.** $m\angle 1 = m\angle 2 = m\angle 3 = m\angle 4 = 90°$, $m\angle 6 = 47°$, $m\angle 5 = m\angle 7 = 133°$ **15.** southwest to northeast **17.** north-northeast to south-southwest **21a.** 90° by vertical angles **b.** 90° by supplementary angles **c.** 90° by vertical angles **d.** All are right angles. **23.** $m\angle 3 = 42°$, $m\angle 4 = 39°$, $m\angle 5 = 99°$, $m\angle 6 = 42°$ **25.** PXT **27.** $P, R, S,$ and T **29.** All 7-sided figures are polygons. **31.** $\frac{1}{3}$

Visualization, p. 135

1. reverse **3.** same

Lesson 4-6

Try This **a.** $\overline{EF}$ **b.** $\overline{YZ}$ **c.** $\triangle XYZ$ **d.** Figures should match when folded over line ℓ. **e.** $\overline{AB} \cong \overline{A'B'}$, $\overline{AC} \cong \overline{A'C'}$, $\overline{BC} \cong \overline{B'C'}$; $\angle A \cong \angle A'$, $\angle B \cong \angle B'$, $\angle C \cong \angle C'$. Also, $\overline{CE} \cong \overline{C'E}$, $\overline{B'D} \cong \overline{BD}$, $\overline{DC'} \cong \overline{DC}$, $\overline{A'E} \cong \overline{AE}$, $\angle A'EC \cong \angle AEC'$, $\angle CDB' \cong \angle C'DB$, etc.

Exercises **1.** H **3.** $\overline{HJ}$ **5.** $\overline{JK}$ **7.** $\triangle HJK$ **13.** DEB **15.** $\angle DBC$ **17.** BF **25.** acute **27.** obtuse **29.** right **31.** 7 **33.** 2

Chapter 4 Review

1. $\overline{AB} \cong \overline{CD}$ **2.** $\overline{YZ} \cong \overline{EF}$ **3.** $\angle A \cong \angle B$, $\angle C \cong \angle E$ **4.** $\angle 2$ and $\angle 3$, $\angle 1$ and $\angle 4$ **5.** $\angle 1$ and $\angle 2$, $\angle 3$ and $\angle 4$, $\angle 2$ and $\angle 3$, $\angle 1$ and $\angle 4$ **6.** 72°, 18°, 108° **7.** $\angle CGD$ and $\angle DGF$, $\angle CGA$ and $\angle AGF$, $\angle CGA$ and $\angle CGE$, $\angle AGF$ and $\angle FGE$, $\angle EGF$ and $\angle EGC$, $\angle DGE$ and $\angle DGA$ **8.** $m\angle 1 = 68°$, $m\angle 2 = 83°$, $m\angle 3 = 68°$, $m\angle 4 = 29°$ **9.** $m\angle ACB = m\angle ECD = 150°$ **10.** $\perp$ **11.** not $\perp$ **12.** not $\perp$ **14.** $\overline{XW}$ **15.** $\angle YXZ$ **16.** $\angle XWY$ **17.** $\triangle ZYW$ **18.** X **19.** $\overline{CA}$

Cumulative Review Chapters 1–4

1a. 14 **2.** $\overleftrightarrow{PQ}$ **3.** $\overline{RV}$ **4.** acute **5.** obtuse **6.** right **8.** 56 has 2 as a factor. **10.** line **11.** plane **12.** If the tomato is ripe, then it is red. **13.** $C = -5, R = 0, T = 6$ **14.** N, K **15.** 9, 9, 15 **16.** -1 **17a.** $(4, 5)$ **b.** $(-4, -5)$ **18a.** $(4, -9)$ **b.** $(-1, -5)$ **19.** $\angle SNC, \angle CNS, \angle N$ **20.** vertex: N, sides: $\overrightarrow{NS}$ and $\overrightarrow{NC}$ **22.** 129° **23.** 52°, 142° **24.** $\angle PQT$ and $\angle RQT$; $\angle PQS$ and $\angle SQR$ **25.** 86°, 66°, 28°, 66°

Chapter 5

Lesson 5-1

Try This **a.** $\angle A$ **b.** $\overline{AC}$ **c.** $\angle C$ **d.** $\overline{AC}$ **e.** equiangular **f.** obtuse **g.** right **h.** isosceles **i.** equilateral **j.** scalene

Exercises   **1.** $\angle Q$ **3.** $\overline{QS}$ **5.** $\overline{RS}$ **7.** acute   **9.** obtuse   **11.** obtuse   **13.** right
15. scalene   **17.** isosceles   **19.** isosceles   **21.** equilateral   **23.** $\triangle QMN$
25. $\triangle TQS$ **27.** $\overline{YZ}$ **29.** $\overline{XY}$ **35.** not possible   **37.** 6 ways   **41.** 42

Lesson 5-2

Try This   **a.** yes   **b.** yes   **c.** no   **d.** 2 lines   **e.** 1 line   **f.** no lines   **g.** 3 lines

Exercises   **1.** 1 line   **3.** no lines   **5.** 1 line   **7.** 4 lines   **17.** 7   **19.** 9.25

Lesson 5-3

Try This   **a.** $\angle A \cong \angle D$, $\angle B \cong \angle E$, $\angle C \cong \angle F$, $\overline{AB} \cong \overline{DE}$, $\overline{BC} \cong \overline{EF}$, $\overline{AC} \cong \overline{DF}$
b. $\angle N \cong \angle P$, $\angle O \cong \angle Q$, $\angle M \cong \angle R$, $\overline{MO} \cong \overline{QR}$, $\overline{NM} \cong \overline{PR}$, $\overline{NO} \cong \overline{PQ}$

Exercises   **1.** $\angle A \cong \angle R$, $\angle B \cong \angle S$, $\angle C \cong \angle T$, $\overline{AB} \cong \overline{RS}$, $\overline{BC} \cong \overline{ST}$, $\overline{AC} \cong \overline{RT}$
3. $\angle X \cong \angle U$, $\angle Y \cong \angle V$, $\angle Z \cong \angle W$, $\overline{XY} \cong \overline{UV}$, $\overline{YZ} \cong \overline{VW}$, $\overline{XZ} \cong \overline{UW}$
5. $\angle A \cong \angle A$, $\angle B \cong \angle B$, $\angle C \cong \angle C$, $\overline{AB} \cong \overline{AB}$, $\overline{BC} \cong \overline{BC}$, $\overline{AC} \cong \overline{AC}$
7. $\angle A$ and $\angle F$, $\angle B$ and $\angle E$, $\angle C$ and $\angle D$, $\overline{AC}$ and $\overline{DF}$, $\overline{AB}$ and $\overline{EF}$, $\overline{BC}$ and $\overline{DE}$ **9.** $\angle M$ and $\angle Q$, $\angle O$ and $\angle S$, $\angle N$ and $\angle P$, $\overline{MN}$ and $\overline{PQ}$, $\overline{OM}$ and $\overline{SQ}$, $\overline{ON}$ and $\overline{SP}$ **11.** $\angle M$ and $\angle U$, $\angle L$ and $\angle T$, $\angle K$ and $\angle V$, $\overline{ML}$ and $\overline{UT}$, $\overline{MK}$ and $\overline{UV}$, $\overline{LK}$ and $\overline{TV}$ **13.** $\angle G$ and $\angle Y$, $\angle K$ and $\angle X$, $\angle H$ and $\angle Z$, $\overline{GK}$ and $\overline{YX}$, $\overline{GH}$ and $\overline{YZ}$, $\overline{KH}$ and $\overline{XZ}$ **15.** $\triangle TPU \cong \triangle SRQ$
19. $\triangle ABC \cong \triangle GHK$ **21.** 0   **23.** 1   **25.** 2   **27.** 12   **29.** 12

Lesson 5-4

Try This   **a.** $\angle A \cong \angle C$, $\angle AEB \cong \angle CED$, $\overline{AB} \cong \overline{CD}$, $\overline{AE} \cong \overline{CE}$, $\overline{AB} \perp \overline{BD}$, $\overline{CD} \perp \overline{BD}$

Exercises   **1.** $\overline{LC} \cong \overline{TB}$, $\overline{VC} \cong \overline{AT}$, $\overline{VL} \cong \overline{AB}$ **3.** $\overline{GQ} \perp \overline{RP}$, $\angle R \cong \angle P$, $\overline{GQ} \cong \overline{GQ}$
5. $\overline{AB} \cong \overline{CD}$, $\angle A \cong \angle C$, $\angle ABD \cong \angle CDB$, $\overline{BD} \cong \overline{BD}$ **7.** $\overline{AB} \perp \overline{BD}$, $\overline{BC} \perp \overline{AD}$, $\overline{AC} \cong \overline{BD}$, $\overline{BC} \cong \overline{BC}$

Lesson 5-5

Try This   **a.** yes   **b.** no   **c.** yes   **d.** no   **e.** yes   **f.** no   **g.** no (by SAS)   **h.** yes
i. no   **j.** no postulate   **k.** SAS   **l.** ASA   **m.** SSS

Exercises   **1.** ASA   **3.** SAS   **5.** SAS or SSS   **7.** SSS   **9.** SAS   **11.** none   **13.** ASA
15. SAS or SSS   **17.** none   **19.** $\angle Q \cong \angle T$, $\angle R \cong \angle V$ **21.** $\overline{QP} \cong \overline{TS}$
25. By the ASA Postulate, the faces of the pyramid are congruent.
27. Because M is the midpoint, two triangles are formed. They are congruent by the SAS Postulate. You can say this of the left and right triangles and the upper and lower triangles.   **29.** yes   **31.** 2   **33.** 2   **35.** 0   **37.** 100

Lesson 5-6

Exercises   **1.** $\triangle TRS$; SAS   **3.** $\triangle MLG$; SAS   **5.** $\triangle AEB$; SSS   **7.** $\triangle GKM$; SAS
9. $\triangle FRS$; SSS   **11.** $\triangle TQP$; SSS   **13.** Yes   **17.** $\triangle ABC \cong \triangle DEF$ **19.** $\angle 1$ and $\angle ABC$ and $\angle 2$ and $\angle DEF$ are supplementary because they both form linear pairs. Because $\angle 1 \cong \angle 2$, $\angle ABC \cong \angle DEF$. $\overline{AB} \cong \overline{DE}$ and $\overline{EF} \cong \overline{BC}$. By SAS, $\triangle ABC \cong \triangle DEF$.   **21.** $\angle CAB \cong \angle DAB$, $\angle CBA \cong \angle DBA$, $\overline{AD} \cong \overline{AC}$
23. $\angle CAB \cong \angle DAB$, $\angle ABC \cong \angle ABD$, $\angle ACB \cong \angle ADB$, $\overline{AB} \cong \overline{AB}$, $\overline{AC} \cong \overline{AD}$, $\overline{BC} \cong \overline{BD}$ **25.** $\angle D \cong \angle U$, $\angle E \cong \angle V$, $\angle F \cong \angle W$, $\overline{DE} \cong \overline{UV}$, $\overline{EF} \cong \overline{VW}$, $\overline{FD} \cong \overline{WU}$ **27.** isosceles   **29.** 14

Lesson 5-7

Try This   **a.** $\triangle SRQ \cong \triangle TRQ$ by SSS, $\angle S \cong \angle T$ as corresponding parts   **b.** She constructed $\triangle DCX$, which is congruent to $\triangle BAX$ by SAS, so $\overline{DC} \cong \overline{BA}$. She measured DC, the length of the corresponding congruent part.   **c.** $\triangle ABT \cong \triangle GFT$ by SSS, so $\angle B \cong \angle F$ as corresponding parts.

Exercises   **1.** SAS, corresponding parts   **3.** SAS, corresponding parts   **5.** SSS, corresponding parts   **7.** ASA, corresponding parts   **9.** The two halves of the kite are congruent by SSS, so $\angle 1$ and $\angle 2$ are congruent.   **11.** $\triangle KGP \cong \triangle TPR$ by ASA, so $\overline{GP} \cong \overline{PR}$ and P is the midpoint of $\overline{GR}$ **13.** $\triangle GKP \cong \triangle RTP$ by ASA, so $\overline{GP} \cong \overline{PR}$ and P is the midpoint of $\overline{GR}$ **15.** $\triangle CAB \cong \triangle DAB$ by ASA, so $\overline{CB} \cong \overline{DB}$.   **17.** $\angle K \cong \angle K$, so $\triangle NKH \cong \triangle GKL$ by ASA.   **19.** $120°$ **21.** $60°$ **23.** 3

Lesson 5-8

Try This   **a.** ASA   **b.** AAS   **c.** HL   **d.** AAS   **e.** no postulate   **f.** SSS   **g.** by AAS and corresponding parts   **h.** By HL, $\triangle RMP \cong \triangle TNP$. Thus, $\overline{MP} \cong \overline{NP}$ as corresponding parts of congruent triangles.

Exercises   **1.** SAS   **3.** AAS   **5.** SSS   **7.** HL   **9.** ASA   **11.** AAS   **13.** AAS, corresponding parts   **15.** HL, corresponding parts   **17.** AAS   **19.** 100 ft high   **21.** $\triangle MQY \cong \triangle SQX$ by AAS, corresponding parts, so $\overline{MQ} \cong \overline{SQ}$ and $\overline{YQ} \cong \overline{QX}$. Q is the midpoint of $\overline{MS}$ and of $\overline{YX}$, so MS and YX bisect each other.   **23.** $\triangle GAE$: right; $\triangle ABE$: obtuse; $\triangle BCE$: acute; $\triangle CDE$: equiangular; $\triangle AEC$: acute   **25.** $\angle EAB$ ($\angle BAE$)   **27.** 35

Algebra Connection, p. 188

1. yes   **3.** yes   **5.** yes   **7.** 0

Chapter 5 Review

1. obtuse   **2.** acute   **3.** right   **4.** isosceles   **5.** equilateral   **6.** scalene   **7.** $\angle G \cong \angle P$, $\angle K \cong \angle T$, $\angle L \cong \angle L$, $\overline{GK} \cong \overline{PT}$, $\overline{KL} \cong \overline{TL}$, $\overline{GL} \cong \overline{PL}$ **8.** 2   **9.** 1   **10.** 0   **11.** $\overline{TG} \cong \overline{FG}$, $\angle TGX \cong \angle FGX$, $\overrightarrow{GX} \perp \overline{TF}$, $\angle GXT$ and $\angle GXF$ are right angles   **13.** no postulate   **14.** SAS   **15.** SSS   **16.** SSS   **17.** SAS   **18.** ASA   **19.** HL   **20.** AAS   **21.** no postulate   **22.** $\triangle MES$, SAS   **23.** ASA, corresponding parts   **24.** SAS, corresponding parts

Chapter 6

Lesson 6-1

Try This   **a.** $m\angle A = 73°$ **b.** $PQ = 16$ **c.** 11   **d.** 65

Exercises   **1.** $70°$ **3.** 62   **5.** 15   **7.** 23   **9.** 35   **11.** 12   **13** $40°$ **15.** 8   **17.** All are $60°$.   **19.** The bisector of the vertex angle (or perpendicular bisector of the base) is a line of symmetry in every isosceles triangle.   **21.** $m\angle A = m\angle C = 45°$, so $AB = BC$. The surveyor should measure $\overline{BC}$.   **23.** AAS   **25.** SSS   **27.** $\overline{AC} \cong \overline{BC}$, $\overline{CD} \cong \overline{CD}$, $\overline{DA} \cong \overline{DB}$, $\angle ACD \cong \angle BCD$, $\angle CDA \cong \angle CDB$, $\angle A \cong \angle B$ **29.** <   **31.** <   **33.** >

Lesson 6-2

Try This **a.** $\angle G$ and $\angle H$ are remote interior angles for both $\angle KFG$ and $\angle JFH$.
b. $m\angle SRT < m\angle RTQ$

Exercises **1.** $\angle ABC, \angle BAC$ **3.** $\angle TQR, \angle TRQ$; $\angle QRT, \angle QTR$ **5.** $m\angle AFC > m\angle D$
7. $m\angle K < m\angle KSH$ **9.** $>$ **11.** $<$ **13.** $>$ **15.** $>$ **17.** always
19. sometimes **21.** $\angle DCE > \angle ABD$ **23.** 12 **25.** 17

Lesson 6-3

Try This **a.** $m\angle T > m\angle S$ **b.** $\angle C, \angle A, \angle B$ **c.** $\overline{NM}, \overline{LN}, \overline{LM}$

Exercises **1.** $m\angle K > m\angle H$ **3.** $\angle T, \angle A, \angle F$ **5.** $\angle F, \angle E, \angle D$ **7.** $\overline{XY}, \overline{WX}, \overline{WY}$
11. $NK > MK$ **13.** $ES > FS$ **15.** If $AB = CD$, then $\overline{AB} \cong \overline{CD}$. **17.** 7
19. 1

Lesson 6-4

Try This **a.** $GK + KH > GH$; $GH + HK > GK$; $HG + GK > HK$ **b.** yes **c.** no
d. no

Exercises **1.** $RS + ST > RT$; $ST + RT > RS$; $RT + SR > ST$ **3.** $QM + QN > NM$;
$QN + NM > QM$; $NM + QM > QN$ **5.** yes **7.** yes **9.** yes **11.** yes
13. yes **15.** $7 < x < 11$ **17.** $9 < x < 17$ **19.** $0.2 < x < 0.8$ **21.** yes
23. 20 mi $< distance <$ 78 mi **25.** The difference of the lengths of two
sides of a triangle is less than the length of the third side. **27.** $ML > MK$
29. $ML > LK > MK$ **31.** $\angle L, \angle M$ **33.** 49

Lesson 6-5

Try This **a.** $EF < ST$ **b.** $KM < XZ$ **c.** $m\angle P < m\angle T$

Exercises **1.** $HK < EF$ **3.** $TS < RP$ **5.** $m\angle D > m\angle A$ **7.** $m\angle X < m\angle T$ **9.** $<$
11. $<$ **13.** $>$ **15.** The circle drawn by the compass with the larger angle
is the larger circle. **17.** $AB + BC > AC$; $BC + AC > AB$; $AC + AB > BC$
19. no **21.** yes **23.** yes **25.** $(-2, -3)$ **27.** $(1, 2)$ **29.** $(-1, 1)$

Lesson 6-6

Exercises **13.** circumcenter **15.** Build it at the incenter. **19.** never **21.** always
23. $EF > RS$ **25.** $AB = 8$; -5 **27.** 10

Geometry in Science, p. 222

5. intersection of diagonals or symmetry lines **7.** intersection of any
2 diameters

Chapter 6 Review

1. 14 **2.** 49 **3.** $\angle BCA, \angle A$ **4.** $\angle JKL, \angle K$ **5.** $m\angle HTJ > m\angle V$
6. $m\angle C > m\angle B$ **7.** $\angle E, \angle M, \angle S$ **8.** $\overline{RS}, \overline{RT}, \overline{ST}$ **9.** $LJ < JK + KL$,
$JK < KL + LJ$, $KL < LJ + JK$ **10.** yes **11.** no **12.** $TS > PM$
13. $m\angle P > m\angle V$

Chapter 7

Lesson 7-1

Try This   **b.** $\overline{CG}, \overline{BF}, \overline{AE}$ **c.** $\overline{BF}, \overline{AE}, \overline{CB}, \overline{DA}$

Exercises   **1.** $\overline{SR}, \overline{LK}, \overline{HJ}$ **3.** $\overline{QR}, \overline{JK}, \overline{HL}$ **5.** $\overline{QJ}, \overline{RK}, \overline{HJ}, \overline{LK}$ **7.** $\overline{PH}, \overline{SL}, \overline{PQ}, \overline{SR}$
13. $\overline{PS}, \overline{PT}$ **15.** $\overline{PQ}, \overline{PR}$ **17.** $\overline{QR}$ and $\overline{TS}, \overline{RS}$ and $\overline{QT}$ **19.** parallel
21. parallel   **23.** parallel   **25.** skew, intersecting, and parallel   **27.** false
29. true   **31.** always   **33.** sometimes   **35.** $SR + RT > ST, SR + ST > RT,$
$ST + RT > SR$ **37.** yes   **39.** $TS < RP$ **41.** $m\angle LNP > m\angle L$ **43.** 5
45. 6

Geometry in Art, p. 233

3. above and to the left

Lesson 7-2

Try This   **a.** alternate interior angles   **b.** interior angles   **c.** corresponding angles
d. alternate exterior angles   **e.** corresponding angles; m and n; p
f. alternate interior angles; p and q; ℓ

Exercises   **1.** corresponding angles   **3.** interior angles   **5.** corresponding
angles   **7.** alternate exterior angles   **9.** alternate interior angles
11. alternate exterior angles   **13.** alternate exterior angles; $m, n; \ell$
15. alternate interior angles; $\ell, p; n$ **17.** interior angles; $\ell, p; m$
19. alternate exterior angles; $\ell, p; n$ **21.** alternate interior angles; $\ell, p; m$
23. $\overrightarrow{SP}$ **25.** $\overline{SQ}$ **27.** $\overrightarrow{SP}$ **31.** $\overline{VS}, \overline{ZW}, \overline{YX}$ **33.** $\overline{ST}, \overline{UV}, \overline{TX}, \overline{UY}$ **37.** yes

---

Algebra Connection, p. 238

1. $3x + 24$ **3.** $30a - 20$ **5.** 9   **7.** 1   **9.** 5

Lesson 7-3

Try This   **a.** $n \parallel q; p \parallel s$ **b.** Interior angles are both right angles, so they are
supplementary.

Exercises   **1.** $m \parallel r; n \parallel q$ **3.** Corresponding angles are congruent.   **5.** Alternate
interior angles are congruent.   **7.** Interior angles are supplementary.
9. Since alternate interior angles are congruent, the table top is parallel to
the ground.   **11.** $\overline{AB} \parallel \overline{CD}, \overline{AC} \parallel \overline{BD}$ **13.** alternate interior angles
15. alternate exterior angles   **17.** 4   **19.** 2   **21.** 5

Lesson 7-4

Try This   **a.** $m\angle 4 = m\angle 5 = m\angle 8 = 130°; m\angle 2 = m\angle 3 = m\angle 6 = m\angle 7 = 50°$
b. $\angle ABC \cong \angle BCD$, and $\angle BAD \cong \angle ADC$; $\angle BAC$ and $\angle ACD$ are
supplementary; $\angle ABD$ and $\angle BDC$ are supplementary   **c.** $\angle XYZ \cong \angle UVW,$
$\angle UVY \cong \angle ZYV$

Exercises   **1.** $m\angle 2 = m\angle 5 = m\angle 6 = 135°; m\angle 1 = m\angle 3 = m\angle 4 = m\angle 7 = 45°$
3. $\angle 1 \cong \angle 2, \angle 3 \cong \angle 4; \angle 1$ and $\angle 5, \angle 5$ and $\angle 2, \angle 3$ and $\angle 6, \angle 6$ and $\angle 4$

are supplementary   **5.** $\angle 1 \cong \angle 4$, $\angle 3 \cong \angle 5$ **7.** $\angle BAE \cong \angle CDE$, $\angle ABE \cong \angle ECD$, $\angle AEB \cong \angle CED$; $\angle ABD$ and $\angle CDB$, $\angle AEB$ and $\angle BED$, $\angle BED$ and $\angle DEC$ are supplementary   **11.** Since the steps are parallel, these angles are all corresponding angles and are therefore congruent.   **13.** yes   **15.** yes   **17.** 3   **19.** 6

Lesson 7-5

Try This **a.** $130°$ **b.** $m\angle D = 30°$, $m\angle E = 20°$ **c.** $\angle 1 \cong \angle 2$ **d.** $m\angle BCD = 55°$, $m\angle B = 70°$, $m\angle CDA = 125°$, $m\angle DCA = 35°$, $m\angle A = 20°$ **e.** $m\angle A = 65°$, $m\angle ACB = 45°$

Exercises **1.** $25°$ **3.** $70°$ **5.** $86°$ **7.** $m\angle E = 100°$, $m\angle F = 32°$ **9.** $m\angle B = 48°$, $m\angle C = 88°$ **11.** $m\angle 1 = 43°$, $m\angle 2 = 47°$ **15.** $m\angle 1 = m\angle 2 = 52.5°$, $m\angle 3 = 82.5°$, $m\angle 4 = 97.5°$ **17.** $x = 18$; $m\angle B = 23°$, $m\angle ACD = 39°$, $m\angle ACB = 141°$ **19.** $m\angle 1 = 36°$, $m\angle 2 = 108°$, $m\angle 3 = 36°$ **21.** yes **23.** Corresponding angles are congruent. **25.** Interior angles are supplementary. **27.** 6

Lesson 7-6

Try This **a.** $\overline{FG}$ **b.** point H **d.** $\angle E$ **e.** $\overline{AC}$

Exercises **1.** point F **3.** $\overline{IJ}$ **5.** $\triangle JHI$ **13.** $\overline{TU}$ **15.** $\angle TWV$ **17.** $\angle PQS$ **19.** yes **21.** translation and reflection; yes **25.** preserve **27.** $m\angle F = 109°$ **29.** $m\angle 1 = 48°$, $m\angle 2 = 42°$ **31.** 1

Chapter 7 Review

1. $\overline{BC}, \overline{AD}, \overline{EH}$ **2.** $\overline{BF}, \overline{CG}, \overline{EF}, \overline{HG}$ **4.** interior angles; $\overleftrightarrow{AD}, \overleftrightarrow{HE}, \overleftrightarrow{BG}$ **5.** alternate interior angles; $\overleftrightarrow{AD}, \overleftrightarrow{HE}, \overleftrightarrow{CG}$ **6.** corresponding angles; $\overleftrightarrow{BG}, \overleftrightarrow{CF}, \overleftrightarrow{AD}$ **7.** $m \parallel p$; $n \parallel q$ **8.** Alternate exterior angles are congruent. **9.** $m\angle EAB = 30°$, $m\angle EBA = 50°$ **10.** $m\angle BED = 40°$, $m\angle ECD = 65°$, $\angle BEC$ and $\angle C$, $\angle DEA$ and $\angle D$, $\angle BEC$ and $\angle AEC$, $\angle AED$ and $\angle DEB$ **11.** $m\angle J = 15°$, $m\angle 2 = 85°$, $m\angle 1 = 95°$ **12.** $m\angle EFG = 65°$, $m\angle E = 25°$, $m\angle EFD = 115°$ **13.** $\overline{UV}$ **14.** $\angle T$

Chapter 8

Lesson 8-1

Try This **a.** $70°$ **b.** $115°$ **c.** rhombus **d.** parallelogram **e.** square

Exercises **1.** $110°$ **3.** $31°$ **5.** $28°$
7. quadrilateral **9.** parallelogram **11.** kite
13. rectangle **15.** square **17.** trapezoid
19. sometimes **21.** never **23.** sometimes
25. never **27.** trapezoid **29.** quadrilateral
31. trapezoid **33.** rhombus **35.** It becomes a rhombus.
37. Theorem 7.1 **39.** $90°$ **41.** $51°$
43. $141°$ **45.** Theorem 7.12 **47.** false
49. $\frac{4}{7}$

Lesson 8-2

Try This **a.** $m\angle B = 153°$, $m\angle C = 27°$, $m\angle D = 153°$ **b.** $m\angle R = m\angle P = 60°$, $m\angle S = m\angle Q = 120°$ **c.** 36 **d.** $DE = GF = 20\frac{3}{4}$, $DG = EF = 13\frac{1}{4}$ **e.** $m\angle D = m\angle F = 68°$, $m\angle E = 112°$ **f.** $AE = EC = 7$, $BE = ED = 4$

Exercises **1.** $m\angle A = 70°$, $m\angle B = m\angle D = 110°$ **3.** $m\angle J = 71°$, $m\angle L = m\angle M = 109°$ **5.** $m\angle R = m\angle P = 67°$, $m\angle S = m\angle Q = 113°$ **7.** $m\angle W = m\angle Y = 110°$, $m\angle X = m\angle Z = 70°$ **9.** 48 **11.** 56.4 **13.** $UT = 17.7$, $RU = ST = 76.9$ **15.** $JM = KL = 9$, $ML = JK = 23$ **17.** $AC = 28$, $ED = 38$ **19.** $AB = 18$, $EB = 8$, $AC = 22$ **21.** $\triangle AED \cong \triangle CEB$, $\triangle AEB \cong \triangle CED$, $\triangle ADC \cong \triangle CBA$, $\triangle ABD \cong \triangle CDB$ **23.** no **25.** $\triangle BAD \cong \triangle DCB$, $\triangle CFD \cong \triangle AEB$, $\triangle CFB \cong \triangle AED$ **27.** false **29.** false **31.** false **33.** $\triangle PQR \cong \triangle STU$ by SSS **35.** 70

Lesson 8-3

Try This **a.** yes, Theorem 8.9 **b.** yes, Theorem 8.8 **c.** Not necessarily. Only 1 pair of opposite sides is congruent.

Exercises **1.** yes; Theorem 8.7 **3.** no; opposite sides not congruent **5.** no; opposite angles not congruent **7.** no; opposite angles not congruent **9.** no; diagonals not bisected **11.** yes; Theorem 8.9 **13.** no; consecutive angles not supplementary **15.** no; not enough information **21.** Diagonals are of constant length and bisect one another where the handles meet. This assures a parallelogram. **23.** $AEFD$ is a parallelogram. **25.** kite **27.** skew

Lesson 8-4

Try This **a.** rectangle **b.** rhombus **c.** parallelogram **d.** It would be a square. **e.** 127° **f.** $\triangle HEJ$, $\triangle HJG$, $\triangle FEJ$, $\triangle FJG$ **g.** It is a right triangle. **h.** It is a square. **i.** 14

Exercises **1.** square **3.** rhombus **5.** rectangle **7.** 6 **9.** 10 **11.** $\triangle CBA$, $\triangle CDA$, $\triangle ABC$ **13.** 90° **15.** 12 **17.** $\triangle QRS$, $\triangle SPQ$, $\triangle RQP$ **19.** isosceles **21.** 90° **23.** Diagonals are congruent and perpendicular; all sides are congruent; all angles are right angles. **25.** All sides are congruent; the diagonals are perpendicular. **29.** $AFED$ is a rhombus **31.** no postulate **33.** EDC; ASA **35.** 9

Lesson 8-5

Try This **a.** bases: $\overline{AB}$ and $\overline{DC}$; legs: $\overline{AD}$ and $\overline{BC}$; median: $\overline{EF}$ **b.** bases: $\overline{AB}$ and $\overline{DC}$; legs: $\overline{AD}$ and $\overline{BC}$; median: $\overline{PS}$ **c.** bases: $\overline{AD}$ and $\overline{BC}$; legs: $\overline{AB}$ and $\overline{DC}$; median: $\overline{EG}$ **d.** bases: $\overline{AD}$ and $\overline{BC}$; legs: $\overline{AB}$ and $\overline{DC}$; median: $\overline{QR}$ **e.** 14.5 **f.** $DE = 30$, $m\angle GFE = 57°$ **g.** $AB = 21$, $DC = 19$, $RS = 20$ **h.** $m\angle A = m\angle D = 95°$, $m\angle B = 85°$ **i.** $m\angle W = 125°$, $m\angle Y = 55°$, $m\angle ZXY = 95°$, $m\angle XZY = 30°$ **j.** $\triangle WXZ \cong \triangle XWY$ by SAS or SSS; $\triangle WZY \cong \triangle XYZ$ by SAS or SSS; $\triangle WPZ \cong \triangle XPY$ by AAS **k.** $m\angle BAC = m\angle ABC = 43°$, $m\angle ACB = m\angle DCE = 94°$, $m\angle ACD = m\angle BCE = 86°$, $m\angle CBE = 76°$, $m\angle DAC = 76°$, $m\angle BDE = m\angle AED = 43°$, $m\angle ADB = 18°$

Exercises **1.** bases: $\overline{AC}$ and $\overline{GE}$; legs: $\overline{AG}$ and $\overline{CE}$; median: $\overline{HD}$ **3.** bases: $\overline{LM}$ and $\overline{KN}$; legs: $\overline{LK}$ and $\overline{MN}$; median: $\overline{PR}$ **5.** 7.8 **7.** 9.5, 3.5, 6.5 **9.** 11, 15, 19 **11.** $m\angle C = 82°$, $m\angle A = m\angle B = 98°$ **13.** $m\angle Y = 80°$, $m\angle W = m\angle X = 100°$ **15.** $m\angle BEC = 62°$, $m\angle AED = 62°$, $m\angle AEB = 118°$; $m\angle BDC = m\angle ACD = 31°$; $m\angle BAC = m\angle ABD = 31°$; $m\angle DAC = 80°$; $m\angle ADB = m\angle BCA = 38°$; $\triangle DAB \cong \triangle CBA$; $\triangle AED \cong \triangle BEC$; $\triangle DAC \cong \triangle CBD$ **19.** $11\frac{1}{2}$ in. **21.** $4\frac{3}{4}$ ft, $5\frac{1}{2}$ ft, $6\frac{1}{4}$ ft **23.** Opposite angles of any isosceles trapezoid are supplementary **25.** adjacent angles **27.** adjacent angles **29.** alternate exterior angles **31.** 16 **33.** 22

Chapter 8 Review

1. opposite angles: $\angle K$ and $\angle M$, or $\angle L$ and $\angle N$; consecutive angles: $\angle K$ and $\angle L$, $\angle L$ and $\angle M$, $\angle M$ and $\angle N$, or $\angle N$ and $\angle K$ **2.** trapezoid **3.** 109° **4.** $m\angle G = 105°$, $m\angle F = m\angle D = 75°$, $DE = 16$, $GF = 16$, $DG = EF = 15$ **5.** $LJ = 4.6$, $KM = 6$ **6.** yes **7.** yes **8.** not necessarily **9.** yes **10.** rhombus **11.** rectangle **12.** $\triangle CED$, $\triangle CEB$, $\triangle AED$ **13.** 9 **14.** 8.5 **15.** $\triangle JLM$, $\triangle LJK$, $\triangle KML$ **16.** $EF = 11$, $DC = 15$ **17.** $m\angle RQV = m\angle QRV = 25°$; $m\angle QVR = 130°$; $m\angle QVT = m\angle RVS = 50°$; $m\angle RTS = m\angle QST = 25°$; $m\angle QTV = m\angle RSV = 52°$; $m\angle TQV = 78°$

Cumulative Review Chapters 5–8

1. equilateral **2.** scalene **3.** isosceles **4.** $\angle P \cong \angle C$, $\overline{PH} \cong \overline{CL}$, $\angle H \cong \angle L$, $\overline{PD} \cong \overline{CF}$, $\angle D \cong \angle F$, $\overline{HD} \cong \overline{LF}$ **5.** $\overline{AB} \cong \overline{BC}$, $\angle ABD \cong \angle CBD$, $\overline{AD} \cong \overline{DC}$, $\overline{BD} \perp \overline{AC}$ **6.** SAS **7.** $\triangle ADC$; SSS **8.** Since $\triangle ABC \cong \triangle ADC$, $\angle D \cong \angle B$ by corresponding parts. **9.** $x = 5$ **10.** $AB = 23$ **11.** $\angle W$, $\angle F$, $\angle P$ **12.** no **13.** corresponding angles **14.** alternate interior angles **15.** interior angles **16.** $\angle 2 \cong \angle 4 \cong \angle 5 \cong \angle 7$, $\angle 1 \cong \angle 3 \cong \angle 6 \cong \angle 8$ **17.** $m\angle C = 56°$, $m\angle BAC = 34°$ **18.** $m\angle A = m\angle C = 40°$, $m\angle B = m\angle D = 140°$ **19.** $PQRS$ is a parallelogram because opposite sides are congruent. **20.** $KLMN$ is a parallelogram because the diagonals bisect each other. **21.** rhombus **22.** bases 4, 10; median 7

Chapter 9

Lesson 9-1

Try This **a.** $\frac{5}{4}$ **b.** $\frac{3}{2}$ **c.** $\frac{2}{3}$ **d.** alto to soprano **e.** $\frac{1}{2}$

Exercises **1.** $\frac{7}{1}$ **3.** $\frac{2}{5}$ **5.** $\frac{4}{1}$ **7.** $\frac{1}{9}$ **9.** $\frac{1}{4}$ **11.** $\frac{7}{8}$ **13.** $\frac{15}{2}$ **15.** $\frac{9}{38}$ **17.** $\frac{2}{5}$ **19.** $\frac{2}{3}$ **21.** $\frac{3}{40}$ **23.** $\frac{14}{25}$ **25.** $\frac{4}{7}$ **27.** $\frac{7}{9}$ **29.** $\frac{15}{1}$ **31.** $\frac{1}{10}$ **35.** $\frac{2}{1}$ **37.** $\frac{7}{3}$ **39.** $\frac{7}{10}$ **41.** $\frac{2}{3}$ **43.** 12 **45.** The number of students must be divisible by $(3 + 2)$. **47.** 5 **49.** 61° **51.** 17, 21; 19

Lesson 9-2

Try This **a.** yes **b.** no **c.** no **d.** yes **e.** 12 **f.** 28 **g.** 18 mi **h.** 5.25 in. **i.** about 12 mi

Exercises **1.** no **3.** yes **5.** yes **7.** yes **9.** 1 **11.** 36 **13.** 64 **15.** 10 **17.** $9\frac{1}{3}$
19. 42 **21.** 25 **23.** 39.05 **25.** 80 km **27.** 2.7 cm **29.** 30 km
31. 2475 mi **35.** 30 mi **37.** 22.5 mi **39.** 36° and 54° **41.** 117° and 63°
43. $\frac{7}{4}$ **45.** $\frac{2}{5}$ **47.** 1:7 **49.** 9 **51.** 4

Lesson 9-3

Try This **a.** no **b.** yes; 2 **c.** yes; $\frac{1}{2}$ **d.** no

Exercises **1.** yes; 3 **3.** yes; $\frac{1}{3}$ **11.** $\frac{1}{4}$ **13.** no **15.** no **17.** yes **19.** 10 **21.** 20
23. 2.6 **25.** 3.8

Lesson 9-4

Try This **a.** $\angle J \cong \angle A$, $\angle K \cong \angle B$, $\angle L \cong \angle C$; $\frac{JK}{AB} = \frac{JL}{AC} = \frac{KL}{BC}$ **b.** $BT = 6\frac{3}{4}$, $CT = 9$
c. $UK = 10$, $EN = 30$

Exercises **1.** $\angle R \cong \angle A$, $\angle T \cong \angle C$, $\angle S \cong \angle B$; $\frac{RS}{AB} = \frac{RT}{AC} = \frac{ST}{BC}$

3. $\angle B \cong \angle J$, $\angle S \cong \angle Z$, $\angle C \cong \angle W$; $\frac{BC}{JW} = \frac{BS}{JZ} = \frac{SC}{ZW}$

5. $\angle D \cong \angle G$, $\angle E \cong \angle I$, $\angle F \cong \angle H$; $\frac{GH}{DF} = \frac{GI}{DE} = \frac{IH}{EF}$
7. $QR = 10$, $PR = 8$ **9.** $WY = 20$, $YZ = 10$ **11.** $AB = 12$, $YZ = 18.5$
13. $KH = 1\frac{7}{8}$, $GK = 3\frac{3}{4}$ **15.** $NP = 4$, $QV = 6$ **17.** 15 cm, 20 cm
19. sometimes **21.** never **23.** yes; 2 **25.** 2

Lesson 9-5

Try This **a.** similar **b.** similar **c.** not similar **d.** 12 **e.** 18.75 m

Exercises **1.** not similar **3.** similar **5.** not similar **7.** 20 **9.** $16\frac{4}{5}$
11. $EM = 16.5$ **13.** 60 ft **17.** 26 ft **19.** 12

Lesson 9-6

Try This **a.** AA **b.** SSS **c.** SAS **d.** AA **e.** 4

Exercises **1.** SSS **3.** SAS **5.** SSS **7.** SAS **9.** AA **11.** 24 **13.** $\triangle PRQ \sim \triangle TRS$
by SAS, so $\angle QRS \cong \angle STR$. $\overline{PQ} \| \overline{ST}$, since alternate interior angles are
congruent. **15.** SSS **17.** 112 ft, 140 ft **19.** 21 ft **21.** 1

Algebra Connection, p. 336

1. 2.65 **3.** 5 **5.** 10 **7.** 3, 4 **9.** 8, 9

Chapter 9 Review

1. 3 to 1 **2.** 3:4 **3.** $\frac{2}{3}$ **4.** 8:3 **5.** 1:3 **6.** no **7.** yes **8.** 15 **9.** 20
10. 57 km **11.** 4.1 cm **12.** 20 km **13.** yes; $\frac{3}{4}$ **15.** $\angle Z \cong \angle G$,
$\angle E \cong \angle T$, $\angle K \cong \angle R$; $\frac{ZE}{GT} = \frac{ZK}{GR} = \frac{EK}{TR}$ **16.** $JA = 6\frac{1}{2}$, $VY = 8\frac{2}{3}$ **17.** 10
18. $12\frac{4}{9}$ **19.** AA **20.** SAS **21.** 12

Chapter 10

Lesson 10-1

Try This **a.** $\triangle PFS \sim \triangle PMF \sim \triangle FMS$ **b.** $\frac{x}{t} = \frac{t}{u}$ **c.** $\frac{x}{r} = \frac{u}{v}$ **d.** 5 **e.** $\sqrt{14} \approx 3.7$
f. 6

Exercises **1.** $\triangle PMD \sim \triangle PHM \sim \triangle MHD$ **3.** $\frac{s}{t} = \frac{t}{e}$ **5.** $\frac{s}{r} = \frac{r}{f}$ **7.** $\frac{m}{d} = \frac{d}{a}$ **9.** 10
11. 21 **13.** $\sqrt{195} \approx 14.0$ **15.** $\sqrt{0.036} \approx 0.2$ **17.** 0.6 **19.** 12
21. $\sqrt{187} \approx 13.7$ **23.** $\sqrt{105} \approx 10.2$ **25.** $\sqrt{\frac{21}{32}} \approx \sqrt{0.65625} \approx 0.81$
27. ≈ 49 in. **29.** 14.6 ft **31.** 10 **33.** $5\frac{6}{23} \approx 5.3$ **35.** SSS **37.** $\frac{10}{1}$
39. $\frac{6}{1}$ **41.** 7 **43.** 10 **45.** 9

Lesson 10-2

Try This **a.** 12 **b.** $\sqrt{51} \approx 7.14$ **c.** 13.9 mi

Exercises **1.** 30 **3.** 15 **5.** 39 **7.** $\sqrt{56} \approx 7.48$ **9.** $\sqrt{149} \approx 12.21$
11. $\sqrt{180} \approx 13.42$ **13.** $\sqrt{1007} \approx 31.73$ **15.** $\sqrt{2125} \approx 46.10$
17. $\sqrt{97} \approx 9.8$ **19.** $\sqrt{16,200} \approx 127.3$ ft **21.** $\sqrt{91} \approx 9.5$ ft **25.** about
158.1 ft **27.** $\sqrt{442} + 9 \approx 30$ ft **29.** $\sqrt{38} \approx 6.2$ ft **31.** ≈ 16.97
33. $\triangle SRP, \triangle SPQ, \triangle PRQ$ **35.** 10.67 ft **37.** $\frac{6}{5}$ or $1\frac{1}{5}$

Probability Connection

1. $\frac{2}{5} = 0.4$ **3.** $\frac{5}{12} \approx 0.42$

Lesson 10-3

Try This **a.** yes **b.** no **c.** no **d.** yes

Exercises **1.** no **3.** no **5.** no **7.** yes **9.** yes **11.** no **13.** no **15.** yes
17. yes **21.** yes **23.** If the distance from home to second base is about
84.9 ft, she can be sure that there are right angles at first and third base.
25. 3 **27.** 6 **29.** not possible **31.** -8 **33.** 1 to 1 **35.** 40.5
37. 8 and 9

Lesson 10-4

Try This **a.** $MF = 3, RF = 3\sqrt{2} \approx 4.2$ **b.** $NP = NS = 10$ **c.** $\frac{15}{\sqrt{2}} \approx 10.6$

Exercises **1.** $s = 2, m = 2\sqrt{2} \approx 2.8$ **3.** $s = 4.5, m = 4.5\sqrt{2} \approx 6.4$ **5.** $r = 14$,
$s = 14$ **7.** $r = 3.5, s = 3.5$ **9.** $r = s = \frac{24}{\sqrt{2}} \approx 17.0$ **11.** $r = s =$
$\frac{3.8}{\sqrt{2}} \approx 2.7$ **13.** $m = 22.6\sqrt{2} \approx 32.0$ **15.** $s = \frac{3}{5}, m = \frac{3}{5}\sqrt{2} \approx 0.8$
17. $r = s = \frac{142}{\sqrt{2}} \approx 100.4$ **21.** $AC = 1, BC = \sqrt{2} \approx 1.4, BD = \sqrt{2} \approx 1.4$,
$CD = 2, CE = 2, DE = 2\sqrt{2} \approx 2.8, EF = 2\sqrt{2} \approx 2.8, DF = 4$
23. 1.522 in. **25.** 23.7

Lesson 10-5

Try This **a.** $ON = 8$, $OR = 16$ **b.** $MQ = 9\sqrt{3} \approx 15.6$, $PQ = 18$ **c.** $TS = 5$, $VT = 5\sqrt{3} \approx 8.7$ **d.** $\frac{12}{\sqrt{3}} \approx 6.9$ ft

Exercises **1.** $t = 5$, $r = 5\sqrt{3} \approx 8.7$ **3.** $t = 13.5$, $r = 13.5\sqrt{3} \approx 23.4$ **5.** $p = 10$, $r = 5\sqrt{3} \approx 8.7$ **7.** $p = 48$, $r = 24\sqrt{3} \approx 41.6$ **9.** $p = 12$, $t = 6$ **11.** $p = 38$, $t = 19$ **13.** $t = \frac{12}{\sqrt{3}} \approx 6.9$, $p = 2t \approx 13.9$ **15.** $t \approx 7.5$, $p = 2t \approx 15.0$ **19.** The ladder is extended to $\frac{24}{\sqrt{3}} \approx 13.9$ ft. It is about 6.9 ft from the wall. **21.** twice the distance from A to C, 9928 ft **23.** $1.8\sqrt{3} \approx 3.1$ m **25.** $h = \frac{s}{2}\sqrt{3}$ **27.** $6\sqrt{3} \approx 10.4$ mm **29.** $12:13$ **31.** 2.25 **33.** 11 and 12

Chapter 10 Review

1. $\triangle TMK \sim \triangle TSM \sim \triangle MSK$ **2.** $\triangle AME \sim \triangle ANM \sim \triangle MNE$ **3.** 8 **4.** $\sqrt{70} \approx 8.37$ **5.** 18 **6.** 13.5 **7.** 25 **8.** 3 **9.** no **10.** no **11.** $PQ = 8.5$, $QR = 8.5\sqrt{3} \approx 14.72$ **12.** $GK = 10$, $GH = 5\sqrt{3} \approx 8.66$ **13.** $MH = NH = 17$ **14.** $RS = 3\sqrt{2} \approx 4.24$, $TR = 6$ **15.** $AR = 5$, $RD = 5\sqrt{2} \approx 7.07$ **16.** $EP = 14$, $EX = 28$

Chapter 11

Lesson 11-1

Try This **a.** Not a polygon; side $\overline{PR}$ intersects five other sides. **b.** polygon **c.** Not a polygon; side $\overline{BC}$ is not a segment. **d.** quadrilateral **e.** heptagon **f.** convex **g.** concave

Exercises **1.** polygon **3.** Not a polygon. Side $\overline{GL}$ intersects three other sides. **5.** polygon **7.** octagon; concave **9.** decagon; concave **11.** decagon; concave **15.** octagon **21.** no **23.** no **25.** 8

Lesson 11-2

Try This **a.** 35 **b.** 170 **c.** 560

Exercises **1.** 90 **3.** 740 **5.** 434 **7.** 2345 **9.** 4850 **11.** 19,700 **13.** 498,500 **15.** 9 **17.** 0 **19.** 66 **21.** 3; 3 **23.** concave quadrilateral **25.** octagon; concave **27.** 3

Problem-Solving Strategies, p. 378

1. 300

Lesson 11-3

Try This **a.** 31 cm **b.** 37 in. **c.** 12.5 cm **d.** 14 cm

Exercises **1.** 20 mm **3.** 41 ft **5.** 9.9 cm **7.** 11.7 cm **9.** $79.65 **13.** 16 **15.** 15 in., 15 in., 18 in. **17.** no **19.** 21 cm **21.** 252 **23.** alternate interior angles **25.** alternate exterior angles **27.** corresponding angles **29.** 5

Lesson 11-4

Try This   **a.** 12,240° **b.** 9900° **c.** 1080° **d.** 94°

Exercises **1.** 1440° **3.** 6840° **5.** 17,640° **7.** 47,700° **9.** 143,640° **11.** 179,640°
13. 88° **15.** 91.5° **19.** 135° each **21.** $x = 9$. The exterior angles measure
18°, 27°, 36°, 45°, 63°, 81°, and 90°. **23.** $x = 12$. The exterior angles
measure 48°, 34°, 94°, 102°, 54°, and 28°. **25.** 168° **27.** 47 cm **29.** 3

Lesson 11-5

Try This   **a.** $4\frac{2}{3}$, $11\frac{2}{3}$

Exercises **1.** 5 **3.** $8\frac{2}{5}$ **5.** $RS = 6\frac{2}{3}$; $UV = 9\frac{1}{3}$ **7.** $DL = 4\frac{1}{6}$; $WV = 4\frac{4}{5}$ **11.** no
13. 15 cm **15.** 2160° **17.** 8640° **19.** 4 **21.** 2

Lesson 11-6

Try This   **a.** 135° **b.** 40° **c.** 72°

Exercises **1.** 120°; 60° **3.** 60°; 120° **5.** $128\frac{4}{7}°$; $51\frac{3}{7}°$ **7.** 162°; 18° **9.** $176\frac{2}{5}°$; $3\frac{3}{5}°$
11. use 60° angles **13.** use 40° angles **15.** use 30° angles **19.** 5250 ft
21. rectangle **25.** 4 **27.** 6 **29.** n **31.** They bisect a pair of opposite
angles or a pair of opposite sides. **33.** 8, 10

Statistics Connection, p. 395

1. 90°; $77\frac{1}{2}°$; the mean is the same as the measure of each angle of a regular
quadrilateral.

Lesson 11-7

Try This   **a.** parallelogram **b.** square

Exercises **1.** parallelogram **3.** rectangle **5.** square **17.** false **19.** false **21.** true
23. 60°; 120° **27.** 5 **29.** 3

Chapter 11 Review

1. polygon **2.** Not a polygon. Sides $\overline{OP}$ and $\overline{TV}$ do not intersect two other
sides. **3.** dodecagon; concave **4.** quadrilateral; convex **5.** 27 **6.** 945
7. 28 in. **8.** 54 cm **9.** 9.1 cm **10.** 11.1 cm **11.** 2880° **12.** 540°
13. 118° **14.** $GN = 15$; $LK = 12\frac{1}{2}$ **15.** 135°; 45° **16.** $157\frac{1}{2}°$; $22\frac{1}{2}°$

Chapter 12

Lesson 12-1

Try This **a.** 12 sq. units **b.** 8 sq. units **c.** 15 sq. units. **d.** 12

Exercises **1.** 23 **3.** 16–17 **5.** 17–18 **7.** 18–19 **9.** 35 **11.** 15 **13.** 38 **15.** 76 **17.** perimeter **19.** area **21.** perimeter **27.** ≈ 15 **31.** 5 **33.** false **35.** false **37.** 24

Algebra Connection, p. 410

1. 10^9 **3.** 2^7 **5.** 100 **7.** 243 **9.** 9,765,625 **11.** 169,744 **13.** yes

Lesson 12-2

Try This **a.** 87 cm^2 **b.** 169 mi^2 **c.** 340 ft^2

Exercises **1.** 98 km^2 **3.** 10.5 cm^2 **5.** 162 yd^2 **7.** 95 in.^2 **9.** 144 m^2 **11.** 384.16 m^2 **13.** 330 ft^2 **15.** 364 m^2 **17.** 3600 ft^2 **19.** 2808 ft^2 **21.** 100 **23.** 10,000 **25.** $b = 8 \text{ cm}$ **27.** $b = 5\sqrt{2} \approx 7.07 \text{ cm}$ **29.** $b = 0.3 \text{ m}$ **31a.** 1.5 **b.** 2.25 **35.** 63 cm^2 **37.** 5984 **39.** 39 m^2 **41.** no **43.** 0 **45.** 15 **47.** 24

Problem-Solving Strategies, p. 416

1. 9025 ft^2 **3.** It is quadrupled. **5.** Create a 22-in. segment using the 11-in. side twice. Measure 17 in. of the segment using the $8\frac{1}{2}$-in. side twice. What remains is $22 - 17 = 5$ in. long.

Lesson 12-3

Try This **a.** 36 mm^2 **b.** 164.9 ft^2 **c.** 13.5 in. **d.** 45 cm^2

Exercises **1.** 91 m^2 **3.** 6 cm^2 **5.** 4060 mm^2 **7.** 8.75 km^2 **9.** 15 ft^2 **11.** $16\frac{7}{8} \text{ in.}^2$ **13.** $h = 1.5 \text{ m}$ **15.** $b \approx 11.67 \text{ cm}$ **17.** 68 m^2 **21.** They have equal area. **23.** 7–8 in.² **25a.** 3.5 **b.** 12.25 **27a.** Does not change. **b.** Does not change. **c.** decreases **d.** decreases **29.** 130 m^2 **31.** $1800°$ **33.** 19 **35.** 7

Lesson 12-4

Try This **a.** 31.08 mm^2 **b.** 18.6 cm^2 **c.** 544 ft^2

Exercises **1.** 37.5 m^2 **3.** 20 cm^2 **5.** 13.5 m^2 **7.** 31.25 in.^2 **9.** $17\frac{3}{16} \approx 17.19 \text{ ft}^2$ **11.** 420 m^2 **13.** about 4680 bricks **15.** \$435 **17.** 18 m **19.** 24.2 cm **21.** 24 m **23.** 12 ft **25.** 7 ft^2 **29a.** 2 **b.** 4 **31.** $\frac{n^2\sqrt{3}}{4}$ **33.** 56 cm^2 **35.** $86\frac{2}{3}$ **37.** 416 ft^2 **39.** $170°$ **41.** 7^2

Lesson 12-5

Try This **a.** 178.5 m^2 **b.** 30 mm^2 **c.** $78\sqrt{3} \approx 135.1 \text{ cm}^2$

Exercises **1.** $26\frac{11}{16}$ ft^2 **3.** 287 cm^2 **5.** 126 cm^2 **7.** 31 ft^2 **9.** $56\frac{7}{16}$ m^2 **11.** 48 ft^2
13. 306 in.^{2} **15.** They have equal area. **17.** 4 m **19.** 3.5 mm
21. $\approx$73 ft^2 **25.** 2 **27.** 48 **29.** 81

Geometry in Technology, p. 430

1. 16 **3.** $13 \times 16 = 208$ ft^2 **5.** $17 \times 17 = 289$ ft^2
7. $12 \times 12.5 = 150$ ft^2 **9.** $8 \times 8 = 64$ ft^2

Chapter 12 Review

1. 18–19 **2.** 78 **3.** 1.8 m **4.** 67 m^2 **5.** 580 **6.** 27 mm^2 **7.** 324 cm^2
8. 6.5 mm **9.** $60\sqrt{3}$ m^2 **10.** 165 **11.** 4.025 **12.** 115.2 dm^2
13. 224 cm^2

Cumulative Review Chapters 9–12

1. $\frac{5}{1}$ **2.** $\frac{4}{11}$ **3.** $\frac{1}{3}$ **4.** yes **5.** no **6.** no **7.** yes; 2 **8.** no
9. $AC = 8$; $BC = 5$ **10.** $RS = 5.25$ **11.** SAS **12.** 16 ft **13.** $\sqrt{48} \approx 6.9$
14. $c = 15$ **15.** $a = \sqrt{3.25} \approx 1.8$ **16.** no **17.** $7, 7\sqrt{2} \approx 9.9$ **18.** 4.5,
$\frac{\sqrt{243}}{2} \approx 7.8$ **19.** octagon, concave **20.** pentagon, convex **21.** 54
22. 84 **23.** 258 cm^2 **24.** 720° **25.** $HJ = 5.25, LM = 3.75$ **26.** 150°
27. 62.9 cm^2 **28.** 9.2 m^2 **29.** 54 cm^2 **30.** 90 dm^2 **31.** 80.5 dkm^2

Chapter 13

Lesson 13-1

Try This **a.** radii: $\overline{JK}, \overline{KH}, \overline{KG}, \overline{KL}$; chords: $\overline{GH}, \overline{GL}, \overline{JH}$; diameters: $\overline{JH}, \overline{GL}$ **b.** 8.5
c. 9.2

Exercises **1.** radii: $\overline{OR}, \overline{OP}, \overline{OG}, \overline{OT}$; chords: $\overline{HJ}, \overline{PT}, \overline{TS}$; diameter: $\overline{PT}$ **3.** 41.3
5. $2\frac{3}{8}$ **7.** 0.1215 **9.** 17.4 **11.** 0.16 **13.** 3.8 cm **17.** It is ≤ 25 mi.
19. The families and the center of the circle form an equilateral triangle.
21. false **23.** rhombus **25.** For any chord AB, $CD = CO + OD = AO + BO > AB$. Thus, $AB < CD$. **27.** 52 in.^{2} **29.** 10

Lesson 13-2

Try This **a.** 16 **b.** 32 **c.** 4.5 **d.** 12.5

Exercises **1.** 24 **3.** 8 **5.** 20 **7.** 13 **11.** Draw two chords on the fragment and
find their perpendicular bisectors. They intersect where the center would
have been. Then a radius can be measured. **13.** 24 cm **15.** radii: $\overline{OA}, \overline{OG}$,
$\overline{OB}, \overline{OT}$; chords: $\overline{AB}, \overline{GT}$; diameter: $\overline{AB}$ **17.** 2.95 in. **19.** 14 yd **21.** 39.6
23. 125

Lesson 13-3

Try This **a.** tangent: n; secant: ℓ, m, p **b.** 13   **c.** 10   **d.** 69°

Exercises **1.** tangent: $\overleftrightarrow{AE}$ and $\overleftrightarrow{DE}$; secant: $\overleftrightarrow{AB}$ and $\overleftrightarrow{CD}$ **3.** 9   **5.** 1.2   **7.** 16
9. 120   **11.** 31°, 118°   **13.** 68°   **15.** 73°   **17.** 1 ft   **19.** $\sqrt{2} + 1 \approx 2.41$ ft
21. 15   **23.** 7   **25.** 10^4

Lesson 13-4

Try This **a.** minor arcs: $\overarc{DF}$, $\overarc{FE}$, $\overarc{DG}$, $\overarc{GE}$, $\overarc{GF}$; major arcs: $\overarc{DGF}$, $\overarc{FDE}$, $\overarc{GED}$, $\overarc{EDG}$,
$\overarc{GDF}$; semicircles: $\overarc{DFE}$, $\overarc{DGE}$ **b.** 110°   **c.** 197°   **d.** $m\overarc{ABC} = 117°$,
$m\overarc{ADC} = 243°$ **e.** $m\angle AOD = 24°$, $m\overarc{AD} = 24°$, $m\overarc{DB} = 156°$

Exercises **1.** minor arcs: $\overarc{XY}$, $\overarc{YZ}$, $\overarc{ZW}$, $\overarc{XW}$, $\overarc{YZW}$; major arcs: $\overarc{YXZ}$, $\overarc{XWY}$, $\overarc{XYW}$, $\overarc{WXZ}$,
$\overarc{YXW}$; semicircles: $\overarc{XYZ}$, $\overarc{XWZ}$ **3.** 47°   **5.** 224°   **7.** 180°
9. $m\overarc{ABC} = 177°$, $m\overarc{ADC} = 183°$ **11.** $m\angle ROU = 125°$, $m\overarc{RS} = 55°$,
$m\overarc{TS} = 125°$ **15.** 72°   **19.** $m\overarc{AB} = 60°$, $m\angle BOF = 30°$, $m\overarc{ABC} = 120°$,
$OF \approx 0.87$ **21.** 92.5°   **23.** 5400°   **25.** 130°   **27.** 38°   **29.** 21°
31. 100,000

Lesson 13-5

Try This **a.** point U **b.** $\overline{TU}$ **c.** 180° clockwise rotation; 180° counterclockwise
rotation   **d.**

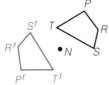

Exercises **1.** H **3.** $\overline{ID}$ **5.** 60° clockwise, 300° counterclockwise
7.   **9.**

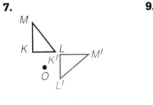
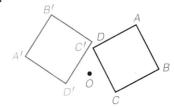

13. $(-2, 3)$ **15.** $(-y, x)$ **17.** true   **19.** false   **21.** false   **23.** $m\overarc{AD} = 19°$,
$m\angle COB = 19°$, $m\overarc{DAB} = 199°$ **25.** 6 cm   **27.** 216 m   **29.** 11

Lesson 13-6

Try This **a.** $\angle GEF$ intercepts $\overarc{GF}$.   **b.** $\angle XWZ$ intercepts $\overarc{XZ}$.   **c.** $\angle PQR$ intercepts $\overarc{PR}$.
d. $m\angle A = 90°$, $m\angle B = 90°$, $m\angle C = 90°$ **e.** $m\angle ADB = 30°$, $m\angle ADC = 60°$,
$m\angle BEC = 30°$ **f.** 59°

Exercises **1.** $\angle CAB$ intercepts $\overarc{CB}$.   **3.** $\angle XYW$ intercepts $\overarc{XW}$; $\angle XYZ$ intercepts $\overarc{XWZ}$;
$\angle YZW$ intercepts $\overarc{YXW}$; $\angle YWZ$ intercepts $\overarc{YZ}$; $\angle WYZ$ intercepts $\overarc{WZ}$.   **5.** 68°
7. 45°   **9.** $m\angle MQP = 110°$, $m\angle QPN = 86°$ **11.** 36°   **13.** 120°

15. 138° **17.** 39° **19.** 121° **21.** 20° **25.** By Th. 13.11, the carpenter's square will fit only if a semicircle has been cut. **27.** 25.42 cm² **29.** 10^5

Lesson 13-7

Exercises **9.** Fold the circle in half, fold the resulting wedge in half, and then fold the resulting wedge in half again. **11.** $m\angle DEG = m\angle DFG = 40°$, $m\widehat{FG} = 100°$ **13.** radii: $\overline{OF}, \overline{OA}, \overline{OB}, \overline{OD}$; chords: $\overline{FD}, \overline{BE}, \overline{AD}$; diameter: $\overline{AD}$ **15.** 3.85 m **17.** $1\frac{3}{4}$ **19.** 81

Lesson 13-8

Try This **a.** 7.5π m ≈ 23.55 m **b.** ≈ 94.2 cm **c.** 24.2 cm **d.** 157 cm

Exercises **1.** 12π m ≈ 37.7 m **3.** 0.5π km ≈ 1.6 km **5.** 25π cm ≈ 78.5 cm **7.** 75π cm ≈ 235.6 cm **9.** 1.72π m $= 5.4$ m **11.** ≈ 13.4 cm **13.** $\approx 942{,}000{,}000$ km **15.** The circumference is halved. **17.** 18,850 cm **19.** 635 cm **21.** You could slip a piece of paper under the band or crawl under it, and you could walk under if you were 1.59 meters tall or less. **23.** 7.5 cm² **25.** 8 **27.** 3

Lesson 13-9

Try This **a.** 64π cm² ≈ 201.06 cm² **b.** 0.0049π in.² ≈ 0.02 in.² **c.** 7.84π km² ≈ 24.63 km² **d.** 173.1 cm² **e.** 150.8 cm²

Exercises **1.** 196π m² ≈ 615.8 m² **3.** 100π mm² ≈ 314.2 mm² **5.** 225π ft² ≈ 706.9 ft² **7.** 81π km² ≈ 254.5 km² **9.** 166.41π m² ≈ 522.8 m² **11.** 81π in.² ≈ 254.5 in.² **13.** 2.25π mm² ≈ 7.1 mm² **15.** 5.29π cm² ≈ 16.6 cm² **17.** 53.0 cm² **19.** 126.1 cm² **21.** ≈ 3.4 m² **25.** 5.4 m² **27.** 101 cm **29.** ≈ 530 ft² **31.** 3342 mi **33.** 4

Chapter 13 Review

1. 31.1 mm **2.** $4\frac{1}{2}$ ft. **3.** 15.8 cm **4.** 0.14 m **5.** 72 **6.** 5 **7.** 5 **8.** 6 **9.** 13 **10.** 100° **11.** minor: $\widehat{AD}, \widehat{CB}, \widehat{AC}, \widehat{BD}$; major $\widehat{CDB}, \widehat{ADC}, \widehat{ABD}, \widehat{BCD}$; semicircles: $\widehat{ACB}, \widehat{CBD}, \widehat{CAD}, \widehat{ADB}$ **12.** 72° **13.** 252° **14.** N **15.** $\overline{QR}$ **16.** 135° clockwise; 225° counterclockwise **17.** 41° **18.** 29°; 58° **19.** 35° **21.** 16π m ≈ 50.3 m **22.** 1.5π m ≈ 4.7 m **23.** 10π in. ≈ 31.4 in. **24.** 36π in.² ≈ 113.1 in.² **25.** 4π m² ≈ 12.6 m² **26.** 10.24π cm² ≈ 32.2 cm² **27.** 52.4 m² **28.** 34.6 m²

Chapter 14

Lesson 14-1

Try This **a.** faces: $ABCDE, GHJKF, ABHG, BCJH, CDKJ, DKFE, EFGA$; edges: $\overline{AB}, \overline{BC}, \overline{CD}, \overline{DE}, \overline{EA}, \overline{GH}, \overline{HJ}, \overline{JK}, \overline{KF}, \overline{FG}, \overline{AG}, \overline{BH}, \overline{CJ}, \overline{DK}, \overline{EF}$; vertices: $A, B, C, D, E, G, H, J, K, F$ **b.** pentahedron

Exercises **1.** faces: $ABCD, \triangle BCF, \triangle ADE, ABFE, DCFE$; edges: $\overline{AD}, \overline{AB}, \overline{BC}, \overline{CD}, \overline{BF}, \overline{CF}, \overline{DE}, \overline{AE}, \overline{EF}$; vertices: A, B, C, D, E, F **3.** $\triangle SXW, \triangle SWP, \triangle SPZ, \triangle SZY, \triangle SYX, XYZPW$; edges: $\overline{SX}, \overline{SW}, \overline{SP}, \overline{SZ}, \overline{SY}, \overline{XW}, \overline{WP}, \overline{PZ}, \overline{ZY}, \overline{YX}$; vertices: S, X, W, P, Z, Y

5. pentahedron **7.** hexahedron **11.** true **13.** false **15.** true **17.** a
19. e **21.** 153.9 m^2 **23.** 63.6 in.^{2} **25.** 4

Lesson 14-2

Try This **a.** right hexagonal prism; bases; *ABCDEF, GHJKLM*; lateral faces: *ABHG,*
BCJH, CDKJ, EDKL, FELM, AFMG; lateral edges: $\overline{AG}, \overline{BH}, \overline{CJ}, \overline{DK}, \overline{EL}, \overline{FM}$,
b. LA = 80; SA = 104 **c.** 84 in.^{2}

Exercises **1.** right triangular prism; bases: $\triangle ABC$, $\triangle DEF$; lateral faces: *ABED, BCFE,*
ACFD; lateral edges: $\overline{AD}, \overline{BE}, \overline{CF}$ **3.** oblique quadrangular prism; bases:
MNPQ, URST; lateral faces: *MNRU, NPSR, QPST, MQTU*; lateral edges: $\overline{MU}$,
$\overline{NR}, \overline{PS}, \overline{QT}$ **5.** LA = 132 mm^2, SA = 188 mm^2 **7.** LA = 120 cm^2,
SA = 132 cm^2 **9.** 120 cm^2 **11.** 144 **13b.** 2611 cm^2 **c.** LA = 2211 cm^2,
SA = 2611 cm^2 **15.** 32 in.^{2} **17.** SA = 48 in.^{2} **19.** octahedron **21.** 27
23. 1,000,000

Lesson 14-3

Try This **a.** regular pentagonal pyramid; base: *FGHIJ*; lateral faces: $\triangle FKG$, $\triangle GKH$,
$\triangle HKI$, $\triangle IKJ$, $\triangle JKF$; lateral edges: $\overline{KF}, \overline{KG}, \overline{KH}, \overline{KI}, \overline{KJ}$ **b.** LA = 60 cm^2,
SA = $(60 + 4\sqrt{3})$ cm$^2 \approx 66.9$ cm^2 **c.** 80 in.^{2}

Exercises **1.** regular quadrangular pyramid; base: *WXYZ*; lateral faces: $\triangle WRX$, $\triangle XRY$,
$\triangle YRZ$, $\triangle ZRW$; lateral edges: $\overline{RW}, \overline{RX}, \overline{RY}, \overline{RZ}$ **3.** regular pentagonal pyramid; base: *ABCDE*; lateral faces: $\triangle AFE$, $\triangle EFD$, $\triangle DFC$, $\triangle CFB$, $\triangle BFA$; lateral
edges: $\overline{FA}, \overline{FB}, \overline{FC}, \overline{FD}, \overline{FE}$ **5.** LA = 216 cm^2, SA = 297 cm^2 **7.** LA =
1600 m^2, SA = 2624 m^2 **9.** 260 cm^2 **11.** 109.44 m^2 **13.** 230 ft^2
15. 79,772 m^2 **17.** always **19.** sometimes **21.** right pentagonal prism;
bases: *ABCDE* and *GHJKF*; lateral faces: *ABHG, BCJH, CDKJ, DEFK, EAGF*;
lateral edges: $\overline{AG}, \overline{BH}, \overline{CJ}, \overline{DK}, \overline{EF}$ **23.** 6

Lesson 14-4

Try This **a.** 8 in.^{3} **b.** 1020 cm^3 **c.** 302.5 cm^3 **d.** 9 ft^3

Exercises **1.** 48 cm^3 **3.** 5600 cm^3 **5.** 960 m^3 **9.** 13.3 ft^3 **11.** 442 in.^{3} **13.** B
15. A **17.** m^3 **19.** cm^3 **21.** 28,800 liters **23.** no **25.** no **27.** 64

Lesson 14-5

Try This **a.** $93\frac{1}{3}$ m^3 **b.** 9625 m^3

Exercises **1.** 24.75 cm^3 **3.** 462 m^3 **5.** 5.875 ft^3 **7.** 525 cm^3 **9.** 180 cm^3
11. $\approx$154.92 m^3 **13.** 7900 cm^3 **15.** The taller one has twice the volume.
17. 9 ft **19.** 166.7 cm^3 **21.** 164 cm^3 **23.** 95 ft^2 **25.** 10^7

Albegra Connection, p. 509

1. 8.4×10^4 **3.** 9.4×10^{10} **5.** 7.3×10^{-7} **7.** 980,000 **9.** 3,900,000
11. 0.0004

Lesson 14-6

Try This **a.** Cylinders and prisms both have congruent bases in parallel planes. The
bases of a prism are polygonal regions; the bases of a cylinder are circular.
b. the vertex of a cone **c.** the height of a cylinder

Exercises **1.** Both are the locus of points a given distance from a given point.
3. a hemisphere **5.** the vertex of a cone **7.** a great circle of a sphere
15. true **17.** true **19.** false **21.** 112 cm^3 **23.** 49π m$^2 \approx 153.9$ m^2
25. 70.56π in.$^2 \approx 221.7$ in.^{2} **27.** 3.7×10^6

Lesson 14-7

Try This **a.** 32π m$^2 \approx 100.5$ m^2 **b.** 76π in.$^2 \approx 238.8$ in.^{2}

Exercises **1.** 60π cm$^2 \approx 188.5$ cm^2 **3.** 2160π cm$^2 \approx 6785.8$ cm^2 **5.** 120π in.$^2 \approx$ 377.0 in.^{2} **7.** 25.5π m$^2 \approx 80.1$ m^2 **11.** 1055.6 cm^2 **13.** \$0.18
15. vertex of a cone **17.** 0.0054 **19.** 7000

Lesson 14-8

Try This **a.** 1872π m$^3 \approx 5881.1$ m^3 **b.** 324π ft$^3 \approx 1017.9$ ft^3 **c.** 36π in.$^3 \approx$ 113.1 in.^{3}

Exercises **1.** 75π cm$^3 \approx 235.6$ cm^3 **3.** 288π m$^3 \approx 904.8$ m^3 **5.** 21.6π ft$^3 \approx 67.9$ ft^3
7. 1488π m$^3 \approx 4674.7$ m^3 **9.** $0.7\overline{6}\pi$ cm$^3 \approx 2.4$ cm^3 **11.** 19.926π ft$^3 \approx$ 62.6 ft^3 **15.** 113.1 in.^{3} **17.** 4 liters **19.** 57,727 liters **21.** 9.3 cm
23. $\frac{\pi}{4}$ **25.** 104π m$^2 \approx 326.7$ m^2 **27.** 4.3×10^3 **29.** 9×10^{-3}

Lesson 14-9

Try This **a.** 256π in.$^2 \approx 804.2$ in.^{2} **b.** 36π cm$^3 \approx 113.1$ cm^3 **c.** 12 cm by 12 cm by 12 cm **d.** 30π m$^3 \approx 94.2$ m^3

Exercises **1.** 268.0 cm^3; 201.1 cm^2 **3.** 14.1 in.^{3}; 28.3 in.^{2} **5.** 1436.8 m^3; 615.8 m^2
7. 38.8 cm^3; 55.4 cm^2 **9.** 179.6 cm^3; 153.9 cm^2 **11.** $(5.5 \times 10^9)\pi$ km^3; $(1.02 \times 10^7)\pi$ km^2 **13.** 3053.6 in.^{3} **15.** $90.\overline{6}\pi$ ft$^3 \approx 284.8$ ft^3
17. 54π m$^3 \approx 169.6$ m^3 **19.** 3 cm **21.** 36π cm^2 **23.** 2 ft^3 **25.** 50.3 ft^2
27. The area of a sphere is equal to 4 times the area of a great circle.
29. 656π m$^3 \approx 2060.9$ m^3 **31.** 1.21π m$^3 \approx 3.8$ m^3 **33.** 12π in. ≈ 37.7 in.
35. 9.4π m ≈ 29.5 m **37.** 4

Lesson 14-10

Try This **a.** $x = 15$, $y = 8$ **b.** 1:3; 1:9; 1:27 **c.** 54π

Exercises **1.** 9; 1:3, 1:9, 1:27 **3.** $\frac{5}{2}$; 2:1, 4:1, 8:1 **5.** 16 **7.** 1000π **9.** 2
11. eight times greater **13.** the larger can **15.** true **17.** false **19.** false
21. Neither; they have the same surface area (216 in.^{2}). **23.** 4188.8 ft^3; 1256.6^2 **25.** interior angles **27.** corresponding angles **29.** $GH + KH >$ GK, $KH + GK > GH$, $GK + GH > KH$ **31.** no **33.** yes **35.** 0.00001

Chapter 14 Review

1. faces: $\triangle FDE$, $\triangle ABC$, FACE, FABD, DBCE; edges; $\overline{FE}$, $\overline{DE}$, $\overline{FD}$, $\overline{AB}$, $\overline{BC}$, $\overline{AC}$, $\overline{FA}$, $\overline{DB}$, $\overline{EC}$; vertices: A, B, C, D, E, F **2.** LA = 117 cm^2, SA = 141 cm^2
3. LA = 198 cm^2; SA = 279 cm^2 **4.** 900 in.^{3} **5.** 1848 m^3 **6.** 75 cm^3
7. 28 in.^{3} **8.** the lateral area of a cylinder **9.** a hemisphere

10. 20π in.$^2 \approx 62.8$ in.^{2} **11.** 44π ft$^2 \approx 138.2$ ft^2 **12.** $18{,}750\pi$ m$^3 \approx$ $58{,}904.9$ m^3 **13.** 1536π m$^3 \approx 4825.5$ m^3 **14.** 2144.7 m^3; 804.2 m^2
15. $3\frac{1}{2}$; $1{:}2$, $1{:}4$; $1{:}8$

Chapter 15

Lesson 15-1

Try This **a.** 5 **b.** $\sqrt{241} \approx 15.52$ **c.** $\sqrt{90} \approx 9.49$ **d.** $\sqrt{122} \approx 11.05$
e. C; $AC = \sqrt{178} \approx 13.3$

Exercises **1.** 5 **3.** $\sqrt{26} \approx 5.1$ **5.** 8 **7.** $\sqrt{45} \approx 6.7$ **9.** $\sqrt{261} \approx 16.2$
11. $\sqrt{5161} \approx 71.8$ **13.** C; $AC = \sqrt{153} \approx 12.4$ **15.** $10 + 5\sqrt{2} +$
$\sqrt{82} \approx 26.1$ **17.** $XZ = 10$, $YW = \sqrt{122} \approx 11.0$ **19.** ≈ 14.4, A to B to C
21. $\sqrt{21^2 + 33^2} = \sqrt{1530} \approx 39.1$ in. **23.** ≈ 15.3, A to B to D to C
25. $300{,}000$ m^3 **27.** 0

Algebra Connection, p. 541

1. 18 **3.** 0.8 **5.** -4 **7.** -3 **9.** -8 **11.** -3.2

Lesson 15-2

Try This **a.** $m = -\frac{3}{2}$ **b.** $m = -\frac{3}{2}$ **c.** $\frac{2}{3}$ **d.** $\frac{8}{6} = \frac{4}{3}$ **e.** $\frac{11}{6}$ **f.** $-\frac{15}{10} = -\frac{3}{2}$
g. 0 **h.** 180 **i.** not defined

Exercises **1.** $\frac{3}{5}$ **3.** $-\frac{7}{2}$ **5.** $-\frac{2}{3}$ **7.** $-\frac{1}{5}$ **9.** 3 **11.** 2 **13.** -2 **15.** $\frac{3}{2}$
17. $-\frac{1}{4}$ **19.** $\frac{1}{6}$ **21.** not defined **23.** 0 **25.** 0 **27.** 0 **29.** $\frac{3}{4}$ **31.** $\frac{9}{10}$
33. 16 **35.** The slope of $\overline{AC}$ is $-\frac{2}{3}$, of $\overline{BC}$ is -2, and of $\overline{AB}$ is not defined.
37. 5, -2, and n, respectively **39.** ≈ 332.4 in.^{3}

Geometry in Science, p. 546

1. isosceles triangle **3.** star **5.** Laser Up, $[1, -3]$, Laser Down, $[6, 0]$, $[0, 6]$, $[-6, 0]$, $[0, -6]$, Laser Up, $[-3, -5]$

Lesson 15-3

Try This **a.** $(6, 4)$ **b.** $(-1, -5)$ **c.** $\sqrt{65} \approx 8.1$

Exercises **1.** $(4, 4)$ **3.** $(1, 5)$ **5.** $\left(3, 2\frac{1}{2}\right)$ **7.** $\left(1\frac{1}{2}, 3\right)$ **9.** $\left(-7\frac{1}{2}, -5\frac{1}{2}\right)$
11. $(-3, -4)$ **13.** $\left(3\frac{1}{2}s, 5t\right)$ **15a.** $10\sqrt{2} \approx 14.1$ **b.** $\sqrt{137} \approx 11.7$
17. $(4, 3)$; 5 **19.** $(-9, -2)$ **21.** $(5, 0)$ **23.** 7

Lesson 15-4

Try This **a.** no **b.** yes **c.** yes **d.** no **e.** no **f.** yes

g.

Exercises **1.** yes **3.** no **5.** yes

11. **13.** **15.** **17.**

19. no **21.** $\sqrt{1800} \approx 42.4$ **23.** $\sqrt{1800} \approx 42.4$ **25.** 1 **27.** -1 **29.** -5

Lesson 15-5

Try This **a.** $-\frac{3}{4}, 3\frac{1}{4}$ **b.** $y = \frac{7}{2}x + \frac{2}{3}$ **c.** perpendicular

Exercises **1.** $8, 3$ **3.** $-5, 4$ **5.** $-6, -9$ **7.** $\frac{2}{3}, 8$ **9.** $-0.9, 0$ **11.** $y = -3x - 2$

13. $y = \frac{5}{3}x - 3$ **15.** $y = 4$ **17.** parallel **19.** neither **21.** perpendicular

23. perpendicular **27.** $(-25, 75)$ **29.** 4 **31.** 1

Chapter 15 Review

1. $\sqrt{109} \approx 10.4$ **2.** C; ≈ 16.97 **3.** -2 **4.** 0 **5.** $-\frac{1}{3}$ **6.** not defined

7. 4 **8.** $(2, 0)$ **9.** $(-5.5, 6.5)$ **10.** no **11.** yes **12.** yes

13. **14.** $5, -2$ **15.** $y = -3x + 3$ **16.** neither

Chapter 16

Lesson 16-1

Try This **a.** $\frac{21}{20}, \frac{21}{29}, \frac{20}{21}, \frac{20}{29}, \frac{29}{20}, \frac{29}{21}$ **b.** $\frac{3}{4}, \frac{4}{5}, \frac{3}{5}, \frac{4}{3}, \frac{5}{4}, \frac{5}{3}$ **c.** hypotenuse is $\overline{ST}$, opposite is $\overline{TU}$, adjacent is $\overline{SU}$ **d.** hypotenuse is $\overline{ST}$, opposite is $\overline{SU}$, adjacent is $\overline{TU}$ **e.** $\frac{9}{15} = 0.6000$ **f.** $\frac{9}{12} = 0.7500$

Exercises **1.** $\frac{3}{4}, \frac{3}{5}, \frac{4}{5}, \frac{4}{3}, \frac{5}{3}, \frac{5}{4}$ **3.** hypotenuse is $\overline{VU}$, opposite is $\overline{TU}$, adjacent is $\overline{VT}$

5. hypotenuse is $\overline{ML}$, opposite is $\overline{KL}$, adjacent is $\overline{KM}$ **7.** $\frac{36}{39} \approx 0.9231$

9. $\frac{15}{39} \approx 0.3846$ **13.** $\triangle RST$ is isosceles; $m\angle R = m\angle S = 45°$. **15.** 28 **17.** -6

Lesson 16-2

Try This   **a.** $\sin R = \dfrac{9}{41}$, $\sin P = \dfrac{40}{41}$ **b.** 0.9231   **c.** 0.4167   **d.** $\cos 60° = \dfrac{1}{2} = $ 0.5 in every right triangle.

Exercises   **1.** $\sin A = \dfrac{4}{5} = 0.8000$, $\sin B = \dfrac{3}{5} = 0.6000$ **3.** $\cos D = \dfrac{48}{52} \approx 0.9231$, $\cos F = \dfrac{20}{52} \approx 0.3846$ **5.** $\tan V = \dfrac{28}{45} \approx 0.6222$, $\tan T = \dfrac{45}{28} \approx 1.6071$ **7.** $\cos 30° = \dfrac{\sqrt{3}}{2} \approx 0.8660$ in every right triangle.   **9.** 0.2925   **11.** 45°   **13.** $\dfrac{5}{12}, \dfrac{5}{13}, \dfrac{12}{13}, \dfrac{12}{5}, \dfrac{13}{12}, \dfrac{13}{5}$ **15.** 696.9 cm^3 **17.** 7   **19.** 4

Lesson 16-3

Try This   **a.** 0.6494   **b.** 0.1564   **c.** 0.1392   **d.** 13.9 cm   **e.** 37.1 m

Exercises   **1.** 0.6428   **3.** 0.2493   **5.** 6.3138   **7.** 0.8660   **9.** 0.6691   **11.** 0.0175   **13.** 0.9986   **15.** 0.4848   **17.** 20.1 cm   **19.** 16.8 cm   **21.** 22.7 ft   **23.** 26 m^2 **25.** 4×10^2 **27.** 9.2×10^{-4}

Lesson 16-4

Try This   **a.** 39°   **b.** 74°   **c.** 55°   **d.** 12°   **e.** 55°; 35°   **f.** 491.5 m

Exercises   **1.** 85°   **3.** 12°   **5.** 67°   **7.** 45°   **9.** 89°   **11.** 45°   **13.** 37°; 53°   **15.** 44°; 46°   **17.** 3.3 km   **19.** 338.5 m   **23.** 5.3 m   **25.** 0.9397   **27.** 1.3764   **29.** 2.4751   **31.** 4   **33.** 7

Probability Connection, p. 576

1. 0.012   **3.** about 0.373   **5.** No, these are not independent events.

Chapter 16 Review

1. $\dfrac{8}{15}, \dfrac{8}{17}, \dfrac{15}{8}, \dfrac{15}{17}, \dfrac{17}{8}, \dfrac{17}{15}$ **2.** hypotenuse is $\overline{PR}$, opposite is $\overline{RQ}$, adjacent is $\overline{PQ}$ **3.** $\dfrac{8}{17} \approx 0.4706$ **4.** $\dfrac{8}{10} = 0.8000$ **5.** $\dfrac{8}{10} = 0.8000$ **6.** $\dfrac{8}{6} \approx 1.3333$ **7.** 0.6157   **8.** 0.1736   **9.** 1.4281   **10.** 0.5000   **11.** 8.8 cm   **12.** 15°   **13.** 38°   **14.** 23°; 67°   **15.** 2.5 km

Cumulative Review Chapters 13–16

1. $CD = 24$ **2.** $RS = 12$, $OW = 5$, $WS = 12$ **3.** $m\widehat{AC} = 180°$, $m\widehat{AB} = 37°$, $m\widehat{BC} = 143°$ **4.** 37° clockwise rotation, 323° counterclockwise rotation   **5.** $m\angle B = 69°$, $m\angle C = 87°$, $m\angle D = 111°$ **6.** 21.0 cm   **7.** 1290.5 in.^{2} **8.** LA = 570 in.^{2}, SA = 630 in.^{2} **9.** LA = 144 cm^2, SA $\approx$ 171.7 cm^2 **10.** LA = 140 cm^2, SA = 189 cm^2 **11.** 195 dkm^3 **12.** 144 cm^3 **13.** 147 m^3 **14.** 895 in.^{2} **15.** 195 dm^3 **16.** 153.9 cm^2, 179.6 cm^3 **17.** $x = 1.5$, $y = 14$ **18.** 2:1, 4:1, 8:1   **19.** $\sqrt{74} \approx 8.6$ **20.** $\sqrt{122} \approx 11.0$ **21.** $-\dfrac{5}{7}$ **22.** $\dfrac{1}{5}$ **23.** not defined   **24.** 0   **25.** $(0.5, 6)$ **26.** no   **27.** yes   **28.** yes   **29.** no   **31.** $y = 6x - 2$ **32.** $\dfrac{1}{2}, -4$ **33.** parallel   **34.** neither   **35.** neither   **36.** $\sin S = 0.5714$, $\cos S = 0.8207$, $\tan S = 0.6963$ **37.** 504 ft   **38.** 39.4°

Dot Paper I

Dot Paper II

Acknowledgments

Photographs

Cover Photograph: Michael Simpson/FPG International

Chapter 1: xvi Stanley Schoenberger/Grant Heilman Photography; 1T NOAA; 1C Donald Specker/Animals, Animals; 1BL Courtesy of the Perkin-Elmer Corp.; 1BR Roger J. Cheng; 5 Frank Siteman/Stock, Boston; 9 Tim Davis*; 17 Hank Morgan/Rainbow; 22 SCALA/Art Resource, NY

Chapter 2: 32 Bill Pierce/Rainbow; 44T Tim Davis*; 46T Ira Kirschenbaum/Stock, Boston; 46B Tim Davis*; 50 David Madison

Chapter 3: 62 Frank Wing/Stock, Boston; 63 Tim Davis*; 67 Tim Davis*; 78 Coco McCoy/Rainbow; 81L Peabody Museum of Salem, photo by M. W. Sexton; 81R Robert Burroughs/Black Star; 84 Tennessee Valley Authority

Chapter 4: 102 NASA; 104 Hal Harrison/Grant Heilman Photography; 116 Arthur C. Parsons/Black Star; 123 Sepp Seitz/Woodfin Camp & Associates; 128 Mount Wilson and Palomar Observatories; 133 Story Litchfield/Stock, Boston; 135B Tim Davis*; 136 Joseph Koudelka/Magnum Photos

Chapter 5: 148 © Harald Sund; 152 George B. Fry III*; 155 Terence A. Gili/Animals, Animals; 156T George B. Fry III*; 156B Carl Zeiss/Monkmeyer; 172T David Burnett/Contact-Woodfin Camp & Associates; 173 Dan Budnik/Woodfin Camp & Associates; 176 Tennessee Valley Authority

Chapter 6: 192 Patrick Ward/Stock, Boston; 205 Gary Milburn/Tom Stack & Associates; 206T Lick Observatory, U. C. Santa Cruz; 209 Sylvia Johnson/Woodfin Camp & Associates; 213 Tim Davis*

Chapter 7: 226 Grant Heilman Photography; 227 George B. Fry III*; 228 George Hall/Woodfin Camp & Associates; 230L Robert Eckert/EKM-Nepenthe; 230R Burk Uzzle/Woodfin Camp & Associates; 234 Tim Davis*; 241T John Colwell/Grant Heilman Photography; 241B Owen Franken/Stock, Boston; 244 Tim Davis*; 255 © 1990 M. C. Escher Heirs/Cordon Art, Baarn, Holland

Chapter 8: 264 Steve Benbow/Woodfin Camp & Associates; 282 Owen Franken/Stock, Boston; 296B Tim Davis*; 296C © Harald Sund

Chapter 9: 304 Ed Young*; 305T Tim Davis*; 309 Runk-Schoenberger/Grant Heilman Photography; 316 George Gardner/The Image Works

Chapter 10: 340 Ed Young*; 345 Fredrik D. Bodin/Stock, Boston; 346 The Bettmann Archive; 349L Peter Menzel/Stock, Boston; 352 Tim Davis*; 363R Clif Garboden/Stock, Boston

Chapter 11: 368 Ed Young*; 369 Roger Malloch/Magnum Photos; 379 Elliott Smith*; 386 Tim Davis*; 387 Tim Davis*; 396 Joseph Flack Weiler/Stock, Boston; 400 © 1990 M. C. Escher Heirs/Cordon Art, Baarn, Holland. Collection Haags Gemeentemuseum, Den Haag

Chapter 12: 404 © Peter Menzel; 405 Dan Chidester/The Image Works; 423L The Bettmann Archive; 423R John Blaustein/Woodfin Camp & Associates; 430 Elliott Smith*

Chapter 13: 436 Ed Young*; 437 George B. Fry III*; 446 Tim Davis*; 451 Tim Davis*; 457 Tim Davis*; 461 © 1990 M. C. Escher Heirs/Cordon Art, Baarn, Holland. Collection Haags Gemeentemuseum, Den Haag; 467 Tim Davis*; 473L Fredrik D. Bodin/Stock, Boston; 473C The New York Historical Society, New York City

Chapter 14: 484 Digital Art/West Light; 485 Mike Yamashita/Woodfin Camp & Associates; 494 Erich Hartmann/Magnum Photos; 500 George B. Fry III*; 510 John Blaustein/Woodfin Camp & Associates; 517 Charles Harbutt/Archive Pictures; 518T Tim Davis*; 523 John Colletti/Stock, Boston; 525TL George B. Fry III*; 525BC Mount Wilson and Las Campanas Observatories, Carnegie Institution of Washington; 525BR NASA; 528 Tim Davis*

Chapter 15: 536 Ed Young*

Chapter 16: 558 Ed Young*; 559 George F. Godfrey/Earth Scenes

Wayland Lee*/Addison-Wesley Publishing Company: 16, 44C, 44B, 49, 70, 103, 109, 122, 124, 125, 126, 135, 172B, 179, 202, 206B, 210, 259, 267, 283, 289, 294, 295, 296T, 305B, 349R, 362, 363L, 370, 374, 415, 438, 454, 455, 470, 473R, 514, 518B, 520, 521, 525TC, 525TR, 525BL

Special thanks to the Chamber of Commerce of the Monterey Peninsula and Beli-Christensen Architects.

Illustration

655 Scott Kim*/Letterforms and Inversions

Calligraphy

John Prestianni*

*Provided expressly for the publisher.

Milestones in Mathematics

c. 30,000+BC   The knucklebones of animals were used as dice in games of chance.

c. 20,000+BC   A wolfbone with 55 notches in two rows divided into groups of five was used for counting (discovered at Vestonice in Czechoslovakia).

c. 8,000 BC   First evidence of recorded counting.

c. 2,000 BC   The Egyptians had arrived at the value for pi of $\pi = 4(8/9)^2$.

c. 1,900 BC   Babylonian scholars used cuneiform numerals to the base 60 in the oldest-known written numeration for place value.

c. 1,700 BC   Sumerian notation was used to solve quadratic equations by the equivalent of the formula we use today.

c.   800 BC   Queen Dido founded the great city of Carthage by solving the geometric "Problem of Dido." A rigorous proof of this problem—what closed curve of specified length will enclose a maximum area—did not come until the nineteenth century.

c.   700 BC   Zero appeared in the Seleucid mathematical tables.

c.   550 BC   Pythagoras developed a logical, deductive proof of the Pythagorean theorem.

c.   300 BC   Euclid wrote the first geometry text, *Elements*.

c.   250 BC   Archimedes wrote *On Mechanical Theorems, Method* for his friend Eratosthenes.

c.   250 AD   An initial-letter shorthand for algebraic equations was developed.

c.   300 AD   Pappus of Alexandria discussed the areas of figures with the same perimeter in the *Mathematical Collection*.

c.   375 AD   Earliest known Mayan Initial Series inscriptions for expressing dates and periods of time.

c.   400 AD   Hypatia, the foremost mathematician in Alexandria, lectured on Diophantine algebra.

       595   Date of an Indian deed on copper plate showing the oldest known use of the nine numerals according to the place value principle: the first written decimal numeration with the structure used today.

825	A treatise on linear and quadratic equations was published by Mohammed Al-Khwarizmi.
850	Mahavira contributed to the development of algebra in India.
1202	Leonardo of Pisa, also called Fibonacci, wrote *Liber abaci*, introducing Arabic numbers to Europe. This book contains his "rabbit problem" involving the numbers we now call Fibonacci.
1261	Yang Hui of China wrote on the properties of the binomial coefficients.
1557	The equal sign (=) came into general use during the 16th century, A.D. (The twin lines as an equal sign were used by the English physician and mathematician Robert Recorde with the explanation that "noe .2. thynges, can be moare equalle.")
1614	John Napier invented logarithms.
1639	René Descartes published his treatise on the application of algebra to geometry (analytic geometry).
1654	Blaise Pascal described the properties of the triangle we now call Pascal's triangle.
1657	Major contributions to number theory were made by Pierre de Fermat including his formulation of the "Pell" equation.
1670	G. Mouton devised a decimal-based measuring system.
1688	The calculus was published by Isaac Newton in *Principia Mathematica*.
1735	Graph theory was originated by Leonard Euler in his paper on the problem, "The Seven Bridges of Konigsberg."
1784	Maria Agnesi developed new ways to deal with problems involving infinite quantities in her book, *Analytical Institutions*.
1799	The fundamental theorem of algebra was delineated by Carl Friederich Gauss, who also developed rigorous proof as the requirement of mathematics.
1816	Sophic Germain published equations which stated the law for vibrating elastic surfaces.
1832	Evariste Galois wrote the theorem stating the conditions under which an equation can be solved.
1854	George Boole developed the postulates of "Boolean Algebra" in *Laws of Thought*.
1854	Mary Fairfax Somerville wrote books to popularize mathematics and extend the influence of the work of mathematicians.
1859	George F. B. Reimann published his work on the distribution of primes; "Reimann's Hypothesis" became one of the famous unsolved problems of mathematics.

1886	Modern combinatorial topology was created by Henri Poincare.
1888	Sonya Kovalesvskaya was awarded the Prix Bordin for her paper "On the Rotation of a Solid Body About a Fixed Point."
1897	David Hilbert published his monumental work on the theory of number fields and later clarified the foundations of geometry.
1906	Grace Chisholm Young and William Young published the first text on set theory.
1914	Srinivasa Ramanujan went to England to collaborate with G. H. Hardy on analytic number theory.
1925	Hermann Weyl published fundamental papers on group theory.
1931	Gödel showed that there must be undecidable propositions in any formal system and that one of those undecidable propositions is consistency.
1932	A completely general theory of ideal numbers was build up, on an axiomatic basis, by Emmy Noether.
1936	The minimax principle in probability and statistics was developed by Abraham Wald.
1937	Goldbach's conjecture that every even number is the sum of two primes ($12 = 5 + 7$, $100 = 3 + 97$) was established by I. M. Vinogradov for every sufficiently large even number that is the sum of, at most, four primes.
1938	Claude E. Shannon discovered the analogy between the truth values of propositions and the states of switches and relays in an electric circuit.
1942	Jacqueline Ferrand created the concept of preholomorphic functions, using these to produce a new methodology for mathematical proofs.
1951	Elizabeth Scott, Jerzy Newyman, and C. D. Shane applied statistical theories to deduce the existence of clusters of galaxies.
1953	Maria Pastori extended the usefulness of the tensor calculus in the pure mathematical investigation of generalized spaces.
1960	Advances in the application of probability and statistics were made by Florence Nightingale David.
1976	Four color problem proved using electronic computing in concert with human deduction.
1985	A new algorithm for factoring large numbers by using elliptic curves was developed by Hendrik W. Lenstra, Jr.
1985	David Hoffman discovered a fourth minimal surface, the first new minimal surface discovered since the 1700s.

Index

A

AA Similarity Theorem, 326–328
AAA Similarity Postulate, 326
AAS congruence, 182–185
absolute coordinates, 546
absolute value, 15, 43, 597
acute angle(s), 12, 150, 563, 597
  of a right triangle, 251
acute triangles, 150, 597
activity *see* Explore, Enrichment,
  Discover, computer enrichment
addition
  of angles, 89–90
  of arc measures, 453
  area, 406
  commutative property of, 188
  of fractions, xiv
  of like terms, 108
  positive and negative numbers,
   43
adjacent angles, 116–117, 597
adjacent side, 560
Algebra Connection
  adding positive and negative
   numbers, 43
  combining like terms and solving
   equations, 108
  distances on the number line, 15
  exponents, 410
  inequalities on a number line,
   212
  multiplying and dividing with
   negative numbers, 541
  number properties, 188
  order of operations and solving
   equations, 88
  radicals, 336
  scientific notation, 509
  solving equations, 82–83
   involving parentheses, 238
  solving equations with variables
   on both sides, 121
  square roots, 336
  subtracting positive and negative
   numbers, 53
alternate exterior angles, 234,
  240, 245, 597
alternate interior angles, 234,
  239, 245, 597
altitude, 597, *see also* height
  of a parallelogram, 417
  of a pyramid, 494
  of a right triangle, 341–343

  of a triangle, 341
Angle Addition Postulate, 89–90,
  453
angle bisector(s), 91, 221–222, 597
  and angles of a triangle, 217
  and incenter, 218
Angle Bisector Theorem, 221
Angle Measure Postulate, 84
angle of incidence, 243
angle of reflection, 243
Angle Sum Theorem, 249
angles(s), 11, 78, 597
  acute, 12, 150, 563, 597
   of a right triangle, 251
  addition of, 89
  adjacent, 116–117, 597
  alternate exterior, 234, 240, 245,
   597
  alternate interior, 234, 239, 245,
   597
  base
   of an isosceles trapezoid, 294
   of an isosceles triangle, 193
  bisecting, 95
  bisector, 91, 217–218, 221–222,
   597
  central, 451–454, 598
  complementary, 109–115, 598
  congruent, 104–105, 158–160,
   163–164, 166, 173–174,
   178–179, 182–185
  consecutive, 265
   of a parallelogram, 274
  copying, 96
  corresponding, 234, 240, 245,
   599
   and similar triangles, 326, 331
  of depression, 572–574, 597
  drawing, 86
  of elevation, 572–574, 597
  of equilateral triangles, 150
  exterior, 200–202, 251
   of a convex polygon, 384
   of an angle, 80
  identifying, 11, 78–80
  of incidence, 243
  included, 149, 600
  inequality and, 204–209
  inscribed, 462–464, 600
  interior, 234, 601
   of a convex polygon, 384
   of an angle, 80, 601
   remote, 200–201, 604
  and intersecting chords, 466
  linear pair, 117–118, 601

  measure(s), 11
   sum of
    of a polygon, 383
    of a quadrilateral, 266
    of a triangle, 249
  measurement, 84–85
  naming, 78–80
  obtuse, 12, 602
  and paper folding, 114
  of parallelograms, 272–275, 281
  of polygons, 383, 390
  of quadrilaterals, 265–294
  of rectangles, 28
  opposite
   of a parallelogram, 272, 281
  rays of, 78
  reference, 560
  of reflection, 243
  of regular polygons, 390
  remote interior, 200–201, 604
  right, 11, 129–130, 604
  of rotation, 457
  sides of, 78–80
  of similar polygons, 321–333
  of squares, 266, 285–288
  straight, 11, 78–80, 605
  sum of measures of a polygon,
   383
  supplementary, 110–115, 605
   interior angles, 240, 245
   and linear pairs, 118
  of triangles, 25, 79
  vertex of, 78–80, 606
  vertical, 122–125, 606
apothem, 415, 597
applications *see* Geometry in Art,
  Geometry in Communication,
  Geometry in Design, Geometry in
  Nature, Geometry in Technology,
  Problem Solving, and Apply in
  each Exercise Set
Arc Addition Theorem, 453
arc(s), 451–454, 597
  addition of, 453
  congruent, 452–454
  degree measure, 452–454
  intercepted, 462–464, 600
  major, 451–454, 601
  minor, 451–454, 602
  naming, 451
  semicircle, 451–454
Archimedean spiral, 128
area, 405–427, 597
  addition of, 406
  approximating, 405–407

of a circle, 476–479
  development of formula using
    a model, 476
estimating, 405–407, 423
of irregular shapes, 414, 420,
  422, 428–429, 478
lateral
  of a cone, 515
  of a cylinder, 514
  of a prism, 490
  of a right prism, 491
of a parallelogram, 417–419
and perimeter, 407
  and reflection, 419, 428
  and rotation, 424, 428
  and scale change, 414, 420,
    424, 428
  and translation, 413, 428
of a rectangle, 411–413
of a regular polygon, 415
of a rhombus, 417–419
of a sector, 478
of a square, 412
of a trapezoid, 426–427
of a triangle, 421–422
  by Heron's formula, 425
surface
  of a cone, 515
  of a cylinder, 514
  of a prism, 490
  of a sphere, 523
unit of, 404
Area Postulate, 405
argument(s)
  convincing, 49, 51, 120, 124, 126,
    152, 173, 242, 252, 257, 372
  logical and flowcharts, 198–199
arithmetic mean, 343
Arithmetic Review, xiv–xv
ASA congruence, 169
associative property, 188, 597
assumption, 33
average, 68, 343, 395
axes, 72
axis
  of a cone, 513
  of a cylinder, 510, 597
  in the x-y plane, 72, 597

B

balance point, 222
bar graphs, 298
base angles
  of an isosceles trapezoid, 294
  of an isosceles triangle, 193, 357
base(s)
  of a cone, 510, 598

of a cylinder, 510, 598
of an exponent, 410
of an isosceles triangle, 193
of an isosceles trapezoid,
  293–294
of a parallelogram, 417, 598
of a prism, 489, 598
of a pyramid, 505–506
of a rectangle, 411, 598
of a trapezoid, 291, 598
of a triangle, 421, 598
BASIC programs
  diagonals of a polygon, 377
  lateral area of a cylinder, 517
  the Midpoint Theorem, 549
bisect, 68
  a segment, 17
  angles, 95–96
bisector(s)
  angle, 91, 597
  and lines of concurrency, 217
  perpendicular, 131, 141, 218–219
    of a chord, 443
    relfection and, 137
  segment, 17, 598
Buffon, Georges, 475

C

calculator
  area by Heron's formula, 425
  circumference, 472
  exponential notation, 410
  geometric mean, 342
  keys

    cos , 570, 572

    π , 472

    sin , 568, 573

    STO and RCL , 111, 425

    tan , 573

    x^2 , 410

    $\sqrt{x}$, 336

  and order of operations, 88
  radicals, 336
  solving right triangles, 572–574
  store and recall keys, 111, 425
  trigonometric ratios, 568
Cartesian coordinate plane *see*
  x-y plane
Cavalieri's Principle, 502, 519
Cavalieri, Bonaventura, 501–502
center

of a circle, 437, 443
finding, 443
of rotation, 457
of a sphere, 511
central angle(s), 451–454, 598
centroid, 222, 598
chord(s), 437–439, 441–445, 598
  intersecting, 445
  lengths of, 442–443
  perpendicular bisector of,
    441–443
  and radii, 441
circle(s)
  arcs of, 452–464
    major, 451–452
    minor, 451–452
  area of, 476
  as a locus of points, 40
  center of, 437–438
  central angles of, 451–454
  chords of, 437–439, 441–444,
    598
  circumscribed, 467–468, 598
  circumference of, 470–472, 598
  congruent, 452
  defined, 437, 598
  diameters of, 437–439, 451–454,
    468, 599
  finding the center of, 443
  great, 511, 600
  inscribed angles of, 462–464
  inscribed polygons, 392, 467–468
  naming, 437
  and π, 471
  radii of, 437–443, 447, 450
  secants of, 446–448
  sectors of, 477, 604
  semi-, 451, 604
  tangents of, 446–448
circumcenter, 219, 598
circumference, 470–472, 598
  development of formula using a
    model, 470
circumscribed circle, 467–468, 598
classifying
  polygons, 370
  polyhedrons, 486
  prisms, 489
  quadrilaterals, 266
  quadrilaterals by diagonals, 286
  triangles
    by angles, 150
    by sides, 150–151
  using a Venn diagram, 271
clockwise rotations, 457
collinear points, 38, 598
  and reflection, 138
combining like terms, 108
commutative property, 188, 598

compass, 16 *see also* constructions
complement, 109
complementary angles, 109–112, 598
compound probability, 576
computer assisted design (CAD), 430
computer exploration *see also*
  computer enrichment
    angle bisectors, 95
    angles and intersecting chords, 466
    angles of a polygon, 383
    angles of regular polygons, 390
    chords and radii, 441
    diagonals of a polygon, 374, 377
    equations and slope, 553
    Exterior Angle Theorem, 200
    exterior angles of a polygon, 384
    inscribed angles and arcs, 462
    lateral area of a cylinder, 517
    Midpoint Theorem, 549
    Opposite Parts Theorem, 204
    parallel lines, 239
    Perpendicular Bisector Theorem, 141
    perspective drawing, 232
    point of concurrency, 217
    Pythagorean Theorem, 346
    SAS and SSS similarity, 331
    similar triangles, 321
computer enrichment
  diagonals of a polygon, 377
  lateral area of a cylinder, 517
  the Midpoint Theorem, 549
computer graphics, 279, 392
concave
  polygons, 371, 598
  quadrilaterals, 265
conclusion, 33, 54–56, 58, 598
concurrent lines, 217–219, 598
conditional sentence, 54–56, 58, 598
  conclusion, 54
  converse of, 56
  hypothesis, 54
cone(s), 510–512, 515–516, 519–520, 530, 599
  axis of, 510
  base of, 510, 598
  height of, 510
  lateral area of, 515
  lateral surface of, 510
  right, 510, 604
  similar, 530
  slant height of, 515, 604
  surface area of, 515
  vertex of, 510
  volume of, 519

congruent, 103, 599
  angles, 104–105, 158–160, 163–164, 166, 173–174, 178–179, 182–185
  arcs, 452
  markings, 163
  regions, 406
  segments, 103, 599
  triangles, 158–160, 168–174
    AAS congruence, 182–185
    ASA congruence, 169–170
    corresponding parts of, 178–179
    HL congruence, 183–185
    SAS congruence, 168–170
    SSS congruence, 167–170
consecutive angles, 265
  of a parallelogram, 274
consecutive sides, 265
constructions
  Archimedean spiral, 128
  bisecting an angle, 95–96, 221
  center of a circle, 443
  the centroid, 222
  circumcenter, 219
  equilateral triangle designs, 154
  flexagons, 373
  geometric devices
    level, 177
    measuring instruments, 365
    pantograph, 320
  golden rectangle, 309
  Greek, 16–18, 95
  incenter, 218
  logarithmic spiral, 128
  parallel lines, 228
  of perpendicular bisector, 131, 141
  of perpendicular lines, 130–131
  polyhedron, 488
  regular inscribed polygons, 467–468
  rhombus, 290
  segments, 16
  spiral, 309
  tangents to points not on a circle, 450
  triangles, 166
converse
  of a conditional statement, 56, 599
  of Hinge Theorem, 214–215
  of Isosceles Triangle Theorem, 194
  of Opposite Parts Theorem, 205
  of Pythagorean Theorem, 352–353
convex
  polygons, 371, 599
  quadrilaterals, 265

convincing argument, 49, 51, 120, 124, 126, 152, 173, 242, 252, 257, 372
coordinate geometry, 72–74, 537–554
  distance between points, 537–539
  graphing equations, 550–551
  graphing inequalities, 540
  line graphs, 77
  midpoints, 547–549
  slope, 542–544
coordinate plane, *see x-y* plane
coordinate(s), 63, 72–74, 599
  absolute, 546
  line graphs, 77
  of a point
    on a number line, 63
    in a plane, 72
  and reflection, 74
  relative, 546
  and translation, 74
coplanar points, 39, 599
copying
  angles, 96
  segments, 16
corresponding angles, 234, 599
  and similar triangles, 326, 331
corresponding parts
  of congruent triangles, 159, 178–179
cosine ratio, 563–565, 568–570, 599
counterclockwise rotations, 457
counterexample, 26–27, 599
cross products, 310, 599
cross section, 37, 599
  area, 502
cube(s), 486, 599
  pattern, 66
  visualizations with, 2, 203, 260
customary units, 9
cylinder(s), 510–518, 599
  axis of, 510
  base of, 510, 598
  lateral area, 514, 517
  right, 510
  surface area, 514
  volume, 518

D

data
  collecting and displaying, 298
decagon, 370, 599
decahedron, 486, 599
deductive reasoning, 33–35, 599
defining a geometric figure, 40
degree, 11, 84, 599

design
computer-assisted, 430
equilateral triangles in, 154
diagonal(s), 34
of an isosceles trapezoid, 293
of a parallelogram, 272, 275, 281
of a polygon, 374–377, 599
of a quadrilateral, 265
diameter
of a circle, 437–439, 599
of a sphere, 528
dimensions, 6
Discover
Angle Bisector Theorem, 221
angles and intersecting chords, 466
cross products of chord lengths, 445
folding midpoints, 20
midsegment of a triangle, 297
paper folding angles, 114
parallel lines and similar triangles, 335
Perpendicular Bisector Theorem, 141
distance
on the coordinate plane, 537
formula, 537–539
on a number line, 64
Distance Formula, 537–539
Distance Postulate, 64
and midpoint, 68
distributive property, 108, 238, 599
division
of negative numbers, 541
DNA molecule, 4
dodecagon, 370, 600
dodecahedron, 486, 600
dot paper, 637, 639
using, 140, 259
perspective, 504
double helix, 4
drawing an angle, 86
drawing(s)
perspective, 232
reading and making, 163–164
the rotation image, 458

E

edge(s), 485
lateral, 489, 601
and perspective drawing, 499
of polyhedrons, 485, 600
endpoint(s), 6, 67–69
and perpendicular bisector, 141
Enrichment
area of a regular polygon, 415
arranging shapes, 71

billiards and angles, 127
constructing tangents to points not on a circle, 450
counterexamples and product claims, 28
finding the height of a tree by similar triangles, 330
flexagons, 373
folding parallel lines, 248
Heron's formula, 425
inequalities on the coordinate plane, 540
isosceles triangles, 197
lateral area of a right prism, 517
making a level, 177
the Midpoint Theorem, 549
Möbius strip, 364
pantograph, 320
paper folding a polygon, 394
periscopes, 243
perspectives and dot paper, 504
polyhedron model, 488
product claims, 28
quadrilaterals and symmetry, 270
rigid objects, 14
shape of DNA molecule, 4
tangrams and quadrilaterals, 284
triangle word web, 153
equations
graphing, 550–551
solving, 82
combining like terms, 108
with variables on both sides, 121
translating to, 115
with parentheses, 238
equiangular triangles, 150, 600
and isosceles triangles, 195
equidistance, 137, 141, 217–218
and angle bisectors, 222
and circumcenter, 218–219
and incenter, 217–218
equilateral triangles, 150, 600
in design, 154
and isosceles triangles, 195
tetrahedron pattern, 162
Escher, M. C., 255, 400, 461
estimation
angle measures, 85–86
and area, 405–407, 423
lengths, 9, 107
perimeter of a polygon, 380
Euclid, 38
event(s), 350
independent, 576
Explore
AA similarity, 326
angle addition, 89
Angle Sum Theory, 249

angles of a polygon, 383
angles of a triangle, 25
angles of quadrilaterals, 265
angles of regular polygons, 390
area of a circle using a parallelogram, 476
area of parallelograms, 417
area of rectangles, 411
area of trapezoids, 426
area of triangles, 421
areas
congruent and noncongruent figures, 406
bisecting an angle, 95
chords and radii, 441
circular objects and π, 470
concurrency, 217
congruent complements and supplements, 111
diagonals of a polygon, 374
the Distance Formula, 537
equations and slope, 553
Exterior Angle theorem, 200
exterior angles of a polygon, 384
geometric shapes, 1
Hinge Theorem, 213
inscribed angles and arcs, 462
Isosceles Right Triangle Theorem, 356
isosceles triangles, 193
lateral area of a cylinder, 514
lateral area of a right prism, 491
linear pairs supplementary, 118
locus of points, 40
median of a trapezoid, 292
Opposite Parts Theorem, 204
paper folding intersecting lines, 49
parallel and perpendicular lines, 554
parallel lines, 239
parallelograms, 272, 280
perpendicular lines, 129
Pythagorean Theorem, 346
reflecting points over the y-axis, 73
reflections, 136
and measures, 138
rhombuses and rectangles, 285
SAS and SSS similarity, 331
SAS, SSS, and ASA congruence, 167
shoe length and measurement, 9
similar right triangles, 341
similar triangles, 321
sine ratios of similar right triangles, 568
SSA and HL, 183

tessellations with nonregular polygons, 397
30°–60° Right Triangle Theorem, 360
translations, 256
trigonometric ratios and similar right triangles, 563
vertical angles and congruence, 122
volume and surface area of similar solids, 529
volume of a pyramid, 505
exponent, 410
exponential notation, 410
exterior
angles
of a triangle, 200–202, 251, 597
of a convex polygon, 384, 600
of an angle, 80
Exterior Angle Theorem, 201–202, 251

F

face(s)
lateral
of prisms, 489–491, 601
of pyramids, 494–496
of polyhedrons, 485, 600
factor, 410
flexagons, 373
flowchart proof, 198–199
45°–45°–90° Triangle Theorem, 356–357

G

gable, 422
Garfield, James A., 429
geometric devices to construct
level, 177
measuring instruments, 365
pantograph, 320
geometric mean, 342, 600
Geometry in Art
computer graphics, 279
perspective drawing, 232
prisms, 499
Geometry in Communication
advertisements, 58
line graphs, 77
Geometry in Design
using equilateral triangles, 154
tessellations with rotations, 461
tiles, 21
Geometry in Nature

golden ratio, 309
spirals, 128
Geometry in Science
centroid, 222
making measuring instruments, 365
numerical control, 546
speed of a raindrop, 567
Geometry in Technology
computer-assisted design, 430
Glossary, 597–606
golden ratio, 309
golden rectangle, 298, 309
grade, 542
graphics
computer, 279, 372, 430
graphing
bar graphs, 298
equations, 550–551
inequalities on a number line, 212
line graphs, 77
pie charts, 455
points on the x-y plane, 72
great circle, 511, 600
Greek constructions, 16–18, 95
see also constructions

H

half-line, 6
height
of a parallelogram, 417
of a prism, 501
of a pyramid, 494
of a rectangle, 411
of a right cone, 510
of a right cylinder, 510
of a right prism, 491
of a right triangle, 341
slant
of a cone, 515, 604
of a pyramid, 494, 604
of a trapezoid, 426
of a triangle, 421–422
hemisphere, 511, 600
volume, 524
heptagon, 370, 600
heptahedron, 486, 600
Heron's formula, 425
hexagon, 370, 600
hexagonal region, 405, 600
hexahedron, 486, 600
Hinge Theorem, 213–215
converse of, 214–215
hip roof, 422
HL congruence, 183–185

horizon line, 232–233
hypotenuse, 183, 560, 600
and isosceles right triangles, 356
and 30°–60° right triangles, 360
hypothesis, 54–56, 58, 600

I

icosahedron, 486, 600
identity element, 188
if-then sentences, 54–56, 58, 600
image, 22, 600
reflection, 136–137, 603
rotation, 457, 604
size transformations, 316
translation, 255
implies, 54
incenter, 218, 600
included angle, 149, 600
included side, 149, 600
independent events, 576
inductive reasoning, 25–27, 33, 600
inequalities
on the coordinate plane, 540
on a number line, 212
in one triangle, 208
in two triangles, 213
inscribed
angles, 462–464, 600
polygons, 392, 467–468, 600
intercepted arcs, 462–464, 600
interior
of an angle, 80, 94, 601
angles, 234, 384, 597, 601
remote, 200–201, 604
intersecting lines, 49, 54, 67
perpendicular lines, 129–132
point of concurrency, 217
intersection, 39, 601
of planes, 46
inverse operations, 82
irrational number, 471
iso-, 153
Isosceles Right Triangle Theorem, 356–357
isosceles trapezoid, 270, 293–294, 601
Isosceles Triangle Theorem, 193–195
isosceles triangle(s), 150, 193–195, 356–357, 601
base of, 193
base angles, 193
legs, 193
right, 356–357
vertex angles, 193

K

Kim, Scott, 653
kite(s), 266, 601

L

lateral area
  of a cone, 515
  of a cylinder, 514, 517
  of a prism, 490, 601
  of a pyramid, 494–496
  of a right prism, 491
lateral edge, 489, 601
lateral faces, 489, 601
lateral surface
  of a cone, 510
  of a cylinder, 510, 601
leg(s)
  of an isosceles triangle, 193, 601
  of a right triangle, 183, 601
  of a trapezoid, 291, 601
length(s)
  of a chord, 442–443
  estimating, 9, 107
  of sides
    of similar polygons, 387–388
    of similar triangles, 321–323,
      326–328, 331–333
  of a triangle, 208, 213
level, 177
like signs, 43
like terms, 108
limit, 470
  and area, 476
limit, 470
line graphs, 77
line of reflection, 136
line of symmetry, 155–156, 601
line(s), 5–6, 44–46
  concurrent, 217–219, 598
  equations of, 550–551
  horizontal
    slopes of, 543
  intersecting, 49, 54
  parallel, 227–229, 239–242,
    244–246, 602
    and slope, 554
  perpendicular, 129–132, 603
    and slope, 554
  of reflection, 136
  secant, 446
  skew, 228, 604
  slope of, 542–544, 605
  of symmetry, 155–156, 601
  tangent, 446
  vertical
    slopes of, 544

linear pair, 117–118, 120, 601
locus, 40, 437, 601
  angle bisector as, 221
  circle as, 40
  perpendicular bisector, 141
  points equidistant from a point,
    40
  points equidistant from sides of
    an angle, 221
logarithmic spiral, 128
logical arguments
  flowcharting, 198–199
look for a pattern, 29, 48, 94
lowest terms, 305

M

major arc, 451, 601
make a drawing, 164–166, 173–174
make an organized list, 416
making geometric devices
  a level, 177
  measuring instruments, 365
  a pantograph, 320
maps
  scale of, 312, 604
Mathematica, 485
mean, 395, 601
  arithmetic, 343
  geometric, 342, 600
measuring instruments, 365
measure(s)
  of angles, 84, 104–105
  arc, 452
  of central angles, 451–454
  of complementary angles,
    109–112
  of inscribed angles, 462–464
  sum of
    of a polygon, 383
    of a quadrilateral, 265
    of a triangle, 25
  of supplementary angles,
    110–112
median(s)
  point of concurrency, 222
  of a set of values, 395, 601
  of a trapezoid, 291–293, 601
  of a triangle, 25–26, 601
metric units, 9
Midpoint Theorem, 547–549
midpoint(s), 17–19, 68–69, 602
  and perpendicular bisector, 131
midsegment, 297, 602
minor arc, 451, 602
model(s)
  area
    of a circle, 476

  of a parallelogram, 417
  of a rectangle, 411
  of a trapezoid, 426
  of a triangle, 421
circumference of a circle, 470
computer
  lateral area of a cylinder, 517
  constructing a polyhedron, 488
  lateral area
    of a pyramid, 495
    of a rectangular prism, 490
  surface area
    of a cone, 515
    of a cylinder, 514
    of a pyramid, 495
    of a rectangular prism, 490
    of a sphere, 523
  volume
    of a cone, 518
    of a cylinder, 518
    of a prism, 500–501
    of a pyramid, 505
    of a sphere, 523
Möbius strip, 364
Möbius, Augustus Ferdinand, 364
mosaic *see* tessellation
multiplication
  of negative numbers, 541

N

n-gon, 370, 602
Napoleon, 179
negative numbers
  multiplying and dividing, 541
nonagon, 370, 602
nonahedron, 486, 602
noncollinear points, 38, 44–45, 602
noncoplanar points, 39, 45, 602
notation
  exponential, 410
  scientific, 509
  standard, 410, 509
number line, 15, 63–64
  and inequalities, 212
number properties, 188
numerical control, 546

O

oblique prism, 489, 602
obtuse angles, 12, 602
obtuse triangles, 150, 602
octagon, 370, 602
octahedron, 486, 602
operations
  order of, 88

opposite
angles of a parallelogram, 272, 281
of a number, 53
parts of a triangle, 149
sides of a parallelogram, 272, 281
Opposite Parts Theorem, 204–206
opposite rays, 67, 602
and linear pairs, 117
and vertical angles, 122
opposite side, 560
opposite sides, 265
order of operations, 88
orientation, 135, 602
and translation, 260
origin, 63, 72, 602
outcome, 350

P

pantograph, 320, 602
paper folding, *see also* Explore, Enrichment, Discover
angle bisectors of a triangle, 217
Angle Sum Theorem, 249
angles, 114
bisecting an angle, 95
center of a circular plate, 444
intersecting lines, 49
making flexagons, 373
parallel lines, 248
perpendicular lines, 129
polygons, 394
quadrilaterals and symmetry, 270
reflection, 73, 136
and symmetry, 155
and tessellations, 397
vertical angles, 122
parallel
lines, 227–229, 239–242, 244–246, 602
and slope, 554
planes, 502
rays, 227, 602
segments, 227, 602
Parallel Postulate, 244
parallelogram(s), 266, 602
area, 417–419
and area of a circle, 476
base of, 417, 598
determining, 280–282
properties of, 272–275
rectangles, 286–288
rhombuses, 285–288
squares, 286–288
parentheses, 238
pentagon, 33, 370, 602
pentagonal prism, 489, 602

pentahedron, 486, 602
perimeter
and area, 407
and reflection, 419, 428
and rotation, 424, 428
and scale change, 414, 420, 424, 428
and translation, 413, 428
of a circle, 470
and circumference, 470
estimating, 380
of polygons, 379, 602
limit of, 470
of a quadrilateral, 273
perpendicular bisector, 131, 141
of a chord, 443
and circumcenter, 219
reflection and, 137
of a segment, 131, 602
construction of, 130–131
of sides of a triangle, 218
Perpendicular Bisector Theorem, 141
perpendicular lines, 129–132, 603
construction of, 130–131
and slope, 554
perspective drawing, 232, 603
prisms, 499
perspectives
and dot paper, 504
π (pi), 471, 603
and area, 477–478
and lateral area, 514–515
and probability, 475
and surface area, 514–515, 523
and volume, 518, 519, 523
pie charts, 455
plane(s), 5, 44–46
coordinate, 72–74, 537–554
intersecting, 46
parallel, 502
and perpendicular lines, 227
tessellations of the, 396–398, 400, 461
x-y, 72–74, 537–554
plotting points, 72–74
point of concurrency, 217
point(s), 5–6
collinear, 38, 598
and reflection, 138
of concurrency, 217
coordinates of, 63, 72–74, 599
coplanar, 39, 599
equidistant, 137, 141
distance between, 64, 537
noncollinear, 38, 44–45, 602
noncoplanar, 45, 602
plotting, 72
reflection of, 74

of tangency, 446
polygon(s), 369, 603
angles of, 383, 390
area, 415
classifying, 370
constructing by inscribing in a circle, 467–468
convex and concave, 371, 598–599
diagonals of, 374
exterior angles of, 384, 391, 600
identifying, 369
inscribed in a circle, 467–468, 600
interior angle measure, 391
limit of areas as area of circle, 476
limit of perimeters and circumference, 470
names of, 370
n-gon, 370, 602
perimeter of, 379, 602
regular, 390–391, 415, 603
similar, 387–388, 604
sum of measures of, 383
vertices of, 369, 606
polygonal region, 405, 603
polyhedron(s), 485–488, 603
constructing, 488
edges of, 485, 600
faces of, 485, 600
vertices of, 485, 606
volume, 500
postulate(s), 44, 586–588, 603
AAA Similarity, 326
Angle Addition, 453
Angle Addition Postulate, 89–90
Angle Measure Postulate, 84
Area Postulate, 405
Cavalieri's Principle, 502, 519
Distance Postulate, 64, 68
Parallel Postulate, 244
power, 410
precision, 9
prism(s), 489–491, 499–502, 603
bases of, 489, 598
classifying, 489
lateral area of, 490, 601
lateral edges of, 489, 601
lateral faces of, 489, 601
oblique, 489, 602
pentagonal, 489
quadrangular, 489, 603
rectangular, 490, 603
and perspective drawing, 499
right, 489
right rectangular
volume, 501
surface area of, 490

triangular, 489, 605
volume of, 500–502
Probability Connection
compound probability, 576
probability and π, 475
probability of a simple event, 351
Problem Solving, 48, 94, 120, 254, 325, 440, 527, 571
Problem-Solving Strategies
draw a Venn diagram, 271
look for a pattern, 29
make a drawing, 166
make an organized list, 416
simplify the problem, 378
use a flowchart, 198–199
write an equation, 115
proportion(s), 310, 603
and similar right triangles, 342
and similar triangles, 321–323
protractor, 16, 84–86 *see also* constructions
pure tessellation, 396, 603
pyramid(s), 494–496, 505–506, 603
altitude of, 494
height of, 494
hexagonal, 494
lateral area of, 495–496
lateral faces of, 494–496
quadrangular, 494
regular, 494, 604
slant height of, 494, 604
surface area of, 495–496
tetrahedron, 494, 605
triangular, 494, 605
vertex of, 494, 606
volume of, 505–506
Pythagorean Theorem, 346–348, 447, 472
and area of a polygon, 415
converse of, 352–353
proof by James Garfield, 429
Pythagorean triple, 359, 603
Pythagorus, 346

Q

quadrangular prism, 489, 603
quadrilateral(s), 34, 265–295, 603
area of, 411–415, 417–419, 421–422, 425–427
classifying, 266
classifying by diagonals, 286
concave, 265
convex, 265
symmetry and, 270

R

radicals, 336
radius (radii)
of a circle, 437–439
of a sphere, 511, 523–524
ratio(s), 305, 603
the golden ratio, 309
π (pi), 471
and proportion, 305, 603
and similar solids, 529
slope as a, 542
trigonometric, 559, 563–565, 568–570
ray(s), 6, 603
concurrent, 217
opposite, 67, 602
parallel, 227, 602
reasoning
deductive, 33–35, 599
inductive, 25–27, 33, 600
rectangle(s), 266, 286–288, 603
area, 411–413
base of, 411, 598
diagonals of, 286
rectangular prism, 490, 603
rectangular region, 405, 603
reference angle, 560
reflection, 22–23, 136–139, 603
and orientation, 135
on the x-y plane, 73–74
reflection image, 136–137, 603
reflectional symmetry, 155
region, 405
regular polygon(s), 390–391, 603
area of, 415
regular pyramid, 494–496, 604
relative coordinates, 546
remote interior angles, 200–201, 604
rhombus(es), 266, 285–288, 604
area of, 417–419
constructing, 290
diagonals of, 285
right angle(s), 11, 604
and perpendicular lines, 129–130
and SAA, 183–185
and triangles, 150
right cone, 510, 604
right cylinder, 510, 604
right prism, 489, 604
lateral area, 491
right triangle(s), 150, 604
altitude of, 341–343
45°–45°, 356–357
HL congruence, 183–185
height of, 341
hypotenuse of, 183, 560, 600

isosceles, 356–357
legs of, 183, 601
similar, 341–343
solving, 572–574
30°–60°, 360–362
and trigonometric ratios, 563
rise, 542
rotation image, 457–458, 604
rotation(s), 22, 604
angle of, 457–458
center of, 457–458
rotation image, 457–458
tessellations with, 461
visualizing a solid figure by, 513
ruler, 9, 16
run, 542

S

SAS congruence, 168–170
SAS Inequality Theorem, 215
SAS Similarity Theorem, 331
scale
of a map, 312, 604
enlargements and reductions, 315–316
of a protractor, 84
scale drawing, 311–312
scale factor
in a plane, 316–318
in space, 528
scalene triangles, 150, 604
scientific notation, 509
secant line, 446
secants, 446–448, 604
sector, 477, 604
area of, 478
segment(s), 6, 67–69, 604
bisector of, 17, 598
concurrent, 217
congruent, 103, 599
copying, 16
endpoints of, 6, 67–69
measuring, 9
midpoint of, 68–69
parallel, 227, 602
perpendicular bisector of, 131, 602
construction of, 130–131
skew, 228, 604
tangent, 447
Selected Answers, 607–640
semi-pure tessellation, 399, 604
semicircle, 451, 604
side(s)
consecutive, 265
included, 149, 600

opposite
   of a parallelogram, 272, 281
   of a polygon, 369
   of a triangle, 149
sign
   of a number, 43
similar(ity)
   figures, 315, 604
   polygons, 387–388, 604
   triangles, 321–323, 326–328,
      331–333
      AA similarity, 326
      SAS similarity, 331
      right, 341–343
   and sine ratios, 568
   solids, 528–530
      surface area, 530
      volume, 530
simple event, 350
simplify the problem, 378
sine ratio, 563–565, 568–570, 604
size transformation, 316, 528
skew lines, 228, 604
slant height
   of a cone, 515, 604
   of a pyramid, 494, 604
slope, 542–544, 605
   of a horizontal line, 543
   and parallel lines, 554
   and perpendicular lines, 554
   of a vertical line, 544
slope-intercept equation, 553
solid(s), 485
   similar, 528–530
solutions
   of equations, 550
solving equations, 82–83
   with like terms, 108
   with variables on both sides, 121
space, 38, 45
space figures, 485, 502
   cones, 510
   cylinders, 510
   polyhedrons, 485–488
   pyramids, 494–496
   spheres, 511
speed of a raindrop, 567
sphere(s), 511, 605
   great circle of, 511
   hemi-, 511, 524, 600
   radius of, 511
   surface area of, 523
   volume of, 523–525
spiral
   Archimedean, 128
   logarithmic, 128
square root(s), 336
   table of, 582
square(s), 266, 286–288, 605

area of, 412
   in design, 21
SSA, 183, 185
SSS congruence, 167–170
SSS Similarity Theorem, 332
standard notation, 410
Statistics Connection
   collecting and displaying data,
      298
   mean and median, 395
straight angle, 11, 78–80, 605
straightedge, 16
subtracting
   positive and negative numbers,
      53
supplement, 110, 124
supplementary angles, 110–115,
   605
   and linear pairs, 118
surface
   lateral
      of a cone, 510
      of a cylinder, 510, 601
surface area
   of a cone, 515
   of a cylinder, 514
   of irregular solids, 492, 516,
      524–527
   of a prism, 490
   of similar solids, 529–530
   of a sphere, 523
Symbols, xiii
symmetry, 155–157, 605
   computer paint programs and,
      279
   line of, 155, 601
   quadrilaterals and, 270

T

tables
   square roots, 582
   times and measures, 584–585
   trigonometric ratios, 583
tangent circles, 448, 605
tangent line, 446, 605
tangent ratio, 563–565, 568–570,
   605
tangent segment, 447
tangents, 446–448
tangrams, 284
terms, 108
tessellation(s), 396–398, 605
   pure, 396, 603
   with rotations, 461
   semi-pure, 399, 604
   with translations, 400
tetrahedron, 162, 486, 494, 605

theorem(s), 49–51, 589–596, 605
   AA Similarity, 326
   Angle Bisector Theorem, 221
   Angle Sum Theorem, 249
   Arc Addition Theorem, 453
   the Distance Formula, 537–539
   Exterior Angle Theorem,
      201–202, 251
   Hinge Theorem, 213–215
   HL Theorem, 447
   Isosceles Right Triangle
      Theorem, 356–357
   Isosceles Triangle Theorem,
      193–195
   Midpoint Theorem, 547–549
   Opposite Parts Theorem,
      204–206
   Perpendicular Bisector Theorem,
      141
   Pythagorean Theorem, 346–348,
      429, 447, 472
      and area of a polygon, 415
      converse of, 352–353
   SAS Inequality Theorem, 215
   SAS Similarity, 331
   SSS Similarity, 332
   30°–60° Right Triangle Theorem,
      360–362
   Triangle Inequality, 208–209
   Vertical Angle Theorem, 123, 453
30°–60° Right Triangle Theorem,
   360–362
tiles, 21
tiling patterns *see* tessellations
transformation(s), 22, 605
   reflections, 22–23, 135–139
   rotations, 22–23, 457–458, 461,
      513
   size, 316, 528
   translations, 22–23, 255
      and tessellations, 400
   on the x-y plane, 73–74
translation image, 255
translation(s), 22, 255, 605
   and tessellations, 400
   on the x-y plane, 74
transversal, 234–236, 238–242, 605
trapezoid(s), 266, 291, 605
   area of, 426–427
   base angles of, 294
   bases of, 291, 598
   isosceles, 270, 294, 601
   legs of, 291, 601
   median of, 291–293, 601
Triangle Inequality, 208–209
triangle(s), 18, 40, 605
   AAS congruence, 182–185
   acute, 150, 597
   area of, 421–422, 425

ASA congruence, 169–170
base of, 421, 598
centroid of, 222, 598
circumcenter of, 219, 598
classifying by angles, 150
classifying by sides, 150–151
congruent, 158–160, 167–174
constructing, 166
equiangular, 150, 600
equilateral, 150, 600
    in design, 154
    pattern for tetrahedron, 162
exterior angles of, 200
HL congruence, 183–185
incenter of, 218, 600
isosceles, 150, 193–195, 601
median of, 25–27, 601
midsegment, 297, 602
obtuse, 150, 602
remote interior angles of, 200
right, 150
    hypotenuse of, 183, 600
    legs of, 183, 600
    similar
        sine ratios and, 568
    solving, 572–574
    trigonometric ratios of, 563
SAS congruence, 168–170
scalene, 150, 604
similar, 321–323, 326–328, 331–333
similar right, 341–343
SSS congruence, 167–170
sum of angle measures of, 25
tessellations with rotations and equilateral, 461
trigonometry and, 559–573
    table of ratios for, 583
triangular prism, 489, 605
triangular pyramid, 494, 605
triangular region, 405, 605
trigonometric ratios, 559–573
    table of, 583
trigonometry, 559–573

U

undecagon, 370, 605
undecahedron, 486, 605
units
    customary, 9
    metric, 9
unlike signs, 43

V

vanishing lines, 232, 605
vanishing point, 232, 499, 606
Venn diagram, 271
vertex (vertices)
    of an angle, 11, 78, 606
    of a cone, 510
    of a polygon, 369, 606
    of a polyhedron, 485, 606
    of a pyramid, 494, 606
    of a quadrilateral, 265
    of a triangle, 18, 149
        order of matching, 159
vertex angle
    of an isosceles triangle, 193
Vertical Angle Theorem, 123, 453
vertical angles, 122–125, 606
views
    of a solid figure, 2
Visualization
    blocks inside a figure, 260
    and coordinates, 552
    cube figure, 2
    cube patterns, 66
        unfolded, 493
    design, pieces from a, 389
    faces of a cube, 203
    holes punched through folded paper, 278
    hidden segments and cross sections, 37
    orientation, 135
    plate, completing a, 562
    rotation and solids, 513

similar figures, 315
squares
    areas of, relative, 409
    figures made from, 350
    tetrahedron pattern, 162
volume, 500–506, 518–520, 523–525, 529–530, 606
    Cavalieri's Principle, 502
    of a cone, 519
    of a cylinder, 518
    of a hemisphere, 524
    of an irregular solid, 507, 521, 524–526
    of a polyhedron, 500
    of a prism, 500–502
    of a pyramid, 505–506
    of a right rectangular prism, 500–501
    of similar solids, 529–530
    of a sphere, 523–525
    unit of, 500

W

word web, 153
write an equation, 115

X

x-axis, 72–74, 606
    and line graphs, 77
x-coordinate, 72–74, 606
x-y plane, 72–74, 537–554, 606
    coordinates on, 72–74
    origin of, 72, 602
    reflection of points, 74
    translation of points, 74

Y

y-axis, 72–74, 606
    and line graphs, 77
y-coordinate, 72–74, 606
y-intercept, 553

Addison
Wesley

Informal Geometry

Addison
Wesley